# WATER AND AQUEOUS SOLUTIONS
Introduction to a Molecular Theory

# WATER AND AQUEOUS SOLUTIONS
## Introduction to a Molecular Theory

Arieh Ben-Naim
*Institute of Chemistry*
*The Hebrew University of Jerusalem*
*Jerusalem, Israel*

PLENUM PRESS · NEW YORK AND LONDON

CHEMISTRY

oging in Publication Data

r. 3. Molecular theory. I. Title.
.22                          74-7325

## Acknowledgments

Thanks are due to the following for permission to reproduce figures from their publications: D. Eisenberg and W. Kauzmann (Figs. 6.4, 6.5, 6.6); A. H. Narten and H. A. Levy (Figs. 6.7, 6.10); A. Rahman and F. H. Stillinger (Figs. 6.32, 6.33); J. A. Barker and R. O. Watts (Fig. 6.30); North-Holland Publishing Company (Figs. 6.19, 6.30, 8.23); The *Journal of Chemical Physics* (Figs. 1.3, 5.2, 5.4, 5.5, 5.6, 5.7, 5.8, 5.9, 6.2, 6.7, 6.9, 6.14, 6.18, 6.19, 6.20, 6.21, 6.22, 6.24, 6.25, 6.26, 6.27, 6.28, 6.29, 6.31, 6.32, 6.33, 8.1, 8.4, 8.7, 8.8, 8.12, 8.13, 8.15, 8.17, 8.18); The Clarendon Press, Oxford (Figs. 6.4, 6.5, 6.6); The *Journal of Solution Chemistry* (Figs. 7.4, 7.5, 7.6, 8.11, 8.20, 8.21, 8.22); *Molecular Physics* (Taylor and Francis, Ltd.) (Figs. 6.21, 6.23); *Chemical Physics Letters* (6.19, 6.30, 8.23); J. Wiley and Sons, Inc. (Figs. 2.6, 2.7, 2.8, 4.3, 6.13, 6.14, 7.1, 7.2).

© 1974 Plenum Press, New York
A Division of Plenum Publishing Corporation
227 West 17th Street, New York, N.Y. 10011

United Kingdom edition published by Plenum Press, London
A Division of Plenum Publishing Company, Ltd.
4a Lower John Street, London W1R 3PD, England

All rights reserved

No part of this book may be reproduced, stored in a retrieval system, or transmitted, in any form or by any means, electronic, mechanical, photocopying, microfilming, recording, or otherwise, without written permission from the Publisher

Printed in the United States of America

**Dedicated to Talma**

# Preface

The molecular theory of water and aqueous solutions has only recently emerged as a new entity of research, although its roots may be found in age-old works. The purpose of this book is to present the molecular theory of aqueous fluids based on the framework of the general theory of liquids. The style of the book is introductory in character, but the reader is presumed to be familiar with the basic properties of water [for instance, the topics reviewed by Eisenberg and Kauzmann (1969)] and the elements of classical thermodynamics and statistical mechanics [e.g., Denbigh (1966), Hill (1960)] and to have some elementary knowledge of probability [e.g., Feller (1960), Papoulis (1965)]. No other familiarity with the molecular theory of liquids is presumed.

For the convenience of the reader, we present in Chapter 1 the rudiments of statistical mechanics that are required as prerequisites to an understanding of subsequent chapters. This chapter contains a brief and concise survey of topics which may be adopted by the reader as the fundamental "rules of the game," and from here on, the development is very slow and detailed.

Excluding the introductory chapter, the book is organized into three parts. The first, Chapters 2–4, presents the general molecular theory of fluids and mixtures. Here the notions of molecular distribution functions are developed with special attention to fluids consisting of nonspherical particles. We have included only those theories judged to be potentially useful in the study of aqueous fluids, so this part may not be considered as an introduction to the theory of fluids *per se*.

With this objective in mind we did not survey the recent developments in the theory of simple fluids. Instead we present ample illustrative examples to stress the contrast between simple fluids, on one hand, and the more complex, aqueous fluids on the other. Of course, the particular choice of topics is a matter of personal taste and has no absolute significance. For instance, the theory of solutions may be developed either along the McMillan–Mayer (1945) theory or along the Kirkwood–Buff (1951) theory. Both are exact and equivalent from the formal point of view. However, the latter is judged to be the more suitable for problems arising in the theory of aqueous fluids.

The second part consists of Chapter 5 alone, which comprises a bridge connecting the formal theory of fluids, on the one hand, and its application to water and aqueous solutions on the other. The construction of this bridge is rendered possible through the generalization of the ideas of molecular distribution functions, which lays the foundation for the so-called mixture-model approach to the theory of fluids. The latter may be viewed as the formal basis for various *ad-hoc* mixture models for water and aqueous solutions that have been suggested by many authors.

The third part, Chapters 6–8, presents the treatment of essentially three systems, namely pure water with zero, one, and two simple solutes, respectively. Chapter 6 includes a brief survey of the properties of water. We have avoided excessive duplication of material which has been fully discussed by Eisenberg and Kauzmann (1969). The emphasis is mainly on the various theoretical approaches, both old and of recent origin, to explain the anomalous properties of this unique fluid. Chapter 7 is concerned with very dilute solutions of simple nonelectrolytes which, from the formal point of view, reflect the properties of pure water with a single solute particle. Both experimental facts and theoretical attempts at interpretation are surveyed. Special attention is devoted to elaboration on the exact meaning and significance of "structural changes" induced by a solute on the solvent.

The last chapter deals with small deviations from very dilute solutions. The problem of hydrophobic interaction, considered to be of crucial importance in biochemical processes, is formulated, and methods of estimating the strength of solute–solute interaction in various solvents are discussed. Preliminary attempts at interpretation, based on concepts developed in the preceding chapters, are also surveyed.

Although the framework of this book could have easily accommodated a chapter on ionic solutions we chose not to include this topic, as several works already exist dealing with it exclusively.

The entire subject of aqueous solutions is still subject to vigorous debate, and many approaches, theories, and interpretations are highly

## Preface

controversial. We have expended a mild effort to represent a reasonable spectrum of opinions advanced by various authors. However, a book on such a subject must inevitably reflect the author's own bias. The common thread linking the subject matter included in Chapters 5–8 is the application of the mixture-model approach to the theory of fluids. It is the author's opinion that this theoretical tool is particularly useful in treating aqueous fluids and, hopefully, will help us to understand these systems on both the molecular and the macroscopic levels.

I am very much indebted to many friends and colleagues who encouraged me in undertaking the writing of this book. Thanks are due to Drs. R. Battino, D. Henderson, H. S. Frank, A. Nitzan, D. Shalitin, F. H. Stillinger, and R. Tenne for reading parts of the manuscript and kindly offering helpful comments and suggestions.

Arieh Ben-Naim

*Jerusalem, Israel*

# Contents

*Chapter 1. Introduction and Prerequisites* . . . . . . . . . 1

1.1. Introduction . . . . . . . . . . . . . . . . . . . . . . . . 1
1.2. Notation . . . . . . . . . . . . . . . . . . . . . . . . . . 2
1.3. Classical Statistical Mechanics . . . . . . . . . . . . . . 6
1.4. Connections between Statistical Mechanics and Thermodynamics . . . . . . . . . . . . . . . . . . . . . . . . . . 9
    1.4.1. $T, V, N$ Ensemble . . . . . . . . . . . . . . . . . . 9
    1.4.2. $T, P, N$ Ensemble . . . . . . . . . . . . . . . . . . 10
    1.4.3. $T, V, \mu$ Ensemble . . . . . . . . . . . . . . . . . . 10
1.5. Basic Distribution Functions in Classical Statistical Mechanics 13
1.6. Ideal Gas . . . . . . . . . . . . . . . . . . . . . . . . . . 15
1.7. Pair Potential and Pairwise Additivity . . . . . . . . . . . 17
1.8. Virial Expansion and van der Waals Equation . . . . . . 25

*Chapter 2. Molecular Distribution Functions* . . . . . . . 29

2.1. Introduction . . . . . . . . . . . . . . . . . . . . . . . . 29
2.2. The Singlet Distribution Function . . . . . . . . . . . . . 30
2.3. Pair Distribution Function . . . . . . . . . . . . . . . . . 36
2.4. Pair Correlation Function . . . . . . . . . . . . . . . . . 39

| | | |
|---|---|---|
| 2.5. | Features of the Radial Distribution Function | 43 |
| | 2.5.1. Ideal Gas | 44 |
| | 2.5.2. Very Dilute Gas | 45 |
| | 2.5.3. Slightly Dense Gas | 45 |
| | 2.5.4. Lennard-Jones Particles at Moderately High Densities | 49 |
| 2.6. | Further Properties of the Radial Distribution Function | 53 |
| 2.7. | Survey of the Methods of Evaluating $g(R)$ | 65 |
| | 2.7.1. Experimental Methods | 65 |
| | 2.7.2. Theoretical Methods | 68 |
| | 2.7.3. Simulation Methods | 69 |
| 2.8. | Higher-Order Molecular Distribution Functions | 75 |
| 2.9. | Molecular Distribution Functions (MDF) in the Grand Canonical Ensemble | 78 |

## Chapter 3. Molecular Distribution Functions and Thermodynamics — 81

| | | |
|---|---|---|
| 3.1. | Introduction | 81 |
| 3.2. | Average Values of Pairwise Quantities | 82 |
| 3.3. | Internal Energy | 85 |
| 3.4. | The Pressure Equation | 88 |
| 3.5. | The Chemical Potential | 91 |
| 3.6. | Pseudo-Chemical Potential | 99 |
| 3.7. | Entropy | 101 |
| 3.8. | Heat Capacity | 102 |
| 3.9. | The Compressibility Equation | 104 |
| 3.10. | Local Density Fluctuations | 109 |
| 3.11. | The Work Required to Form a Cavity in a Fluid | 114 |
| 3.12. | Perturbation Theories of Liquids | 120 |

## Chapter 4. Theory of Solutions — 123

| | | |
|---|---|---|
| 4.1. | Introduction | 123 |
| 4.2. | Molecular Distribution Functions in Mixtures; Definitions | 124 |
| 4.3. | Molecular Distribution Functions in Mixtures; Properties | 127 |

| | | |
|---|---|---|
| 4.4. | Mixtures of Very Similar Components . . . . . . . . . . . | 135 |
| 4.5. | The Kirkwood–Buff Theory of Solutions . . . . . . . . . . | 137 |
| 4.6. | Symmetric Ideal Solutions; Necessary and Sufficient Conditions | 145 |
| 4.7. | Small Deviations from Symmetric Ideal (SI) Solutions . . . | 153 |
| 4.8. | Dilute Ideal (DI) Solutions . . . . . . . . . . . . . . . . . | 155 |
| 4.9. | Small Deviations from Dilute Ideal Solutions . . . . . . . | 159 |
| 4.10. | A Completely Solvable Example . . . . . . . . . . . . . | 164 |
| | 4.10.1. Ideal Gas Mixture as a Reference System . . . . . . . . . | 167 |
| | 4.10.2. Symmetric Ideal Solution as a Reference System . . . . . . | 167 |
| | 4.10.3. Dilute Ideal Solution as a Reference System . . . . . . . . | 168 |
| 4.11. | Standard Thermodynamic Quantities of Transfer . . . . . . | 170 |
| | 4.11.1. Entropy . . . . . . . . . . . . . . . . . . . . . . . . . . . | 174 |
| | 4.11.2. Enthalpy . . . . . . . . . . . . . . . . . . . . . . . . . . . | 176 |
| | 4.11.3. Volume . . . . . . . . . . . . . . . . . . . . . . . . . . . | 176 |

## Chapter 5. Generalized Molecular Distribution Functions and the Mixture-Model Approach to Liquids    177

| | | |
|---|---|---|
| 5.1. | Introduction . . . . . . . . . . . . . . . . . . . . . . . | 177 |
| 5.2. | The Singlet Generalized Molecular Distribution Function . . | 179 |
| | 5.2.1. Coordination Number (CN) . . . . . . . . . . . . . . . . | 180 |
| | 5.2.2. Binding Energy (BE) . . . . . . . . . . . . . . . . . . . | 183 |
| | 5.2.3. Volume of the Voronoi Polyhedron (VP) . . . . . . . . . . | 184 |
| | 5.2.4. Combination of Properties . . . . . . . . . . . . . . . . | 186 |
| 5.3. | Illustrative Examples of GMDF's . . . . . . . . . . . . . | 187 |
| 5.4. | Pair and Higher-Order GMDF's . . . . . . . . . . . . . | 194 |
| 5.5. | Relations between Thermodynamic Quantities and GMDF's . | 195 |
| | 5.5.1. Heat Capacity at Constant Volume . . . . . . . . . . . . | 197 |
| | 5.5.2. Heat Capacity at Constant Pressure . . . . . . . . . . . . | 198 |
| | 5.5.3. Coefficient of Thermal Expansion . . . . . . . . . . . . . | 200 |
| | 5.5.4. Isothermal Compressibility . . . . . . . . . . . . . . . . | 200 |
| 5.6. | The Mixture-Model (MM) Approach; General Considerations | 201 |
| 5.7. | The Mixture-Model Approach to Liquids; Classifications Based on Local Properties of the Molecules . . . . . . . . | 208 |
| 5.8. | General Relations between Thermodynamics and Quasicomponent Distribution Functions (QCDF) . . . . . . . | 211 |

5.9. Reinterpretation of Some Thermodynamic Quantities Using the Mixture-Model Approach . . . . . . . . . . . . . 215
5.10. Some Thermodynamic Identities in the Mixture-Model Approach  217

## Chapter 6. Liquid Water . . . . . . . . . . . . . . . . . 223

6.1. Introduction . . . . . . . . . . . . . . . . . . . . . 223
6.2. Survey of Properties of Water . . . . . . . . . . . . . 225
   6.2.1. The Single Water Molecule . . . . . . . . . . . . . 226
   6.2.2. The Structure of Ice and Ice Polymorphs . . . . . . . . 227
   6.2.3. Liquid Water . . . . . . . . . . . . . . . . . . . 230
6.3. The Radial Distribution Function of Water . . . . . . . . 233
6.4. Effective Pair Potential for Water . . . . . . . . . . . . 238
   6.4.1. Construction of a HB Potential by Gaussian Functions . . . . 241
   6.4.2. Construction of a HB Potential Function Based on the Bjerrum Model for Water . . . . . . . . . . . . . . . . . . 243
6.5. Virial Coefficients of Water . . . . . . . . . . . . . . 245
6.6. Survey of Theories of Water . . . . . . . . . . . . . . 248
6.7. A Prototype of an Interstitial Model for Water . . . . . . . 252
   6.7.1. Description of the Model . . . . . . . . . . . . . . 252
   6.7.2. General Features of the Model . . . . . . . . . . . . 255
   6.7.3. First Derivatives of the Free Energy . . . . . . . . . . 258
   6.7.4. Second Derivatives of the Free Energy . . . . . . . . . 261
6.8. Application of an Exact Two-Structure Model (TSM) . . . . 265
6.9. Embedding ad hoc Models for Water in the General Framework of the Mixture-Model Approach . . . . . . . . . . 276
6.10. A Possible Definition of the "Structure of Water" . . . . . 280
6.11. Waterlike Particles in Two Dimensions . . . . . . . . . . 283
   6.11.1. The Physical Model . . . . . . . . . . . . . . . . 284
   6.11.2. Application of the Percus–Yevick Integral Equation . . . . 288
   6.11.3. Application of Monte Carlo Method . . . . . . . . . . 292
6.12. Waterlike Particles in Three Dimensions . . . . . . . . . 299
   6.12.1. Application of the Monte Carlo Technique . . . . . . . 299
   6.12.2. Application of the Percus–Yevick Equation . . . . . . . 301
   6.12.3. Application of the Molecular Dynamics Method . . . . . 302
6.13. Conclusion . . . . . . . . . . . . . . . . . . . . . 306

## Chapter 7. Water with One Simple Solute Particle ... 309

- 7.1. Introduction . . . . . . . . . . . . . . . . . . . . . 309
- 7.2. Survey of Experimental Observations . . . . . . . . . . 312
- 7.3. "Hard" and "Soft" Parts of the Dissolution Process . . . . 321
- 7.4. Application of a Two-Structure Model . . . . . . . . . . 328
- 7.5. Application of an Interstitial Model . . . . . . . . . . . 337
- 7.6. The Problem of Stabilization of the Structure of Water . . . 343
- 7.7. Application of a Continuous Mixture-Model Approach . . . 354
- 7.8. Conclusion . . . . . . . . . . . . . . . . . . . . . . 360

## Chapter 8. Water with Two or More Simple Solutes. Hydrophobic Interaction (HI) . . . . . . . . 363

- 8.1. Introduction . . . . . . . . . . . . . . . . . . . . . 363
- 8.2. Survey of Experimental Evidence on Hydrophobic Interaction 367
- 8.3. Formulation of the Problem of Hydrophobic Interaction (HI) 373
- 8.4. A Possible Connection between HI and Experimental Quantities 377
- 8.5. An Approximate Connection between HI and Experimental Quantities . . . . . . . . . . . . . . . . . . . . . . 384
- 8.6. Further Experimental Data on HI . . . . . . . . . . . . 392
- 8.7. Generalizations . . . . . . . . . . . . . . . . . . . . 398
  - 8.7.1. HI among a Set of $M$ Identical Spherical, Nonpolar Solute Particles . . . . . . . . . . . . . . . . . . . . . 398
  - 8.7.2. Attaching a Methyl Group to Various Molecules . . . . . . 404
  - 8.7.3. Attaching an Ethyl Group to Various Molecules . . . . . . 406
- 8.8. Hydrophobic Interaction at Zero Separation . . . . . . . . 407
- 8.9. Hydrophobic Interaction at More Realistic Distances . . . . 414
  - 8.9.1. Extrapolation from $\delta A^{HI}(0)$ and $\delta A^{HI}(\sigma_1)$ . . . . . . . . . 414
  - 8.9.2. Hydrophobic Interaction between Bulky Molecules . . . . . 416
- 8.10. Entropy and Enthalpy of Hydrophobic Interaction . . . . . 422
- 8.11. Hydrophobic Interaction and Structural Changes in the Solvent 428
- 8.12. Simulations . . . . . . . . . . . . . . . . . . . . . 436
- 8.13. Conclusion . . . . . . . . . . . . . . . . . . . . . . 438

*Chapter 9. Appendix* . . . . . . . . . . . . . . . . . . 441

9-A. Rotational Partition Function for a Rigid Asymmetric Molecule 441
9-B. Functional Derivative and Functional Taylor Expansion . . 443
9-C. The Ornstein–Zernike Relation . . . . . . . . . . . . . . 447
9-D. The Percus–Yevick Integral Equation . . . . . . . . . . . 450
9-E. Solution of the Percus–Yevick (PY) Equation . . . . . . . 453
9-F. The Chemical Potential in the $T$, $P$, $N$ Ensemble . . . . . . 454
9-G. The Chemical and the Pseudo-Chemical Potential in a Lattice Model . . . . . . . . . . . . . . . . . . . . . . . . . . . 457

*Glossary of Abbreviations* . . . . . . . . . . . . . . . . 459

*References* . . . . . . . . . . . . . . . . . . . . . . . . 461

*Index* . . . . . . . . . . . . . . . . . . . . . . . . . . 471

Chapter 1

# Introduction and Prerequisites

## 1.1. INTRODUCTION

A molecular theory of matter provides a bridge between the molecular properties of the individual particles constituting the system under observation and the macroscopic properties of the bulk matter as revealed in an actual experiment. The mathematical tools to establish such a bridge are contained within the realm of statistical mechanics.

In this book, we restrict our considerations to systems at equilibrium. It is well known that the statistical theory of nonequilibrium systems has not reached the same level of perfection as the theory of systems at thermodynamic equilibrium. It is also well known that development of the molecular theory of liquids has lagged well behind the corresponding work in gases and solids. It is not difficult to trace the reasons for this. Gases are characterized by a relatively low density. Hence, simultaneous interactions among a large group of molecules may be neglected. In fact, one can include in a systematic fashion interactions between pairs, triplets, etc., to get a plausible theory of real gases (e.g., Mayer and Mayer, 1940; Hill, 1956; Uhlenbeck and Ford, 1962). The other extreme case is the perfect solid, which is characterized by regularity of the lattice structure. This feature again helps to simplify the mathematical apparatus required to handle the solid state.

A liquid is neither a rarefied nor an ordered collection of molecules. Therefore, it is no wonder that progress in the theory of liquids has been slow and difficult. Nevertheless, in the past two decades, remarkable progress has been achieved in the theory of simple liquids, i.e., liquids composed of simple molecules or atoms.

This chapter surveys some of the basic elements of statistical mechanics necessary for the development of the subject matter in the subsequent chapters. Most of the material in this chapter is presumed to be known to the reader. The main reason for presenting it here is to establish a unified system of notation which will be employed throughout the book.

## 1.2. NOTATION

The location of the center of the molecule is denoted by the vector $\mathbf{R} = (x, y, z)$, where $x$, $y$, and $z$ are the Cartesian coordinates of a specific point in the molecule, chosen as its center. For instance, for a water molecule, it will be convenient to choose the center of the oxygen atom as the center of the molecule. We shall often use the symbol $\mathbf{R}$ to denote the location of a specific point in the system, not necessarily occupied by the center of a molecule. If there are several molecules, then $\mathbf{R}_i$ denotes the location of the center of the $i$th molecule.

An infinitesimal element of volume is denoted by

$$d\mathbf{R} = dx\, dy\, dz \qquad (1.1)$$

This represents the *volume* of a small cube defined by the edges $dx$, $dy$, and $dz$, as illustrated in Fig. 1.1. Some texts use the notation $d^3R$ for the element of volume to distinguish it from the infinitesimal vector, denoted by $d\mathbf{R}$. In this book, we never need to use an infinitesimal vector; thus, $d\mathbf{R}$ will always signify an element of volume.

We usually make no distinction between the notation for the *region* (defined, say, by the cube of edges $dx$, $dy$, and $dz$) and its *volume* (given by the product $dx\, dy\, dz$). For instance, we may say that the center of a particle falls within $d\mathbf{R}$, meaning that the center is contained in the *region* of space defined by the cube of edges $dx$, $dy$, and $dz$. On the other hand, if $dx = dy = dz = 2$ Å, then $d\mathbf{R}$ is also the *volume* (8 Å$^3$) of this region.

The element of volume $d\mathbf{R}$ is understood to be located at the point $\mathbf{R}$. In some cases, it will be convenient to choose an element of volume other than a cubic one. For instance, an infinitesimal spherical shell of radius $R$ and width $dR$ has the volume

$$d\mathbf{R} = 4\pi R^2\, dR \qquad (1.2)$$

# Introduction and Prerequisites

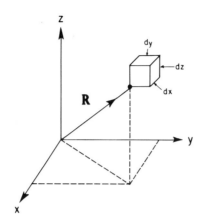

Fig. 1.1. An infinitesimal element of volume $d\mathbf{R} = dx\, dy\, dz$ located at the point $\mathbf{R}$.

For simple particles such as hard spheres or argon atoms, the designation of the locations of all the centers of the particles in the system is a sufficient description of the *configuration* of the system. A more detailed description of both location and orientation is needed for more complex molecules. For a rigid, nonspherical molecule, we use $\mathbf{R}_i$ to designate the location of its center, and $\mathbf{\Omega}_i$ the orientation of the whole molecule. As an example, consider a water molecule as being a rigid particle.[1] Figure 1.2 shows one possible set of angles used to describe the orientation of a water molecule. Let $\boldsymbol{\mu}$ be the vector originating from the center and bisecting the H–O–H angle. Then, two angles, say $\phi$ and $\theta$, are required to fix the orientation of this vector. In addition, a third angle $\psi$ is needed to describe the rotation of the whole molecule about the axis $\boldsymbol{\mu}$. The choice of the triplet of angles included in $\mathbf{\Omega}$ is by no means a unique one. One very commonly used set is the Euler angles (Goldstein, 1950). Although we shall never need to specify the choice of angles, we shall always presume that the Euler angles have been chosen.

In general, integration over the variable $\mathbf{R}_i$ means integration over the whole volume of the system, i.e.,

$$\int d\mathbf{R}_i = \int_V d\mathbf{R}_i = \int_0^L dx_i \int_0^L dy_i \int_0^L dz_i \qquad (1.3)$$

---

[1] The assumption is made that the geometry of the molecule is fixed. This is consistent with the well-known fact that deviations from the equilibrium values for the O–H distance and the H–O–H angle are almost negligible at room temperature (see also Chapter 6).

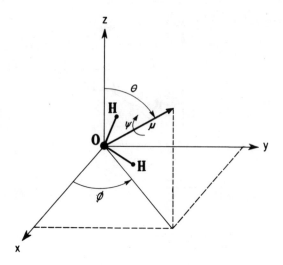

Fig. 1.2. One possible choice of orientational angles for a water molecule. The vector $\boldsymbol{\mu}$, originating from the center of the oxygen and bisecting the H–O–H angle, can be specified by the two polar angles $\theta$ and $\phi$. In addition, the angle of rotation of the molecule about the axis $\boldsymbol{\mu}$ is denoted by $\psi$.

where for simplicity we have assumed that the region of integration is a cube of length $L$. The integration over $\boldsymbol{\Omega}_i$ will be understood to be over all possible orientations of the molecule. Using, for instance, the set of Euler angles, we have

$$\int d\boldsymbol{\Omega}_i = \int_0^{2\pi} d\phi_i \int_0^{\pi} \sin\theta_i \, d\theta_i \int_0^{2\pi} d\psi_i \qquad (1.4)$$

The *configuration* of a rigid molecule is thus specified by a six-dimensional vector, including the *location* and the *orientation* of the molecule, namely,

$$\mathbf{X}_i = \mathbf{R}_i, \boldsymbol{\Omega}_i = (x_i, y_i, z_i, \phi_i, \theta_i, \psi_i) \qquad (1.5)$$

The configuration of a system of $N$ rigid molecules is denoted by

$$\mathbf{X}^N = \mathbf{X}_1, \mathbf{X}_2, \ldots, \mathbf{X}_N \qquad (1.6)$$

Similarly, the infinitesimal element of the configuration of a single molecule is denoted by

$$d\mathbf{X}_i = d\mathbf{R}_i \, d\boldsymbol{\Omega}_i \qquad (1.7)$$

# Introduction and Prerequisites

and for $N$ molecules,

$$d\mathbf{X}^N = d\mathbf{X}_1 d\mathbf{X}_2 \cdots d\mathbf{X}_N \tag{1.8}$$

For a general molecule, the specification of the configuration by six coordinates may be insufficient. A simple example is $n$-butane, a schematic description of which is given in the Fig. 1.3. Here, the "center" of the molecule is chosen as the position of one of the carbon atoms. The orientation of the triplet of carbon atoms, say, 1, 2, and 3, is denoted by $\mathbf{\Omega}$. This specification leaves one degree of freedom, which is referred to as internal rotation, and is denoted by $\phi_{23}$. This completes our specification of the configuration of the molecule (it is presumed that bond lengths and angles are fixed at their equilibrium values). In the general case, the set of internal rotational angles will comprise the vector $\mathbf{P}$ and will be referred to as the *conformation* of the molecule. Thus, the total *configuration* in the general case includes the coordinates of *location*, *orientation*, and *internal rotation*, and is denoted by

$$\mathbf{Y} = \mathbf{R}, \mathbf{\Omega}, \mathbf{P} \tag{1.9}$$

In most of the book, we shall be dealing with rigid molecules, in which case a specification of internal rotations is not required.

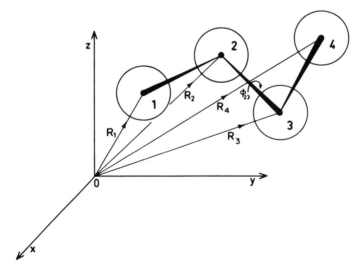

Fig. 1.3. Schematic description of $n$-butane. The total configuration of the molecule consists of the location of, say, the carbon numbered 1 ($\mathbf{R}_1$), the orientation of the (rigid) triplet of carbons, say 1, 2, and 3, and an internal rotational angle, designated by $\phi_{23}$.

## 1.3. CLASSICAL STATISTICAL MECHANICS

A fundamental quantity in statistical mechanics is the partition function (PF). There are various forms of the partition function, each having its advantage in application to a particular problem. The most common one is the so-called canonical PF, and applies to a system having a fixed number of molecules $N$ contained in a vessel of volume $V$ and maintained at a constant temperature $T$. (For a multicomponent system, $N$ is replaced by the vector $\mathbf{N} = N_1, N_2, \ldots, N_c$ specifying the composition of the system, i.e., $N_i$, $i = 1, 2, \ldots, c$ is the number of molecules of the $i$th species.)

The PF for such a system is defined by

$$Q(T, V, N) = \sum_{i=1}^{\infty} \exp(-\beta E_i) \tag{1.10}$$

Here, $\beta = (kT)^{-1}$, with $k$ the Boltzmann constant ($k = 1.38 \times 10^{-16}$ erg·deg$^{-1}$) and $T$ the absolute temperature. $E_i$ is the $i$th eigenvalue, or energy level, of the Hamiltonian of the system. The summation is carried out over all the eigenstates of this Hamiltonian.

Although the partition function defined in (1.10) is a more fundamental starting point in statistical mechanics, we shall always use the classical analog of the PF which, for a system of $N$ simple particles (i.e., spherical particles having no internal structure), takes the form

$$Q(T, V, N) = (1/h^{3N} N!) \int \cdots \int d\mathbf{p}^N \, d\mathbf{R}^N \exp(-\beta H) \tag{1.11}$$

Here, $h$ is the Planck constant ($h = 6.625 \times 10^{-27}$ erg·sec) and $H$ is the classical Hamiltonian of the system, given by

$$H(\mathbf{p}^N, \mathbf{R}^N) = \sum_{i=1}^{N} (\mathbf{p}_i^2/2m) + U_N(\mathbf{R}^N) \tag{1.12}$$

with $\mathbf{p}_i$ the momentum vector of the $i$th particle (presumed to possess only translational degrees of freedom), and $m$ is the mass of each particle. The total potential energy of the system at the specified configuration $\mathbf{R}^N$ is denoted by $U_N(\mathbf{R}^N)$.

We shall not be concerned with the details of the transition from the quantum mechanical to the classical version of the partition function. This can be found in standard texts (e.g., Hill, 1956; Huang, 1963). We note only that the expression (1.11) is not purely classical insofar as it contains

# Introduction and Prerequisites

two remnants of quantum mechanical origin: the Planck constant and the $N$-factorial.

The integration over the momenta in (1.11) can be performed to obtain

$$h^{-3N} \int_{-\infty}^{\infty} \cdots \int_{-\infty}^{\infty} d\mathbf{p}^N \exp\left[-\beta \sum_{i=1}^{N} (\mathbf{p}_i^2/2m)\right]$$

$$= \left[h^{-1} \int_{-\infty}^{\infty} dp \exp(-\beta p^2/2m)\right]^{3N} = \left[h^{-1}(2m/\beta)^{1/2} \int_{-\infty}^{\infty} \exp(-x^2)\, dx\right]^{3N}$$

$$= [(2\pi mkT)^{3/2}/h^3]^N = \Lambda^{-3N} \quad (1.13)$$

where we have introduced the momentum partition function

$$\Lambda = h/(2\pi mkT)^{1/2} \quad (1.14)$$

which is often referred to as the "thermal de Broglie wavelength" of the particles at temperature $T$.

Another important quantity is the configurational partition function, defined by

$$Z_N = \int \cdots \int d\mathbf{R}^N \exp[-\beta U_N(\mathbf{R}^N)] \quad (1.15)$$

which can be used to rewrite the canonical PF in (1.11) as

$$Q(T, V, N) = Z_N/(N!\,\Lambda^{3N}) \quad (1.16)$$

Clearly, the most difficult problem in statistical mechanics is the evaluation of the multidimensional integral (1.15). Unfortunately, the integrand does not factorize into $N$ factors, each depending on one of the vectors $\mathbf{R}_i$. Major efforts have been aimed at finding approximate ways of evaluating integrals of the type in (1.15). Some of these will be described in subsequent chapters.

For a system of $N$ nonspherical molecules, the PF is modified as follows:

$$Q(T, V, N) = \frac{q^N}{(8\pi^2)^N \Lambda^{3N} N!} \int \cdots \int d\mathbf{X}^N \exp[-\beta U_N(\mathbf{X}^N)] \quad (1.17)$$

Here, $\Lambda$ has the same significance as in (1.14) and (1.16). There are essentially two new features in the PF given in (1.17). First, we have acknowledged the fact that the total potential energy of the system depends upon the locations and orientations of the molecules. Hence, integration is carried out over all possible configurations of the system. Second, the factor $q$

includes the rotational PF as well as any internal PF that the molecule may possess. (Further discussion on the rotational PF of a rigid molecule is included in Appendix 9-A.)

It is important to realize that factoring out the internal PF ($q$) as in (1.17) involves the assumption that the internal degrees of freedom of a single molecule do not depend on the nature of the environment in which the molecule may be found. That is, different configurations $\mathbf{X}^N$ produce different environments for the molecules, which in turn may affect the internal partition function of the molecule. Such an assumption may not be granted automatically and we shall return to discuss this topic in relation to water in Chapter 6. Until then, we shall assume the validity of the PF as written in (1.17).

A second PF that we shall use often is the isothermal–isobaric PF, defined by

$$\Delta(T, P, N) = C \int_0^\infty dV\, Q(T, V, N) \exp(-\beta PV) \tag{1.18}$$

where $P$ is the pressure and the integration extends over all the possible values of $V$. This partition function is multiplied by a constant $C$ having the dimensions of a reciprocal volume, so that the whole expression is rendered dimensionless. The exact specification of $C$ will be of no concern to us since we shall only be interested in ratios of such quantities, in which case any additional constant will eventually be cancelled out. [For more details, see Hill (1956), Brown (1958), and Münster (1969).] The isothermal–isobaric PF is more convenient than the canonical PF for establishing relations with experimental quantities, which are more commonly obtained in a system under constant pressure than constant volume. This aspect will be demonstrated in some of the applications to aqueous solutions.

Finally, we introduce the so-called grand partition function, which is characterized by the thermodynamic variables $T$, $V$, and $\mu$, i.e., for an open system at constant $T$, $V$, and $\mu$. The formal definition is

$$\Xi(T, V, \mu) = \sum_{N=0}^\infty Q(T, V, N) \exp(\beta \mu N) \tag{1.19}$$

where $\mu$ is the chemical potential and it is understood that the first term in the sum is unity, i.e.,

$$Q(T, V, N = 0) = 1 \tag{1.20}$$

In this book, we shall use the grand PF mainly for the study of the thermodynamics of solutions in Chapter 4.

# Introduction and Prerequisites

## 1.4. CONNECTIONS BETWEEN STATISTICAL MECHANICS AND THERMODYNAMICS

In this section, we present a list of the most useful relations between the various partition functions and the corresponding thermodynamic quantities. In each case, we first cite the "fundamental relation" which may be viewed as an axiomatic assumption, then follow with further relations obtained by purely thermodynamic arguments. The relations are grouped according to the set of variables employed in the characterization of the system.

### 1.4.1. T, V, N Ensemble

The fundamental relation in the $T$, $V$, $N$ ensemble is that between the Helmholtz free energy and the canonical partition function:

$$A(T, V, N) = -kT \ln Q(T, V, N) \tag{1.21}$$

If the PF $Q(T, V, N)$ of a system can be evaluated, either exactly or approximately, relation (1.21) gives an explicit dependence of $A$ on the variables $T$, $V$, and $N$ (as well as on the molecular properties of the particles constituting our system). Such a relation is fundamental in the sense that all thermodynamic information on the system can be extracted from it by the application of standard thermodynamic relations. For instance, one can use the relation

$$dA = -S\, dT - P\, dV + \mu\, dN \tag{1.22}$$

with $S$ the entropy, $P$ the pressure, and $\mu$ the chemical potential [for a multicomponent system, the last term on the rhs of (1.22) should be interpreted as a scalar product $\boldsymbol{\mu} \cdot d\mathbf{N} = \sum_{i=1}^{c} \mu_i \, dN_i$], to get

$$S = -\left(\frac{\partial A}{\partial T}\right)_{V,N} = k \ln Q + kT\left(\frac{\partial \ln Q}{\partial T}\right)_{V,N} \tag{1.23}$$

$$P = -\left(\frac{\partial A}{\partial V}\right)_{T,N} = kT\left(\frac{\partial \ln Q}{\partial V}\right)_{T,N} \tag{1.24}$$

$$\mu = \left(\frac{\partial A}{\partial N}\right)_{T,V} = -kT\left(\frac{\partial \ln Q}{\partial N}\right)_{T,V} \tag{1.25}$$

Other relations can be obtained readily; for instance, the internal energy is obtained from (1.21) and (1.23):

$$E = A + TS = kT^2 \left(\frac{\partial \ln Q}{\partial T}\right)_{V,N} \tag{1.26}$$

The constant-volume heat capacity can be obtained by differentiating (1.26) with respect to temperature.

### 1.4.2. T, P, N Ensemble

The fundamental relation in the $T, P, N$ ensemble is between the Gibbs free energy and the isothermal–isobaric PF, namely,

$$G(T, P, N) = -kT \ln \Delta(T, P, N) \tag{1.27}$$

Using the thermodynamic relation

$$dG = -S\, dT + V\, dP + \mu\, dN \tag{1.28}$$

we get

$$S = -\left(\frac{\partial G}{\partial T}\right)_{P,N} = k \ln \Delta + kT \left(\frac{\partial \ln \Delta}{\partial T}\right)_{P,N} \tag{1.29}$$

$$V = \left(\frac{\partial G}{\partial P}\right)_{T,N} = -kT \left(\frac{\partial \ln \Delta}{\partial P}\right)_{T,N} \tag{1.30}$$

$$\mu = \left(\frac{\partial G}{\partial N}\right)_{T,P} = -kT \left(\frac{\partial \ln \Delta}{\partial N}\right)_{T,P} \tag{1.31}$$

The enthalpy can be obtained from (1.27) and (1.29):

$$H = G + TS = kT^2 \left(\frac{\partial \ln \Delta}{\partial T}\right)_{P,N} \tag{1.32}$$

The constant-pressure heat capacity can be obtained by differentiating (1.32) with respect to temperature. The isothermal compressibility is obtained by differentiating (1.30) with respect to pressure, etc.

### 1.4.3. T, V, μ Ensemble

The fundamental relation in the $T, V, \mu$ ensemble is

$$P(T, V, \mu)V = kT \ln \Xi(T, V, \mu) \tag{1.33}$$

## Introduction and Prerequisites

Using the thermodynamic relation

$$d(PV) = S\, dT + P\, dV + N\, d\mu \tag{1.34}$$

we get

$$S = \left(\frac{\partial (PV)}{\partial T}\right)_{V,\mu} = k \ln \Xi + kT\left(\frac{\partial \ln \Xi}{\partial T}\right)_{V,\mu} \tag{1.35}$$

$$P = \left(\frac{\partial (PV)}{\partial V}\right)_{T,\mu} = kT\left(\frac{\partial \ln \Xi}{\partial V}\right)_{T,\mu} = kT\frac{\ln \Xi}{V} \tag{1.36}$$

$$N = \left(\frac{\partial (PV)}{\partial \mu}\right)_{T,V} = kT\left(\frac{\partial \ln \Xi}{\partial \mu}\right)_{T,V} \tag{1.37}$$

This completes our general survey of the relations between thermodynamic quantities and molecular properties (through the partition function) of the system. In the above list, we adopted the common practice in thermodynamics of not using a special notation for exact or average quantities. For instance, in the $T$, $V$, $N$ ensemble, $N$ is a fixed number, whereas in the $T$, $V$, $\mu$ ensemble, the number of particles fluctuates. Therefore, in relation (1.37), for example, one should use a special symbol such as $\bar{N}$ or $\langle N \rangle$ to denote the *average* number of particles in the system. In the following sections, we shall not elaborate on such a distinction in the notation if the meaning is quite apparent from the context. The discussion of fluctuation contains such a distinction, and we use the notation $\langle \rangle$ to denote an average quantity in the appropriate ensemble.

We close this section with a few relations expressing the extent of fluctuation of various thermodynamic quantities. In the $T$, $V$, $N$ ensemble, the most important relation is that giving the fluctuation in the internal energy, i.e.,

$$\langle (E - \langle E \rangle)^2 \rangle = \langle E^2 \rangle - \langle E \rangle^2 = kT^2 C_V \tag{1.38}$$

where the constant-volume heat capacity is defined by

$$C_V = (\partial \langle E \rangle / \partial T)_{V,N} \tag{1.39}$$

Similarly, in the $T$, $P$, $N$ ensemble, we have

$$\langle (H - \langle H \rangle)^2 \rangle = \langle H^2 \rangle - \langle H \rangle^2 = kT^2 C_P \tag{1.40}$$

where

$$C_P = (\partial \langle H \rangle / \partial T)_{P,N} \tag{1.41}$$

Note that we use the same symbol $\langle \rangle$ for different averages in (1.38) and

(1.40), the former standing for an average in the $T, V, N$ ensemble, the latter in the $T, P, N$ ensemble (see also Section 1.5).

Two other relations of importance in the $T, P, N$ ensemble are

$$\langle (V - \langle V \rangle)^2 \rangle = \langle V^2 \rangle - \langle V \rangle^2 = kT\varkappa_T \langle V \rangle \tag{1.42}$$

where the isothermal compressibility is defined by

$$\varkappa_T = -\frac{1}{\langle V \rangle}\left(\frac{\partial \langle V \rangle}{\partial P}\right)_{T,N} \tag{1.43}$$

The second involves the cross fluctuations of volume and enthalpy, which are related to the thermal expansivity

$$\langle (V - \langle V \rangle)(H - \langle H \rangle) \rangle = \langle VH \rangle - \langle V \rangle \langle H \rangle = kT^2 \alpha \langle V \rangle \tag{1.44}$$

where

$$\alpha = \frac{1}{\langle V \rangle}\left(\frac{\partial \langle V \rangle}{\partial T}\right)_{P,N} \tag{1.45}$$

In the $T, V, \mu$ ensemble, of foremost importance is the fluctuation in the number of particles, which, for a one-component system, reads

$$\langle (N - \langle N \rangle)^2 \rangle = \langle N^2 \rangle - \langle N \rangle^2 = kT\left(\frac{\partial \langle N \rangle}{\partial \mu}\right)_{T,V} = kTV\left(\frac{\partial \varrho}{\partial \mu}\right)_T \tag{1.46}$$

where we have used $\varrho = \langle N \rangle / V$ for the average density.

Furthermore, using the Gibbs–Duhem relation

$$-S\,dT + V\,dP = N\,d\mu \tag{1.47}$$

we get

$$\left(\frac{\partial P}{\partial \mu}\right)_T = \frac{N}{V} = \varrho \tag{1.48}$$

and also

$$\left(\frac{\partial \varrho}{\partial \mu}\right)_T = \left(\frac{\partial \varrho}{\partial P}\right)_T \left(\frac{\partial P}{\partial \mu}\right)_T = \varkappa_T \varrho^2 \tag{1.49}$$

Combining (1.46) and (1.49), we get

$$\langle N^2 \rangle - \langle N \rangle^2 = kTV\varrho^2 \varkappa_T \tag{1.50}$$

Further relations involving cross fluctuations in number of particles in a multicomponent system are discussed in Chapter 4.

# Introduction and Prerequisites

## 1.5. BASIC DISTRIBUTION FUNCTIONS IN CLASSICAL STATISTICAL MECHANICS

Most of the subject matter of statistical mechanics is concerned with the evaluation of average quantities. The averaging process differs according to the choice of the set of thermodynamic variables which characterize the system under consideration.

Let $Q(\mathbf{X}^N)$ be any function of the configuration $\mathbf{X}^N$. The average of this function is then defined by

$$\langle Q \rangle = \int \cdots \int d\mathbf{X}^N \, P(\mathbf{X}^N) Q(\mathbf{X}^N) \qquad (1.51)$$

where $P(\mathbf{X}^N)$ is the distribution function,[2] i.e., $P(\mathbf{X}^N) \, d\mathbf{X}^N$ is the probability of finding particle 1 in the element $d\mathbf{X}_1, \ldots,$ particle $N$ in the element $d\mathbf{X}_N$.

We shall often say that $P(\mathbf{X}^N)$ is the probability density of observing the "event" $\mathbf{X}^N$. One should remember, however, that the *probability* of an exact event $\mathbf{X}^N$ is always zero, and that the probabilistic meaning is assigned only to an element of "volume" $d\mathbf{X}^N$ at $\mathbf{X}^N$.

In the $T, V, N$ ensemble, the basic distribution function is the probability density for observing the configuration $\mathbf{X}^N$,

$$P(\mathbf{X}^N) = \frac{\exp[-\beta U_N(\mathbf{X}^N)]}{\int \cdots \int d\mathbf{X}^N \exp[-\beta U_N(\mathbf{X}^N)]} \qquad (1.52)$$

Clearly, the normalization condition for $P(\mathbf{X}^N)$ is

$$\int \cdots \int d\mathbf{X}^N \, P(\mathbf{X}^N) = 1 \qquad (1.53)$$

which is consistent with the probabilistic meaning of $P(\mathbf{X}^N) \, d\mathbf{X}^N$. In other words, the probability of finding the system in any possible configuration is unity.

In the $T, P, N$ ensemble, the basic distribution function is the probability density of finding a system with a volume $V$ *and* a configuration $\mathbf{X}^N$:

$$P(\mathbf{X}^N, V) = \frac{\exp[-\beta PV - \beta U_N(\mathbf{X}^N)]}{\int dV \int \cdots \int d\mathbf{X}^N \exp[-\beta PV - \beta U_N(\mathbf{X}^N)]} \qquad (1.54)$$

---

[2] The concept of "distribution function" as used in statistical mechanics differs from that employed in the mathematical theory of probability. For instance, in a one-dimensional system, the distribution function $F(x)$ is defined as $F(x) = \int_{-\infty}^{x} f(y) \, dy$, where $f(y)$ is referred to as the density function. Most of the quantities we shall be using in this book are actually density functions and not distribution functions, in the mathematical sense.

The integration over $V$ extends from zero to infinity. The probability density of observing a system with volume $V$ independently of the configuration is obtained from (1.54) by integrating over all configurations, i.e.,

$$P(V) = \int \cdots \int d\mathbf{X}^N \, P(\mathbf{X}^N, V) \tag{1.55}$$

Finally, the conditional distribution function defined by

$$P(\mathbf{X}^N/V) = \frac{P(\mathbf{X}^N, V)}{P(V)} = \frac{\exp[-\beta PV - \beta U_N(\mathbf{X}^N)]}{\int \cdots \int d\mathbf{X}^N \exp[-\beta PV - \beta U_N(\mathbf{X}^N)]}$$

$$= \frac{\exp[-\beta U_N(\mathbf{X}^N)]}{\int \cdots \int d\mathbf{X}^N \exp[-\beta U_N(\mathbf{X}^N)]} \tag{1.56}$$

is the conditional probability density of finding a system in the configuration $\mathbf{X}^N$, given that the system has the volume $V$. Note the similarity between (1.52) and (1.56).

In the $T, V, \mu$ or grand ensemble, the basic distribution function defined by

$$P(\mathbf{X}^N, N) = \frac{(q^N/N!) \exp[\beta \mu N - \beta U_N(\mathbf{X}^N)]}{\sum_{N=0}^{\infty} (q^N/N!)[\exp(\beta \mu N)] \int \cdots \int d\mathbf{X}^N \exp[-\beta U_N(\mathbf{X}^N)]} \tag{1.57}$$

is the probability density of observing a system with precisely $N$ particles and the configuration $\mathbf{X}^N$.

The probability of finding a system in the $T, V, \mu$ ensemble with exactly $N$ particles is obtained from (1.57) by integrating over all possible configurations, namely,

$$P(N) = \int \cdots \int d\mathbf{X}^N \, P(\mathbf{X}^N, N) \tag{1.58}$$

which can be rewritten, using the notation of Section 1.3, as

$$P(N) = Q(T, V, N)[\exp(\beta \mu N)]/\Xi(T, V, \mu) \tag{1.59}$$

The conditional distribution function defined by

$$P(\mathbf{X}^N/N) = \frac{P(\mathbf{X}^N, N)}{P(N)} = \frac{\exp[-\beta U_N(\mathbf{X}^N)]}{\int \cdots \int d\mathbf{X}^N \exp[-\beta U_N(\mathbf{X}^N)]} \tag{1.60}$$

is the probability density of observing a system in the configuration $\mathbf{X}^N$, given that the system contains precisely $N$ particles. Note again the similarity between (1.60) and (1.52).

## Introduction and Prerequisites

The above expressions are the most fundamental distribution functions; from these we shall compute average quantities in the various ensembles. In the next chapter, we introduce some additional distribution functions, referred to as molecular distribution functions. The latter are derived from the basic distributions in a rather straightforward manner.

### 1.6. IDEAL GAS

The ideal gas is an important example for demonstrating the power of the methods of statistical mechanics. The reason for this stems from the fact that relatively simple and explicit expressions can be derived for the thermodynamic properties of the gas in terms of its molecular parameters. An ideal gas is characterized by the absense of intermolecular forces between the molecules, i.e.,

$$U_N(\mathbf{X}^N) \equiv 0 \tag{1.61}$$

for each configuration $\mathbf{X}^N$. Of course, no real system obeys relation (1.61). Nevertheless, the results obtained for ideal gases, using assumption (1.61), are closely attained by real gases at very low densities when the occurrence of pairs, triplets, etc. of particles at short interparticle separation becomes an exceedingly rare event.

Substituting (1.61) in (1.17), we realize that the crux of the difficulty involved in the evaluation of the configurational integral is immediately removed, and we have

$$\begin{aligned} Q(T, V, N) &= \frac{q^N}{(8\pi^2)^N \Lambda^{3N} N!} \int \cdots \int d\mathbf{X}^N \\ &= \frac{q^N}{(8\pi^2)^N \Lambda^{3N} N!} \left[ \int_V d\mathbf{R} \int_0^{2\pi} d\phi \int_0^{\pi} \sin\theta\, d\theta \int_0^{2\pi} d\psi \right]^N \\ &= \frac{q^N V^N}{\Lambda^{3N} N!} \end{aligned} \tag{1.62}$$

For simple spherical particles we get, from either (1.15) and (1.16) or from (1.62) (with $q = 1$)

$$Q(T, V, N) = V^N/\Lambda^{3N} N! \tag{1.63}$$

Note that $q$ and $\Lambda$ depend on the temperature and not on the volume $V$ or on $N$. An important consequence of this is that the equation of the state of an ideal gas is independent of the particular molecules constituting the system. To see this, we derive the expression for the pressure (1.24) by differentiating (1.62) with respect to volume:

$$P = kT\left(\frac{\partial \ln Q}{\partial V}\right)_{T,N} = \frac{kTN}{V} = \varrho kT \qquad (1.64)$$

Evidently, this equation of state does not depend on the properties of the molecules. This behavior is not shared by other thermodynamic relations of the ideal gas. For instance, the chemical potential obtained from (1.25) and (1.62) and using the Stirling approximation[3] is

$$\mu = -kT\left(\frac{\partial \ln Q}{\partial N}\right)_{T,V} = kT \ln(\Lambda^3 q^{-1}) + kT \ln \varrho$$
$$= \mu^{og}(T) + kT \ln \varrho \qquad (1.65)$$

where $\varrho = N/V$ is the number density and $\mu^{og}(T)$ is the standard chemical potential. The latter conveys the properties of the individual molecules in the system. We note in passing that (1.65) exemplifies the merits of statistical mechanics. Once we know the details of the properties of a single molecule, we can get, through (1.65), an explicit expression for the chemical potential of the system.

Another useful expression is that for the entropy of an ideal gas, which can be obtained from (1.23) and (1.62):

$$S = k \ln Q + kT\left(\frac{\partial \ln Q}{\partial T}\right)_{V,N} = \tfrac{5}{2}kN - Nk \ln(\varrho\Lambda^3 q^{-1}) + kTN \frac{\partial \ln q}{\partial T} \qquad (1.66)$$

For simple particles, this reduces to the well-known relation

$$S = \tfrac{5}{2}kN - Nk \ln \varrho\Lambda^3 \qquad (1.67)$$

---

[3] In this book, we always use the Stirling approximation in the form

$$\ln N! = N \ln N - N$$

A better approximation for small values of $N$ is (Feller, 1960)

$$\ln N! = N \ln N - N + \tfrac{1}{2} \ln(2\pi N)$$

However, for $N$ of the order $10^{23}$, the first approximation is sufficient. It is also useful to note that, within the Stirling approximation, we have

$$(\partial/\partial N)(\ln N!) = \ln N$$

Similarly, the energy of a system of simple particles is obtained from (1.63) and (1.67) as

$$E = A + TS = kTN \ln \varrho \Lambda^3 - kTN + T(\tfrac{5}{2}kN - Nk \ln \varrho \Lambda^3) = \tfrac{3}{2}kTN \tag{1.68}$$

which, in this case, is entirely due to the kinetic energy of the particles.

The heat capacity for a system of simple particles is obtained directly from (1.68) as

$$C_V = (\partial E/\partial T)_V = \tfrac{3}{2}kN \tag{1.69}$$

which may be viewed as originating from the accumulation of $\tfrac{1}{2}k$ per translational degree of freedom of a particle. For molecules also having rotational degrees of freedom, we get

$$C_V = 3kN \tag{1.70}$$

which is built up of $\tfrac{3}{2}kN$ from the translational and $\tfrac{3}{2}kN$ from the rotational degrees of freedom. If other internal degrees of freedom are present, there are additional contributions to $C_V$.

## 1.7. PAIR POTENTIAL AND PAIRWISE ADDITIVITY

The total potential energy of interaction $U_N(\mathbf{X}^N)$ appearing in the classical partition function (1.17) is defined as the work required to bring $N$ molecules from infinite separation to the configuration $\mathbf{X}^N$. This work may, in general, be a very complicated function of the configuration $\mathbf{X}^N$. In order to achieve progress in the theory of liquids, one needs to make some simplifying assumptions. The most fruitful one has been the so-called pairwise additivity assumption, which states that the total work required to bring $N$ molecules from infinite separation to the configuration $\mathbf{X}^N$ is equal to the sum of the work required to bring each pair of molecules, say $i$ and $j$, from infinite separation to the final configuration $\mathbf{X}_i$, $\mathbf{X}_j$. Thus, one writes for the total potential energy

$$U_N(\mathbf{X}^N) = \sum_{1 \leq i < j \leq N} U(\mathbf{X}_i, \mathbf{X}_j) \tag{1.71}$$

where $U(\mathbf{X}_i, \mathbf{X}_j)$ is called the pair potential and the summation in (1.71) extends over all of the $N(N-1)/2$ pairs of different molecules in the system.

Most of the progress in the molecular theory of liquids has been achieved for systems obeying the pairwise additivity assumption. One system for which (1.71) is obeyed is a system composed of hard spheres. Hard spheres (HS) are idealized particles defined by a pair potential which is zero when the separation is greater than their diameter $\sigma$, and infinity when the separation is less than $\sigma$, i.e.,

$$U^{\text{HS}}(\mathbf{R}_i, \mathbf{R}_j) = \begin{cases} 0 & \text{for } |\mathbf{R}_i - \mathbf{R}_j| > \sigma \\ \infty & \text{for } |\mathbf{R}_i - \mathbf{R}_j| < \sigma \end{cases} \quad (1.72)$$

A schematic illustration of the hard-sphere pair potential is depicted in Fig. 1.4.

Hard spheres are characterized by the single parameter $\sigma$, their diameter. Of course, the definition of the *pair* potential (1.72) does not automatically provide a definition of $U_N(\mathbf{R}^N)$ for a system of hard spheres. The most reasonable definition for this would be the following:

$$U_N(\mathbf{R}^N) = \begin{cases} 0 & \text{if } |\mathbf{R}_i - \mathbf{R}_j| > \sigma \text{ for all } i,j = 1, 2, \ldots, N; i \neq j \\ \infty & \text{if at least one } |\mathbf{R}_i - \mathbf{R}_j| \text{ is less than } \sigma \end{cases} \quad (1.73)$$

The total potential energy is zero if no two particles are closer to each other than $\sigma$, and is infinity once at least a single pair is at a closer distance than $\sigma$. Clearly, using the definition (1.72), we obtain

$$\sum_{1 \leq i < j \leq N} U^{\text{HS}}(\mathbf{R}_i, \mathbf{R}_j) = \begin{cases} 0 & \text{if } |\mathbf{R}_i - \mathbf{R}_j| > \sigma \\ & \text{for all } i,j = 1, 2, \ldots, N; i \neq j \\ \infty & \text{if at least one } |\mathbf{R}_i - \mathbf{R}_j| \\ & \text{is less than } \sigma \end{cases} \quad (1.74)$$

Comparing (1.73) and (1.74), we get the equality

$$U_N(\mathbf{R}^N) = \sum_{1 \leq i < j \leq N} U^{\text{HS}}(\mathbf{R}_i, \mathbf{R}_j) \quad (1.75)$$

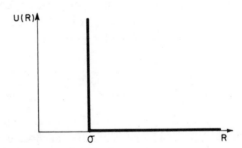

Fig. 1.4. Pair potential for hard spheres of diameter $\sigma$.

# Introduction and Prerequisites

One should remember that hard spheres are not real particles, and (1.75) is valid by virtue of the definition (1.73). Therefore, the pairwise additivity assumption must be viewed as being a built-in feature of the *definition* of a system of hard spheres.[4]

By a simple generalization, one can define nonspherical hard particles for which (1.71) is fulfilled. Other systems for which the pairwise additivity assumption is presumed to hold are systems of idealized point charges, point dipoles, point quadrupoles, etc.

A system of real particles such as argon is believed to obey relation (1.71) approximately. Although it is now well known that even the simplest molecules do not obey (1.71) exactly, it is still considered a useful approximation without which little, if any, progress in the theory of liquids could have been achieved.

Let us present a general argument underlying the adoption of a particular pair potential for real systems. We begin by noting that the pair potential function $U(R)$ is not known exactly for any pair of real particles. Nevertheless, we know that at a large distance there is a net attraction between the two particles, whereas at a very short distance they exert very strong repulsive forces on each other. The general form of the pair potential for two simple atoms such as argon is depicted in Fig. 1.5. [Note that since the energy of the system is defined up to an arbitrary additive constant, we can make the choice of fixing $U(R) = 0$ at $R = \infty$. The important features of the pair potential are the slopes, rather than the values, at each point.]

Knowing the general *form* of $U(R)$ is not sufficient for developing a theory of liquids; we must translate the form into a mathematical function. Of course, there are infinitely many possible functions which have forms similar to the one depicted in Fig. 1.5. The most common one, which partially rests on theoretical grounds, is the so-called Lennard-Jones (LJ) pair potential, which reads

$$U_{\text{LJ}}(R) = 4\varepsilon[(\sigma/R)^{12} - (\sigma/R)^6] \qquad (1.76)$$

This is a two-parameter function which conveys the general form of the expected "real" pair potential of a pair of simple atoms. The parameter $\sigma$ can be conveniently assigned the meaning of an effective diameter of the

---

[4] Of course, one can define $U_N(\mathbf{R}^N)$ for a system of hard spheres by (1.72) and (1.75) and then obtain the property (1.73). This point of view is stressed in the following for Lennard-Jones particles.

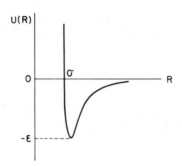

Fig. 1.5. General form of the pair potential function for simple and spherical molecules.

particles, whereas $\varepsilon$ can serve as a measure of the strength of the interaction between the two particles.

Let us examine a few features of the LJ function (1.76). Clearly,

$$U_{\text{LJ}}(R = \sigma) = 0 \tag{1.77}$$

For $R < \sigma$, the two particles exert strong repulsive forces on each other and therefore this region is effectively impenetrable, which justifies the meaning of $\sigma$ as an effective diameter. The minimum of $U_{\text{LJ}}(R)$ occurs at $R = 2^{1/6}\sigma$ and its value is

$$U_{\text{LJ}}(R = 2^{1/6}\sigma) = -\varepsilon \tag{1.78}$$

Two LJ curves with parameters for neon and argon are shown in Fig. 1.6. The force exerted on one particle by the other is given by

$$F(R) = -\partial U_{\text{LJ}}(R)/\partial R \tag{1.79}$$

This is negative (attraction) for $R > 2^{1/6}\sigma$ and positive (repulsion) for $R < 2^{1/6}\sigma$.

The leading, long-range, $\sim R^{-6}$ behavior of the pair potential has some theoretical basis. On the other hand, the $\sim R^{-12}$ behavior at short distances is simply a convenient analytical way of expressing the strong repulsive forces [for more details, see Hirschfelder et al. (1954)]. Therefore, the overall LJ function should be viewed basically as a "model" for the pair potential operating between two real particles. We can now define a system of "LJ particles" as a system of imaginary particles for which the pair potential is the LJ function (1.76) and for which the total potential function for a system of $N$ such particles obeys the assumption of pairwise additivity.

# Introduction and Prerequisites

It is important to stress that knowledge even of the exact pair potential does not furnish any information on the potential energy of a system containing more than two particles.

In most theories of the liquid state, one replaces the system of real particles by "model particles" such as LJ particles. The crucial assumption is that the behavior of the model system will faithfully represent the behavior of the real system. Very similar arguments will guide us in choosing an adequate model system of particles to represent the water molecules.

There are various methods of fixing the "best" parameters $\sigma$ and $\varepsilon$ for various molecules. Table 1.1 gives an idea of the order of magnitude of these parameters as obtained from the second virial coefficient (see next section).

The values of $\sigma$ and $\varepsilon$ do not have any absolute significance. First, because of the arbitrariness of the LJ function, one could have chosen

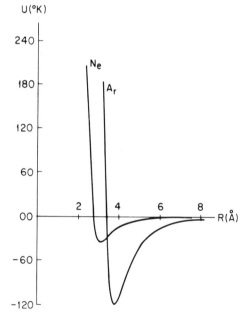

Fig. 1.6. The form of the Lennard-Jones curves for neon (with $\varepsilon/k = 34.9°K$ and $\sigma = 2.78$ Å) and for argon (with $\varepsilon/k = 119.8°K$ and $\sigma = 3.405$ Å). The parameters are obtained from data on the second virial coefficient for these gases. Note that the minimum of each curve occurs at $R = 2^{1/6}\sigma$.

### Table 1.1

**Values of $\sigma$ and of $\varepsilon/k$ for the Lennard-Jones Function Fitted to Obtain the Second Virial Coefficient of Some Simple Gases**[a]

| Gas | $\sigma$, Å | $\varepsilon/k$, °K |
|---|---|---|
| Ne | 2.78 | 34.9 |
| A | 3.40 | 119.8 |
| Kr | 3.60 | 171.0 |
| Xe | 4.10 | 221.0 |
| $CH_4$ | 3.82 | 148.2 |

[a] From Hirschfelder et al. (1954).

an analytical function quite different from (1.76) and characterized by different parameters. Second, even if the LJ function were the best one to represent the pair potential of the real molecules, there is as yet no experimental method of determining the parameters $\sigma$ and $\varepsilon$ uniquely. More details on this problem can be found in the book by Hirschfelder et al. (1954). For recent discussion see also Maitland and Smith (1973) and Barker et al. (1971).

We now turn to a brief consideration of the origin of nonadditivity effects of the potential function. We demonstrate the idea with a simple example. Consider a system of three particles (Fig. 1.7) whose pair potential is a superposition of a hard-sphere and a dipole–dipole interaction:

$$U_2(\mathbf{R}_1, \mathbf{\Omega}_1, \mathbf{R}_2, \mathbf{\Omega}_2) = U^{\text{HS}}(|\mathbf{R}_2 - \mathbf{R}_1|) + U^{\text{DD}}(\mathbf{R}_1, \mathbf{\Omega}_1, \mathbf{R}_2, \mathbf{\Omega}_2) \quad (1.80)$$

where $U^{\text{HS}}(R)$ is defined in (1.72) (note that $U^{\text{HS}}$ is a function of the scalar distance $|\mathbf{R}_2 - \mathbf{R}_1|$ only), and the dipole–dipole interaction is given by

$$U^{\text{DD}}(\mathbf{R}_1, \mathbf{\Omega}_1, \mathbf{R}_2, \mathbf{\Omega}_2) = \frac{\boldsymbol{\mu}_1 \cdot \boldsymbol{\mu}_2 - 3(\boldsymbol{\mu}_1 \cdot \mathbf{u}_{12})(\boldsymbol{\mu}_2 \cdot \mathbf{u}_{12})}{|\mathbf{R}_2 - \mathbf{R}_1|^3} \quad (1.81)$$

Here, $\boldsymbol{\mu}_i$ is the dipole moment vector of the $i$th particle and $\mathbf{u}_{12}$ is a unit vector along $\mathbf{R}_2 - \mathbf{R}_1$. For any given configuration of the three particles, we have in this case the exact relation of additivity

$$U_3(\mathbf{X}_1, \mathbf{X}_2, \mathbf{X}_3) = U_2(\mathbf{X}_1, \mathbf{X}_2) + U_2(\mathbf{X}_1, \mathbf{X}_3) + U_2(\mathbf{X}_2, \mathbf{X}_3) \quad (1.82)$$

where $\mathbf{X}_i$ stands for $\mathbf{R}_i, \mathbf{\Omega}_i$.

## Introduction and Prerequisites

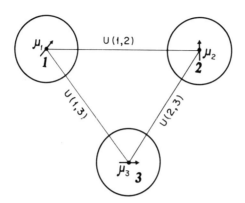

Fig. 1.7. A configuration of three hard spheres with a point dipole embedded at their centers. The dipole vectors are denoted by $\mathbf{\mu}_1$, $\mathbf{\mu}_2$, and $\mathbf{\mu}_3$. The pair potentials are $U(1, 2)$, $U(1, 3)$, and $U(2, 3)$.

Next, given the same system of three particles, suppose that we were not aware of the detailed structure of the particles and therefore would like to describe its pair potential by a function of the separation only. The pair potential that we will get is the average over all the orientations of the two particles, namely,

$$V_2(\mathbf{R}_i, \mathbf{R}_j) \equiv \langle U_2(\mathbf{X}_i, \mathbf{X}_j)\rangle_2$$
$$= \frac{\int\int d\mathbf{\Omega}_i\, d\mathbf{\Omega}_j\, U_2(\mathbf{X}_i, \mathbf{X}_j) \exp[-\beta U_2(\mathbf{X}_i, \mathbf{X}_j)]}{\int\int d\mathbf{\Omega}_i\, d\mathbf{\Omega}_j\, \exp[-\beta U_2(\mathbf{X}_i, \mathbf{X}_j)]} \quad (1.83)$$

The symbol $\langle\ \rangle_2$ stands for an average over all the orientations of the pair of particles. The average interaction energy for the triplet of particles at positions $\mathbf{R}_1$, $\mathbf{R}_2$, $\mathbf{R}_3$ is defined similarly as

$$V_3(\mathbf{R}_1, \mathbf{R}_2, \mathbf{R}_3) = \langle U_3(\mathbf{X}_1, \mathbf{X}_2, \mathbf{X}_3)\rangle_3 \quad (1.84)$$

where here the symbol $\langle\ \rangle_3$ means an average over all the orientations of the three particles, namely

$$\langle U_3(\mathbf{X}_1, \mathbf{X}_2, \mathbf{X}_3)\rangle_3$$
$$= \frac{\int\int\int d\mathbf{\Omega}_1\, d\mathbf{\Omega}_2\, d\mathbf{\Omega}_3\, U_3(\mathbf{X}_1, \mathbf{X}_2, \mathbf{X}_3) \exp[-\beta U_3(\mathbf{X}_1, \mathbf{X}_2, \mathbf{X}_3)]}{\int\int\int d\mathbf{\Omega}_1\, d\mathbf{\Omega}_2\, d\mathbf{\Omega}_3\, \exp[-\beta U_3(\mathbf{X}_1, \mathbf{X}_2, \mathbf{X}_3)]} \quad (1.85)$$

We recall that if we record all of the positions and orientations of the particles, we have a perfect additivity as given in (1.82). However, if we disregard the orientation, we have a new pair potential (1.83) and a new triplet potential (1.84), which in this case does not necessarily obey the additivity assumption. To see this more clearly, let us substitute (1.82) in (1.84) to obtain

$$\begin{aligned} V_3(\mathbf{R}_1, \mathbf{R}_2, \mathbf{R}_3) &= \langle U_3(\mathbf{X}_1, \mathbf{X}_2, \mathbf{X}_3) \rangle_3 \\ &= \langle U_2(\mathbf{X}_1, \mathbf{X}_2) \rangle_3 + \langle U_2(\mathbf{X}_1, \mathbf{X}_3) \rangle_3 + \langle U_2(\mathbf{X}_2, \mathbf{X}_3) \rangle_3 \\ &\neq \langle U_2(\mathbf{X}_1, \mathbf{X}_2) \rangle_2 + \langle U_2(\mathbf{X}_1, \mathbf{X}_3) \rangle_2 + \langle U_2(\mathbf{X}_2, \mathbf{X}_3) \rangle_2 \\ &= V_2(\mathbf{R}_1, \mathbf{R}_2) + V_2(\mathbf{R}_1, \mathbf{R}_3) + V_2(\mathbf{R}_2, \mathbf{R}_3) \end{aligned} \qquad (1.86)$$

The inequality in (1.86) arises because, in general, the averages $\langle \rangle_2$ and $\langle \rangle_3$ do not produce the same results. In other words, the average interaction for a pair of particles may be different if, in the process of averaging, a third particle is present in the system.

The above example illustrates a general principle. If we have a very detailed specification of our system, then we hope to get complete additivity. Once, however, we average over part of the degrees of freedom of the system, we may get a simpler description of the system, but with the sacrifice of additivity.

With a little imagination, we can infer from the above example the source of nonadditivity of interaction in real molecules. Let us observe three argon atoms at positions $\mathbf{R}_1$, $\mathbf{R}_2$, and $\mathbf{R}_3$. If, instead of this description, we record at each moment the locations of all the electrons and nuclei, then we have an exact additive potential (assuming that the electrons and nuclei are point charges[5]). Such a description would be formidably complicated and hardly useful. Instead, we average over all the positions of the electrons, getting a relatively simplified function (of fewer degrees of freedom). The resulting averaging process will, however, in general produce nonadditivity. Again we see the same principle: Ignoring part of the degrees of freedom has the virtue of simplifying the description of our system, but at the same time we have sacrificed the property of additivity.

In the next chapter, we encounter another example of nonadditivity produced by averaging over part of the degrees of freedom of the system.

---

[5] Of course, the interaction between the nuclei themselves may not be additive and we shall have to specify, say, the positions and orientations of more elementary particles.

## 1.8. VIRIAL EXPANSION AND VAN DER WAALS EQUATION

In Section 1.6, we noted that the equation of the state of an ideal gas is "universal," i.e., it does not depend on any particular property of the gas under observation. For any gas at sufficiently low density, we have

$$P/kT = N/V = \varrho \tag{1.87}$$

The universal character of the equation of state ceases to hold for nonideal gases. This fact is already revealed by the well-known van der Waals equation, which can be written as

$$(P + \varrho^2 a)(1 - \varrho b) = \varrho kT \tag{1.88}$$

where the constants $a$ and $b$ were first introduced by van der Waals to account for the "attraction" between the particles and the finite volume of the particles, respectively. As we shall soon see, these two constants are actually two features of the intermolecular potential operating between the particles, and any division between "attraction" and "size" of the particles is quite arbitrary. Before showing this, we note that modern theories of real gases provide evidence that the van der Waals equation is actually valid only in a certain limited range of densities. Let us rewrite (1.88) as an expansion of $P/kT$ in the density $\varrho$,

$$\frac{P}{kT} = \varrho - \varrho^2 \frac{a}{kT} + \frac{P\varrho b}{kT} + \varrho^3 \frac{ab}{kT} \tag{1.89}$$

It is now known that this particular expansion is valid up to terms of order $\varrho^2$. Since at low densities, $P/kT \sim \varrho$, we see that the term $P\varrho b/kT$ is of order $\varrho^2$. Therefore, we can replace $P/kT$ by $\varrho$ if we are going to retain the expansion up to the second order in the density, and we get

$$\frac{P}{kT} = \varrho + \left(b - \frac{a}{kT}\right)\varrho^2 + \cdots \tag{1.90}$$

The term of order $\varrho^3$ has been dropped in (1.90). We already see here that $a$ and $b$ appear in a certain combination as a coefficient of the second power of the density. Statistical mechanics gives a general procedure for identifying this coefficient, as well as higher ones in the density expansion of the pressure.

The general expansion of the pressure can be written as

$$P/kT = \varrho + B_2(T)\varrho^2 + B_3(T)\varrho^3 + \cdots \quad (1.91)$$

where the $B_k(T)$ are called the virial coefficients of the gas. Statistical mechanics provides general expressions for $B_k(T)$ in terms of the molecular properties of the gas. For systems of simple spherical particles, we have the well-known relations [see, for example, Hill (1956)]

$$B_2(T) = -\tfrac{1}{2} \int_0^\infty \{\exp[-\beta U(R)] - 1\} 4\pi R^2 \, dR \quad (1.92)$$

$$B_3(T) = -(1/3V) \iiint_V d\mathbf{R}^3 \, \{\exp[-\beta U(\mathbf{R}_1, \mathbf{R}_2)] - 1\}$$
$$\times \{\exp[-\beta U(\mathbf{R}_1, \mathbf{R}_3)] - 1\}\{\exp[-\beta U(\mathbf{R}_2, \mathbf{R}_3)] - 1\} \quad (1.93)$$

Let us further elaborate on the content of $B_2(T)$ for a Lennard-Jones gas. We know that for $R \leq \sigma$, the potential function is positive and increases steeply to infinity as we reduce the distance. Therefore, we can approximate the integral in (1.92) by

$$B_2(T) = -\tfrac{1}{2} \left[ \int_0^\sigma \{\exp[-\beta U(R)] - 1\} 4\pi R^2 \, dR \right.$$
$$\left. + \int_\sigma^\infty \{\exp[-\beta U(R)] - 1\} 4\pi R^2 \, dR \right]$$
$$\approx \tfrac{1}{2} \left[ \int_0^\sigma 4\pi R^2 \, dR + \beta \int_\sigma^\infty U(R) 4\pi R^2 \, dR \right]$$

i.e., we put $\exp[-\beta U(R)] = 0$ for $R \leq \sigma$ and expand to the first order the exponent $\exp[-\beta U(R)] \sim 1 - \beta U(R)$ for $R > \sigma$. If we now make the identification

$$b = \frac{1}{2} \int_0^\sigma 4\pi R^2 \, dR = \frac{1}{2} \frac{4\pi \sigma^3}{3} \quad (1.94)$$

$$\frac{a}{kT} = \frac{-\int_\sigma^\infty U(R) 4\pi R^2 \, dR}{2kT} \quad (1.95)$$

we see that the constant $b$ is related to the "volume" of the particles, whereas the constant $a$ is roughly an average "attraction" between them. These two features appear in (1.92) in the integral over the full pair potential $U(R)$. The splitting into two terms is of course arbitrary; we can choose $\sigma$ at any point for which $\exp[-\beta U(R)] \approx 0$ for all $R \leq \sigma$. The rest of the

# Introduction and Prerequisites

integral in (1.95) will, in general, include both an attractive and a repulsive part of the potential function. For completeness, we also cite the expressions for $B_2(T)$ and $B_3(T)$ for nonspherical particles:

$$B_2(T) = - [1/2(8\pi^2)] \int d\mathbf{X}_2 f(\mathbf{X}_1, \mathbf{X}_2) \tag{1.96}$$

$$B_3(T) = - [1/3(8\pi^2)^2] \int d\mathbf{X}_2 \, d\mathbf{X}_3 f(\mathbf{X}_1, \mathbf{X}_2) f(\mathbf{X}_1, \mathbf{X}_3) f(\mathbf{X}_2, \mathbf{X}_3) \tag{1.97}$$

where we have introduced the notation

$$f(\mathbf{X}_i, \mathbf{X}_j) = \exp[-\beta U(\mathbf{X}_i, \mathbf{X}_j)] - 1 \tag{1.98}$$

Note that in (1.93) and (1.97) we have assumed the pairwise additivity of the total potential function for three particles.

## Chapter 2
# Molecular Distribution Functions

## 2.1. INTRODUCTION

The notions of molecular distribution functions (MDF) command a central role in the theory of fluids. Of foremost importance among these are the singlet and the pair distribution functions. This chapter is mainly devoted to describing and surveying the fundamental features of these two functions. At the end of the chapter, we briefly mention the general definitions of higher-order MDF's. These are rarely incorporated into actual applications, since very little is known about their properties.

The pair correlation function, introduced in Section 2.4, conveys information on the mode of packing of the molecules in the liquid. This information is often referred to as representing a sort of "order," or amount of "structure" that persists in the liquid. Moreover, the pair correlation function forms an important bridge between molecular properties and thermodynamic quantities. This topic is discussed in Chapter 3.

We start by defining the singlet and pair distribution functions for a system of *rigid*, not necessarily spherical particles. This is a rather simple generalization of the corresponding functions for spherical particles which are discussed in current texts on the theory of liquids.

All of the MDF's are easily derivable from the fundamental distribution functions introduced in Chapter 1. They may be assigned either a probabilistic meaning or interpreted as averages of number densities in an appropriate ensemble. Both aspects will be presented in Sections 2.2 and 2.3.

In order for the reader to acquire a good familiarity with the pair correlation functions, ample illustrations are presented in Sections 2.4–2.6. Both theoretical and experimental methods of evaluating this function are surveyed in Section 2.7.

## 2.2. THE SINGLET DISTRIBUTION FUNCTION

All the results obtained in this section are very simple and almost self-evident. Nevertheless, we have taken the liberty to make the subject matter appear more complicated than is really necessary. The reason is that we would like to stress some subtle points which are particularly clear in the context of this simple case, but are more difficult to grasp in other cases that will appear throughout the book.

The system under consideration consists of $N$ rigid particles at a given temperature and contained in a vessel of volume $V$. The basic probability density for such a system is (see Section 1.5)

$$P(\mathbf{X}^N) = \frac{\exp[-\beta U_N(\mathbf{X}^N)]}{\int \cdots \int d\mathbf{X}^N \exp[-\beta U_N(\mathbf{X}^N)]} \quad (2.1)$$

An average of any function of the configuration $F(\mathbf{X}^N)$ in the $T, V, N$ ensemble is defined by

$$\bar{F} = \int \cdots \int d\mathbf{X}^N \, P(\mathbf{X}^N) F(\mathbf{X}^N) \quad (2.2)$$

In some cases, we use a special symbol, such as $\langle F \rangle$, for an average quantity. However, we refrain from using this notation whenever the meaning of that quantity as an average is quite evident.

As a very simple example, let us express the average number of particles in a certain region $S$ within the vessel. (A particle is said to be in the region $S$ whenever its *center* falls within that region.) Let $N(\mathbf{X}^N, S)$ be the number of particles in $S$ at a particular configuration $\mathbf{X}^N$ of the system. One may imagine taking a snapshot of the system at some instant and counting the number of particles that happen to fall in $S$ at that particular configuration. An illustration is given in Fig. 2.1.

The average number of particles in $S$ is, according to (2.2),

$$N(S) = \int \cdots \int d\mathbf{X}^N \, P(\mathbf{X}^N) N(\mathbf{X}^N, S) \quad (2.3)$$

# Molecular Distribution Functions

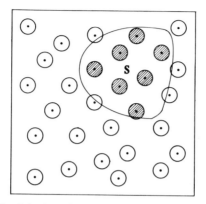

Fig. 2.1. A region $S$ within the vessel containing the system. Shaded circles indicate particles whose center falls within $S$.

This relation can be rewritten in a somewhat more complicated form, which will turn out to be useful for later applications.

Let us define the characteristic function

$$A_i(\mathbf{R}_i, S) = \begin{cases} 1 & \text{if } \mathbf{R}_i \in S \\ 0 & \text{if } \mathbf{R}_i \notin S \end{cases} \quad (2.4)$$

The symbol $\in$ means "belongs to" or "contained in." The quantity $N(\mathbf{X}^N, S)$ now can be expressed as

$$N(\mathbf{X}^N, S) = \sum_{i=1}^{N} A_i(\mathbf{R}_i, S) \quad (2.5)$$

which simply means that in order to *count* the number of particles within $S$, we have to check the *location* of each particle separately. Each particle whose center falls within $S$ will contribute unity to the sum on the rhs of (2.5); hence, the sum counts the number of particles in $S$. Introducing (2.5) into (2.3), we get

$$\begin{aligned} N(S) &= \int \cdots \int d\mathbf{X}^N \, P(\mathbf{X}^N) \sum_{i=1}^{N} A_i(\mathbf{R}_i, S) \\ &= \sum_{i=1}^{N} \int \cdots \int d\mathbf{X}^N \, P(\mathbf{X}^N) A_i(\mathbf{R}_i, S) \\ &= N \int \cdots \int d\mathbf{X}^N \, P(\mathbf{X}^N) A_1(\mathbf{R}_1, S) \end{aligned} \quad (2.6)$$

In (2.6), we have exploited the fact that all the particles are equivalent; hence, the sum over the index $i$ produces $N$ integrals having the same magnitude. We may therefore select one of these integrals, say $i = 1$, and replace the sum by a product of $N$ equal integrals.

The "mole fraction" of particles within $S$ is defined as

$$x(S) = N(S)/N = \int \cdots \int d\mathbf{X}^N P(\mathbf{X}^N) A_1(\mathbf{R}_1, S) \tag{2.7}$$

This quantity may also be assigned a probabilistic meaning that is often useful. To see this, we recall that the function $A_1(\mathbf{R}_1, S)$ employed in (2.7) has the effect of reducing the range of integration from $V$ to a restricted range which fulfills the condition of $\mathbf{R}_1$ being located in $S$. Symbolically, this can be rewritten as[1]

$$\int_V \cdots \int_V d\mathbf{X}^N P(\mathbf{X}^N) A_1(\mathbf{R}_1, S) = \int \cdots \int_{R_1 \in S} d\mathbf{X}^N P(\mathbf{X}^N) = P_1(S) \tag{2.8}$$

We recall that $P(\mathbf{X}^N)$ is the probability density of the occurrence of the "event" $\mathbf{X}^N$. Therefore, integration over all the "events" $\mathbf{X}^N$ for which the condition $\mathbf{R}_1 \in S$ is fulfilled gives the probability of the occurrence of the condition, i.e., $P_1(S)$ is the probability that a specific particle, say number 1, will be found in $S$. Of course, we could have chosen in (2.6) any specific particle other than 1. From (2.7) and (2.8), we arrive at an important relation:

$$x(S) = P_1(S) \tag{2.9}$$

which states that the mole fraction of particles in $S$ equals the probability that a *specific* particle, say 1, will be found in $S$. Chapter 5 gives some other relations similar to (2.9).

We now introduce the singlet molecular distribution function, which is obtained from $N(S)$ in the limit of a very small region $S$. First we note

---

[1] The content of (2.8) may be more apparent in a one-dimensional example. Let $S$ be a segment $(a, b)$ on the $x$ axis; then we have

$$\int_{-\infty}^{\infty} P(x) A(x, S) \, dx = \int_{x \in S} P(x) \, dx = \int_a^b P(x) \, dx$$

where $A(x, S)$ is defined as unity whenever $x \in S$ and zero elsewhere. Note that in (2.8) we have integration over both locations and orientations of the particles. In the present context only the integration over the locations is affected by $A_1(\mathbf{R}_1, S)$.

## Molecular Distribution Functions

that $A_i(\mathbf{R}_i, S)$ can also be written as

$$A_i(\mathbf{R}_i, S) = \int_S \delta(\mathbf{R}_i - \mathbf{R}') \, d\mathbf{R}' \qquad (2.10)$$

where we have exploited the basic property of the Dirac delta function: The integral is unity if $\mathbf{R}_i \in S$ and zero otherwise.

If $S$ is an infinitesimally small region $d\mathbf{R}'$, then we have

$$A_i(\mathbf{R}_i, d\mathbf{R}') = \delta(\mathbf{R}_i - \mathbf{R}') \, d\mathbf{R}' \qquad (2.11)$$

Hence, from (2.6), we get

$$N(dR') = d\mathbf{R}' \int \cdots \int d\mathbf{X}^N P(\mathbf{X}^N) \sum_{i=1}^{N} \delta(\mathbf{R}_i - \mathbf{R}') \qquad (2.12)$$

The average local (number) density of particles in the element of volume $d\mathbf{R}'$ is defined by

$$\varrho^{(1)}(\mathbf{R}') = N(d\mathbf{R}')/d\mathbf{R}' = \int \cdots \int d\mathbf{X}^N P(\mathbf{X}^N) \sum_{i=1}^{N} \delta(\mathbf{R}_i - \mathbf{R}') \qquad (2.13)$$

which is also referred to as the singlet molecular distribution function.

The meaning of $\varrho^{(1)}(\mathbf{R}')$ as local density will prevail in all of our applications. However, in some cases, one also may assign to $\varrho^{(1)}(\mathbf{R}')$ the meaning of probability density. This must be done with some care, as will be shown. First, we can rewrite (2.13) as we did in (2.6), in the form

$$\varrho^{(1)}(\mathbf{R}') = N \int \cdots \int d\mathbf{X}^N P(\mathbf{X}^N) \, \delta(\mathbf{R}_1 - \mathbf{R}') = N P^{(1)}(\mathbf{R}') \qquad (2.14)$$

The interpretation of $P^{(1)}(\mathbf{R}') \, d\mathbf{R}'$ follows from the same argument as in the case of $P_1(S)$ in (2.8). Namely, this is the probability of finding a specific particle, say 1, in $d\mathbf{R}'$ at $\mathbf{R}'$. Hence, $P^{(1)}(\mathbf{R}')$ is often referred to as the *specific* singlet distribution function.

The next question is: What is the probability of finding *any* particle in $d\mathbf{R}'$? To answer this question, we consider the following events:

| event | probability of the event |
|---|---|
| particle 1 in $d\mathbf{R}'$ | $P^{(1)}(\mathbf{R}') \, d\mathbf{R}'$ |
| particle 2 in $d\mathbf{R}'$ | $P^{(1)}(\mathbf{R}') \, d\mathbf{R}'$ |
| ⋮ | ⋮ |
| particle $N$ in $d\mathbf{R}'$ | $P^{(1)}(\mathbf{R}') \, d\mathbf{R}'$ |

Clearly, by virtue of the equivalence of the particles, we have exactly the same probability for all the events listed above.

The event "*any* particle in $d\mathbf{R}'$" means either "particle 1 in $d\mathbf{R}'$," or "particle 2 in $d\mathbf{R}'$," ..., or "particle $N$ in $d\mathbf{R}'$." In probability language, this event is called the *union* of all the events listed above, and is written symbolically as

$$\{\text{any particle in } d\mathbf{R}'\} = \bigcup_{i=1}^{N} \{\text{particle } i \text{ in } d\mathbf{R}'\} \qquad (2.15)$$

In general, there exists no simple relation between the probability of a union of events and the probabilities of the individual events. However, if we choose $d\mathbf{R}'$ to be small enough so that no more than a single particle may be found in it at any given time, then all the events listed above become disjointed (i.e., occurrence of one event excludes the possibility of simultaneous occurrence of any other event). In this case, we have the additivity relation for the probability of the union, namely,

$$\begin{aligned}\Pr\{\text{any particle in } d\mathbf{R}'\} &= \sum_{i=1}^{N} \Pr\{\text{particle } i \text{ in } d\mathbf{R}'\} \\ &= \sum_{i=1}^{N} P^{(1)}(\mathbf{R}')\, d\mathbf{R}' = NP^{(1)}(\mathbf{R}')\, d\mathbf{R}' \\ &= \varrho^{(1)}(\mathbf{R}')\, d\mathbf{R}' \qquad (2.16)\end{aligned}$$

(The symbol Pr stands for "probability of.") Relation (2.16) gives the probabilistic meaning of the quantity $\varrho^{(1)}(\mathbf{R}')\, d\mathbf{R}'$, which is contingent upon the choice of a sufficiently small element of volume $d\mathbf{R}'$. The quantity $\varrho^{(1)}(\mathbf{R}')$ is often referred to as the *generic* singlet distribution function.

Some care must be exercised when using the probabilistic meaning of $\varrho^{(1)}(\mathbf{R}')\, d\mathbf{R}'$. For instance, the probability of finding a *specific* particle, say 1, in a region $S$ is obtained from the *specific* singlet distribution function simply by integration:

$$P_1(S) = \int_S P^{(1)}(\mathbf{R}')\, d\mathbf{R}' \qquad (2.17)$$

which follows from the fact that the events "particle 1 in $d\mathbf{R}'$" and "particle 1 in $d\mathbf{R}''$" are disjoint events (i.e., a specific particle cannot be in two different elements $d\mathbf{R}'$ and $d\mathbf{R}''$ simultaneously). Hence, the probability of the union (particle 1 in $S$) is obtained as the sum (or integral) of the probabilities of the individual events.

## Molecular Distribution Functions

This property is not shared by the *generic* singlet distribution function, and the integral

$$\int_S \varrho^{(1)}(\mathbf{R}') \, d\mathbf{R}' \tag{2.18}$$

does not have the meaning of the probability of the event "any particle in $S$." The reason is that the events "a particle in $d\mathbf{R}'$" and "a particle in $d\mathbf{R}''$" are not disjoint, hence one cannot obtain the probability of their union in a simple fashion. It is for this reason that the meaning of $\varrho^{(1)}(\mathbf{R}')$ as a local density at $\mathbf{R}'$ should be preferred. If $\varrho^{(1)}(\mathbf{R}') \, d\mathbf{R}'$ is viewed as the average number of particles in $d\mathbf{R}'$, then, clearly, (2.18) is the average number of particles in $S$. The meaning of $\varrho^{(1)}(\mathbf{R}') \, d\mathbf{R}'$ as an average number of particles is preserved upon integration; the probabilistic meaning is not.

A particular example of (2.18) occurs when $S$ is chosen as the total volume of the vessel, i.e.,

$$\int_V \varrho^{(1)}(\mathbf{R}') \, d\mathbf{R}' = N \int_V P^{(1)}(\mathbf{R}') \, d\mathbf{R}' = N \tag{2.19}$$

The last equality follows from the normalization of $P^{(1)}(\mathbf{R}')$; i.e., the probability of finding particle 1 in any place in $V$ is unity. The normalization condition (2.19) can also be obtained directly from (2.13).

In a homogeneous fluid, we expect that $\varrho^{(1)}(\mathbf{R}')$ will have the same value at any point $\mathbf{R}'$ within the system. (This is true apart from a very small region near the surface of the system, which we always neglect in considering macroscopic systems). Therefore, we put

$$\varrho^{(1)}(\mathbf{R}') = \text{const} \tag{2.20}$$

and from (2.19) we get

$$\text{const} \times \int_V d\mathbf{R}' = N \tag{2.21}$$

Hence

$$\varrho^{(1)}(\mathbf{R}') = N/V = \varrho \tag{2.22}$$

The last relation is quite a trivial result for homogeneous systems. It states that the local density at each point $\mathbf{R}'$ is equal to the bulk density $\varrho$. This is, of course, not true in an inhomogeneous system.

In a similar fashion, we may introduce the singlet distribution function for location and orientation, which by analogy to (2.14) is defined as

$$\varrho^{(1)}(\mathbf{X}') = \int \cdots \int d\mathbf{X}^N \, P(\mathbf{X}^N) \sum_{i=1}^{N} \delta(\mathbf{X}_i - \mathbf{X}')$$

$$= N \int \cdots \int d\mathbf{X}^N \, P(\mathbf{X}^N) \, \delta(\mathbf{X}_1 - \mathbf{X}')$$

$$= NP^{(1)}(\mathbf{X}') \qquad (2.23)$$

where $P^{(1)}(\mathbf{X}')$ is the probability density of finding a specific particle at a given configuration $\mathbf{X}'$.

Again, in a homogeneous and isotropic fluid, we expect that

$$\varrho^{(1)}(\mathbf{X}') = \text{const} \qquad (2.24)$$

Hence, using the normalization condition

$$\int \varrho^{(1)}(\mathbf{X}') \, d\mathbf{X}' = N \int P^{(1)}(\mathbf{X}') \, d\mathbf{X}' = N \qquad (2.25)$$

we get

$$\varrho^{(1)}(\mathbf{X}') = N/V(8\pi^2) = \varrho/8\pi^2 \qquad (2.26)$$

The connection between $\varrho^{(1)}(\mathbf{R}')$ and $\varrho^{(1)}(\mathbf{X}')$ is obtained simply by integration over all the orientations[2]

$$\varrho^{(1)}(\mathbf{R}') = \int \varrho^{(1)}(\mathbf{X}') \, d\mathbf{\Omega}' \qquad (2.27)$$

## 2.3. PAIR DISTRIBUTION FUNCTION

In this section, we introduce the pair distribution function. We first present its meaning as a probability density and then show how it can be reinterpreted as an average quantity.

The starting point is the basic probability density $P(\mathbf{X}^N)$, which was defined in Chapter 1, Eq. (1.52), for the $T, V, N$ ensemble (a similar procedure can be employed for introducing the molecular distribution function in other ensembles). The *specific* pair distribution function is defined as the probability density of finding molecule 1 at $\mathbf{X}'$ *and* molecule 2 at $\mathbf{X}''$. This

---

[2] Note that we are using the same symbol $\varrho^{(1)}$ to denote two different functions, a function of $\mathbf{R}'$ and a function of $\mathbf{X}'$. However, no confusion should arise from this notation since the argument of the function is always indicated.

# Molecular Distribution Functions

can be obtained from $P(\mathbf{X}^N)$ by integrating over all the configurations of the remaining $N-2$ molecules[3]

$$P^{(2)}(\mathbf{X}', \mathbf{X}'') = \int \cdots \int d\mathbf{X}_3 \cdots d\mathbf{X}_N \, P(\mathbf{X}', \mathbf{X}'', \mathbf{X}_3, \ldots, \mathbf{X}_N) \quad (2.28)$$

Clearly, $P^{(2)}(\mathbf{X}', \mathbf{X}'') \, d\mathbf{X}' \, d\mathbf{X}''$ is the probability of finding a specific particle, say 1, in $d\mathbf{X}'$ at $\mathbf{X}'$ and another particle, say 2, in $d\mathbf{X}''$ at $\mathbf{X}''$. The same probability applies for any specific pair of two different particles.

Consider the following list of events and their corresponding probabilities.

| event | probability of the event |
|---|---|
| particle 1 in $d\mathbf{X}'$ *and* particle 2 in $d\mathbf{X}''$ | $P^{(2)}(\mathbf{X}', \mathbf{X}'') \, d\mathbf{X}' \, d\mathbf{X}''$ |
| particle 1 in $d\mathbf{X}'$ *and* particle 3 in $d\mathbf{X}''$ | $P^{(2)}(\mathbf{X}', \mathbf{X}'') \, d\mathbf{X}' \, d\mathbf{X}''$ |
| ⋮ | |
| particle 1 in $d\mathbf{X}'$ *and* particle $N$ in $d\mathbf{X}''$ | $P^{(2)}(\mathbf{X}', \mathbf{X}'') \, d\mathbf{X}' \, d\mathbf{X}''$ |
| particle 2 in $d\mathbf{X}'$ *and* particle 1 in $d\mathbf{X}''$ | $P^{(2)}(\mathbf{X}', \mathbf{X}'') \, d\mathbf{X}' \, d\mathbf{X}''$ |
| ⋮ | |
| particle $N$ in $d\mathbf{X}'$ *and* particle $N-1$ in $d\mathbf{X}''$ | $P^{(2)}(\mathbf{X}', \mathbf{X}'') \, d\mathbf{X}' \, d\mathbf{X}''$ |

(2.29)

The event

$$\{\text{a particle in } d\mathbf{X}' \text{ and another particle in } d\mathbf{X}''\} \quad (2.30)$$

is clearly the union of all the $N(N-1)$ events listed in (2.29). However, the probability of the event (2.30) is the sum of all the probabilities of the events listed in (2.29) *only* if the latter are disjoint. This condition can be realized when the elements of volume $d\mathbf{R}'$ and $d\mathbf{R}''$ (contained in $d\mathbf{X}'$ and $d\mathbf{X}''$, respectively) are small enough so that no more than one of the events in (2.29) may occur at any given time. For this case, we define the *generic* pair distribution function as

$$\varrho^{(2)}(\mathbf{X}', \mathbf{X}'') \, d\mathbf{X}' \, d\mathbf{X}''$$
$$= \Pr\{\text{a particle in } d\mathbf{X}' \text{ and a different particle in } d\mathbf{X}''\}$$
$$= \sum_{i \neq j} \Pr\{\text{particle } i \text{ in } d\mathbf{X}' \text{ and particle } j \text{ in } d\mathbf{X}''\}$$
$$= \sum_{i \neq j} P^{(2)}(\mathbf{X}', \mathbf{X}'') \, d\mathbf{X}' \, d\mathbf{X}'' = N(N-1) P^{(2)}(\mathbf{X}', \mathbf{X}'') \, d\mathbf{X}' \, d\mathbf{X}'' \quad (2.31)$$

---

[3] We use primed vectors like $\mathbf{X}', \mathbf{X}'', \ldots$ to distinguish them from the vectors $\mathbf{X}_3, \mathbf{X}_4, \ldots$ whenever the two sets of vectors have a different "status." For instance, in (2.28), the primed vectors are fixed in the integrand. Such a distinction is not essential, although it may help to avoid confusion.

The last equality in (2.31) follows from the equivalence of all the $N(N-1)$ pairs of specific different particles. Using the definition of $P^{(2)}(\mathbf{X}', \mathbf{X}'')$ in (2.28), we can transform the definition of $\varrho^{(2)}(\mathbf{X}', \mathbf{X}'')$ into an expression which will be interpreted as an average quantity:

$$\varrho^{(2)}(\mathbf{X}', \mathbf{X}'')\, d\mathbf{X}'\, d\mathbf{X}''$$
$$= N(N-1)\, d\mathbf{X}'\, d\mathbf{X}'' \int \cdots \int d\mathbf{X}_3 \cdots d\mathbf{X}_N\, P(\mathbf{X}', \mathbf{X}'', \mathbf{X}_3, \ldots, \mathbf{X}_N)$$
$$= N(N-1)\, d\mathbf{X}'\, d\mathbf{X}'' \int \cdots \int d\mathbf{X}_1 \cdots d\mathbf{X}_N\, P(\mathbf{X}_1 \cdots \mathbf{X}_N)$$
$$\times \delta(\mathbf{X}_1 - \mathbf{X}')\, \delta(\mathbf{X}_2 - \mathbf{X}'')$$
$$= d\mathbf{X}'\, d\mathbf{X}'' \int \cdots \int d\mathbf{X}^N\, P(\mathbf{X}^N) \sum_{\substack{i=1 \\ i \neq j}}^{N} \sum_{j=1}^{N} \delta(\mathbf{X}_i - \mathbf{X}')\, \delta(\mathbf{X}_j - \mathbf{X}'') \quad (2.32)$$

In the second form on the rhs of (2.32), we employ the basic property of the Dirac delta function, so that integration is now extended over all the vectors $\mathbf{X}_1, \ldots, \mathbf{X}_N$. In the third form on the rhs of (2.32), we have used the equivalence of the $N$ particles, as was done in (2.31), to get an average of the quantity

$$d\mathbf{X}'\, d\mathbf{X}'' \sum_{\substack{i=1 \\ i \neq j}}^{N} \sum_{j=1}^{N} \delta(\mathbf{X}_i - \mathbf{X}')\, \delta(\mathbf{X}_j - \mathbf{X}'') \quad (2.33)$$

which can be viewed as a "counting function," i.e., for any specific configuration $\mathbf{X}^N$, this quantity counts the number of *pairs* of particles occupying the elements $d\mathbf{X}'$ and $d\mathbf{X}''$. Hence, the integral (2.32) counts the average number of *pairs* occupying $d\mathbf{X}'$ and $d\mathbf{X}''$. The normalization of $\varrho^{(2)}(\mathbf{X}', \mathbf{X}'')$ follows directly from (2.32):

$$\int\int d\mathbf{X}'\, d\mathbf{X}''\, \varrho^{(2)}(\mathbf{X}', \mathbf{X}'') = N(N-1) \quad (2.34)$$

which is the *exact* number of pairs in $V$. As in the previous section, we note that the meaning of $\varrho^{(2)}(\mathbf{X}', \mathbf{X}'')$ as an average quantity is preserved upon integration over any region $S$. This is not the case, however, when its probabilistic meaning is adopted. For instance, the quantity

$$\int_S \int_S d\mathbf{X}'\, d\mathbf{X}''\, \varrho^{(2)}(\mathbf{X}', \mathbf{X}'') \quad (2.35)$$

is the average number of pairs occupying the region $S$ (a factor of $\tfrac{1}{2}$ should

# Molecular Distribution Functions

be included if we are interested in *different* pairs). This quantity is in general not a probability.

It is also useful to introduce the *locational* (or the spatial) pair distribution function, defined by

$$\varrho^{(2)}(\mathbf{R}', \mathbf{R}'') = \int \int d\mathbf{\Omega}'\, d\mathbf{\Omega}''\, \varrho^{(2)}(\mathbf{X}', \mathbf{X}'') \tag{2.36}$$

where integration is carried out over the orientations of the two particles. $\varrho^{(2)}(\mathbf{R}', \mathbf{R}'')\, d\mathbf{R}'\, d\mathbf{R}''$ is the average number of pairs occupying $d\mathbf{R}'$ and $d\mathbf{R}''$, or alternatively, for infinitesimal elements $d\mathbf{R}'$ and $d\mathbf{R}''$, the probability of finding one particle in $d\mathbf{R}'$ at $\mathbf{R}'$ and a second particle in $d\mathbf{R}''$ at $\mathbf{R}''$. It is sometimes convenient to denote the quantity defined in (2.36) by $\bar{\varrho}^{(2)}(\mathbf{R}', \mathbf{R}'')$, to distinguish it from the different function $\varrho^{(2)}(\mathbf{X}', \mathbf{X}'')$. However, since we specify the arguments of the functions, there should be no reason for confusion as to this notation.

## 2.4. PAIR CORRELATION FUNCTION

In most applications, it has been found more useful to employ the pair correlation function, defined below, rather than the pair distribution function itself.

Consider the two elements of "volume" $d\mathbf{X}'$ and $d\mathbf{X}''$ and the intersection of the two events:

$$\{\text{a particle in } d\mathbf{X}'\} \cap \{\text{a particle in } d\mathbf{X}''\} \tag{2.37}$$

The symbol $\cap$ may be read as "and," i.e., the combination in (2.37) means that the first *and* the second events occur.[4]

Two events are called *independent* whenever the probability of their intersection is equal to the product of the probabilities of the two events. In general, the two separate events given in (2.37) are not independent; the occurrence of one of them may influence the likelihood, or the probability, of occurrence of the other. For instance, if the separation $R = |\mathbf{R}'' - \mathbf{R}'|$ between the two elements is very small (compared to the molecular diameter of the particles), then fulfilling one event strongly reduces the chances of the second.

We now invoke the following physically plausible contention. In a fluid, if the separation $R$ between two elements is very large, then the two

---

[4] It is presumed that the two particles referred to in (2.37) are different.

events in (2.37) become independent. Therefore, we can write for the probability of their intersection

$$\varrho^{(2)}(\mathbf{X}', \mathbf{X}'') \, d\mathbf{X}' \, d\mathbf{X}''$$
$$= \Pr[\{\text{a particle in } d\mathbf{X}'\} \cap \{\text{a particle in } d\mathbf{X}''\}]$$
$$= \Pr\{\text{a particle in } d\mathbf{X}'\} \cdot \Pr\{\text{a particle in } d\mathbf{X}''\}$$
$$= \varrho^{(1)}(\mathbf{X}') \, d\mathbf{X}' \, \varrho^{(1)}(\mathbf{X}'') \, d\mathbf{X}'', \quad \text{for } R \to \infty \quad (2.38)$$

Or, in short,

$$\varrho^{(2)}(\mathbf{X}', \mathbf{X}'') \stackrel{R \to \infty}{=} \varrho^{(1)}(\mathbf{X}')\varrho^{(1)}(\mathbf{X}'') = (\varrho/8\pi^2)^2 \quad (2.39)$$

the last equality holding for a homogeneous and isotropic fluid. If (2.39) holds, it is often said that the local densities at $\mathbf{X}'$ and $\mathbf{X}''$ are uncorrelated.[5] (The condition $R \to \infty$ should be implemented only after the thermodynamic limit, $N \to \infty$, $V \to \infty$, $N/V = \text{const}$, has been taken.)

For any finite distance $R$, the factorization of $\varrho^{(2)}(\mathbf{X}', \mathbf{X}'')$ into a product may not be valid. We now introduce the pair *correlation function*, which measures the extent of deviation from (2.39), and is defined by

$$\varrho^{(2)}(\mathbf{X}', \mathbf{X}'') = \varrho^{(1)}(\mathbf{X}')\varrho^{(1)}(\mathbf{X}'')g(\mathbf{X}', \mathbf{X}'')$$
$$= (\varrho/8\pi^2)^2 g(\mathbf{X}', \mathbf{X}'') \quad (2.40)$$

The second equality holds for a homogeneous and isotropic fluid. A related quantity is the *locational* pair correlation function, defined in terms of the locational pair distribution function

$$\varrho^{(2)}(\mathbf{R}', \mathbf{R}'') = \varrho^2 g(\mathbf{R}', \mathbf{R}'') \quad (2.41)$$

The relation between $g(\mathbf{R}', \mathbf{R}'')$ and $g(\mathbf{X}', \mathbf{X}'')$ follows from (2.36), (2.40), and (2.41):

$$g(\mathbf{R}', \mathbf{R}'') = [1/(8\pi^2)^2] \int \int d\mathbf{\Omega}' \, d\mathbf{\Omega}'' \, g(\mathbf{X}', \mathbf{X}'') \quad (2.42)$$

which can be viewed as the angle average of $g(\mathbf{X}', \mathbf{X}'')$. Note that again we use the same symbol, $g$, for two different functions in (2.42). It is also worth noting that the more detailed notation, such as $g^{(2)}(\mathbf{X}', \mathbf{X}'')$, should

---

[5] In probability theory, there is a precise distinction between the concepts of independence and uncorrelation. The latter follows from the former, but the reverse, however, is not true.

# Molecular Distribution Functions

be used whenever a distinction from higher-order correlation functions is made. Here, however, we shall not need to use the latter.

In this book, we shall be interested only in the homogeneous and isotropic fluids. In such a case, there is a redundancy in specifying the full configuration of the pair of particles by 12 coordinates $(\mathbf{X}', \mathbf{X}'')$. It is clear that for any configuration of the pair $\mathbf{X}', \mathbf{X}''$, the correlation $g(\mathbf{X}', \mathbf{X}'')$ will be invariant to translation and rotation of the pair as a unit, keeping the relative configuration of one particle toward the other fixed. Therefore, we can reduce to six the number of independent variables necessary for the full description of the pair correlation function. For instance, we may choose the location of one particle at the origin of the coordinate system $\mathbf{R}' = 0$, and fix its orientation, say, at $\phi' = \theta' = \psi' = 0$. Hence, the pair correlation function is a function of only the six variables $\mathbf{X}'' = \mathbf{R}'', \mathbf{\Omega}''$.

Similarly, the function $g(\mathbf{R}', \mathbf{R}'')$ is a function of only the scalar distance $R = |\mathbf{R}'' - \mathbf{R}'|$. (For instance, $\mathbf{R}'$ may be chosen at the origin $\mathbf{R}' = 0$ and, because of the isotropy of the fluid, the orientation of $\mathbf{R}''$ is of no importance. Therefore, only the separation $R$ is left as the independent variable.) The function $g(R)$, i.e., the pair correlation function expressed explicitly as a function of the distance $R$, is often referred to as the radial distribution function. This function has played a central role in the theory of fluids.

We now turn to a somewhat different interpretation of the pair correlation function. We define the conditional probability of observing a particle in $d\mathbf{X}''$ at $\mathbf{X}''$, given a particle at $\mathbf{X}'$:

$$\varrho(\mathbf{X}''/\mathbf{X}')\,d\mathbf{X}'' = \frac{\varrho^{(2)}(\mathbf{X}', \mathbf{X}'')\,d\mathbf{X}'\,d\mathbf{X}''}{\varrho^{(1)}(\mathbf{X}')\,d\mathbf{X}'} = \varrho^{(1)}(\mathbf{X}'')g(\mathbf{X}', \mathbf{X}'')\,d\mathbf{X}'' \qquad (2.43)$$

The last equality follows from the definition of $g(\mathbf{X}', \mathbf{X}'')$ in (2.40).

It is important to note that the probability of finding a particle at an exact configuration $\mathbf{X}''$ is zero, which is the reason for taking an infinitesimal element of "volume" at $\mathbf{X}''$. On the other hand, the conditional probability may be defined for an exact *condition*: "given a particle at $\mathbf{X}'$." This may formally be seen from (2.43), where $d\mathbf{X}'$ cancels out once we form the ratio of the two distribution functions. Hence, one can actually take the limit $d\mathbf{X}' \to 0$ in the definition of the conditional probability. [For more details, see Papoulis (1965).]

We recall that the quantity $\varrho^{(1)}(\mathbf{X}'')\,d\mathbf{X}''$ is the local density of particles at $\mathbf{X}''$. We now show that the quantity defined in (2.43) is the (conditional) local density at $\mathbf{X}''$, given a particle at $\mathbf{X}'$. In other words, we fix a particle at $\mathbf{X}'$ and view the rest of the $N-1$ particles as a system subjected to the field of force produced by the first particle. Clearly, the new system is no longer homogeneous, nor isotropic; therefore, the local density may be different at each point.

Let us work out the formal argument for a system obeying the pairwise additivity of the total potential function, i.e., we assume that

$$U_N(\mathbf{X}_1, \ldots, \mathbf{X}_N) = \sum_{i<j} U_{ij}(\mathbf{X}_i, \mathbf{X}_j)$$
$$= U_{N-1}(\mathbf{X}_2, \ldots, \mathbf{X}_N) + \sum_{j=2}^{N} U_{1j}(\mathbf{X}_1, \mathbf{X}_j) = U_{N-1} + B_1 \quad (2.44)$$

In (2.44), we have split the total potential energy of the system of $N$ particles into two parts: the potential energy of interaction among the $N-1$ particles, and the interaction of one particle, chosen as particle 1, with the $N-1$ particles. Once we fix the configuration of particle 1 at $\mathbf{X}_1$, the rest of the system can be looked upon as a system in an "external" field given by $B_1$.

From the definitions (2.1), (2.23), (2.32), and (2.43), we get

$$\varrho(\mathbf{X}''/\mathbf{X}')$$
$$= \frac{N(N-1)\int \cdots \int d\mathbf{X}^N \exp[-\beta U_N(\mathbf{X}^N)]\, \delta(\mathbf{X}_1 - \mathbf{X}')\, \delta(\mathbf{X}_2 - \mathbf{X}'')}{N \int \cdots \int d\mathbf{X}^N \exp[-\beta U_N(\mathbf{X}^N)]\, \delta(\mathbf{X}_1 - \mathbf{X}')}$$
$$= \frac{(N-1)\int \cdots \int d\mathbf{X}_2 \cdots d\mathbf{X}_N \exp[-\beta U_N(\mathbf{X}', \mathbf{X}_2, \ldots, \mathbf{X}_N)]\, \delta(\mathbf{X}_2 - \mathbf{X}'')}{\int \cdots \int d\mathbf{X}_2 \cdots d\mathbf{X}_N \exp[-\beta U_N(\mathbf{X}', \mathbf{X}_2, \ldots, \mathbf{X}_N)]}$$
$$= (N-1)\int \cdots \int d\mathbf{X}_2 \cdots d\mathbf{X}_N\, P^*(\mathbf{X}', \mathbf{X}_2, \ldots, \mathbf{X}_N)\, \delta(\mathbf{X}_2 - \mathbf{X}'') \quad (2.45)$$

where $P^*(\mathbf{X}', \mathbf{X}_2, \ldots, \mathbf{X}_N)$ is the basic probability density of a system of $N-1$ particles placed in an "external" field produced by a particle fixed at $\mathbf{X}'$, i.e.,

$$P^*(\mathbf{X}', \mathbf{X}_2, \ldots, \mathbf{X}_N) = \frac{\exp(-\beta U_{N-1} - \beta B_1)}{\int \cdots \int d\mathbf{X}_2 \cdots d\mathbf{X}_N \exp(-\beta U_{N-1} - \beta B_1)} \quad (2.46)$$

with $B_1 = \sum_{j=2}^{N} U_{1j}(\mathbf{X}_1, \mathbf{X}_j)$. [Compare with the probability density defined in (2.1).]

We now observe that relation (2.45) has the same structure as relation (2.23) but with two minor differences. First, (2.45) refers to a system of $N-1$ instead of $N$ particles; second, the system of $N-1$ particles is placed in an external field. Hence, (2.45) is simply the local density at $\mathbf{X}''$

# Molecular Distribution Functions

of a system of $N - 1$ particles placed in the external field $B_1$. This is an example of a singlet molecular distribution function which is not everywhere constant.

Similarly, for the locational pair correlation function, we have

$$\varrho(\mathbf{R}''/\mathbf{R}') = \varrho g(\mathbf{R}', \mathbf{R}'') \tag{2.47}$$

which is the conditional average density at $\mathbf{R}''$ given a particle at $\mathbf{R}'$. This interpretation of the pair correlation function will be most useful in the forthcoming applications.

As noted above, the function $g(\mathbf{R}', \mathbf{R}'')$ is a function of the scalar distance $R = |\mathbf{R}'' - \mathbf{R}'|$. (This is true both for spherical particles and for molecular fluids for which an orientational average has been carried out.) Because of the spherical symmetry of the locational pair correlation function, the local density has the same value for any point on the spherical shell of radius $R$ from the center of the fixed particle at $\mathbf{R}'$. It is also convenient to choose as an element of volume a spherical shell of width $dR$ and radius $R$. The average number of particles in this element of volume is

$$\varrho g(\mathbf{R}', \mathbf{R}'') \, d\mathbf{R}'' = \varrho g(R) 4\pi R^2 \, dR \tag{2.48}$$

Sometimes, the function $g(R)4\pi R^2$ rather than $g(R)$ is referred to as the radial distribution function. In this book, we use this term only for $g(R)$ itself.

## 2.5. FEATURES OF THE RADIAL DISTRIBUTION FUNCTION

In this section, we illustrate the general features of the radial distribution function (RDF), $g(R)$, for a system of simple spherical particles. From the definitions (2.32) and (2.40) (applied to spherical particles), we get

$$g(\mathbf{R}', \mathbf{R}'') = \frac{N(N-1)}{\varrho^2}$$

$$\times \frac{\int \cdots \int d\mathbf{R}_3 \cdots d\mathbf{R}_N \exp[-\beta U_N(\mathbf{R}', \mathbf{R}'', \mathbf{R}_3, \ldots, \mathbf{R}_N)]}{\int \cdots \int d\mathbf{R}_1 \cdots d\mathbf{R}_N \exp[-\beta U_N(\mathbf{R}_1, \ldots, \mathbf{R}_N)]} \tag{2.49}$$

This general relation will be used to extract information on the behavior of $g(R)$ for some simple systems. A more useful expression, which we shall need only for demonstrative purposes, is the density expansion of $g(\mathbf{R}', \mathbf{R}'')$, which reads (Hill, 1956)

$$g(\mathbf{R}', \mathbf{R}'') = \{\exp[-\beta U(\mathbf{R}', \mathbf{R}'')]\}\{1 + B(\mathbf{R}', \mathbf{R}'')\varrho + C(\mathbf{R}', \mathbf{R}'')\varrho^2 + \cdots\} \quad (2.50)$$

where the coefficients $B(\mathbf{R}', \mathbf{R}'')$, $C(\mathbf{R}', \mathbf{R}'')$, etc., are given in terms of integrals over the so-called Mayer $f$-function, defined by

$$f(\mathbf{R}', \mathbf{R}'') = \exp[-\beta U(\mathbf{R}', \mathbf{R}'')] - 1 \quad (2.51)$$

For instance,

$$B(\mathbf{R}', \mathbf{R}'') = \int_V f(\mathbf{R}', \mathbf{R}_3) f(\mathbf{R}'', \mathbf{R}_3)\, d\mathbf{R}_3 \quad (2.52)$$

We now turn to some specific examples.

### 2.5.1. Ideal Gas

The RDF for an ideal gas can be obtained directly from definition (2.49). Putting $U_N \equiv 0$ for all configurations, the integrations become trivial and we get

$$g(\mathbf{R}', \mathbf{R}'') = \frac{N(N-1)}{\varrho^2} \frac{\int \cdots \int d\mathbf{R}_3 \cdots d\mathbf{R}_N}{\int \cdots \int d\mathbf{R}_1 \cdots d\mathbf{R}_N} = \frac{N(N-1)V^{N-2}}{\varrho^2 V^N} \quad (2.53)$$

or

$$g(R) = 1 - (1/N) \quad (2.54)$$

As we expect, $g(R)$ is practically unity for any value of $R$. This is an obvious reflection of the basic property of an ideal gas, i.e., absence of correlation follows from absence of interaction. The term $N^{-1}$ is typical of constant-volume systems. At the thermodynamic limit $N \to \infty$, $V \to \infty$, $N/V = \text{const}$, this term may, for most purposes, be dropped.[6] Of course, in order to get the correct normalization of $g(R)$, one should use the exact relation (2.54), which yields

$$\varrho \int_V g(\mathbf{R}', \mathbf{R}'')\, d\mathbf{R}'' = \varrho \int_V [1 - (1/N)]\, d\mathbf{R}'' = N - 1 \quad (2.55)$$

which is exactly the total number of particles in the system, excluding the one fixed at $\mathbf{R}'$.

---

[6] In some instances, care must be employed to take the proper limit of infinite systems. For an example, see the next chapter, Section 3.9.

## 2.5.2. Very Dilute Gas

At very low densities, $\varrho \to 0$, we may neglect all powers of $\varrho$ in the density expansion of $g(R)$, in which case we get from (2.50)

$$g(R) = \exp[-\beta U(R)], \quad \varrho \to 0 \tag{2.56}$$

where $U(R)$ is the pair potential operating between two particles. Relation (2.56) is essentially the Boltzmann distribution law. Since at low densities, encounters in which more than two particles are involved are very rare, the pair distribution function is determined solely by the pair potential.

One direct way of obtaining (2.56) from the definition (2.49) (and not through the density expansion) is to consider the case of a system containing only two particles. Letting $N = 2$ in (2.49), we get

$$g(R) = \frac{2}{\varrho^2} \frac{\exp[-\beta U(R)]}{Z_2} \tag{2.57}$$

where $Z_2$ is the configurational partition function for the case $N = 2$.

Since by definition $U(R) \to 0$ as $R \to \infty$, we can use (2.57) to form the ratio

$$g(R)/g(\infty) = \exp[-\beta U(R)] \tag{2.58}$$

Assuming that at $R \to \infty$, $g(\infty)$ is practically unity, we get from (2.58)

$$g(R) = \exp[-\beta U(R)] \tag{2.59}$$

which is the same as (2.56). Note that (2.56) and (2.59) have been obtained for two apparently different conditions ($\varrho \to 0$ on one hand and $N = 2$ on the other). The identical results for $g(R)$ in the two cases reflect the fact that at very low densities, only interactions between *pairs* determine the behavior of $g(R)$. In other words, a pair of interacting particles does not "feel" the presence of the other particles. The form of $g(R)$ as $\varrho \to 0$ for a system of hard spheres and Lennard-Jones particles is depicted in Fig. 2.2. It is seen that for HS particles as $\varrho \to 0$, correlation exists only for $R < \sigma$. For $R > \sigma$, the function $g(R)$ is identically unity. For LJ particles, we observe a single peak in $g(R)$ at the same point for which $U(R)$ has a minimum, namely at $R = 2^{1/6}\sigma$.

## 2.5.3. Slightly Dense Gas

In the context of this section, a slightly dense gas is a gas properly described by the first-order expansion in density, such as the linear term

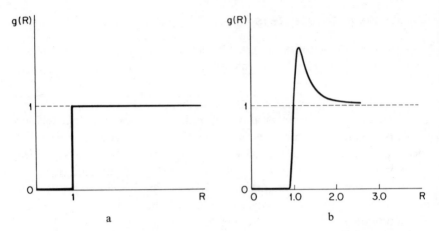

Fig. 2.2. The form of $g(R)$ at very low densities ($\varrho \to 0$). (a) For hard spheres with $\sigma = 1$; (b) for Lennard-Jones particles with parameters $\varepsilon/kT = 0.5$ and $\sigma = 1.0$.

in (2.50). Before analyzing the content of the coefficient $B(\mathbf{R}', \mathbf{R}'')$ in the expansion of $g(R)$, let us demonstrate its origin by considering a system of three particles. Putting $N = 3$ in the definition (2.49), we get

$$g(\mathbf{R}', \mathbf{R}'') = \frac{6}{\varrho^2} \frac{\int d\mathbf{R}_3 \exp[-\beta U_3(\mathbf{R}', \mathbf{R}'', \mathbf{R}_3)]}{Z_3} \qquad (2.60)$$

where $Z_3$ is the configurational partition function for a system of three particles. Assuming pairwise additivity of the potential energy $U_3$ and using the definition of the function $f$ in (2.51), we can transform (2.60) into

$$g(\mathbf{R}', \mathbf{R}'') = \frac{6}{\varrho^2} \exp[-\beta U(\mathbf{R}', \mathbf{R}'')]$$
$$\times \frac{\int d\mathbf{R}_3 [f(\mathbf{R}', \mathbf{R}_3) f(\mathbf{R}_3, \mathbf{R}'') + f(\mathbf{R}', \mathbf{R}_3) + f(\mathbf{R}'', \mathbf{R}_3) + 1]}{Z_3} \qquad (2.61)$$

Noting again that $U(\mathbf{R}', \mathbf{R}'') = 0$ for $R = |\mathbf{R}'' - \mathbf{R}'| \to \infty$, we form the ratio[7]

$$\frac{g(R)}{g(\infty)} = \exp[-\beta U(R)]$$
$$\times \frac{\int d\mathbf{R}_3 f(\mathbf{R}', \mathbf{R}_3) f(\mathbf{R}_3, \mathbf{R}'') + 2 \int d\mathbf{R}_3 f(\mathbf{R}', \mathbf{R}_3) + V}{\lim_{R \to \infty}[\int d\mathbf{R}_3 f(\mathbf{R}', \mathbf{R}_3) f(\mathbf{R}_3, \mathbf{R}'') + 2 \int d\mathbf{R}_3 f(\mathbf{R}', \mathbf{R}_3) + V]} \qquad (2.62)$$

[7] By $R \to \infty$, we mean here a very large distance compared with the molecular diameter.

## Molecular Distribution Functions

Clearly, the two integrals over $f(\mathbf{R}', \mathbf{R}_3)$ and $f(\mathbf{R}'', \mathbf{R}_3)$ are equal and independent of the separation $R$, i.e.,

$$C \equiv \int_V d\mathbf{R}_3 f(\mathbf{R}', \mathbf{R}_3) = \int_V d\mathbf{R}_3 f(\mathbf{R}'', \mathbf{R}_3) \tag{2.63}$$

[Note that since $f(R)$ is a short-range function of $R$, the integral in (2.63) does not depend on $V$.]

On the other hand, we have the limiting behavior

$$\lim_{R \to \infty} \int d\mathbf{R}_3 f(\mathbf{R}', \mathbf{R}_3) f(\mathbf{R}_3, \mathbf{R}'') = 0 \tag{2.64}$$

which follows from the fact that the two factors in the integrand contribute to the integral only if $\mathbf{R}_3$ is close simultaneously to both $\mathbf{R}'$ and $\mathbf{R}''$, a situation that cannot be attained if $R = |\mathbf{R}'' - \mathbf{R}'| \to \infty$.

Using (2.63), (2.64), and the definition (2.52), we now rewrite (2.62) as

$$\frac{g(R)}{g(\infty)} = \exp[-\beta U(R)] \frac{B(\mathbf{R}', \mathbf{R}'') + 2C + V}{2C + V} \tag{2.65}$$

Since $C$ is constant, it may be neglected, as compared with $V$, in the thermodynamic limit. Also, assuming that $g(\infty)$ is practically unity, we get the final form of $g(R)$ for this case:

$$g(R) = \{\exp[-\beta U(R)]\}[1 + (1/V)B(\mathbf{R}', \mathbf{R}'')], \quad R = |\mathbf{R}'' - \mathbf{R}'| \tag{2.66}$$

It is interesting to note that $1/V$ appearing in (2.66) replaces the density $\varrho$ in (2.50). In fact, the quantity $1/V$ may be interpreted as the density of the "free" particles (i.e., the particles in excess over the two fixed at $\mathbf{R}'$ and $\mathbf{R}''$) for the case $N = 3$.

The foregoing derivation of (2.66) illustrates the origin of the coefficient $B(\mathbf{R}', \mathbf{R}'')$, which, in principle, results from the simultaneous interaction of three particles [compare this result with (2.59)]. This is actually the meaning of the term "slightly dense gas." Whereas in a very dilute gas we take account of interactions between pairs only, here we also consider the effect of interactions among three particles, but not more.

Next we examine the content of the function $B(\mathbf{R}', \mathbf{R}'')$ and its effect on $g(R)$ for a system of hard spheres (HS). In this case, one has the property

$$f(R) = \begin{cases} -1 & \text{for } R < \sigma \\ 0 & \text{for } R > \sigma \end{cases} \tag{2.67}$$

Therefore, the only contribution to the integral in (2.52) comes from regions in which both $f(\mathbf{R}', \mathbf{R}_3)$ and $f(\mathbf{R}'', \mathbf{R}_3)$ are equal to $-1$. This may occur for $R < 2\sigma$. The integrand vanishes either when $|\mathbf{R}' - \mathbf{R}_3| > \sigma$ or $|\mathbf{R}'' - \mathbf{R}_3| > \sigma$. Furthermore, for $|\mathbf{R}'' - \mathbf{R}'| < \sigma$, the exponential factor in (2.66), $\exp[-\beta U(\mathbf{R}', \mathbf{R}'')]$, vanishes. Thus, the only region of interest is $\sigma \leq R \leq 2\sigma$. Figure 2.3 depicts such a situation. Since the value of the integrand in the region where it is nonzero equals $(-1) \times (-1) = 1$, the integration problem reduces to the geometric problem of computing the volume of the intersection of the two spheres of radius $\sigma$ (note that the *diameter* of the hard spheres is $\sigma$). The solution can be obtained by transforming to bipolar coordinates. We write

$$u = |\mathbf{R}_3 - \mathbf{R}'|, \quad v = |\mathbf{R}_3 - \mathbf{R}''| \quad (2.68)$$

Then the element of volume $d\mathbf{R}_3$ can be chosen (noting the axial symmetry of our problem) as a ring of radius $y_3$ and cross section $dx_3\, dy_3$ (see Fig. 2.4), i.e.,

$$d\mathbf{R}_3 = 2\pi y_3\, dx_3\, dy_3 \quad (2.69)$$

The transformation of variables and its Jacobian are

$$u^2 = x_3^2 + y_3^2, \quad v^2 = y_3^2 + (R - x_3)^2, \quad \frac{\partial(x_3, y_3)}{\partial(u, v)} = \frac{uv}{y_3 R} \quad (2.70)$$

Hence

$$d\mathbf{R}_3 = (2\pi/R) uv\, du\, dv \quad (2.71)$$

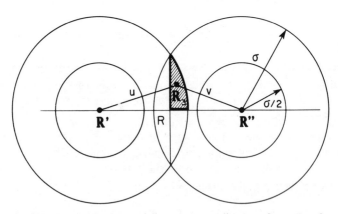

Fig. 2.3. Two hard spheres of diameter $\sigma$ at a distance of $\sigma < R < 2\sigma$ from each other. The intersection of the two spheres of *radius* $\sigma$ about $\mathbf{R}'$ and $\mathbf{R}''$ is twice the volume of revolution of the shaded area in the figure.

# Molecular Distribution Functions

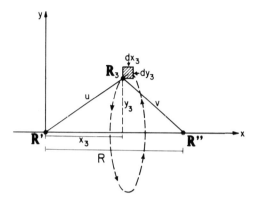

Fig. 2.4. Bipolar coordinates used for the integration over the intersection region of Fig. 2.3.

Using this transformation, we rewrite (2.52) as

$$B(R) = 2 \frac{2\pi}{R} \int_{R/2}^{\sigma} u \, du \int_{R-u}^{u} v \, dv = \frac{4\pi\sigma^3}{3} \left[ 1 - \frac{3}{4} \frac{R}{\sigma} + \frac{1}{16} \left( \frac{R}{\sigma} \right)^3 \right] \tag{2.72}$$

The factor 2 is included in (2.72) since we compute only half of the intersection of the two spheres. (This corresponds to the volume of revolution of the shaded area in Fig. 2.3.) Using (2.72), we can now write explicitly the form of the radial distribution function for hard spheres at "slightly dense" concentration:

$$g(R) = \begin{cases} 0 & \text{for } R < \sigma \\ 1 + \varrho \, \frac{4\pi\sigma^3}{3} \left[ 1 - \frac{3}{4} \frac{R}{\sigma} + \frac{1}{16} \left( \frac{R}{\sigma} \right)^3 \right] & \text{for } \sigma < R < 2\sigma \\ 1 & \text{for } R > 2\sigma \end{cases} \tag{2.73}$$

The form of this function is depicted in Fig. 2.5. Further implications of (2.73) will be discussed in Section 2.6, where we reinterpret the correction term $B(R)$ as an effective "attraction" between a pair of hard spheres.

## 2.5.4. Lennard-Jones Particles at Moderately High Densities

Lennard-Jones (LJ) particles are supposed to closely resemble in behavior real simple spherical particles such as argon. In this section, we

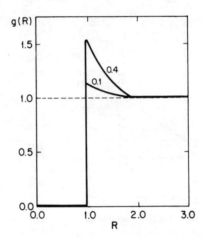

Fig. 2.5. The form of $g(R)$ for hard-sphere particles ($\sigma = 1$), using the first-order expansion in the density. The two curves correspond to $\varrho = 0.1$ and $\varrho = 0.4$ [see Eq. (2.73)]. [Note that Eq. (2.73) may not be valid for these densities.]

present some further information on the behavior of $g(R)$ and its dependence on density and on temperature. The LJ particles are defined through their pair potential as

$$U_{\text{LJ}}(R) = 4\varepsilon[(\sigma/R)^{12} - (\sigma/R)^6] \qquad (2.74)$$

Figure 2.6 demonstrates the dependence of $g(R)$ on the density. The dimensionless densities $\varrho\sigma^3$ are recorded next to each curve.[8] One observes that at very low densities, there is a single peak, corresponding to the minimum in the potential function (2.74). At successively higher densities, new peaks develop which become more and more pronounced as the density increases. The location of the first peak is essentially unchanged, even though its height increases steadily. The locations of the new peaks occur roughly at integral multiples of $\sigma$, i.e., at $R \sim \sigma, 2\sigma, 3\sigma, \ldots$ . This feature reflects the propensity of the spherical molecules to pack, at least locally, in concentric and nearly equidistant spheres about a given molecule. This is

---

[8] In this book, $\sigma$ is expressed in arbitrary units of length. Hence, $\varrho$ is the number density per the corresponding unit of volume. Note that the only significant parameter is the dimensionless quantity $\varrho\sigma^3$.

# Molecular Distribution Functions

a very fundamental property of the fluid and deserves further attention.

Consider a random arrangement (a configuration) of spherical particles in the fluid. (An illustration in two dimensions is provided in Fig. 2.7.) Now consider a spherical shell of width $d\sigma$ and radius $\sigma$, and inquire as to the average number of particles in this "element of volume." If the center of the spherical shell has been chosen at random, as was done on the rhs of the figure, we find that on the average, the number of particles is $\varrho 4\pi\sigma^2 \, d\sigma$. On the other hand, if we choose the center of the spherical shell so that it coincides with the center of a particle, then on the average, we find more particles in this element of volume. The drawing on the left illustrates this case for one configuration. One sees that in this particular example, there are six particles in the element of volume on the left as compared with two particles on the right. Similarly, we could have drawn spherical shells of width $d\sigma$ at $2\sigma$ and again have found excess particles in the element of volume, the origin of which has been chosen at the center of one of the particles. The excess of particles at the distances $\sigma, 2\sigma, 3\sigma$, etc., from the center of a particle is manifested in the various peaks of the function $g(R)$. Clearly, this effect decays rapidly as the distance from the center increases. We see from Fig. 2.6 that $g(R)$ is almost unity for $R \gtrsim 5\sigma$. This means that correlation between the local densities at two points $\mathbf{R}'$ and $\mathbf{R}''$ extends over a relatively short range.

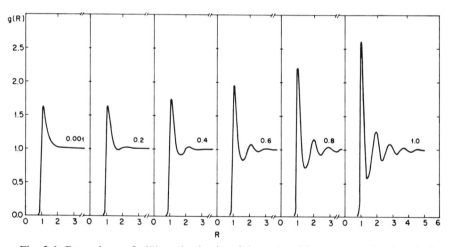

Fig. 2.6. Dependence of $g(R)$ on the density of the system. The corresponding (number) densities are indicated next to each curve. The functions $g(R)$ for this illustration were computed by numerical solution of the Percus–Yevick equation for Lennard-Jones particles with parameters $\sigma = 1.0$ and $\varepsilon/kT = 0.5$ (for details, see Appendix 9-E).

At short distances, say $\sigma \lesssim R \lesssim 5\sigma$, in spite of the random distribution of the particles, there is a sort of "order" revealed by the form of the RDF. This order is often referred to as the local structure of the liquid. The local character of this structure should be stressed, since it contrasts with the long-range order typical of a perfect lattice.

Figure 2.8 demonstrates the dependence of $g(R)$ on the parameter $\varepsilon/kT$ (that is, the LJ energy parameter $\varepsilon$ for a system at fixed temperature; or, alternatively, for a given system with a fixed parameter $\varepsilon$, we get the dependence on temperature). The illustration in Fig. 2.8 corresponds to a fixed density of $\varrho\sigma^3 = 0.6$. One sees that as $\varepsilon/kT$ increases, the peaks become more pronounced, indicating an increase in the local "order" in the liquid. The values of $\varepsilon/kT$ for which the computations have been carried out are indicated next to each curve. Some details on the method used to compute the various functions $g(R)$ in this section can be found in Appendix 9-E.

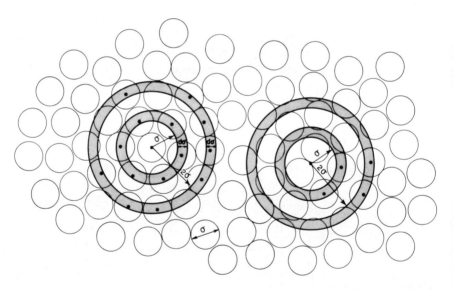

Fig. 2.7. A random distribution of spheres in two dimensions. Two spherical shells of width $d\sigma$ with radius $\sigma$ and $2\sigma$ are drawn (the diameter of the spheres is $\sigma$). On the left, the center of the spherical shell coincides with the center of one particle, whereas on the right, the center of the spherical shell has been chosen at a random point. It is clearly observed that the two shells on the left are filled, by centers of particles, to a larger extent than the corresponding shells on the right. The average excess of particles in these shells, drawn from the center of a given particle, is manifested by the various peaks of $g(R)$.

# Molecular Distribution Functions

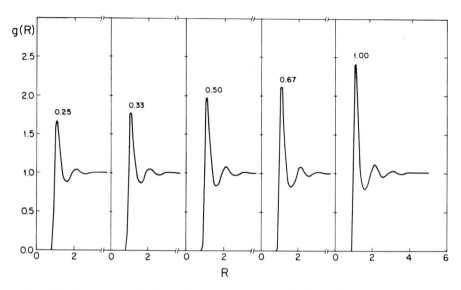

Fig. 2.8. Dependence of $g(R)$ on the energy parameter for Lennard-Jones particles with $\sigma = 1.0$ and (number) density $\varrho = 0.6$. The value of $\varepsilon/kT$ is indicated next to each curve. (For details of the method of computation of these curves, see Appendix 9-E.)

## 2.6. FURTHER PROPERTIES OF THE RADIAL DISTRIBUTION FUNCTION

From the definition of $g(R)$, it follows that the average number of particles in a spherical shell of radius $R$ (from a center of a given particle) and width $dR$ is

$$N(dR) = \varrho g(R) 4\pi R^2 \, dR \tag{2.75}$$

Hence, the average number of particles in a sphere of radius $R_M$ (excluding the particle at the center) is

$$N_{\mathrm{CN}}(R_M) = \varrho \int_0^{R_M} g(R) 4\pi R^2 \, dR \tag{2.76}$$

The quantity $N_{\mathrm{CN}}(R_M)$ may be referred to as the coordination number of particles, computed for the particular sphere of radius $R_M$. A choice of $\sigma \lesssim R_M \lesssim 2\sigma$ will give a coordination number that conforms to the common usage of this concept. There exist other methods of defining the concept of coordination number, which are summarized and discussed by Pings (1968).

A function related to the pair correlation function is the potential of average force (and torque), defined by[9]

$$g(\mathbf{X'}, \mathbf{X''}) = \exp[-\beta W(\mathbf{X'}, \mathbf{X''})] \qquad (2.77)$$

We henceforth specialize to spherical particles. A similar treatment can be carried out for the more general case of rigid particles, but this will be of no use to us.

From the definition of $g(\mathbf{R'}, \mathbf{R''})$ in (2.49), we have

$$\exp[-\beta W(\mathbf{R'}, \mathbf{R''})]$$
$$= \frac{N(N-1)}{\varrho^2} \frac{\int \cdots \int d\mathbf{R}_3 \cdots d\mathbf{R}_N \exp[-\beta U_N(\mathbf{R'}, \mathbf{R''}, \mathbf{R}_3, \ldots, \mathbf{R}_N)]}{Z_N} \qquad (2.78)$$

We now take the gradient of $W(\mathbf{R'}, \mathbf{R''})$ with respect to the vector $\mathbf{R'}$, and get

$$-\beta \nabla' W(\mathbf{R'}, \mathbf{R''})$$
$$= \nabla' \left[ \ln \int \cdots \int d\mathbf{R}_3 \cdots d\mathbf{R}_N \exp[-\beta U_N(\mathbf{R'}, \mathbf{R''}, \mathbf{R}_3, \ldots, \mathbf{R}_N)] \right] \qquad (2.79)$$

where the symbol $\nabla'$ stands for the gradient with respect to $\mathbf{R'} = (x', y', z')$, i.e.,

$$\nabla' = \left( \frac{\partial}{\partial x'}, \frac{\partial}{\partial y'}, \frac{\partial}{\partial z'} \right)$$

We also assume that the total potential energy is pairwise additive; hence, we may write

$$U_N(\mathbf{R'}, \mathbf{R''}, \mathbf{R}_3, \ldots, \mathbf{R}_N)$$
$$= U_{N-2}(\mathbf{R}_3, \ldots, \mathbf{R}_N) + \sum_{i=3}^{N} [U(\mathbf{R}_i, \mathbf{R'}) + U(\mathbf{R}_i, \mathbf{R''})] + U(\mathbf{R'}, \mathbf{R''}) \qquad (2.80)$$

The gradient of $U_N$ is then

$$\nabla' U_N(\mathbf{R'}, \mathbf{R''}, \mathbf{R}_3, \ldots, \mathbf{R}_N) = \sum_{i=3}^{N} \nabla' U(\mathbf{R}_i, \mathbf{R'}) + \nabla' U(\mathbf{R'}, \mathbf{R''}) \qquad (2.81)$$

---

[9] As in the notation of $g(\mathbf{X'}, \mathbf{X''})$, we omit a superscript to indicate the order of the potential of average force. The one defined in (2.77) is often denoted by $W^{(2)}(\mathbf{X'}, \mathbf{X''})$, to distinguish it from higher-order potentials of average force.

## Molecular Distribution Functions

Performing the differentiation in (2.79), we get

$$-\nabla' W(\mathbf{R}', \mathbf{R}'')$$
$$= \frac{\int \cdots \int d\mathbf{R}_3 \cdots d\mathbf{R}_N \exp(-\beta U_N)[-\sum_{i=3}^{N} \nabla' U(\mathbf{R}_i, \mathbf{R}') - \nabla' U(\mathbf{R}', \mathbf{R}'')]}{\int \cdots \int d\mathbf{R}_3 \cdots d\mathbf{R}_N \exp(-\beta U_N)} \quad (2.82)$$

Note that the integration in the numerator of (2.82) is over $\mathbf{R}_3 \cdots \mathbf{R}_N$; hence, the quantity $\nabla'(\mathbf{R}', \mathbf{R}'')$ can be placed in front of the integral sign. We also introduce the conditional probability density of finding the $N - 2$ particles at a specified configuration, given that two particles are at $\mathbf{R}', \mathbf{R}''$:

$$P(\mathbf{R}_3, \ldots, \mathbf{R}_N / \mathbf{R}', \mathbf{R}'')$$
$$= \frac{P(\mathbf{R}', \mathbf{R}'', \mathbf{R}_3, \ldots, \mathbf{R}_N)}{P(\mathbf{R}', \mathbf{R}'')} = \frac{\exp[-\beta U_N(\mathbf{R}', \mathbf{R}'', \mathbf{R}_3, \ldots, \mathbf{R}_N)]}{Z_N}$$
$$\times \frac{Z_N}{\int \cdots \int d\mathbf{R}_3 \cdots d\mathbf{R}_N \exp[-\beta U_N(\mathbf{R}', \mathbf{R}'', \mathbf{R}_3, \ldots, \mathbf{R}_N)]}$$
$$= \frac{\exp[-\beta U_N(\mathbf{R}', \mathbf{R}'', \mathbf{R}_3, \ldots, \mathbf{R}_N)]}{\int \cdots \int d\mathbf{R}_3 \cdots d\mathbf{R}_N \exp[-\beta U_N(\mathbf{R}', \mathbf{R}'', \mathbf{R}_3, \ldots, \mathbf{R}_N)]} \quad (2.83)$$

Using (2.83), we rewrite (2.82) as

$$-\nabla' W(\mathbf{R}', \mathbf{R}'') = -\nabla' U(\mathbf{R}', \mathbf{R}'')$$
$$+ \int \cdots \int d\mathbf{R}_3 \cdots d\mathbf{R}_N P(\mathbf{R}_3, \ldots, \mathbf{R}_N / \mathbf{R}', \mathbf{R}'')$$
$$\times \sum_{i=3}^{N} [-\nabla' U(\mathbf{R}_i, \mathbf{R}')]$$
$$= -\nabla' U(\mathbf{R}', \mathbf{R}'') + \left\langle -\sum_{i=3}^{N} \nabla' U(\mathbf{R}_i, \mathbf{R}') \right\rangle_{N-2} \quad (2.84)$$

In (2.84), we have expressed $-\nabla' W(\mathbf{R}', \mathbf{R}'')$ as a sum of two terms. The first term is simply the "direct" force exerted on the particle at $\mathbf{R}'$ when the second particle is at $\mathbf{R}''$. The second term is the *conditional* average force [note that the average has been carried out using the conditional probability density (2.83)] exerted on the particle at $\mathbf{R}'$ by all the other particles present in the system. It is an average over all the configurations of the $N - 2$ particles (as indicated by the notation $\langle \ \rangle_{N-2}$) keeping $\mathbf{R}'$ and $\mathbf{R}''$ fixed. The latter may be referred to as the "indirect" force operating on the particle at $\mathbf{R}'$, which originates from all the other particles, excluding the one at $\mathbf{R}''$. The foregoing discussion justifies the designation of $W(\mathbf{R}', \mathbf{R}'')$ as the "potential of average force." Its gradient gives the average force, including direct and indirect contributions, operating on the particle at $\mathbf{R}'$.

The form of the function $W(R)$, with $R = |\mathbf{R}'' - \mathbf{R}'|$ for LJ particles, and its density dependence are depicted in Fig. 2.9. Clearly, at very low densities, the potential of average force is identical with the pair potential; this follows from the negligible effect of all the other particles present in the system. At higher densities, the function $W(R)$ shows successive maxima and minima [corresponding exactly to the minima and maxima of $g(R)$, by virtue of the definition (2.77)]. The interesting point worth noting is that the indirect force at, say $R > \sigma$, can be either attractive or repulsive even in the region where the direct force is purely attractive.

There is a fundamental analogy between the concept of the potential of average force and the average interaction potential between two dipoles or atoms, as discussed in Section 1.7. In all of these cases, we average over part of the degrees of freedom of the system and thereby produce new features of the interaction which did not exist in the "bare" (or original) potential function. Here, we have obtained alternating regions of repulsion and attraction at $R > \sigma$, whereas in the same region, the original direct potential was everywhere attractive. Another feature which results from the averaging process is the possible loss of the pairwise additivity of the interaction. Some examples were discussed in Section 1.7. Further elaboration on this question will be presented in Section 2.8.

It is instructive to examine in more detail the origin of the attractive region of $W(R)$ for a system of hard spheres. This may be demonstrated already for a system of hard spheres at low density, for which we had an explicit expression for the RDF. Thus, using relation (2.73) and the definition (2.77), we get

$$W(R) = \begin{cases} \infty & \text{for } R < \sigma \\ -kT \ln[1 + \varrho B(R)] \sim -kT\varrho \dfrac{4\pi\sigma^3}{3}\left[1 - \dfrac{3}{4}\dfrac{R}{\sigma} + \dfrac{1}{16}\left(\dfrac{R}{\sigma}\right)^3\right] & \text{for } \sigma \leq R \leq 2\sigma \\ 0 & \text{for } R > 2\sigma \end{cases}$$
(2.85)

This function is depicted in Fig. 2.10. Note that the force is attractive everywhere in the range $\sigma \leq R \leq 2\sigma$, i.e.,

$$-\frac{\partial}{\partial R}W(R) = kT\varrho \frac{4\pi\sigma^3}{3}\left(-\frac{3}{4\sigma} + \frac{3}{16}\frac{R^2}{\sigma^3}\right) < 0, \qquad \sigma \leq R \leq 2\sigma$$
(2.86)

Molecular Distribution Functions

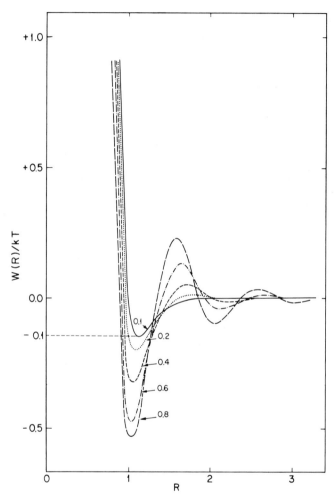

Fig. 2.9. Schematic illustration of the form of the function $W(R)$ and its density dependence. (The curves in this figure were computed for a system of Lennard-Jones particles in two dimensions with parameters $\sigma = 1.0$ and $\varepsilon/kT = 0.1$.) The number density is indicated next to each curve. Note that for $\varrho = 0.1$, the curve of $W(R)/kT$ is almost identical with $U(R)/kT$, with a minimum energy of $W(R)/kT \approx -0.1$. The first minimum becomes deeper and new minima develop as the density increases.

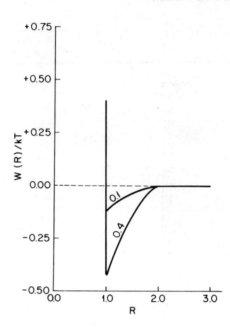

Fig. 2.10. The forms of $W(R)/kT$ for hard spheres at low densities. The curves are computed for spheres of diameter $\sigma = 1.0$ and two densities, $\varrho = 0.1$ and $\varrho = 0.4$. Note that for $1 \leq R \leq 2$, we have an attractive potential of average force. [Computed from Eq. (2.85); note that Eq. (2.85) is valid at very low densities, and may not hold for these particular densities.]

This is a correct formal result, yet it is quite surprising if we remember that we are dealing with hard-sphere particles. It is important to try to get an insight into the origin of this "attractive force" before accepting the formal result unquestioningly. We now examine this phenomenon for a system of three "almost" hard spheres, two of them fixed at $\mathbf{R}_1$ and $\mathbf{R}_2$ (with $R = |\mathbf{R}_2 - \mathbf{R}_1|$), and a third particle serving as a "solvent." Clearly, this is the simplest solvent we can envisage besides a vacuum. The choice of "almost hard spheres" has been made essentially for convenience and will become clearer later. The pair potential for our system is depicted in Fig. 2.11. The idea is that for $R > \sigma$, the potential is zero, as it is for hard spheres, whereas for $R < \sigma$, the potential is very strongly repulsive, yet the slope is finite, in contrast to the infinite slope for hard spheres.

# Molecular Distribution Functions

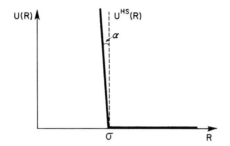

Fig. 2.11. An almost-hard-sphere potential function. As the angle $\alpha$ tends to zero, one gets the hard-sphere potential $U^{HS}(R)$.

Figure 2.12 shows the system under consideration. The two fixed particles are denoted by 1 and 2 and are placed at a distance $\sigma \leq R \leq 2\sigma$ (which is the only region of interest for the present discussion). The third particle, which may wander about, is denoted by 3, and a few possible positions of this particle are shown in the figure.

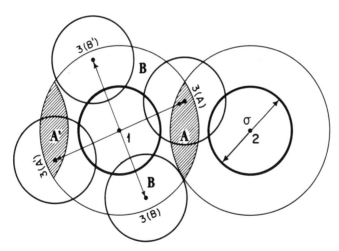

Fig. 2.12. Two almost hard spheres, numbered 1 and 2, at a distance of $\sigma < R < 2\sigma$. A third particle, numbered 3, is shown in various positions: In $3(B)$ and $3(B')$, particle 3 exerts repulsive forces on 1 of equal magnitudes but opposite directions. The probabilities of these two positions are equal since 3 interacts only with 1. On the other hand, in the two positions $3(A)$ and $3(A')$, particle 3 exerts equal forces on 1 and in opposite directions, but now the probability of position $3(A)$ is smaller than that of $3(A')$.

The total force operating on particle 1 is, according to (2.84),

$$-\nabla_1 W(\mathbf{R}_1, \mathbf{R}_2) = \int_V d\mathbf{R}_3\, P(\mathbf{R}_3/\mathbf{R}_1, \mathbf{R}_2)[-\nabla_1 U(\mathbf{R}_1, \mathbf{R}_3)]$$

$$= \int_A + \int_{A'} + \int_B \tag{2.87}$$

(Note that for $\sigma < R \leq 2\sigma$, the direct force is zero.) In the last form on the rhs of (2.87), we have split the integral over $V$ into three integrals over three nonoverlapping regions. The region $A$ includes all points in space for which $|\mathbf{R}_3 - \mathbf{R}_1| < \sigma$ *and* $|\mathbf{R}_3 - \mathbf{R}_2| < \sigma$, i.e., the region in which particle 3 "touches" both particles 1 and 2. It is obtained by the intersection of the two spheres of *radius* $\sigma$ drawn about the centers 1 and 2. Region $A'$ is obtained by reflecting each point in $A$ through the center of particle 1. Region $B$ comprises all the remaining points for which particle 3 interacts with 1 but not with 2. [Other regions in space are of no importance since they do not contribute to the integral (2.87).]

Now, for each position of particle 3 in $B$, there exists a "complementary" position obtained by reflection through the center of 1. These pairs of positions will have the same probability density [since $P(\mathbf{R}_3/\mathbf{R}_1, \mathbf{R}_2)$ depends only on $U(\mathbf{R}_1, \mathbf{R}_3)$ in these regions]. Furthermore, the forces exerted on 1 by such a pair of positions will be equal in magnitude but opposite in direction. An example of such a pair is shown in Fig. 2.12 and is denoted by $3(B)$ and $3(B')$. It is therefore clear that the contribution to the integral over region $B$ in (2.87) will be zero.

Turning to regions $A$ and $A'$, we see again that for each point in $A$, we have a point in $A'$ in which the force exerted by 3 on 1 will have the same magnitude but opposite direction. An example is the pair denoted by $3(A)$ and $3(A')$. But now the probability of finding the particle 3 in $A$ will always be smaller than the probability of finding the particle in the complementary point in $A'$. The reason is that $P(\mathbf{R}_3/\mathbf{R}_1, \mathbf{R}_2)$ is proportional to $\exp[-\beta U(\mathbf{R}_1, \mathbf{R}_3) - \beta U(\mathbf{R}_2, \mathbf{R}_3)]$ in $A$, but proportional to $\exp[-\beta U(\mathbf{R}_1, \mathbf{R}_3)]$ in $A'$. Since the potential is positive in these regions, the probability of finding particle 3 at a point in $A$ will always be smaller than the probability of finding it at the complementary point in $A'$. Hence, the integral over $A$ is larger in absolute magnitude than the integral over $A'$, and the overall direction of the force in (2.87) will be from $\mathbf{R}_1$ toward $\mathbf{R}_2$, i.e., the indirect force operating between 1 and 2, due to the presence of 3, is attractive.

To summarize, we have started with particles having no *direct* attractive forces and have arrived at an *indirect* attractive force, which origin-

Molecular Distribution Functions 61

ates from the averaging process over all possible positions of a third particle. The latter, though also exerting only repulsive forces, operates on particle 1 in an asymmetric manner, which leads to a net average attraction between particles 1 and 2.

Of course, we could have chosen the potential function in Fig. 2.11 to be as steep as we wish, and hence approach very closely the potential function for hard spheres. The latter is didactically inconvenient, however, since we have to deal with the awkward product of infinite forces with zero probabilities under the integral sign in (2.87).[10]

A different and often useful function related to $g(R)$ is defined by

$$y(R) = g(R) \exp[+\beta U(R)] \quad (2.88)$$

A glance at (2.50) shows that $y(R)$ is simply the second factor in (2.50), which includes the density expansion of $g(R)$. The elimination of the factor $\exp[-\beta U(R)]$ appearing in (2.50) is often useful, since the remaining function $y(R)$ becomes everywhere an analytic function of $R$ even for hard spheres. We illustrate this point by considering only the first-order term in the expansion (2.50) for hard-sphere particles, where we have

$$y(R) = 1 + B(R)\varrho \quad (2.89)$$

Note that for $R \leq \sigma$, $g(R)$ is zero because of the factor $\exp[-\beta U(R)]$ in (2.50) (see also Fig. 2.5) and clearly has a singularity at the point $R = \sigma$. Once we get rid of the singular factor in (2.50), we are left with the function $y(R)$, which changes smoothly even in the range $0 \leq R \leq \sigma$. Let us examine the specific example given in (2.89).

The form of $y(R)$ for $R > \sigma$ is identical to the form of $g(R)$ in this region, which has been given in (2.73). The only region for which we have to compute $y(R)$ is thus $R \leq \sigma$. Using arguments similar to those in Section 2.5., we get, from (2.89) and the definition of $B(R)$ in (2.52),

$$y(R) = 1 + \frac{4\pi\varrho}{R} \int_{R/2}^{\sigma} u\, du \int_{|R-u|}^{u} v\, dv$$

$$= 1 + \varrho \frac{4\pi\sigma^3}{3} \left[ 1 - \frac{3}{4} \frac{R}{\sigma} + \frac{1}{16} \left(\frac{R}{\sigma}\right)^3 \right] \quad \text{for} \quad 0 \leq R \leq \sigma \quad (2.90)$$

[10] The arguments presented here are based on the definition of $g(R)$ and hence $W(R)$ in the *configurational* space. In a real system, where particles possess kinetic energy of translation, one may argue that collisions of particle 3 with 1 are asymmetric because of partial shielding by particle 2. Hence, particle 3 will collide with 1 more often from the left-hand side than from the right-hand side, producing the same net effect of attraction between particles 1 and 2.

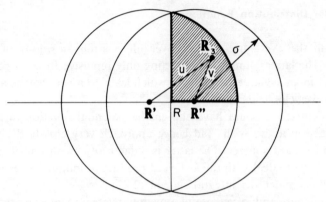

Fig. 2.13. Region of integration [Eq. (2.90)] for two hard spheres of *diameter* σ at a distance $0 \leq R \leq \sigma$. The integration extends over the volume of revolution of the intersection between the two circles centered at **R'** and **R''**. Half of this intersection is darkened in the figure.

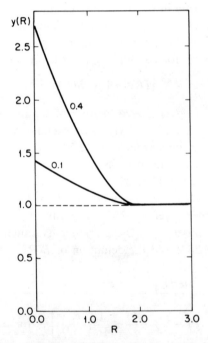

Fig. 2.14. The form of the function $y(R)$ for hard spheres, using Eq. (2.90). Note that the function $y(R)$ changes smoothly as we cross $R = \sigma = 1$ and that it is a monotonically decreasing function of $R$ in the region $0 \leq R \leq \sigma$. The two curves correspond to the same systems as in Fig. 2.5, i.e., for $\sigma = 1.0$ and $\varrho = 0.1$ and $\varrho = 0.4$.

## Molecular Distribution Functions

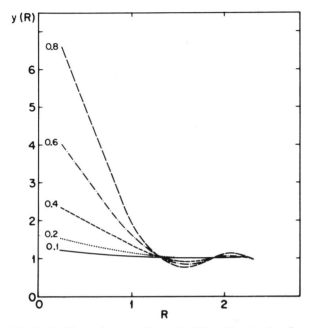

Fig. 2.15. Schematic variation of $y(R)$ with density for Lennard-Jones particles (these computations correspond to Lennard-Jones particles in two dimensions with $\sigma = 1.0$ and $\varepsilon/kT = 0.1$. The number density is indicated next to each curve). Note again the monotonic behavior of $y(R)$ in the region $0 \leq R \leq \sigma$.

The region of integration in (2.90) differs from that in (2.72) only in the lower limit, $|R - u|$ instead of $R - u$ in (2.72). The region of integration is shown in Fig. 2.13 (compare with Fig. 2.3 for $\sigma < R < 2\sigma$), and it consists of twice the volume of revolution of the shaded area in the figure. Note that since the integration over $v$ in (2.90) gives $v^2/2$, the sign of $R - u$ in the lower limit is of no importance, hence the integral in (2.90) gives the same result as in (2.72). Figure 2.14 shows the form of $y(R)$ for HS particles using (2.90) and Fig. 2.15 shows $y(R)$ for LJ particles at various densities. Note in particular that the function $y(R)$ is continuous in crossing the point $R = \sigma$ into the physically inaccessible region $R < \sigma$. This feature renders this function of some convenience in the theory of integral equations, where one first solves for $y(R)$ and then constructs $g(R)$ by (2.88) (for more details, see Appendices 9-D and 9-E). We exploit the information on $y(R)$ at $0 \leq R \leq \sigma$ in Chapter 8 in our discussion on hydrophobic interactions.

Finally, we now relate the potential of average force to the Helmholtz free energy. This relation will also be found useful for the study of hydrophobic interactions (Chapter 8).

Consider a system of $N$ simple spherical particles in a volume $V$ at temperature $T$. The Helmholtz free energy for such a system is

$$\exp[-\beta A(T, V, N)] = (1/N!\,\Lambda^{3N}) \int \cdots \int d\mathbf{R}^N \exp[-\beta U(\mathbf{R}^N)] \quad (2.91)$$

Now consider a slightly modified system in which two particles, say 1 and 2, have been fixed at the points $\mathbf{R}'$ and $\mathbf{R}''$, respectively. The free energy for such a system is denoted by $A(\mathbf{R}', \mathbf{R}'')$ and we have

$$\exp[-\beta A(\mathbf{R}', \mathbf{R}'')] = \frac{1}{(N-2)!\,\Lambda^{3(N-2)}} \int \cdots \int d\mathbf{R}_3 \cdots d\mathbf{R}_N$$
$$\times \exp[-\beta U_N(\mathbf{R}', \mathbf{R}'', \mathbf{R}_3, \ldots, \mathbf{R}_N)] \quad (2.92)$$

Note the differences between (2.91) and (2.92). Let us denote by $A(R)$, with $R = |\mathbf{R}'' - \mathbf{R}'|$, the free energy of such a system when the separation between the two particles is $R$, and form the difference

$$\Delta A(R) = A(R) - A(\infty) \quad (2.93)$$

which is the work required to bring the two particles from fixed position at infinite separation, to fixed position where the separation is $R$. The process is carried out at constant volume and temperature. From (2.49), (2.77), (2.91), and (2.93), we get

$$\exp[-\beta\,\Delta A(R)]$$
$$= \frac{\int \cdots \int d\mathbf{R}_3 \cdots d\mathbf{R}_N \exp[-\beta U_N(\mathbf{R}', \mathbf{R}'', \mathbf{R}_3, \ldots, \mathbf{R}_N)]}{\lim_{R\to\infty}\{\int \cdots \int d\mathbf{R}_3 \cdots d\mathbf{R}_N \exp[-\beta U_N(\mathbf{R}', \mathbf{R}'', \mathbf{R}_3, \ldots, \mathbf{R}_N)]\}}$$
$$= g(R)/g(\infty) = \exp\{-\beta[W(R) - W(\infty)]\} \quad (2.94)$$

This relation is important. It is a connection between the *work* required to bring two particles from one configuration to another, and the ratio of the *probabilities* of the two events. A similar relation between the probability of occurrence of an event and the work involved in the creation of that event is discussed in Section 3.11.

## 2.7. SURVEY OF THE METHODS OF EVALUATING g(R)

There exist essentially three sources of information on the form of the function $g(R)$. The first is experimental, based on X-ray, or neutron, diffraction by liquids. The second is theoretical, and usually involves integral equations for $g(R)$ or some related function. The third method may be classified as an intermediate between theory and experiment. It consists of various simulation techniques that have recently become very important for investigating the liquid state.

In this section, we survey the fundamental principles of the various methods. The details are highly technical in character and are not essential for understanding the subject matter in the subsequent chapters. We refer the reader to the literature for recent reviews on each of the methods that will be discussed.

### 2.7.1. Experimental Methods

The original and most extensively employed method of evaluating $g(R)$ experimentally is the study of the X-ray diffraction pattern by liquids. Recently, diffraction of neutrons has been found increasingly useful for this purpose. The principal idea of converting diffraction patterns into pair distribution functions is common to both methods, though they differ both in experimental detail and scope of information that they provide.

Consider a monochromatic beam of electromagnetic waves with a given wave number vector $\mathbf{k}_0$ ($\mathbf{k}_0$ is a vector in the direction of propagation of the plane wave with absolute magnitude $|\mathbf{k}_0| = 2\pi/\lambda$, where $\lambda$ is the wavelength), scattered from a system of $N$ molecules.[11] The scattered wave number vector is denoted by $\mathbf{k}$. Two important assumptions are made regarding this process. (1) The incident and the scattered beam have the same wavelength, i.e.,

$$|\mathbf{k}_0| = |\mathbf{k}| = 2\pi/\lambda \qquad (2.95)$$

---

[11] Actually, the X rays are scattered by the electrons, and therefore one should start by considering the ensemble of electrons as the fundamental medium for the scattering experiment. However, since all the atoms (or molecules) are considered to be equivalent, one may group together all the electrons belonging to the same atom (or molecule). This grouping procedure is equivalent to a virtual replacement of each atom (or molecule) by a single electron situated at some chosen "center" of the molecule. To compensate for this replacement, one introduces the molecular structure factor (which is the same for all the molecules in the system). In the present treatment, this factor is absorbed in the proportionality constant $\alpha'$ in Eq. (2.97).

This means that the scattering is "elastic," often referred to as "Rayleigh scattering." (2) Each ray entering the system is scattered only once. This assumption is essential for obtaining the required relation between the intensity of the scattered beam and the *pair* distribution function. If multiple scattering occurs, then such a relationship would involve higher-order molecular distribution functions.

Let $E_j$ be the complex amplitude of the wave scattered by the $j$th particle, measured at the point of observation (which is presumed to be far from the system). The total amplitude measured at the point of observation is

$$E(\mathbf{R}^N) = \sum_{j=1}^{N} E_j = E_0 \sum_{j=1}^{N} \exp(i\phi_j) \qquad (2.96)$$

where in $E_0$ we have included all the factors which have equal values for all the waves scattered, and $\phi_j$ is the phase of the wave scattered from the $j$th particle (here, $i = \sqrt{-1}$). We have also explicitly denoted the dependence of $E$ on the configuration $\mathbf{R}^N$.

The intensity of the scattered wave at the point of observation is proportional to the square of the amplitude; hence

$$I(\mathbf{R}^N) = \alpha E(\mathbf{R}^N) E(\mathbf{R}^N)^* = \alpha' \sum_{j=1}^{N} \exp(i\phi_j) \sum_{n=1}^{N} \exp(-i\phi_n)$$

$$= \alpha' \left\{ N + \sum_{j \neq n} \exp[i(\phi_j - \phi_n)] \right\} \qquad (2.97)$$

where $\alpha$ and $\alpha'$ are numerical factors. In the last form on the rhs of (2.97), we have split the sum into two terms, those with $j = n$ and those with $j \neq n$. The important observation is that $I(\mathbf{R}^N)$ is expressed in (2.97) as a sum over terms each of which depends on one *pair* of particles. According to a general theorem (which will be derived in the next chapter, Section 3.2), the average of such an expression over all the configurations of the system can be expressed as an integral involving the pair distribution function. Before taking the average over all the configurations of the system, it is convenient to express the phase difference $\phi_j - \phi_n$ in terms of the positions of the $j$th and $n$th particles (the relevant geometry is given in Fig. 2.16):

$$(a + b)2\pi/\lambda = \mathbf{R}_{jn} \cdot (\mathbf{k}_0 - \mathbf{k}) = \phi_j - \phi_n \qquad (2.98)$$

The difference in the wave vectors $\mathbf{k}_0 - \mathbf{k}$ is related to the scattering angle $\theta$ through

$$|\mathbf{S}|/2|\mathbf{k}| = \sin(\theta/2), \qquad \mathbf{S} = \mathbf{k}_0 - \mathbf{k} \qquad (2.99)$$

## Molecular Distribution Functions

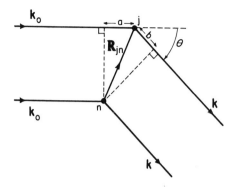

Fig. 2.16. Schematic illustration of the scattering of an electromagnetic wave by two particles $j$ and $n$. The wave vectors of the incident and the scattered rays are $\mathbf{k}_0$ and $\mathbf{k}$, respectively. The difference between the optical distance of the two rays scattered from $j$ and $n$ is $a + b$.

Hence, (2.97) can be rewritten as

$$I(\mathbf{R}^N, \mathbf{S}) = \alpha' \left\{ N + \sum_{j \neq n} \exp(i\mathbf{R}_{jn} \cdot \mathbf{S}) \right\} \quad (2.100)$$

where we have also explicitly introduced the dependence of $I$ on the vector $\mathbf{S}$. Taking the average over all the configurations, say in the $T, V, N$ ensemble, we get

$$I(\mathbf{S}) = \langle I(\mathbf{R}^N, S) \rangle = \alpha' \left[ N + \left\langle \sum_{j \neq n} \exp(i\mathbf{R}_{jn} \cdot \mathbf{S}) \right\rangle \right]$$

$$= \alpha' \left[ N + \int \int d\mathbf{R}_1 \, d\mathbf{R}_2 \, \varrho^{(2)}(\mathbf{R}_1, \mathbf{R}_2) \exp(i\mathbf{R}_{12} \cdot \mathbf{S}) \right] \quad (2.101)$$

In the last step of (2.101), we have used a theorem that will be proven, in more general terms, in the next chapter. One may notice that since the pair distribution function depends only on the distance $R_{12}$, and the exponential function depends only on the scalars $R_{12}$, $\theta$, and $\lambda$, one can actually rewrite the integral in (2.101) as a one-dimensional integral. An important feature that should be noticed is that the integral in (2.101) is essentially a Fourier transform of the pair distribution function. (The constant $\alpha'$ may also depend on $S$ through the structure factor that has been included in it.) Hence, one can, in principle, take the inverse transform and express $\varrho^{(2)}(R)$ as an integral involving $I(S)$. This process requires some approximations which follow from the incomplete knowledge of the function $I(S)$ over the full range of values of $S$ (or the scattering angles $\theta$). Further details may be found in the work of Hirschfelder et al. (1954), Fisher (1964), Kruh (1962), and Pings (1968).

### 2.7.2. Theoretical Methods

By theoretical methods, one usually refers to the devising of integral equations, the solutions of which give $g(R)$ or a function related to $g(R)$ in a simple manner. Integral equations were originally suggested by Yvon (1935) and Kirkwood (1935), who have been followed by many researchers [see also Hill (1956), Rice and Gray (1965), and Münster (1969)]. The most promising and successful integral equation is the so-called Percus–Yevick (PY) equation. This has been used extensively to compute $g(R)$ for simple liquids such as argon, achieving remarkable agreement with experimental results. The derivation of the PY equation, though of interest in itself, is deferred to the appendices, since it involves some mathematical techniques which are rather intricate. Here we discuss its general structure.

An integral equation for $g(R)$ can be written symbolically as

$$g(R) = F[g(r), U(r); \varrho, T] \qquad (2.102)$$

where $F$ stands for a functional of $g(r)$ and $U(r)$ which is also dependent on thermodynamic parameters, such as $\varrho, T$ or $P, T$, etc.

Once we have chosen a model potential to describe our system of real particles, e.g., Lennard-Jones particles (see Section 1.7), the functional equation is viewed as an equation involving $g(R)$ as the unknown function. This equation can be solved either by analytical methods or by numerical methods (see, for instance, Appendix 9-E). One version of the PY equation is

$$y(\mathbf{R}_1, \mathbf{R}_2) = 1 + \varrho \int [y(\mathbf{R}_1, \mathbf{R}_3)f(\mathbf{R}_1, \mathbf{R}_3)]$$
$$\times [y(\mathbf{R}_3, \mathbf{R}_2)f(\mathbf{R}_3, \mathbf{R}_2) + y(\mathbf{R}_3, \mathbf{R}_2) - 1]\, d\mathbf{R}_3 \qquad (2.103)$$

where

$$y(\mathbf{R}_1, \mathbf{R}_2) = g(\mathbf{R}_1, \mathbf{R}_2) \exp[+\beta U(\mathbf{R}_1, \mathbf{R}_2)] \qquad (2.104)$$

$$f(\mathbf{R}_1, \mathbf{R}_2) = \exp[-\beta U(\mathbf{R}_1, \mathbf{R}_2)] - 1 \qquad (2.105)$$

Equation (2.103) is an integral equation for $y$, provided we have chosen a pair potential $U$ (and, hence, $f$ is presumed to be a given function). Once we have solved (2.103) for $y$, we can compute $g$ through (2.104). A simpler

## Molecular Distribution Functions

version of Eq. (2.103) for spherical particles is given in Appendix 9-E. As an illustration of the content of the integral equation, let us substitute $y \equiv 1$ on the rhs of (2.103) (which is the solution for $y$ at extremely low densities, $\varrho \to 0$; see also discussion in Section 2.6). This is also a common first step in an iterational procedure for solving Eq. (2.103):

$$y(\mathbf{R}_1, \mathbf{R}_2) = 1 + \varrho \int f(\mathbf{R}_1, \mathbf{R}_3) f(\mathbf{R}_3, \mathbf{R}_2)\, d\mathbf{R}_3 \qquad (2.106)$$

which can be identified with the first-order expansion of $y$ in the density, as discussed in Section 2.6. We can continue this process and substitute $y$ from (2.106) into the rhs of (2.103) to obtain successively higher-order terms in the density expansion of $y$. This expansion is not exact since it is based on an approximate integral equation. Nevertheless, the solution of (2.103) by iteration is considered to give reliable results for simple fluids up to considerably high densities. It is for this reason that we have chosen to illustrate the properties of $g(R)$ by solutions obtained from this equation (for more details, see Appendix 9-E).

### 2.7.3. Simulation Methods

As noted before, simulation methods may be considered to be intermediate between theory and experiment. In fact, various kinds of simulations have been devised for the purpose of computing $g(R)$ as well as other properties of the liquid. We now make a distinction between three of these methods.

#### 2.7.3.1. Experimental Simulations

Morrell and Hildebrand (1934) described a simple and interesting method of computing $g(R)$ for a system of balls of macroscopic size. They used gelatin balls suspended in liquid gelatin that were mechanically shaken in their vessel. The coordinates of some of the balls could be measured by taking photographs of the system at various times. From these data, they were able to compute the distribution of distances between the balls, and from this, the radial distribution function (see Hildebrand *et al.*, 1970).

A similar method was employed by Bernal and his collaborators (Bernal and King, 1968), who used steel balls in a random configuration to compute $g(R)$ (as well as other distribution functions of interest in understanding the random character of the liquid state). The form of $g(R)$ obtained from these methods was remarkably similar to the RDF of argon obtained from the X-ray diffraction pattern. One important conclusion that may be drawn from these experiments is that the mode of packing of the atoms in the liquid follows essentially from geometric considerations of the packing of hard spheres. The fact that atoms have an effective hard-core diameter already dictates the type of packing of the molecules in the liquid. The additional attractive forces operating between real molecules have a relatively small effect on the mode of packing. This conclusion has been confirmed by many authors using considerably different arguments.

### 2.7.3.2. Monte Carlo Simulation

A closely related "experiment" can be carried out on a computer. Instead of shaking a system of steel balls by mechanical means, one can generate random configurations of particles by computer. The latter method can be used to compute the RDF as well as other thermodynamic quantities of a system.

The specific method devised by Metropolis *et al.* (1953) to compute the properties of liquids is now known as the Monte Carlo (MC) method. In fact, this is a special procedure to compute multidimensional integrals numerically.

Consider the computation of any average quantity, say in the $T$, $V$, $N$ ensemble:

$$\langle F \rangle = \int \cdots \int d\mathbf{R}^N \, P(\mathbf{R}^N) F(\mathbf{R}^N) \qquad (2.107)$$

where $F(\mathbf{R}^N)$ is any function of the configuration $\mathbf{R}^N$. In particular, the pair distribution function (and, hence, the RDF) can be viewed as an average of the form (2.107) (see Section 2.3).

From the definition of the Riemann integral, the quantity $\langle F \rangle$ can be approximated by a sum over discrete events (configurations)

$$\langle F \rangle \approx \sum_{i=1}^{n} P_i F_i \qquad (2.108)$$

where $P_i$ is the probability of the $i$th configuration and $F_i$ the value of the function $F(\mathbf{R}^N)$ at the $i$th configuration.

A glance at (2.107) reveals that two very severe difficulties arise in any attempt to approximate the average $\langle F \rangle$ by a sum of the form (2.108). First, "a configuration" means a specification of $3N$ coordinates (for spherical particles), and, clearly, such a number of configurations cannot be

# Molecular Distribution Functions

handled in a computer if $N$ is of the order of $10^{23}$. This limitation compels us to choose $N$ of the order of a few hundred. The question of the suitability of such a small sample of molecules to represent a macroscopic system immediately arises. The second difficulty concerns the convergence of the sum in (2.108). Suppose we have already chosen $N$. The question is: How many configurations $n$ should we select in order to ensure the validity of the approximation in (2.108)? Again, time limitations impose restrictions on the number $n$ that we can afford in an actual "experiment."

Suppose now that we are given the opportunity of carrying out a computation with, say, $N$ of the order of $10^2$–$10^3$ and with $n$ of the order of $10^5$–$10^6$. What should be an intelligent way of exploiting this opportunity to get meaningful results which will be relevant to the study of the liquid state?

Metropolis *et al.* (1953) suggested two ingenious ways to overcome the two difficulties mentioned above. First, a system with $N \approx 10^2$ possesses surface effects that may dominate the behavior of the sample and therefore may not be suitable to represent a macroscopic property of the liquid. To reduce the surface effects, it has been suggested to impose periodic boundary conditions. This trick is simpler to visualize in one or two dimensions (Fig. 2.17), where the system is converted into a ring or a torus, respectively. In three dimensions, one may imagine that a particle leaving the surface $A$ in Fig. 2.17 enters through the surface $A'$. If this is done for each of the pairs of opposing surfaces, the system may be viewed as "closed on itself," in the sense of the two simpler examples given in the figure.

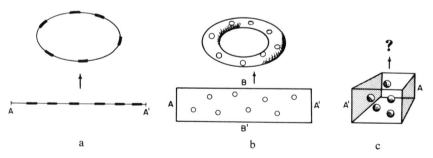

Fig. 2.17. Illustration of the periodic boundary conditions. (a) A one-dimensional system of "spheres" (rods). The periodic boundary condition is obtained by identifying the end $A$ with $A'$. The result is a closed circle. (b) A two-dimensional system of "spheres" (disks). The periodic boundary condition is obtained by identifying the edges $A$ with $A'$, and $B$ with $B'$. The result is a torus. (c) A three-dimensional system of spheres. The periodic boundary condition is obtained by identifying opposite faces, such as $A$ and $A'$, with each other. No pictorial description of the resulting system is possible.

The second idea is both more profound and more interesting. Suppose we have a system of $N$ particles at a relatively high density. A choice of a *random* configuration for the centers of all particles will almost always lead to the event that at least two particles overlap. Hence, $U(\mathbf{R}^N)$ of that configuration will be large and positive and therefore $P(\mathbf{R}^N)$ will be very small. Merely picking a sequence of $n$ random configurations would therefore be a very inefficient way of using computer time. Most of the time, we would be handling configurations which are physically almost unattainable. The suggested trick was expressed very succinctly by Metropolis *et al.* (1953): "Instead of choosing configurations *randomly*, then weighting them with $\exp[-\beta U_N(\mathbf{R}^N)]$, we choose configurations with *probability* $\exp[-\beta U_N(\mathbf{R}^N)]$ and weight them *evenly*."

Let us illustrate the essence of this idea for an exaggerated situation in a one-dimensional integral (see Fig. 2.18)

$$\int_a^b P(x)F(x)\,dx \approx \sum_{i=1}^n P_i F_i \qquad (2.109)$$

Suppose that $P(x)$ is almost zero in most of the interval $(a, b)$. In choosing a series of $n$ points at random, we shall be spending most of our time computing the values of the integrand that are nearly zero. A very small fraction of the configurations will hit the important region under the sharp peak.

The modified procedure is to construct a sequence of "events" (or configurations) in which the probability of the newly chosen event depends on that of the former event. Such a scheme of events is referred to as a Markov chain. Here we have a specific example of such a chain, which we shall very briefly describe.

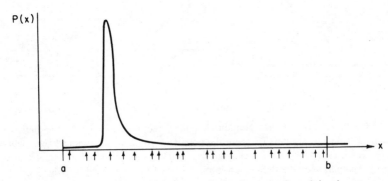

Fig. 2.18. A function $P(x)$ which has a single sharp peak and is almost zero everywhere else. The arrow indicates random numbers $x_i$ ($a \leq x_i \leq b$) at which the integrand is evaluated.

# Molecular Distribution Functions

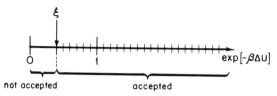

Fig. 2.19. Schematic illustration of the accepted and unaccepted regions in the Monte Carlo computation. If $\exp[-\beta \Delta U] > 1$, the new configuration is accepted. If $\exp[-\beta \Delta U] < 1$, then one selects a random number $0 \leq \xi \leq 1$ and make a further test; if $\exp[-\beta \Delta U] > \xi$, the new configuration is accepted; if $\exp[-\beta \Delta U] < \xi$, it is not accepted.

Consider a given configuration designated by the index $i$ and total potential energy $U_i$. We now take a new configuration $i + 1$ (say, by randomly moving one particle) and check the difference in the potential energy

$$\Delta U = U_{i+1} - U_i$$

If $\Delta U < 0$, the new configuration is accepted. If $\Delta U > 0$, we make another check. We select a random number $0 \leq \xi \leq 1$ and make the following decision: If $\exp[-\beta \Delta U] > \xi$, the new configuration is accepted; if $\exp[-\beta \Delta U] < \xi$, the new configuration is rejected. All these possibilities are shown schematically in Fig. 2.19. The net effect of this particular recipe is to let the choice of the next configuration be biased by its probability. As can be seen from Fig. 2.18, the above procedure consists of a series of points which, with large probability, "senses its way" toward the region that has the larger contribution to the integral. Therefore, in the long run, we will be spending more time working with configurations in the "important" regions of the integrand. Of course, we must have some representatives from the "unimportant" region, but this is done with extreme parsimony.

The Monte Carlo method has been used extensively to investigate the properties of simple fluids such as noble gases. The results of such computations are considered quite reliable even for a system consisting of $N \approx 10^2$ particles. For hard-sphere fluids, where no (real) experimental results may be obtained, this kind of (computer) experiment is an important source of information with which theoretical results (say, from integral equations) may be compared. A thorough review of all the technical and mathematical questions concerning this method has been presented by Wood (1968).

### 2.7.3.3. Molecular Dynamics

The molecular dynamics method is conceptually simpler than the Monte Carlo method. Here again, we can compute various averages of the form (2.107) and hence the RDF as well. The method consists in a direct solution of the equations of motion of a sample of $N$ ($\approx 10^2$) particles. In principle, the method amounts to computing *time* averages rather than *ensemble* averages, and was first employed for simple liquids by Alder and Wainwright (1957) [see review by Alder and Hoover (1968)]. The problem of surface effects is dealt with as in the Monte Carlo method. The sequence of events is now not *random*, but follows the trajectory which is dictated to the system by the equations of motion. In this respect, this method is of a more general scope, since it permits the computation of equilibrium as well as transport properties of the system.

Consider a given configuration of the system $\mathbf{R}^N$ and a given distribution of momenta at some particular time $t$. The total force acting on the $k$th particle is given by the sum of the forces exerted by all the other particles. The equations of motion for the system of $N$ particles are thus

$$m\, d^2\mathbf{R}_k/dt^2 = \sum_{j=1, j \neq k}^{N} \mathbf{F}(\mathbf{R}_{kj}) = -\sum_{j=1, j \neq k}^{N} \nabla_k U(\mathbf{R}_{kj}) \qquad (2.110)$$

Here, $m$ is the mass of each particle. We now let the $k$th particle move under the influence of the *constant* force given in (2.110) during a short time $\Delta t$. Continuation of this process produces a series of configurations over which average quantities may be computed. The essential approximation involves the assumption that the force is constant during $\Delta t$. Therefore, it is essential to choose a $\Delta t$ short enough to validate this approximation. Yet $\Delta t$ must not be too small, otherwise the system will tend to equilibrium very sluggishly, and the whole process will require excessive computation time.[12] (There is no absolute criterion for determining when equilibrium has been reached. One practical test is to start with different initial conditions and see if we end up with the same computed average quantities.)

---

[12] A similar consideration occurs in the Monte Carlo method. To proceed from one configuration to the next, one usually makes a small translation $\Delta \mathbf{R}$ to a given particle selected at random. $\Delta \mathbf{R}$ must be small enough, otherwise the moved particle will, with large probability, enter a position already occupied by another particle, and the new configuration will be rejected. On the other hand, if $\Delta \mathbf{R}$ is too small, the configuration will hardly change from one step to another, and excessive time will be needed to make significant changes of the configuration.

Both the Monte Carlo and the molecular dynamic methods have been applied successfully to investigate the properties of the liquid state. In spite of their inherent limitations (finiteness of $N$ and the number of steps that may be handled by the computer), these methods are considered to provide "exact" results for a model-particle system. This statement deserves further elaboration. We recall that our knowledge of the analytical form of the pair potential is incomplete even for the simplest atoms such as the noble gases. Furthermore, even if we knew the exact form of $U(R)$, say for argon, it would still be difficult to estimate to what extent high-order potentials are important to determine the properties of the liquid. Therefore, if we compare a theoretical result for $g(R)$ (say, by solving an integral equation) against an experimental result, we are actually comparing a property of a model with a property of a real substance. Such a comparison is, of course, the ultimate aim of the theory. Nevertheless, the extent of agreement between the theory and the experiment is often not a fair test of the theory. A better test is to compare the theory with an experiment carried out on *exactly the same* system. Such a comparison is made possible by the computer experiments, where we may deal with the same particles (say hard spheres or Lennard-Jones particles) on which the theoretical tools are applied. In this sense, the output of the simulation techniques provides an "exact" experimental source of information. To be sure, the information applies strictly to systems of model particles and not to real particles.

## 2.8. HIGHER-ORDER MOLECULAR DISTRIBUTION FUNCTIONS

Molecular distribution functions (MDF) of order higher than two rarely appear in actual computations of thermodynamic quantities of liquids. This is not because they are not needed there, but because nearly nothing is known of their analytical properties. The importance of higher-order MDF's is threefold. First, they appear in some relations connecting thermodynamic quantities and molecular properties, such as the relation for the heat capacity or partial molar quantities. Second, some of the common relations between thermodynamic quantities and the *pair* distribution function hold only by virtue of the assumption of pairwise additivity of the total potential energy. If we give up this assumption, and include higher-order potentials, we must use higher-order MDF's (more details are given in the next chapter). Finally, MDF's of higher order do appear in some intermediate steps in various derivations of integral equations. Some knowledge about their properties is therefore a necessity.

Consider again the basic distribution function in the $T, V, N$ ensemble (2.1). The *specific* $n$th-order MDF is defined as the probability density of finding particles 1, 2, ..., $n$ in the configuration $\mathbf{X}^n = \mathbf{X}_1, \mathbf{X}_2, \ldots, \mathbf{X}_n$:

$$P^{(n)}(\mathbf{X}^n) = \int \cdots \int d\mathbf{X}_{n+1} \cdots d\mathbf{X}_N \, P(\mathbf{X}^N) \tag{2.111}$$

The corresponding *generic* $n$th-order MDF is defined by

$$\varrho^{(n)}(\mathbf{X}^n) = [N!/(N-n)!]P^{(n)}(\mathbf{X}^n) \tag{2.112}$$

which is obtained by generalizing the arguments used in previous sections to define $\varrho^{(1)}$ and $\varrho^{(2)}$.

In some applications, it is more advantageous to rewrite (2.112) as an average quantity, similar to Eqs. (2.13) and (2.32), namely

$$\begin{aligned}\varrho^{(n)}(\mathbf{X}_1', \ldots, \mathbf{X}_n') &= [N!/(N-n)!] \int \cdots \int d\mathbf{X}_{n+1} \cdots d\mathbf{X}_N \\ &\quad \times P(\mathbf{X}_1', \mathbf{X}_2', \ldots, \mathbf{X}_n'; \mathbf{X}_{n+1}, \ldots, \mathbf{X}_N) \\ &= \int \cdots \int d\mathbf{X}_1 \cdots d\mathbf{X}_N \, P(\mathbf{X}^N) \sum_{i_1=1}^{N} \sum_{\substack{i_2=1 \\ i_1 \neq i_2 \neq i_3 \cdots \neq i_n}}^{N} \cdots \sum_{i_n=1}^{N} \\ &\quad \times \delta(\mathbf{X}_{i_1} - \mathbf{X}_1') \cdots \delta(\mathbf{X}_{i_n} - \mathbf{X}_n')\end{aligned} \tag{2.113}$$

Note that we have used primed vectors for quantities that are fixed. In the second form on the rhs of (2.113), we have used the basic property of the Dirac delta function, and the fact that there are altogether $N!/(N-n)!$ terms in the sum under the integral sign. The meaning of $\varrho^{(n)}$ as a probability density, or as an average density, can be obtained by direct generalization of the arguments employed for $\varrho^{(1)}$ and $\varrho^{(2)}$.

Various conditional MDF's can be introduced; for instance, the singlet conditional distribution function, for finding a particle at $\mathbf{X}_3$ given two particles at $\mathbf{X}_1$ and $\mathbf{X}_2$, is

$$\varrho(\mathbf{X}_3/\mathbf{X}_1, \mathbf{X}_2) = \varrho^{(3)}(\mathbf{X}_1, \mathbf{X}_2, \mathbf{X}_3)/\varrho^{(2)}(\mathbf{X}_1, \mathbf{X}_2) \tag{2.114}$$

This function will appear in the study of hydrophobic interaction in Chapter 8. It is important to realize that the quantity defined in (2.114) is the average local density of particles at $\mathbf{R}_3$ (with orientation $\mathbf{\Omega}_3$) of a system subjected to an "external" field of force produced by fixing two particles

# Molecular Distribution Functions

at $X_1$ and $X_2$. The arguments for this interpretation are exactly the same as those given in Section 2.4 [following Eq. (2.45)].

Higher-order correlation functions and potentials of average forces are defined in analogy with previous definitions for pairs. For instance, for $n = 3$, we define

$$\varrho^{(3)}(X_1, X_2, X_3) = \varrho^{(1)}(X_1)\varrho^{(1)}(X_2)\varrho^{(1)}(X_3)g^{(3)}(X_1, X_2, X_3) \quad (2.115)$$

and

$$g^{(3)}(X_1, X_2, X_3) = \exp[-\beta W^{(3)}(X_1, X_2, X_3)] \quad (2.116)$$

Using an interpretation similar to that for $W(X_1, X_2)$ (Section 2.6), we call $W^{(3)}(X_1, X_2, X_3)$ the potential of average force for the triplet of particles in the configuration $X_1, X_2, X_3$, the average being carried out over all the configurations of the remaining $N - 3$ particles. This function has played an important role in the derivation of integral equations for the pair correlation function. Although we shall not be interested in this topic in the remainder of the book, it is appropriate to point out one analogy between the expected behavior of $W^{(3)}$ and other average potentials discussed in Section 1.7. In Section 1.7, we demonstrated that the interaction among three dipoles is pairwise additive as long as we specify the *locations* and the *orientations* of the three dipoles. Once we average over part of the degrees of freedom (the orientations), we get, first, new features for the pair potential, and, in addition, we may lose the pairwise additivity. There exists a similar situation here, although some imagination is needed to see the analogy. We start with a system of $N$ particles. We select three of them and average over all the configurations of the $N - 3$ particles. Thus, we have averaged over part of the degrees of freedom of the system, and thereby obtained new properties for the average potential. Considering the potential of average force for pairs, we have already shown in Section 2.6 that new properties arise as result of the averaging process. In a similar way, we expect that even if our original system obeys the pairwise additivity assumption for the total potential energy, this property will be lost once we average over part of the degrees of freedom. Hence, we expect that, in particular, $W^{(3)}(X_1, X_2, X_3)$ will not be pairwise additive. In the historical development of the theory of liquids, the latter assumption was actually employed, and is known as the Kirkwood superposition approximation. For $W^{(3)}$, one writes

$$W^{(3)}(X_1, X_2, X_3) = W^{(2)}(X_1, X_2) + W^{(2)}(X_1, X_3) + W^{(2)}(X_2, X_3) \quad (2.117)$$

(here, $W^{(2)}$ is the same as $W$ in Section 2.6). It is important to recognize that the assumption of pairwise additivity for $W^{(3)}$ is on a different "level" than the assumption of pairwise additivity for $U_3$.

## 2.9. MOLECULAR DISTRIBUTION FUNCTIONS (MDF) IN THE GRAND CANONICAL ENSEMBLE

In discussing closed systems, we have stressed the advantage of the *generic* over the *specific* MDF's. The latter presupposes the possibility of identification of the individual particles, which of course we cannot achieve in practice. Nevertheless, the *specific* MDF's were found useful at least as intermediary concepts in the process of formulating the definitions of the generic MDF's. When we turn to open systems, it becomes meaningless to discuss the specific MDF's. Hence, we proceed directly to define the generic MDF's in the grand canonical, or the $T$, $V$, $\mu$, ensemble.

We recall that the probability of finding a system in the $T$, $V$, $\mu$ ensemble with exactly $N$ particles is (Section 1.5)

$$P(N) = Q(T, V, N)[\exp(\beta\mu N)]/\Xi(T, V, \mu) \qquad (2.118)$$

where $Q(T, V, N)$ and $\Xi(T, V, \mu)$ are the canonical and the grand canonical partition functions, respectively.

The conditional $n$th-order MDF of finding the configuration $\mathbf{X}^n$, given that the system has $N$ particles, is

$$\varrho^{(n)}(\mathbf{X}^n/N) = \frac{N!}{(N-n)!} \frac{\int \cdots \int d\mathbf{X}_{n+1} \cdots d\mathbf{X}_N \exp[-\beta U_N(\mathbf{X}^N)]}{\int \cdots \int d\mathbf{X}^N \exp[-\beta U_N(\mathbf{X}^N)]} \qquad (2.119)$$

Clearly, this quantity is defined for $n \leq N$ only. The $n$th-order MDF in the $T$, $V$, $\mu$ ensemble is defined as the average of (2.119) with the weight given in (2.118), i.e.,

$$\varrho^{(n)}(\mathbf{X}^n) = \sum_{N \geq n} P(N) \varrho^{(n)}(\mathbf{X}^n/N)$$

$$= \frac{1}{\Xi} \sum_{N \geq n} \frac{N!}{(N-n)!}$$

$$\times \frac{Q(T, V, N)[\exp(\beta\mu N)] \int \cdots \int d\mathbf{X}_{n+1} \cdots d\mathbf{X}_N \exp[-\beta U_N(\mathbf{X}^N)]}{Z_N} \qquad (2.120)$$

Recalling that

$$Q(T, V, N) = (q^N/N!)Z_N \qquad (2.121)$$

## Molecular Distribution Functions

and denoting

$$\lambda = \exp(\beta\mu) \qquad (2.122)$$

we can rewrite (2.120) as

$$\varrho^{(n)}(\mathbf{X}^n) = \frac{1}{\Xi} \sum_{N \geq n} \frac{(\lambda q)^N}{(N-n)!} \int \cdots \int d\mathbf{X}_{n+1} \cdots d\mathbf{X}_N \exp[-\beta U_N(\mathbf{X}^N)]$$
$$(2.123)$$

The normalization condition for $\varrho^{(n)}(\mathbf{X}^n)$ is obtained from (2.120) by integrating over all the configurations $\mathbf{X}^n$

$$\int \cdots \int d\mathbf{X}^n \, \varrho^{(n)}(\mathbf{X}^n) = \sum_{N \geq n} P(N) \frac{N!}{(N-n)!} = \left\langle \frac{N!}{(N-n)!} \right\rangle \qquad (2.124)$$

where here the symbol $\langle \, \rangle$ stands for an average in the $T, V, \mu$ ensemble. Let us work out two simple and important cases. For $n = 1$, we have

$$\int d\mathbf{X}_1 \, \varrho^{(1)}(\mathbf{X}_1) = \langle N!/(N-1)! \rangle = \langle N \rangle \qquad (2.125)$$

which is simply the average number of particles in a system in the $T, V, \mu$ ensemble. [Compare with (2.19) in the $T, V, N$ ensemble.] Using essentially the same arguments as in Section 2.2, we get for a homogeneous and isotropic system

$$\varrho^{(1)}(\mathbf{X}) = \langle N \rangle / 8\pi^2 V = \varrho / 8\pi^2 \qquad (2.126)$$

which is the same as in Eq. (2.26), but with the replacement of the exact $N$ by the average $\langle N \rangle$.

For $n = 2$, we get from (2.124)

$$\int \int d\mathbf{X}_1 \, d\mathbf{X}_2 \, \varrho^{(2)}(\mathbf{X}_1, \mathbf{X}_2) = \langle N!/(N-2)! \rangle = \langle N(N-1) \rangle = \langle N^2 \rangle - \langle N \rangle$$
$$(2.127)$$

Relations (2.125) and (2.127) will be very important for applications in Chapters 3 and 4.

In a similar fashion, one may introduce correlation functions in the $T, V, \mu$ ensemble. Of particular importance is the pair correlation function defined by[13]

$$\varrho^{(2)}(\mathbf{X}_1, \mathbf{X}_2) = \varrho^{(1)}(\mathbf{X}_1) \varrho^{(1)}(\mathbf{X}_2) g(\mathbf{X}_1, \mathbf{X}_2) \qquad (2.128)$$

---

[13] In our notation, we shall make no distinction between MDF's defined in various ensembles. The appropriate meaning in each case should be clear from the context.

One important property of $g(\mathbf{X}_1, \mathbf{X}_2)$, defined in the $T$, $V$, $\mu$ ensemble, is its limiting behavior at low densities:

$$g(\mathbf{X}_1, \mathbf{X}_2) \xrightarrow{\varrho \to 0} \exp[-\beta U(\mathbf{X}_1, \mathbf{X}_2)] \qquad (2.129)$$

which is strictly true without additional terms of the order of $\langle N \rangle^{-1}$. For more details, see Hill (1956) and Münster (1968).

*Chapter 3*

# Molecular Distribution Functions and Thermodynamics

## 3.1. INTRODUCTION

This chapter summarizes the most important relations between thermodynamic quantities and molecular distribution functions (MDF). The majority of these relations apply to systems obeying the assumption of pairwise additivity for the total potential energy. We shall indicate, however, how to modify the relations when higher-order potentials are to be incorporated in the formal theory. In general, higher-order potentials bring in higher-order MDF's. Since very little is known about the analytical behavior of the latter, such relationships are rarely useful in applications.

Most of the specific derivations carried out in this section apply to systems of simple spherical particles. We shall also point out the appropriate generalizations for nonspherical particles, which do not possess internal rotations.

There is a step common to most of the procedures leading to the relations between thermodynamic quantities and the pair distribution function. Therefore, in the next section we derive a general theorem connecting averages of pairwise quantities and the pair distribution function. In fact, we have already quoted one application of this theorem in Section 2.7. Further applications will appear in subsequent sections of this chapter.

## 3.2. AVERAGE VALUES OF PAIRWISE QUANTITIES

Consider an average of a general quantity $F(\mathbf{X}^N)$ in the $T, V, N$ ensemble

$$\langle F \rangle = \int \cdots \int d\mathbf{X}^N \, P(\mathbf{X}^N) F(\mathbf{X}^N) \tag{3.1}$$

with

$$P(\mathbf{X}^N) = \{\exp[-\beta U_N(\mathbf{X}^N)]\}/Z_N \tag{3.2}$$

A pairwise quantity is a function that is expressible as a sum of terms, each of which depends on the configuration of a *pair* of particles, namely

$$F(\mathbf{X}^N) = \sum_{i \neq j} f(\mathbf{X}_i, \mathbf{X}_j) \tag{3.3}$$

where we have assumed that the sum is over all different pairs. In the present treatment, all of the $N$ particles are presumed to be equivalent, so that the function $f$ is the same for each pair of indices. (The extension to mixtures will be discussed at the end of this section.)

Substituting (3.3) in (3.1), we get

$$\begin{aligned}
\langle F \rangle &= \int \cdots \int d\mathbf{X}^N \, P(\mathbf{X}^N) \sum_{i \neq j} f(\mathbf{X}_i, \mathbf{X}_j) \\
&= \sum_{i \neq j} \int \cdots \int d\mathbf{X}^N \, P(\mathbf{X}^N) f(\mathbf{X}_i, \mathbf{X}_j) \\
&= N(N-1) \int \cdots \int d\mathbf{X}^N \, P(\mathbf{X}^N) f(\mathbf{X}_1, \mathbf{X}_2) \\
&= \int d\mathbf{X}_1 \int d\mathbf{X}_2 \, f(\mathbf{X}_1, \mathbf{X}_2) \left[ N(N-1) \int \cdots \int d\mathbf{X}_3 \cdots d\mathbf{X}_N \, P(\mathbf{X}^N) \right] \\
&= \int d\mathbf{X}_1 \int d\mathbf{X}_2 \, f(\mathbf{X}_1, \mathbf{X}_2) \varrho^{(2)}(\mathbf{X}_1, \mathbf{X}_2)
\end{aligned} \tag{3.4}$$

All the steps taken in (3.4) are quite simple. It is instructive, however, to go through them carefully since this is a standard procedure in the theory of classical fluids. In the first step on the rhs of (3.4), we have merely interchanged the signs of summation and integration. The second step is important. We exploit the fact that all particles are equivalent; thus each term in the sum has the same numerical value, independent of the indices $i, j$. Hence, we replace the sum over $N(N-1)$ terms by $N(N-1)$ times one integral. In the latter, we have chosen the (arbitrary) indices 1 and 2,

## Molecular Distribution Functions and Thermodynamics

but, clearly, due to the equivalence of the particles, we could have chosen any other two indices. The third and fourth steps make use of the definition of the pair distribution function defined in Section 2.3.

We can rewrite the final result of (3.4) as

$$\langle F \rangle = \int d\mathbf{X}' \int d\mathbf{X}'' f(\mathbf{X}', \mathbf{X}'') \varrho^{(2)}(\mathbf{X}', \mathbf{X}'') \tag{3.5}$$

where we have changed to primed vectors to stress the fact that we do not refer to any specific pair of particles [as might be inferred erroneously from the final form of the rhs of (3.4)].

A simpler version of (3.5) occurs when the particles are spherical so that each configuration $\mathbf{X}$ consists of only the locational vector $\mathbf{R}$. This is the most frequent case in the theory of simple fluids. The corresponding expression for the average in this case is

$$\langle F \rangle = \int d\mathbf{R}' \int d\mathbf{R}'' f(\mathbf{R}', \mathbf{R}'') \varrho^{(2)}(\mathbf{R}', \mathbf{R}'') \tag{3.6}$$

A common case that often occurs is when the function $f(\mathbf{R}', \mathbf{R}'')$ depends only on the separation between the two points $R = |\mathbf{R}'' - \mathbf{R}'|$. In addition, for homogeneous and isotropic fluids, $\varrho^{(2)}(\mathbf{R}', \mathbf{R}'')$ also depends only on $R$. This permits the transformation of (3.6) into a one-dimensional integral, which, in fact, is the most useful form of the result (3.6). To do this, we first transform to relative coordinates

$$\bar{\mathbf{R}} = \mathbf{R}', \quad \mathbf{R} = \mathbf{R}'' - \mathbf{R}' \tag{3.7}$$

Hence,

$$\langle F \rangle = \int d\bar{\mathbf{R}} \int d\mathbf{R}\, f(\mathbf{R}) \varrho^{(2)}(\mathbf{R}) = V \int d\mathbf{R}\, f(\mathbf{R}) \varrho^{(2)}(\mathbf{R}) \tag{3.8}$$

Next, we transform to polar coordinates

$$d\mathbf{R} = dx\, dy\, dz = R^2 \sin\theta\, d\theta\, d\phi\, dR \tag{3.9}$$

Since the integrand in (3.8) depends on the scalar $R$, we can integrate over all the orientations to get the final form

$$\begin{aligned}\langle F \rangle &= V \int_0^\infty f(R) \varrho^{(2)}(R) 4\pi R^2\, dR \\ &= \varrho^2 V \int_0^\infty f(R) g(R) 4\pi R^2\, dR \end{aligned} \tag{3.10}$$

It is clear from (3.10) that a knowledge of the pairwise function $f(R)$ in (3.3), together with the radial distribution function $g(R)$, is sufficient to evaluate the average quantity $\langle F \rangle$. Note that we have taken as infinity the upper limit of the integral in (3.10), which is not always permitted. In most practical cases, however, $f(R)$ will be of finite range. Since $g(R)$ tends to unity at distances of a few molecular diameters (excluding the region near the critical point), the upper limit of the integral in (3.10) can be taken to be of the order of a few molecular diameters only. Hence, extension to infinity would not affect the value of the integral.

We now briefly mention two straightforward extensions of the theorem (3.5).

(a) For mixtures of, say, two components, a pairwise function is defined as

$$F(\mathbf{X}^{N_A+N_B}) = \sum_{i \neq j} f_{AA}(\mathbf{X}_i, \mathbf{X}_j) + \sum_{i \neq j} f_{BB}(\mathbf{X}_i, \mathbf{X}_j)$$
$$+ \sum_{i=1}^{N_A} \sum_{j=1}^{N_B} f_{AB}(\mathbf{X}_i, \mathbf{X}_j) + \sum_{i=1}^{N_B} \sum_{j=1}^{N_A} f_{BA}(\mathbf{X}_i, \mathbf{X}_j) \quad (3.11)$$

where $\mathbf{X}^{N_A+N_B}$ stands for the configuration of the whole system of $N_A + N_B$ particles of species $A$ and $B$. Here, $f_{\alpha\beta}$ is the pairwise function for the pair of species $\alpha$ and $\beta$ ($\alpha = A, B$ and $\beta = A, B$). Altogether, we have in (3.11) $N_A(N_A - 1) + N_B(N_B - 1) + 2N_AN_B$ terms which correspond to all of the $(N_A + N_B)(N_A + N_B - 1)$ pairs in the system.

Note that in (3.11) we have assumed, as in (3.3), the summation over $i \neq j$ for pairs of the same species. This is not required for pairs of different species. Using exactly the same procedure as for the one-component system, we get for the average quantity

$$\langle F \rangle = \int d\mathbf{X}' \int d\mathbf{X}'' f_{AA}(\mathbf{X}', \mathbf{X}'') \varrho_{AA}^{(2)}(\mathbf{X}', \mathbf{X}'')$$
$$+ \int d\mathbf{X}' \int d\mathbf{X}'' f_{BB}(\mathbf{X}', \mathbf{X}'') \varrho_{BB}^{(2)}(\mathbf{X}', \mathbf{X}'')$$
$$+ \int d\mathbf{X}' \int d\mathbf{X}'' f_{AB}(\mathbf{X}', \mathbf{X}'') \varrho_{AB}^{(2)}(\mathbf{X}', \mathbf{X}'')$$
$$+ \int d\mathbf{X}' \int d\mathbf{X}'' f_{BA}(\mathbf{X}', \mathbf{X}'') \varrho_{BA}^{(2)}(\mathbf{X}', \mathbf{X}'') \quad (3.12)$$

where $\varrho_{\alpha\beta}^{(2)}$ are the pair distribution functions for species $\alpha$ and $\beta$, which will be introduced in Chapter 4. The extension to multicomponent systems

# Molecular Distribution Functions and Thermodynamics

does not involve any new ideas. Note also that $f_{AB}(\mathbf{X'}, \mathbf{X''})$ may be different from $f_{BA}(\mathbf{X'}, \mathbf{X''})$, but the last two integrals will have the same value.

(b) For functions $F$ that depend on pairs as well as on triplets of particles of the form

$$F(\mathbf{X}^N) = \sum_{i \neq j} f(\mathbf{X}_i, \mathbf{X}_j) + \sum_{i \neq j, j \neq k} h(\mathbf{X}_i, \mathbf{X}_j, \mathbf{X}_k) \quad (3.13)$$

we get the result

$$\langle F \rangle = \int d\mathbf{X'} \int d\mathbf{X''} f(\mathbf{X'}, \mathbf{X''}) \varrho^{(2)}(\mathbf{X'}, \mathbf{X''})$$

$$+ \int d\mathbf{X'} \int d\mathbf{X''} \int d\mathbf{X'''} h(\mathbf{X'}, \mathbf{X''}, \mathbf{X'''}) \varrho^{(3)}(\mathbf{X'}, \mathbf{X''}, \mathbf{X'''}) \quad (3.14)$$

The arguments leading to (3.14) are the same as those for (3.4). The new element that enters into (3.14) is the triplet distribution function. Similarly, we can write down formal relations for average quantities which depend on larger numbers of particles. The result would be integrals involving successively higher-order molecular distribution functions. Unfortunately, even (3.14) is rarely useful since we do not have sufficient information on $\varrho^{(3)}$. Functions of the form (3.13) may occur, for instance, if we include three-body potentials to describe the total potential energy.

## 3.3. INTERNAL ENERGY

Consider a system in the $T$, $V$, $N$ ensemble and assume that the total potential energy of interaction is pairwise additive, namely

$$U_N(\mathbf{X}^N) = \tfrac{1}{2} \sum_{i \neq j} U(\mathbf{X}_i, \mathbf{X}_j) \quad (3.15)$$

The factor $\tfrac{1}{2}$ is included in (3.15) since the sum over $i \neq j$ counts each pair interaction twice.

The partition function for this system is

$$Q(T, V, N) = \frac{q^N}{N!} Z_N = \frac{q^N}{N!} \int \cdots \int d\mathbf{X}^N \exp[-\beta U_N(\mathbf{X}^N)] \quad (3.16)$$

where we have included in $q^N$ the momentum partition function (compare with the notation in Section 1.3).

The internal energy of the system is given by[1]

$$E = -T^2 \frac{\partial (A/T)}{\partial T} = kT^2 \frac{\partial \ln Q}{\partial T}$$

$$= NkT^2 \frac{\partial \ln q}{\partial T} + kT^2 \frac{\partial \ln Z_N}{\partial T} \qquad (3.17)$$

The first term on the rhs includes the internal and the kinetic energy of the individual molecules. For instance, for spherical and structureless molecules (see Section 1.3) we have $q = \Lambda^{-3}$ and, hence,

$$N\varepsilon^K = NkT^2(\partial \ln q/\partial T) = \tfrac{3}{2}NkT \qquad (3.18)$$

which is the average translational kinetic energy of the molecules.

The second term on the rhs of (3.17) is the average energy of interaction among the particles. This can be seen immediately by performing the derivative of the configurational partition function:

$$kT^2 \frac{\partial \ln Z_N}{\partial T} = \frac{\int \cdots \int d\mathbf{X}^N \exp[-\beta U_N(\mathbf{X}^N)] U_N(\mathbf{X}^N)}{Z_N}$$

$$= \int \cdots \int d\mathbf{X}^N P(\mathbf{X}^N) U_N(\mathbf{X}^N) \qquad (3.19)$$

Hence, the total internal energy is

$$E = N\varepsilon^K + \langle U_N \rangle \qquad (3.20)$$

where $N\varepsilon^K$ denotes the first term on the rhs of (3.17).

The average potential energy in (3.20), with the assumption of pairwise additivity (3.15), fulfills the conditions of the previous section; hence, we can immediately apply theorem (3.5) to obtain

$$E = N\varepsilon^K + \tfrac{1}{2} \int d\mathbf{X}' \int d\mathbf{X}'' \, U(\mathbf{X}', \mathbf{X}'') \varrho^{(2)}(\mathbf{X}', \mathbf{X}'') \qquad (3.21)$$

For spherical particles, we can transform relation (3.21) into a one-dimensional integral. Using the same arguments as in Section 3.2 to derive (3.10), we get, from (3.21),

$$E = N\varepsilon^K + \tfrac{1}{2} N\varrho \int_0^\infty U(R) g(R) 4\pi R^2 \, dR \qquad (3.22)$$

---

[1] Note that $E$ is referred to as the internal energy in the thermodynamic sense. $\varepsilon^K$ designates the internal energy of a *single* molecule.

## Molecular Distribution Functions and Thermodynamics    87

For nonpolar particles, $U(R)$ will usually have a range of a few molecular diameters; hence, the main contribution to the integral on the rhs of (3.22) comes from the finite region around the origin.

It is instructive to dwell on the direct interpretation of the second term on the rhs of (3.22). The average potential energy of interaction can be computed as follows. We select a particle and compute its total interaction with the rest of the system. Since the local density of particles at a distance $R$ from the center of the selected particle is $\varrho g(R)$, the average number of particles in the spherical element of volume $4\pi R^2\, dR$ is $\varrho g(R) 4\pi R^2\, dR$. Hence, the average interaction of a given particle with the rest of the system is

$$\int_0^\infty U(R)\varrho g(R) 4\pi R^2\, dR \qquad (3.23)$$

We now repeat the same computation for each particle. Since the $N$ particles are identical, the average interaction of each particle with the medium is the same. However, if we multiply (3.23) by $N$, we will be counting each pair interaction twice. Hence, we must multiply by $N$ and divide by two to obtain the average interaction energy for the whole system:

$$\tfrac{1}{2} N\varrho \int_0^\infty U(R) g(R) 4\pi R^2\, dR \qquad (3.24)$$

Relation (3.22) is evidently simpler than the original form of the total energy in (3.20). Once we have presumed an analytical form for $U(R)$ and acquired information (from either theoretical or experimental sources) on $g(R)$, we can compute the energy of the system by a one-dimensional integration. Remember that the basic simplification achieved in (3.21) is due to the assumption of pairwise additivity of the total potential energy. In addition, to get (3.22), we have assumed that the particles are spherical. For nonspherical particles, we can still simplify (3.21) by transforming to relative coordinates and integrating over the configuration of one particle:

$$\begin{aligned} E &= N\varepsilon^K + \tfrac{1}{2} V 8\pi^2 \int d\mathbf{X}\, U(\mathbf{X}) \varrho^{(2)}(\mathbf{X}) \\ &= N\varepsilon^K + (N\varrho/16\pi^2) \int d\mathbf{X}\, U(\mathbf{X}) g(\mathbf{X}) \end{aligned} \qquad (3.25)$$

Here, we have a six-dimensional integral instead of a one-dimensional integral as in (3.22).

## 3.4. THE PRESSURE EQUATION

The pressure equation is computed in this section for a system of spherical particles. This choice is made only because of notational convenience. We shall quote the corresponding equation for nonspherical particles at the end of this section.

The pressure is obtained from the Helmholtz free energy by

$$P = -(\partial A/\partial V)_{T,N} \qquad (3.26)$$

where

$$A = -kT \ln Q(T, V, N) \qquad (3.27)$$

Note that the dependence of $Q$ on the volume comes only through the configurational partition function, hence

$$P = kT(\partial \ln Z_N/\partial V)_{T,N} \qquad (3.28)$$

In order to perform this derivative, we first express $Z_N$ explicitly as a function of $V$. For macroscopic systems, we assume that the pressure is independent of the geometric form of the vessel. Hence, for convenience, we choose a cube of edge $V^{1/3}$ so that the configurational partition function is written as

$$Z_N = \int_0^{V^{1/3}} \cdots \int_0^{V^{1/3}} dx_1\, dy_1\, dz_1 \cdots dx_N\, dy_N\, dz_N \exp(-\beta U_N(\mathbf{R}^N))] \qquad (3.29)$$

We now make the following transformation of variables:

$$x_i' = V^{-1/3} x_i, \qquad y_i' = V^{-1/3} y_i, \qquad z_i' = V^{-1/3} z_i \qquad (3.30)$$

so that the limits of the integral in (3.29) become independent of $V$:

$$Z_N = V^N \int_0^1 \cdots \int_0^1 dx_1'\, dy_1'\, dz_1' \cdots dx_N'\, dy_N'\, dz_N' \exp(-\beta U_N) \qquad (3.31)$$

With the new set of variables, the total potential becomes a function of the volume:

$$U_N = \tfrac{1}{2} \sum_{i \neq j} U(R_{ij}) = \tfrac{1}{2} \sum_{i \neq j} U(V^{1/3} R_{ij}') \qquad (3.32)$$

## Molecular Distribution Functions and Thermodynamics

The relation between the distances expressed by the two sets of variables is

$$R_{ij} = [(x_j - x_i)^2 + (y_j - y_i)^2 + (z_j - z_i)^2]^{1/2}$$
$$= V^{1/3}[(x_j' - x_i')^2 + (y_j' - y_i')^2 + (z_j' - z_i')^2]^{1/2}$$
$$= V^{1/3} R_{ij}' \qquad (3.33)$$

We may now perform the differentiation of (3.31) with respect to the volume:

$$\left(\frac{\partial Z_N}{\partial V}\right)_{T,N} = NV^{N-1} \int_0^1 \cdots \int_0^1 dx_1' \cdots dz_N' \exp(-\beta U_N)$$
$$+ V^N \int_0^1 \cdots \int_0^1 dx_1' \cdots dz_N' [\exp(-\beta U_N)]\left(-\beta \frac{\partial U_N}{\partial V}\right)$$
(3.34)

From (3.32), we also have

$$\frac{\partial U_N}{\partial V} = \frac{1}{2} \sum_{i \neq j} \frac{\partial U(R_{ij})}{\partial R_{ij}} \frac{\partial R_{ij}}{\partial V} = \frac{1}{2} \sum_{i \neq j} \frac{\partial U(R_{ij})}{\partial R_{ij}} \frac{1}{3} V^{-2/3} R_{ij}'$$
$$= \frac{1}{6V} \sum_{i \neq j} \frac{\partial U(R_{ij})}{\partial R_{ij}} R_{ij} \qquad (3.35)$$

Combining (3.34) and (3.35) and transforming back to the original variables, we obtain

$$\left(\frac{\partial \ln Z_N}{\partial V}\right)_{T,N} = \frac{N}{V} - \frac{\beta}{6V} \int \cdots \int d\mathbf{R}^N P(\mathbf{R}^N) \sum_{i \neq j} \frac{\partial U(R_{ij})}{\partial R_{ij}} R_{ij} \qquad (3.36)$$

The second term on the rhs of (3.36) is an average of a pairwise quantity. Therefore, we can apply the general theorem (3.5) of Section 3.2 to obtain

$$P = kT\left(\frac{\partial \ln Z_N}{\partial V}\right)_{T,N} = kT\varrho - \frac{\varrho^2}{6} \int_0^\infty R \frac{\partial U(R)}{\partial R} g(R) 4\pi R^2 \, dR \qquad (3.37)$$

This is the final form of the pressure equation for a system of spherical particles obeying the assumption of pairwise additivity for the total potential energy. Note that the first term is the "ideal gas" pressure (i.e., starting with the ideal gas partition function with $U_N \equiv 0$). The second term carries the effect of the intermolecular forces on the pressure. Note that, in general, $g(R)$ is a function of density so that this term is not the second-order term in the density expansion of the pressure. There is a simple case where $g(R)$ is independent of $\varrho$ and this is for very low densities, where we have (see Section 2.5)

$$g(R) = \exp[-\beta U(R)], \qquad \varrho \to 0 \qquad (3.38)$$

Noting that

$$\frac{\partial}{\partial R}\{\exp[-\beta U(R)] - 1\} = -\beta \frac{\partial U(R)}{\partial R} \exp[-\beta U(R)] \qquad (3.39)$$

we can integrate (3.37) by parts to obtain

$$P = kT\varrho - \tfrac{1}{6}\varrho^2 \int_0^\infty -kT(\partial/\partial R)\{\exp[-\beta U(R)] - 1\}4\pi R^3 \, dR$$

$$= kT\varrho - \tfrac{1}{2}kT\varrho^2 \int_0^\infty \{\exp[-\beta U(R)] - 1\}4\pi R^2 \, dR$$

$$= kT\varrho + kT\varrho^2 B_2(T) \qquad (3.40)$$

which is the correct form for the second-order term of the density expansion of the pressure (see Section 1.8).

The general relation (3.37) is very useful in computing the equation of state of a system once we know the form of the function $g(R)$. Indeed, such computations have been performed to test theoretical methods of evaluating $g(R)$ (see also Section 3.9).

In a mixture of $c$ components, the generalization of (3.37) is straightforward and leads to

$$P = \sum_{\alpha=1}^{c} kT\varrho_\alpha - \frac{1}{6} \sum_{\alpha,\beta=1}^{c} \varrho_\alpha \varrho_\beta \int_0^\infty \frac{\partial U_{\alpha\beta}(R)}{\partial R} g_{\alpha\beta}(R) 4\pi R^3 \, dR \qquad (3.41)$$

where $\varrho_\alpha$ is the density of the $\alpha$ species and $g_{\alpha\beta}(R)$ is the pair correlation function for the pair of species $\alpha$ and $\beta$.

For a system of rigid, nonspherical molecules, the derivation of the pressure equation is essentially the same as that for spherical molecules, but with somewhat more notational complication, the result being

$$P = kT\varrho - (1/6V) \int d\mathbf{X}' \int d\mathbf{X}'' \, [\mathbf{R}\cdot\nabla_\mathbf{R} U(\mathbf{X}', \mathbf{X}'')]\varrho^{(2)}(\mathbf{X}', \mathbf{X}'') \qquad (3.42)$$

where

$$\mathbf{R} = \mathbf{R}'' - \mathbf{R}' \qquad (3.43)$$

Molecular Distribution Functions and Thermodynamics            91

## 3.5. THE CHEMICAL POTENTIAL

The chemical potential is defined, in the $T$, $V$, $N$ ensemble, by

$$\mu = (\partial A/\partial N)_{T,V} \qquad (3.44)$$

For reasons that will become clear in the following paragraphs, the chemical potential cannot be expressed as a simple integral involving the pair correlation function.

Consider, for example, the pressure equation that we have derived in the previous section, which we denote by

$$P = P\{g(R); \varrho, T\} \qquad (3.45)$$

By this notation, we simply mean that we have expressed the pressure as a function of $\varrho$ and $T$, and also in terms of $g(R)$ (which is itself a function of $\varrho$ and $T$).

Since

$$P = -\left(\frac{\partial A}{\partial V}\right)_{T,N} = -\left(\frac{\partial a}{\partial(\varrho^{-1})}\right)_T \qquad (3.46)$$

where $a = A/N$ and $\varrho^{-1} = V/N$, we can integrate (3.46) to obtain

$$a = -\int P\{g(R); \varrho, T\}\, d(\varrho^{-1}) \qquad (3.47)$$

We see that in order to express $a$ in terms of $g(R)$, we must know the explicit dependence of $g(R)$ on the density. Thus, even if we have used the pressure equation in the integrand of (3.47), we need a second integration, over the density, to get the Helmholtz free energy per particle. The chemical potential follows from the relation

$$\mu = a + pv \qquad (3.48)$$

with $v = V/N$.

As a second route to computing $a$, consider the energy equation derived in Section 3.3, which we denote by

$$E = E\{g(R); \varrho, T\} \qquad (3.49)$$

The relation between the energy per particle and the Helmholtz free energy is

$$e = \frac{E}{N} = -T^2 \left\{ \frac{\partial(a/T)}{\partial T} \right\}_\varrho \qquad (3.50)$$

which can be integrated to obtain

$$\frac{a}{T} = \frac{-1}{N} \int \frac{E\{g(R); \varrho, T\}}{T^2} dT \tag{3.51}$$

Again we see that if we use the energy expression [in terms of $g(R)$] in the integrand of (3.51), we must also know the dependence of $g(R)$ on the temperature.

The above two illustrations show that in order to obtain a relation between $\mu$ and $g(R)$, it is not sufficient to know the function $g(R)$ at a given $\varrho$ and $T$; one needs the more detailed knowledge of $g(R)$ *and* its dependence on either $\varrho$ or $T$. This difficulty is not removable. As we shall soon see, it follows from the fact that the chemical potential is *not* an average of a pairwise quantity, and therefore the general theorem of Section 3.2 is not applicable here. Let us further elaborate on this point.

The chemical potential can be written as

$$\mu = \left(\frac{\partial A}{\partial N}\right)_{T,V} = \lim_{dN \to 0} \frac{A(N + dN) - A(N)}{dN} = \frac{A(N + 1) - A(N)}{1} \tag{3.52}$$

All the equalities in (3.52) hold for macroscopic systems where the addition of one particle ($dN = 1$) can be considered to be an infinitesimal change in the variable $N$.[2]

Relation (3.52) simply means that in order to compute the chemical potential, it is sufficient to observe the change of the Helmholtz free energy upon the addition of one particle. We now use the connection between the

---

[2] The validity of (3.52) is guaranteed because of the extensive character of the free energy. Instead of taking the limit $dN \to 0$ as is required in the conventional definition of a derivative, we keep $dN$ fixed and take the limit $N \to \infty$, $V \to \infty$, and $N/V = \text{const}$, i.e.,

$$\lim_{dN \to 0} \frac{A(N + dN) - A(N)}{dN} = \lim_{dN \to 0} \left[ A\left(\frac{N}{dN} + 1\right) - A\left(\frac{N}{dN}\right) \right]$$

$$= \lim_{M \to \infty} [A(M + 1) - A(M)]$$

where we put $M = N/dN$. Clearly, the above manipulations do not hold for all differentiable functions. For instance,

$$\lim_{dN \to 0} \frac{\sin(N + dN) - \sin N}{dN} \neq \lim_{N \to \infty} [\sin(N + 1) - \sin N]$$

## Molecular Distribution Functions and Thermodynamics

Helmholtz free energy and the canonical partition function to obtain

$$\exp(-\beta\mu) = \exp\{-\beta[A(T, V, N+1) - A(T, V, N)]\} = \frac{Q(T, V, N+1)}{Q(T, V, N)}$$

$$= \frac{[q^{N+1}/\Lambda^{3(N+1)}(N+1)!]\int\cdots\int d\mathbf{R}_0\cdots d\mathbf{R}_N \exp[-\beta U_{N+1}]}{[q^N/\Lambda^{3N}N!]\int\cdots\int d\mathbf{R}_1\cdots d\mathbf{R}_N \exp[-\beta U_N]}$$

(3.53)

Note that the added particle has been given the index zero. Using the assumption of pairwise additivity of the total potential, we may split $U_{N+1}$ into two terms

$$U_{N+1}(\mathbf{R}_0, \ldots, \mathbf{R}_N) = U_N(\mathbf{R}_1, \ldots, \mathbf{R}_N) + \sum_{j=1}^{N} U(\mathbf{R}_0, \mathbf{R}_j)$$

$$= U_N(\mathbf{R}_1, \ldots, \mathbf{R}_N) + B(\mathbf{R}_0, \ldots, \mathbf{R}_N) \quad (3.54)$$

where we have included all the interactions of the zeroth particle with the rest of the system into the quantity $B(\mathbf{R}_0, \ldots, \mathbf{R}_N)$. Using (3.54) and the general expression for the basic probability density in the $T$, $V$, $N$ ensemble, we rewrite (3.53) as

$$\exp(-\beta\mu) = [q/\Lambda^3(N+1)]\int\cdots\int d\mathbf{R}_0\, d\mathbf{R}_1\cdots d\mathbf{R}_N$$

$$\times P(\mathbf{R}_1, \ldots, \mathbf{R}_N) \exp[-\beta B(\mathbf{R}_0, \ldots, \mathbf{R}_N)]$$

We now transform to relative coordinates

$$\mathbf{R}_i' = \mathbf{R}_i - \mathbf{R}_0, \quad i = 1, 2, \ldots, N \quad (3.55)$$

and note that $B(\mathbf{R}_0, \ldots, \mathbf{R}_N)$ is actually a function only of the relative coordinates $\mathbf{R}_1', \ldots, \mathbf{R}_N'$. [For instance, $U(\mathbf{R}_0, \mathbf{R}_j)$ is a function of $\mathbf{R}_j'$ and not of both $\mathbf{R}_0$ and $\mathbf{R}_j$.]

Hence,

$$\exp(-\beta\mu) = [q/\Lambda^3(N+1)]\int d\mathbf{R}_0 \int\cdots\int d\mathbf{R}_1'\cdots d\mathbf{R}_N'$$

$$\times P(\mathbf{R}_1', \ldots, \mathbf{R}_N') \exp[-\beta B(\mathbf{R}_1', \ldots, \mathbf{R}_N')] \quad (3.56)$$

Now the integrand is independent of $\mathbf{R}_0$, so that we may integrate over $\mathbf{R}_0$ to obtain the volume. The inner integral is simply the average, in the $T$, $V$, $N$ ensemble, of the quantity $\exp(-\beta B)$, i.e.,

$$\exp(-\beta\mu) = [qV/(N+1)\Lambda^3]\langle\exp(-\beta B)\rangle \quad (3.57)$$

Putting $\varrho = N/V \simeq (N+1)/V$ (macroscopic system), we can rearrange (3.57) to obtain the relation

$$\mu = kT\ln(\varrho\Lambda^3 q^{-1}) - kT\ln\langle\exp(-\beta B)\rangle \qquad (3.58)$$

This is an important expression. The first term on the rhs of (3.58) includes the density of the system and internal properties of the single molecules. The second term is an average of $\exp(-\beta B)$, where $B$ may be referred to as the "binding energy" of a given particle to the rest of the system. We may also view it as an "external" field that acts on the system of $N$ particles by a given particle fixed at some point in the system. It is important to observe that

$$\phi(\mathbf{R}_0 \cdots \mathbf{R}_N) \equiv \exp[-\beta B(\mathbf{R}_0, \ldots, \mathbf{R}_N)] = \prod_{j=1}^{N}\exp[-\beta U(\mathbf{R}_0, \mathbf{R}_j)] \qquad (3.59)$$

is not a pairwise quantity in the sense of (3.3), i.e., it is not a *sum*, but a *product* of pairwise functions. This is the inherent reason why we cannot express the chemical potential as a simple integral involving only the pair distribution function.

In (3.47) and (3.51), we have demonstrated that the chemical potential [through relation (3.48)] can be written formally as an integral over $g(R)$ provided we know its dependence on either $T$ or $\varrho$. We now derive a third expression due to Kirkwood, which employs the idea of a coupling parameter $\xi$. The ultimate expression for the chemical potential would be an integral over both $R$ and $\xi$ involving the function $g(R, \xi)$.

Let us define an auxiliary potential function as follows:

$$U(\xi) = U_N(\mathbf{R}_1, \ldots, \mathbf{R}_N) + \xi\sum_{j=1}^{N}U(\mathbf{R}_0, \mathbf{R}_j) \qquad (3.60)$$

which can be compared with (3.54). Clearly,

$$U(\xi = 0) = U_N(\mathbf{R}_1, \ldots, \mathbf{R}_N) \qquad (3.61)$$

$$U(\xi = 1) = U_{N+1}(\mathbf{R}_0, \ldots, \mathbf{R}_N) \qquad (3.62)$$

The idea is that by changing $\xi$ from zero to unity, the function $U(\xi)$ changes continuously from $U_N$ to $U_{N+1}$.

Associated with $U(\xi)$ we also define an auxiliary configurational partition function by

$$Z(\xi) = \int\cdots\int d\mathbf{R}_0\,d\mathbf{R}_1\cdots d\mathbf{R}_N\exp[-\beta U(\xi)] \qquad (3.63)$$

## Molecular Distribution Functions and Thermodynamics

where, clearly, we have

$$Z(\xi = 0) = \int \cdots \int d\mathbf{R}_0 \, d\mathbf{R}_1 \ldots d\mathbf{R}_N \exp(-\beta U_N) = V Z_N \quad (3.64)$$

and

$$Z(\xi = 1) = Z_{N+1} \quad (3.65)$$

The expression (3.53) for the chemical potential can be rewritten using the above notation as

$$\mu = kT \ln(\varrho \Lambda^3 q^{-1}) - kT \ln Z(\xi = 1) + kT \ln Z(\xi = 0) \quad (3.66)$$

or using the identity

$$kT \ln Z(\xi = 1) - kT \ln Z(\xi = 0) = kT \int_0^1 \frac{\partial \ln Z(\xi)}{\partial \xi} d\xi \quad (3.67)$$

we get

$$\mu = kT \ln(\varrho \Lambda^3 q^{-1}) - kT \int_0^1 \frac{\partial \ln Z(\xi)}{\partial \xi} d\xi \quad (3.68)$$

We now differentiate $Z(\xi)$ directly to obtain

$$\begin{aligned}
kT & \frac{\partial \ln Z(\xi)}{\partial \xi} \\
&= \frac{kT}{Z(\xi)} \int \cdots \int d\mathbf{R}_0 \cdots d\mathbf{R}_N \{\exp[-\beta U(\xi)]\} \left[ -\beta \sum_{j=1}^N U(\mathbf{R}_0, \mathbf{R}_j) \right] \\
&= -\int \cdots \int d\mathbf{R}_0 \cdots d\mathbf{R}_N \, P(\mathbf{R}^{N+1}, \xi) \sum_{j=1}^N U(\mathbf{R}_0, \mathbf{R}_j) \\
&= -\sum_{j=1}^N \int \cdots \int d\mathbf{R}_0 \cdots d\mathbf{R}_N \, P(\mathbf{R}^{N+1}, \xi) U(\mathbf{R}_0, \mathbf{R}_j) \\
&= -N \int \int d\mathbf{R}_0 \, d\mathbf{R}_1 \, U(\mathbf{R}_0, \mathbf{R}_1) \int \cdots \int d\mathbf{R}_2 \cdots d\mathbf{R}_N \, P(\mathbf{R}^{N+1}, \xi) \\
&= -\frac{1}{N+1} \int \int d\mathbf{R}_0 \, d\mathbf{R}_1 \, U(\mathbf{R}_0, \mathbf{R}_1) \varrho^{(2)}(\mathbf{R}_0, \mathbf{R}_1, \xi) \\
&= -\varrho \int_0^\infty U(R) g(R, \xi) 4\pi R^2 \, dR \quad (3.69)
\end{aligned}$$

The formal steps in (3.69) are very similar to those in Section 3.2, and there is no need to go through them in detail. The only new feature in (3.69) is the appearance of the parameter $\xi$, employed throughout in the distribution functions.

We now combine (3.69) with (3.68) to obtain the final expression for the chemical potential:

$$\mu = kT \ln(\varrho \Lambda^3 q^{-1}) + \varrho \int_0^1 d\xi \int_0^\infty U(R)g(R, \xi)4\pi R^2 \, dR \qquad (3.70)$$

We can also define the "standard chemical potential" by

$$\mu^{\circ g} = kT \ln(\Lambda^3 q^{-1}) \qquad (3.71)$$

and

$$kT \ln \gamma^{\text{ideal gas}} = \varrho \int_0^1 d\xi \int_0^\infty U(R)g(R, \xi)4\pi R^2 \, dR \qquad (3.72)$$

to rewrite (3.70) in the form

$$\mu = \mu^{\circ g} + kT \ln(\varrho \gamma^{\text{ideal gas}}) \qquad (3.73)$$

where, in (3.72), we have an explicit expression for the activity coefficient $\gamma^{\text{ideal gas}}$ which measures the extent of deviation of the chemical potential from its ideal gas form. The quantity $\varrho g(R, \xi)$ is the local density of particles around a given particle that is "coupled," to the extent $\xi$, to the rest of the system. The last sentence means that the total potential energy of the system is given by (3.60) with a fixed value of $\xi$. It is important to observe that (3.70) is not a simple integral involving the function $g(R)$, since a more detailed knowledge of the function $g(R, \xi)$ is required. In this respect, it is equivalent to the other two possible expressions for $\mu$ given in the beginning of this section.

In order to appreciate the significance of (3.70), let us compare it with another expression for the chemical potential. Suppose that we have chosen a very low density $\varrho_0$, for which the chemical potential has the ideal gas form

$$\mu(\varrho_0) = \mu^{\circ g} + kT \ln \varrho_0 \qquad (3.74)$$

We now write the chemical potential at the final density $\varrho$ as

$$\mu(\varrho) - \mu(\varrho_0) = \int_{\varrho_0}^{\varrho} (\partial \mu / \partial \varrho)_T \, d\varrho \qquad (3.75)$$

Combining (3.74) with (3.75), we get

$$\mu(\varrho) = [\mu^{\circ g} + kT \ln \varrho_0] + \int_{\varrho_0}^{\varrho} (\partial \mu / \partial \varrho)_T \, d\varrho \qquad (3.76)$$

## Molecular Distribution Functions and Thermodynamics

In (3.76), we have split the chemical potential into two terms: The first corresponds to the work required to add a single particle (at $T$, $V$ constant) to an ideal system of very low density; the second is the work required to "build up" the *density* from the initial value $\varrho_0$ to the final value $\varrho$. The essence of Kirkwood's method is to achieve a different splitting of $\mu(\varrho)$ into two parts. First, we introduce a hypothetical particle which is not coupled to the rest of the system ($\xi = 0$). Then, the *particle* is continuously "built up" by changing $\xi$ from zero to unity. This is the meaning of the splitting of $\mu$ in (3.70). The first term corresponds to the work of adding a particle uncoupled to the system (with density $\varrho$). The second term is the work required to "build up" the added particle in the system, using the parameter $\xi$ as a coupling device.

One important property of the coupling integral in (3.70) should be noted. For simple particles, where the pair potential drops to zero at long distances as $R^{-6}$, the upper limit of the integral over $R$ can be replaced by some finite value $R_{\max}$ which is of the order of a few molecular diameters [note that $g(R)$ also decays to unity at a distance of the order of few molecular diameters]. Therefore, we conclude that the coupling work gets its principal contribution only from the local environment of the added particle.

As an illustrative application of (3.70), consider the limiting form of $g(R)$ at low densities (Section 2.5)

$$g(R) = \exp[-\beta U(R)] \qquad (3.77)$$

and hence

$$g(R, \xi) = \exp[-\beta \xi U(R)] \qquad (3.78)$$

Substituting (3.78) into (3.70), we get an immediate integral over $\xi$:

$$\int_0^1 d\xi \int_0^\infty U(R) \exp[-\beta \xi U(R)] 4\pi R^2 \, dR$$
$$= -kT \int_0^\infty \{\exp[-\beta U(R)] - 1\} 4\pi R^2 \, dR \qquad (3.79)$$

Using the definition (see Section 1.8)

$$B_2(T) = -\tfrac{1}{2} \int_0^\infty \{\exp[-\beta U(R)] - 1\} 4\pi R^2 \, dR \qquad (3.80)$$

we can write (3.70) for this case as

$$\mu = \mu^{\circ g} + kT \ln \varrho + 2kT B_2(T) \varrho \qquad (3.81)$$

The last term on the rhs of (3.81) can be viewed as the first-order term in the expansion of the activity coefficient (3.72) in the density.

In fact, the virial expansion can be recovered from (3.81) by using the thermodynamic relation

$$dP = \varrho \, d\mu \quad (T \text{ constant}) \tag{3.82}$$

From (3.81), we get

$$d\mu = (kT/\varrho) \, d\varrho + 2kTB_2(T) \, d\varrho \tag{3.83}$$

Combining (3.82) and (3.83) yields

$$dP = [kT + 2kTB_2(T)\varrho] \, d\varrho \tag{3.84}$$

which, upon integration between $\varrho = 0$ and the final density $\varrho$, yields

$$P = kT\varrho + kTB_2(T)\varrho^2 \tag{3.85}$$

which is the leading form of the virial expansion of the pressure.

We now briefly summarize the modifications that must be introduced into the equation for the chemical potential for more complex systems.

(a) For systems that do not obey the assumption of pairwise additivity for the potential energy, Eq. (3.70) becomes invalid. In a formal way, one can derive an analogous relation involving higher-order molecular distribution functions, which, as noted before, may hardly be of practical value.

(b) For rigid, nonspherical particles whose potential energy obeys the assumption of pairwise additivity, a relation similar to (3.70) holds. However, one must now integrate over the location as well as the orientation of the particle. The generalized relation is

$$\mu = kT \ln(\varrho \Lambda^3 q^{-1}) + \int_0^1 d\xi \int d\mathbf{X}'' \, U(\mathbf{X}', \mathbf{X}'') \varrho(\mathbf{X}''/\mathbf{X}', \xi) \tag{3.86}$$

Here, $q$ includes the rotational as well as the internal partition function of a single molecule. The quantity $\varrho(\mathbf{X}''/\mathbf{X}', \xi)$ is the local density of particles at $\mathbf{X}''$, given a particle at $\mathbf{X}'$ coupled to the extent $\xi$. Clearly, the whole integral on the rhs of (3.86) does not depend on the choice of $\mathbf{X}'$ (for instance, we can take $\mathbf{R}' = 0$ and $\mathbf{\Omega}' = 0$ and measure $\mathbf{X}''$ relative to this choice).

Molecular Distribution Functions and Thermodynamics     99

## 3.6. PSEUDO-CHEMICAL POTENTIAL

The chemical potential is the work (here, at $T$, $V$ constant) associated with the addition of one particle to a macroscopically large system:

$$\mu = A(T, V, N+1) - A(T, V, N) \qquad (3.87)$$

The pseudo-chemical potential refers to the work associated with the addition of one particle to a *fixed* position in the system,[3] say at $\mathbf{R}_0$:

$$\bar{\mu} = A(T, V, N+1; \mathbf{R}_0) - A(T, V, N) \qquad (3.88)$$

The statistical mechanical expression for the pseudo-chemical potential can be expressed, similarly to (3.53), as a ratio between two partition functions corresponding to the difference in the Helmholtz free energies in (3.88), i.e.,

$$\exp(-\beta\bar{\mu}) = \frac{(q^{N+1}/\Lambda^{3N}N!)\int\cdots\int d\mathbf{R}_1\cdots d\mathbf{R}_N \exp[-\beta U_{N+1}(\mathbf{R}_0,\ldots,\mathbf{R}_N)]}{(q^N/\Lambda^{3N}N!)\int\cdots\int d\mathbf{R}_1\cdots d\mathbf{R}_N \exp[-\beta U_N(\mathbf{R}_1,\ldots,\mathbf{R}_N)]}$$
$$(3.89)$$

It is instructive to observe the differences between (3.53) and (3.89). Since the added particle in (3.89) is devoid of the translational degree of freedom, it will not bear a momentum partition function. Hence, we have $\Lambda^{3N}$ instead of $\Lambda^{3(N+1)}$ as in (3.53). For the same reason, the integration in the numerator of (3.89) is over the $N$ locations $\mathbf{R}_1,\ldots,\mathbf{R}_N$ and not over $\mathbf{R}_0,\ldots,\mathbf{R}_N$ as in (3.53). Furthermore, since we have added a particle to a fixed position, it is distinguishable from the other particles, hence we have $N!$ instead of $(N+1)!$ as in (3.53).

Once we have set up the statistical mechanical expression (3.89), the following formal steps are nearly the same as in the previous section. Relation (3.89) can be rewritten, using the notation of Section 3.5, as

$$\exp(-\beta\bar{\mu}) = q\int\cdots\int d\mathbf{R}^N P(\mathbf{R}^N) \exp[-\beta B(\mathbf{R}_0,\ldots,\mathbf{R}_N)]$$
$$= q\langle\exp(-\beta B)\rangle \qquad (3.90)$$

---

[3] In this section, we restrict ourselves to processes at $T$, $V$ constant. The analogous treatment in the $T$, $P$, $N$ ensemble will be discussed in Appendix 9-F. Also, for simplicity, we assume that the particles are spherical; hence, only the *location* of the added particle is specified in (3.88). For more complex particles, one can also specify the orientation of the added molecule.

or

$$\bar{\mu} = kT \ln q^{-1} - kT \ln \langle \exp(-\beta B) \rangle$$
$$= kT \ln q^{-1} + \varrho \int_0^1 d\xi \int_0^\infty U(R) g(R, \xi) 4\pi R^2 \, dR \qquad (3.91)$$

which should be compared with (3.58) and (3.70). Note that we have added the particle to a fixed position $\mathbf{R}_0$; therefore, from the formal point of view, $\bar{\mu}$ depends on $\mathbf{R}_0$. However, in a homogeneous fluid, all the points of the system are presumed to be equivalent (except for a small region near the boundaries, which is negligible for our present purposes), and therefore $\bar{\mu}$ is effectively independent of $\mathbf{R}_0$.

Combining (3.91) with either (3.58) or (3.70), we get the expression for the chemical potential

$$\mu = \bar{\mu} + kT \ln(\varrho \Lambda^3) \qquad (3.92)$$

This relation has a simple and important interpretation. The work $\mu$ required to add a particle to the system is split into two parts. First, we add the particle to a *fixed* position, say $\mathbf{R}_0$, the corresponding work being $\bar{\mu}$. Next, we remove the constraint imposed by fixing the position of the particle; the corresponding work is the second term on the rhs of (3.92). In the second step, the particle "gains" the factor $\Lambda^3$ due to the recovery of the translational kinetic energy. The factor $V$ is gained since the total volume is now accessible to the particle. Finally, since the released particle is no longer distinguishable from the other particles, there is a gain of a factor $N$ (more precisely, $N+1$) which, together with the volume $V$, produces the density $\varrho$ in (3.92). Of course, all these details are already embedded in the difference between expressions (3.53) and (3.89).

The significance of the above interpretation may be better understood with the aid of the following example. Suppose we start with a system of $N$ identical particles and add a new particle only slightly different from the other particles (say a different isotope). If we repeat all the arguments of this and the previous sections, we end up with the following expression for the chemical potential of this particle:

$$\mu' = \bar{\mu}' + kT \ln[(1/V)(\Lambda')^3] \qquad (3.93)$$

Here, $\mu'$ is the work of introducing the particle to the system. The term $\bar{\mu}'$ is the corresponding pseudo-chemical potential and $(\Lambda')^3$ is its momentum partition function. The latter two quantities differ very slightly from the corresponding (unprimed) quantities in (3.92). The essential major difference is the replacement of the density $\varrho = N/V$ in (3.92) by $1/V$ in (3.93). The reason is that the added particle is presumed to be distinguishable from the remaining particles, whether free or fixed at $\mathbf{R}_0$. Hence, no gain of a factor $N$ is achieved by releasing the particle from its fixed position. A further illustration for a lattice model is presented in Appendix 9-G.

## 3.7. ENTROPY

With the expressions derived thus far for the energy, pressure, and chemical potential, one can easily obtain an expression for the entropy, using thermodynamic relationships. For instance, we can use the relation

$$TS = E + PV - N\mu \tag{3.94}$$

or the differential relations

$$\left(\frac{\partial S}{\partial T}\right)_{V,N} = \frac{1}{T}\left(\frac{\partial E}{\partial T}\right)_{V,N} \tag{3.95}$$

$$\left(\frac{\partial S}{\partial V}\right)_{T,N} = \left(\frac{\partial P}{\partial T}\right)_{V,N} \tag{3.96}$$

to obtain the entropy. As in the case of the chemical potential, we shall soon learn that the entropy is not expressible in terms of $g(R)$ in a simple manner. Let us look more closely at the reason for this.

The entropy is obtained from the Helmholtz free energy by differentiation with respect to temperature, i.e.,

$$-S = \left(\frac{\partial A}{\partial T}\right)_{V,N} = \frac{\partial}{\partial T}\left\{-kT\ln\left[\left(\frac{q^N}{N!\Lambda^{3N}}\right)Z_N\right]\right\} \tag{3.97}$$

If $U_N \equiv 0$, then $Z_N = V^N$, and we have the entropy of the ideal gas system, which we denote by $S^{\text{id}}$. Hence

$$S = S^{\text{id}} + (\partial/\partial T)(kT\ln Z_N) \tag{3.98}$$

Differentiating the configurational partition function (for a system of spherical particles) with respect to temperature yields

$$(\partial/\partial T)(\ln Z_N) = Z^{-1}\int\cdots\int d\mathbf{R}^N\,\{\exp[-\beta U_N(\mathbf{R}^N)]\}[U_N(\mathbf{R}^N)/kT^2]$$

$$= (1/kT^2)\int\cdots\int d\mathbf{R}^N\,P(\mathbf{R}^N)U_N(\mathbf{R}^N) \tag{3.99}$$

where we have used the relation

$$P(\mathbf{R}^N) = \{\exp[-\beta U_N(\mathbf{R}^N)]\}/Z_N \qquad (3.100)$$

Using (3.99) and (3.100), we can rewrite (3.98) as

$$S = S^{\text{id}} + k \ln Z_N + kT(\partial/\partial T)(\ln Z_N)$$
$$= S^{\text{id}} + k \ln Z_N + (1/T)\left\{\int \cdots \int d\mathbf{R}^N\, P(\mathbf{R}^N)[-kT \ln P(\mathbf{R}^N) - kT \ln Z_N]\right\}$$
$$= S^{\text{id}} - k \int \cdots \int d\mathbf{R}^N\, P(\mathbf{R}^N) \ln P(\mathbf{R}^N)$$
$$= S^{\text{id}} - k \langle \ln P(\mathbf{R}^N) \rangle \qquad (3.101)$$

This cannot be expressed as a simple integral involving $g(R)$ only. The reason is the same as for the chemical potential.[4]

## 3.8. HEAT CAPACITY

In this section, we show that the heat capacity (here, at constant volume) is not expressible in terms of the pair distribution function. In fact, we shall see that molecular distribution functions of up to order four are required for this purpose. In Chapter 5, we discuss a different possibility of expressing the heat capacity in terms of generalized molecular distribution functions.

Consider a system characterized by the variables $T$, $V$, $N$. The internal energy (see Section 3.3) is given by

$$E = N\varepsilon^K + \langle U_N \rangle \qquad (3.102)$$

where

$$\langle U_N \rangle = \int \cdots \int d\mathbf{R}^N\, P(\mathbf{R}^N) U_N(\mathbf{R}^N) \qquad (3.103)$$

The heat capacity is obtained by direct differentiation with respect to temperature

$$C_V = \left(\frac{\partial E}{\partial T}\right)_{V,N} = NC^K + \int \cdots \int d\mathbf{R}^N \left[\frac{\partial P(\mathbf{R}^N)}{\partial T}\right]_{V,N} U_N(\mathbf{R}^N) \qquad (3.104)$$

---

[4] Here, we have expressed only the nonideal part of the entropy as an average of $\ln P(\mathbf{R}^N)$. In fact, this is a general form for the total entropy of a system in any ensemble, i.e., if $P_i$ is the probability of observing the state $i$, then the entropy is given by $S = -k \sum_i P_i \ln P_i$. [See, for example, Hill (1960).]

## Molecular Distribution Functions and Thermodynamics

From (3.100), we get

$$\frac{\partial P(\mathbf{R}^N)}{\partial T}$$

$$= \frac{\{\exp[-\beta U_N(\mathbf{R}^N)]\} U_N(\mathbf{R}^N)/kT^2}{Z_N}$$

$$- \frac{\{\exp[-\beta U_N(\mathbf{R}^N)]\} \int \cdots \int d\mathbf{R}^N \{\exp[-\beta U_N(\mathbf{R}^N)]\}[U_N(\mathbf{R}^N)/kT^2]}{Z_N^2} \tag{3.105}$$

Substituting (3.105) into (3.104), we get

$$C_V = NC^K + (1/kT^2)[\langle U_N(\mathbf{R}^N)^2 \rangle - \langle U_N(\mathbf{R}^N) \rangle^2] \tag{3.106}$$

which is similar to the general relation between heat capacity and fluctuations in the total energy as discussed in Chapter 1. Here, however, we have explicitly expressed the fluctuations only in the potential energy. Now, assuming that

$$U_N(\mathbf{R}^N) = \tfrac{1}{2} \sum_{i \ne j} U(\mathbf{R}_i, \mathbf{R}_j) \tag{3.107}$$

we immediately see that the average quantity $\langle U_N(\mathbf{R}^N)^2 \rangle$ is not an average of a pairwise quantity. Substituting (3.107) into (3.106) and collecting integrals of equal values, we get

$$\begin{aligned}
C_V &= NC^K + (1/4kT^2)[2N(N-1)\langle U(\mathbf{R}_1, \mathbf{R}_2)^2 \rangle \\
&\quad + 4N(N-1)(N-2)\langle U(\mathbf{R}_1, \mathbf{R}_2)U(\mathbf{R}_2, \mathbf{R}_3) \rangle \\
&\quad + N(N-1)(N-2)(N-3)\langle U(\mathbf{R}_1, \mathbf{R}_2)U(\mathbf{R}_3, \mathbf{R}_4) \rangle \\
&\quad + \{N(N-1)\langle U(\mathbf{R}_1, \mathbf{R}_2) \rangle\}^2] \\
&= NC^K + (1/4kT^2)\bigg[2 \int d(1,2)\, U(1,2)\varrho^{(2)}(1,2) \\
&\quad + 4 \int d(1,2,3)\, U(1,2)U(2,3)\varrho^{(3)}(1,2,3) \\
&\quad + \int d(1,2,3,4)\, U(1,2)U(3,4)\varrho^{(4)}(1,2,3,4) \\
&\quad + \bigg(\int d(1,2)\, U(1,2)\varrho^{(2)}(1,2)\bigg)^2 \bigg]
\end{aligned} \tag{3.108}$$

In the last form on the rhs of (3.108), we have used a common shorthand notation, where the numbers in brackets stand for the configuration of the corresponding particle. We see that the heat capacity is expressible in terms of the molecular distribution function up to order four. Because of lack of information on these functions, relation (3.108) has had, thus far, no practical value. A different expression for $C_V$ in the grand canonical ensemble has been derived by Buff and Brout (1955).

## 3.9. THE COMPRESSIBILITY EQUATION

The compressibility relation is one of the simplest and most useful relations between a thermodynamic quantity and the pair correlation function. In this section, we derive this relation and point out some of its outstanding features.

We consider here a system of rigid, nonspherical particles in the $T$, $V$, $\mu$ ensemble. We stress from the outset that no assumption of additivity of the potential energy is invoked at any stage of the derivation.

We recall the normalization conditions for $\varrho^{(1)}(\mathbf{X}_1)$ and for $\varrho^{(2)}(\mathbf{X}_1, \mathbf{X}_2)$ in the $T$, $V$, $\mu$ ensemble (see Section 2.9)

$$\int d\mathbf{X}_1 \, \varrho^{(1)}(\mathbf{X}_1) = \langle N \rangle \tag{3.109}$$

$$\int d\mathbf{X}_1 \, d\mathbf{X}_2 \, \varrho^{(2)}(\mathbf{X}_1, \mathbf{X}_2) = \langle N^2 \rangle - \langle N \rangle \tag{3.110}$$

where the symbol $\langle \ \rangle$ stands for an average in the $T$, $V$, $\mu$ ensemble. Squaring (3.109) and subtracting from (3.110) yields

$$\int d\mathbf{X}_1 \, d\mathbf{X}_2 \, [\varrho^{(2)}(\mathbf{X}_1, \mathbf{X}_2) - \varrho^{(1)}(\mathbf{X}_1)\varrho^{(1)}(\mathbf{X}_2)] = \langle N^2 \rangle - \langle N \rangle^2 - \langle N \rangle \tag{3.111}$$

Using the relations (homogeneous and isotropic fluid)

$$\varrho^{(1)}(\mathbf{X}_1) = \varrho/8\pi^2 \tag{3.112}$$

$$g(\mathbf{X}_1, \mathbf{X}_2) = \varrho^{(2)}(\mathbf{X}_1, \mathbf{X}_2)/\varrho^{(1)}(\mathbf{X}_1)\varrho^{(1)}(\mathbf{X}_2) \tag{3.113}$$

$$g(\mathbf{R}_1, \mathbf{R}_2) = [1/(8\pi^2)^2] \int d\mathbf{\Omega}_1 \, d\mathbf{\Omega}_2 \, g(\mathbf{X}_1, \mathbf{X}_2) \tag{3.114}$$

we can rearrange (3.111) to obtain

$$\varrho^2 \int d\mathbf{R}_1 \, d\mathbf{R}_2 \, [g(\mathbf{R}_1, \mathbf{R}_2) - 1] = \langle N^2 \rangle - \langle N \rangle^2 - \langle N \rangle \tag{3.115}$$

Since $g(\mathbf{R}_1, \mathbf{R}_2)$ depends only on the scalar separation $R = |\mathbf{R}_2 - \mathbf{R}_1|$,

## Molecular Distribution Functions and Thermodynamics

we can rewrite (3.115) as

$$\varrho \int_V d\mathbf{R}\, [g(R) - 1] + 1 = (\langle N^2 \rangle - \langle N \rangle^2)/\langle N \rangle$$

$$= \varrho \int_0^\infty [g(R) - 1] 4\pi R^2\, dR + 1 \quad (3.116)$$

The last equality holds since we know that $g(R) - 1$ tends rapidly to zero as $R \to \infty$, so that convergence of the integral is ensured (see below for further discussion). Relation (3.116) is an important connection between the radial distribution function and fluctuations in the number of particles. In Section 1.4, we obtained the relation

$$\langle N^2 \rangle - \langle N \rangle^2 = kTV\varrho^2 \varkappa_T \quad (3.117)$$

where $\varkappa_T$ is the isothermal compressibility of the system. Combining (3.117) with (3.116), we get the final result

$$\varkappa_T = \frac{1}{kT\varrho} + \frac{1}{kT} \int_V d\mathbf{R}\, [g(R) - 1] = \frac{1}{kT\varrho} + \frac{1}{kT} \int_0^\infty [g(R) - 1] 4\pi R^2\, dR \quad (3.118)$$

which is known as the compressibility equation. Note that the first term on the rhs of (3.118) is the compressibility of an ideal gas. That is, for a system obeying the relation

$$P = \varrho kT \quad (3.119)$$

we get

$$\varkappa_T = -\frac{1}{V}\left(\frac{\partial V}{\partial P}\right)_{T,N} = \left(\frac{\partial \ln \varrho}{\partial P}\right)_T = \frac{1}{kT\varrho} \quad (3.120)$$

Hence, the second term on the rhs of (3.118) conveys the contribution to the compressibility due to the existence of interaction (and therefore correlation) among the particles.

Let us survey some of the salient properties of the compressibility equation (3.118).

1. We recall that no assumption on the total potential energy has been introduced to obtain (3.118). In the previous sections, we found relations between some thermodynamic quantities and the pair correlation functions which were based explicitly on the assumption of pairwise additivity of the total potential energy. We recall, also, that higher-order molecular distribution functions must be introduced if higher-order potentials are not

negligible. Relation (3.118) does not depend on the additivity assumption, hence it suffers no modification should high-order potentials be of importance. In this respect, the compressibility equation is far more general than the previously obtained relations (e.g., the energy or the pressure relation).

2. The compressibility equation involves the knowledge of the *radial* distribution function, even when the system consists of nonspherical particles. We recall that previously obtained relations between, say, the energy or the pressure, and the pair correlation function were dependent on the type of particle under consideration [compare, for instance, relations (3.22) and (3.25)]. The compressibility of the system depends only on the *spatial* pair correlation function. If nonspherical particles are considered, it is understood that $g(R)$ is obtained by averaging over all orientations (3.114).

3. The compressibility equation is a simple integral over $g(R)$. It does not require explicit knowledge of $U(R)$ (or higher-order potentials). It is true that $g(R)$ is a function of $U(R)$. However, once we have obtained the former, we can use it directly to compute the compressibility by means of (3.118). This is not possible for the computation of, say, the energy.

One of the most important applications of the compressibility equation is to test various theories of computation of $g(R)$. We recall that the pressure equation (3.37) has been found useful for computing the equation of state of a substance, and hence can be used as a sensitive test of the theory that has furnished $g(R)$. Similarly, by integrating the compressibility equation, we obtain the equation of state of the system, which may serve as a different test of the theory. Clearly, if we use the *exact* function $g(R)$ in the pressure or in the compressibility equations, we must end up with the same equation of state. However, since we usually employ only approximations for $g(R)$, the results of the two equations may be different. Therefore, the discrepancy between the two results obtained with the same $g(R)$ using the pressure and the compressibility equations can serve as a very sensitive test of the accuracy of the method of computing $g(R)$.

In applying the compressibility equation (3.118), care must be exercised in order to ensure that we have taken the limiting form of $g(R)$ for infinitely large systems. Failing to take the proper limit may lead to spurious conclusions.

Let us first demonstrate the source of difficulty by a simple example. Consider an ideal gas in the $T, V, N$ ensemble. In Section 2.5, we saw that $g(R)$ in this case has the form

$$g(R) = 1 - (1/N) \quad \text{(ideal gas; } T, V, N \text{ ensemble)} \quad (3.121)$$

Molecular Distribution Functions and Thermodynamics    107

On the other hand, if we take the infinite-system-size limit of (3.121), we get

$$g(R) = 1 \qquad (3.122)$$

which is also the form of $g(R)$ obtained for the ideal gas in the $T$, $V$, $\mu$ system. [For more details, see Hill (1956) and Münster (1969).]

The difference between (3.121) and (3.122) is admittedly small (for large $N$). This is true as long as we are interested in $g(R)$ itself, so that a term of order $N^{-1}$ can be considered as negligible. However, if we use $g(R)$ in the compressibility equation, different results are obtained from (3.121) and (3.122), namely[5]

$$\varkappa_T = \frac{1}{kT\varrho} + \frac{1}{kT} \int_V d\mathbf{R} \left(\frac{-1}{N}\right)$$

$$= \frac{1}{kT\varrho} - \frac{1}{kT\varrho} = 0 \qquad \text{[using } g(R) \text{ from (3.121)]} \qquad (3.123)$$

$$\varkappa_T = \frac{1}{kT\varrho} \qquad \text{[using } g(R) \text{ from (3.122)]} \qquad (3.124)$$

Clearly, only the second result gives the correct compressibility of the ideal gas (3.120).

Relations (3.121) and (3.122) hold for the ideal gas. For the general case, the important feature of $g(R)$ that causes the trouble is the limiting behavior of $g(R)$ as $R \to \infty$. The result for the general case is [for more details, see Lebowitz and Percus (1961)]

$$g(R) \xrightarrow{R \to \infty} 1 - (\varrho kT\varkappa_T/N) \qquad (T, V, N \text{ ensemble}) \qquad (3.125)$$

$$g(R) \xrightarrow{R \to \infty} 1 \qquad (T, V, \mu \text{ ensemble}) \qquad (3.126)$$

Clearly, (3.126) can be obtained from (3.125) by taking the infinite-system-size limit ($N \to \infty$). Note also that (3.121) is a particular case of (3.125).

The origin of the discrepancy between the two results in the $T$, $V$, $N$ and in the $T$, $V$, $\mu$ ensembles is the difference in the normalization conditions for the molecular distribution functions. In particular, in the $T$, $V$, $N$ ensemble, we have

$$\langle N^2 \rangle = \langle N \rangle^2 = N^2 \qquad (3.127)$$

---

[5] Note that in (3.123) we have used the first form on the rhs of (3.118), so that integration over $\mathbf{R}$ gives the total volume $V$.

Hence, the normalization condition is

$$\int d\mathbf{X}_1 \, d\mathbf{X}_2 \, [\varrho^{(2)}(\mathbf{X}_1, \mathbf{X}_2) - \varrho^{(1)}(\mathbf{X}_1)\varrho^{(1)}(\mathbf{X}_2)] = -N \quad (3.128)$$

which is equivalent to the normalization condition

$$\varrho \int_0^\infty [g(R) - 1] 4\pi R^2 \, dR = -1 \quad (T, V, N \text{ ensemble}) \quad (3.129)$$

The last result simply means that the total number of particles *around* a *given* particle minus the total number of particles in the system is exactly minus one. This simple calculation does not hold for the open system, where $N$ is not a fixed number.

The reader may wonder why we have only now dealt with the question of the limiting behavior of $g(R)$ as $R \to \infty$. The reason is quite simple. In all of our previous integrals, $g(R)$ appeared with another function in the integrand. For instance, in the equation for the energy, we have an integral of the form

$$\int_0^\infty U(R) g(R) 4\pi R^2 \, dR \quad (3.130)$$

Clearly, since $U(R)$ is presumed to tend to zero, as, say, $R^{-6}$ as $R \to \infty$, it is of no importance whether the limiting behavior of $g(R)$ is given by (3.125) or (3.126); in both cases, the integrand will become practically zero as $R$ becomes large enough so that $U(R) \approx 0$. The unique feature of the compressibility relation is that only $g(R)$ appears under the sign of integration. Therefore, different results may be anticipated according to the different limiting behavior of $g(R)$ as $R \to \infty$.

As a subsidiary result of this section, we derive a relation between the density derivative of the chemical potential and an integral involving $g(R)$.

Recall the thermodynamic relation (see Section 1.4)

$$(\partial \mu / \partial \varrho)_T = 1/\varkappa_T \varrho^2 \quad (3.131)$$

Combining (3.131) with (3.118) yields

$$\left(\frac{\partial \mu}{\partial \varrho}\right)_T = \frac{kT}{\varrho + \varrho^2 G} = kT \left(\frac{1}{\varrho} - \frac{G}{1 + \varrho G}\right) \quad (3.132)$$

where we have denoted

$$G = \int_0^\infty [g(R) - 1] 4\pi R^2 \, dR \quad (3.133)$$

Molecular Distribution Functions and Thermodynamics    109

Relation (3.132) will be generalized in the next chapter to mixtures. In the context of this section, we note that by integrating (3.132) with respect to the density, we can get the chemical potential, i.e.,

$$\mu = \int \frac{kT\, d\varrho}{\varrho + \varrho^2 G} + \text{const} \tag{3.134}$$

Thus, once we have $g(R)$ and its density dependence, we can determine $\mu$ from (3.134). The constant of integration is evaluated as follows: We choose a very low density $\varrho_0$ for which we know that

$$\mu(\varrho_0) = \mu^{og} + kT \ln \varrho_0 \tag{3.135}$$

The chemical potential at the final density $\varrho$ is written as

$$\mu(\varrho) = \mu(\varrho_0) + \int_{\varrho_0}^{\varrho} \left(\frac{\partial \mu}{\partial \varrho}\right)_T d\varrho \tag{3.136}$$

Hence, from (3.134), (3.135), and (3.136) we get

$$\mu(\varrho) = [\mu^{og} + kT \ln \varrho_0] + \int_{\varrho_0}^{\varrho} \frac{kT\, d\varrho}{\varrho + \varrho^2 G} \tag{3.137}$$

This relation holds for any sufficiently low density $\varrho_0$.

## 3.10. LOCAL DENSITY FLUCTUATIONS

The content of this section is closely related to that of the previous section. We shall be interested in the fluctuations of the density in a given region $S$ within the system.

Consider, for example, a system in the $T$, $V$, $N$ ensemble. We select a region $S$ within the vessel containing the system and inquire as to the number of particles that fall within $S$ for a given configuration[6] $\mathbf{R}^N$:

$$N(\mathbf{R}^N, S) = \sum_{i=1}^{N} \int_S \delta(\mathbf{R}_i - \mathbf{R}')\, d\mathbf{R}' \tag{3.138}$$

Each term in the sum over $i$ is unity whenever $\mathbf{R}_i$ is in $S$, and is zero otherwise. Therefore, the sum over $i$ counts all the particles that are within $S$

[6] The results of this section apply to a system of particles which are not necessarily spherical. Here, we use $\mathbf{R}^N$ for describing the locations of all the *centers* of the particles. The orientations are of no importance for the present considerations.

at a given configuration $\mathbf{R}^N$. The average number of particles in $S$ is

$$\begin{aligned}\langle N(S)\rangle &= \sum_{i=1}^{N} \int_V \cdots \int d\mathbf{R}^N\, P(\mathbf{R}^N) \int_S \delta(\mathbf{R}_i - \mathbf{R}')\, d\mathbf{R}' \\ &= N \int_V \cdots \int d\mathbf{R}^N\, P(\mathbf{R}^N) \int_S \delta(\mathbf{R}_1 - \mathbf{R}')\, d\mathbf{R}' \\ &= \int_S d\mathbf{R}'\, N \int_V \cdots \int d\mathbf{R}^N\, P(\mathbf{R}^N)\, \delta(\mathbf{R}_1 - \mathbf{R}') \\ &= \int_S d\mathbf{R}'\, \varrho^{(1)}(\mathbf{R}') \\ &= \varrho V(S) \end{aligned} \qquad (3.139)$$

The last relation holds for a homogeneous fluid, where $V(S)$ is the volume of the region $S$.

Now consider the average of the square of $N(\mathbf{R}^N, S)$:

$$\langle N(S)^2\rangle = \int_V \cdots \int d\mathbf{R}^N\, P(\mathbf{R}^N) \int_S \sum_{i=1}^{N} \delta(\mathbf{R}_i - \mathbf{R}')\, d\mathbf{R}' \int_S \sum_{j=1}^{N} \delta(\mathbf{R}_j - \mathbf{R}'')\, d\mathbf{R}'' \qquad (3.140)$$

Rearranging (3.140), we get

$$\begin{aligned}\langle N(S)^2\rangle &= \sum_{i=1}^{N}\sum_{j=1}^{N} \int_S d\mathbf{R}' \int_S d\mathbf{R}'' \int_V \cdots \int d\mathbf{R}^N\, P(\mathbf{R}^N)\, \delta(\mathbf{R}_i - \mathbf{R}')\, \delta(\mathbf{R}_j - \mathbf{R}'') \\ &= \sum_{i=1}^{N} \int_S d\mathbf{R}' \int_S d\mathbf{R}''\, \delta(\mathbf{R}' - \mathbf{R}'') \int_V \cdots \int d\mathbf{R}^N\, P(\mathbf{R}^N)\, \delta(\mathbf{R}_i - \mathbf{R}') \\ &\quad + \sum_{i\neq j} \int_S d\mathbf{R}' \int_S d\mathbf{R}'' \int_V \cdots \int d\mathbf{R}^N\, P(\mathbf{R}^N)\, \delta(\mathbf{R}_i - \mathbf{R}')\, \delta(\mathbf{R}_j - \mathbf{R}'') \\ &= \int_S d\mathbf{R}' \int_S d\mathbf{R}''\, \delta(\mathbf{R}' - \mathbf{R}'')\varrho^{(1)}(\mathbf{R}') + \int_S d\mathbf{R}' \int_S d\mathbf{R}''\, \varrho^{(2)}(\mathbf{R}', \mathbf{R}'') \\ &= \langle N(S)\rangle + \int_S d\mathbf{R}' \int_S d\mathbf{R}''\, \varrho^{(2)}(\mathbf{R}', \mathbf{R}'') \end{aligned} \qquad (3.141)$$

In the second step on the rhs of (3.141), we have split the double sum over $i$ and $j$ into two sums; the first consists of all the terms with $i = j$ and the second includes the terms with $i \neq j$. We have also used the identity of the product of two Dirac delta functions[7]

$$\delta(\mathbf{R}_i - \mathbf{R}')\, \delta(\mathbf{R}_i - \mathbf{R}'') = \delta(\mathbf{R}_i - \mathbf{R}')\, \delta(\mathbf{R}' - \mathbf{R}'') \qquad (3.142)$$

[7] The analog of (3.142) in the discrete case is $\delta_{ik}\delta_{ij} = \delta_{ik}\delta_{kj}$.

## Molecular Distribution Functions and Thermodynamics 111

In the third step of (3.141), we have used the definition of the pair distribution function.

The fluctuations in the number of particles within $S$ are given by

$$\langle \Delta N(S)^2 \rangle = \langle (N(S) - \langle N(S) \rangle)^2 \rangle = \langle N(S)^2 \rangle - \langle N(S) \rangle^2$$

$$= \langle N(S) \rangle + \int_S d\mathbf{R}' \int_S d\mathbf{R}'' \, \varrho^{(2)}(\mathbf{R}', \mathbf{R}'')$$

$$- \int_S d\mathbf{R}' \, \varrho^{(1)}(\mathbf{R}') \int_S d\mathbf{R}'' \, \varrho^{(1)}(\mathbf{R}'')$$

$$= \langle N(S) \rangle + \varrho^2 \int_S d\mathbf{R}' \int_S d\mathbf{R}'' \, [g(R) - 1] \qquad (3.143)$$

where $R = |\mathbf{R}'' - \mathbf{R}'|$.

Relation (3.143) resembles (3.115) of the previous section. Care must be exercised in applying (3.143) for the whole volume $V$. If we are in the $T, V, \mu$ ensemble, then, taking $S$ as the whole volume of the system, we get the compressibility relation (3.115). However, if we are in the $T, V, N$ ensemble, then $N$ is *fixed* and we have

$$\langle N(V) \rangle = N(V) = N \qquad (3.144)$$

and (3.143) reduces to

$$N + \varrho^2 \int_V d\mathbf{R}' \int_V d\mathbf{R}'' \, [g(R) - 1] = 0 \qquad (3.145)$$

which is the correct normalization condition for $g(R)$ in the $T, V, N$ ensemble [compare with (3.129)], but is different from the compressibility relation.

Another limiting case is that in which $S$ is an infinitesimally small region, such that at most one particle can occupy $S$ at any given time. For instance, if the maximum diameter of $S$ is smaller than $\sigma$ (the effective diameter of the particles), then $g(R)$ will be zero in the integrand of (3.143) and we get

$$\langle \Delta N(S)^2 \rangle = \langle N(S) \rangle - \varrho^2 V(S)^2 = \varrho V(S)[1 - \varrho V(S)] \qquad (3.146)$$

Thus, if $V(S)$ is infinitesimally small, then fluctuations in the number of particles are dominated by $\varrho V(S)$.

Next, consider two regions $S_1$ and $S_2$ in the vessel containing the system (Fig. 3.1). The cross fluctuations in the number of particles in the

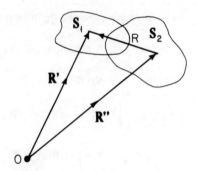

Fig. 3.1. Two regions $S_1$ and $S_2$ in the vessel containing the system. $\mathbf{R}'$ and $\mathbf{R}''$ are two points within $S_1$ and $S_2$, respectively.

two regions are given by

$$\langle \Delta N(S_1)\, \Delta N(S_2) \rangle$$
$$= \langle [N(S_1) - \langle N(S_1) \rangle][N(S_2) - \langle N(S_2) \rangle] \rangle$$
$$= \langle N(S_1) N(S_2) \rangle - \langle N(S_1) \rangle \langle N(S_2) \rangle$$
$$= \int \cdots \int_V d\mathbf{R}^N\, P(\mathbf{R}^N) \int_{S_1} \sum_{i=1}^N \delta(\mathbf{R}_i - \mathbf{R}')\, d\mathbf{R}' \int_{S_2} \sum_{j=1}^N \delta(\mathbf{R}_j - \mathbf{R}'')\, d\mathbf{R}''$$
$$\quad - \int_{S_1} d\mathbf{R}'\, \varrho^{(1)}(\mathbf{R}') \int_{S_2} d\mathbf{R}''\, \varrho^{(1)}(\mathbf{R}'')$$
$$= \sum_{i=1}^N \int_{S_1} d\mathbf{R}' \int_{S_2} d\mathbf{R}''\, \delta(\mathbf{R}' - \mathbf{R}'') \int \cdots \int_V d\mathbf{R}^N\, P(\mathbf{R}^N)\, \delta(\mathbf{R}_i - \mathbf{R}')$$
$$\quad + \sum_{i \neq j} \int_{S_1} d\mathbf{R}' \int_{S_2} d\mathbf{R}'' \int \cdots \int_V d\mathbf{R}^N\, P(\mathbf{R}^N)\, \delta(\mathbf{R}_i - \mathbf{R}')\, \delta(\mathbf{R}_j - \mathbf{R}'')$$
$$\quad - \int_{S_1} d\mathbf{R}'\, \varrho^{(1)}(\mathbf{R}') \int_{S_2} d\mathbf{R}''\, \varrho^{(1)}(\mathbf{R}'')$$
$$= \int_{S_1} d\mathbf{R}' \int_{S_2} d\mathbf{R}''\, \varrho^{(1)}(\mathbf{R}')\, \delta(\mathbf{R}' - \mathbf{R}'')$$
$$\quad + \int_{S_1} d\mathbf{R}' \int_{S_2} d\mathbf{R}''\, [\varrho^{(2)}(\mathbf{R}', \mathbf{R}'') - \varrho^{(1)}(\mathbf{R}')\varrho^{(1)}(\mathbf{R}'')] \qquad (3.147)$$

We now distinguish between two cases:

(1) If the two regions $S_1$ and $S_2$ do not overlap, the first term on the rhs of (3.147) is zero. This follows from the property of the Dirac

## Molecular Distribution Functions and Thermodynamics

delta function:

$$\int_{S_2} d\mathbf{R}'' \, \delta(\mathbf{R}' - \mathbf{R}'') = 0 \quad \text{if } \mathbf{R}' \notin S_2 \quad (3.148)$$

Since $\mathbf{R}_1'$ is always within $S_1$, (3.148) will hold whenever $S_1$ and $S_2$ do not intersect.

(2) If $S_1$ and $S_2$ do intersect, we shall have for the first term on the rhs of (3.147)

$$\int_{S_2} d\mathbf{R}'' \, \delta(\mathbf{R}' - \mathbf{R}'') = \begin{cases} 0 & \text{if } \mathbf{R}' \notin S_2 \\ 1 & \text{if } \mathbf{R}' \in S_2 \end{cases} \quad (3.149)$$

But $\mathbf{R}'$ is always within $S_1$; thus, we have

$$\int_{S_1} d\mathbf{R}' \, \varrho^{(1)}(\mathbf{R}') \int_{S_2} d\mathbf{R}'' \, \delta(\mathbf{R}' - \mathbf{R}'') = \int_{S_1 \cap S_2} d\mathbf{R}' \, \varrho^{(1)}(\mathbf{R}') = \varrho V(S_1 \cap S_2) \quad (3.150)$$

where $S_1 \cap S_2$ stands for the intersection region between $S_1$ and $S_2$. The last equality on the rhs of (3.150) holds for homogeneous fluids. For this case, we write the final form of (3.147) as

$$\langle \Delta N(S_1) \, \Delta N(S_2) \rangle = \varrho V(S_1 \cap S_2) + \varrho^2 \int_{S_1} d\mathbf{R}' \int_{S_2} d\mathbf{R}'' \, [g(R) - 1] \quad (3.151)$$

Some special cases are the following:

(1) If $S_1$ and $S_2$ are identical, i.e., $S_1 = S_2 = S$, then (3.151) reduces to the previous result (1.143).

(2) If $S_1$ and $S_2$ are infinitesimal, nonoverlapping elements of volume, say $d\mathbf{R}'$ and $d\mathbf{R}''$, then

$$\langle \Delta N(d\mathbf{R}') \, \Delta N(d\mathbf{R}'') \rangle = \varrho^2 [g(R) - 1] \, d\mathbf{R}' \, d\mathbf{R}'' \quad (3.152)$$

On the other hand, dividing (3.152) by $d\mathbf{R}' \, d\mathbf{R}''$ and using the second equality on the rhs of (3.147), we get

$$\frac{\langle \Delta N(d\mathbf{R}') \, \Delta N(d\mathbf{R}'') \rangle}{d\mathbf{R}' \, d\mathbf{R}''} = \frac{\langle N(d\mathbf{R}')N(d\mathbf{R}'') \rangle - \langle N(d\mathbf{R}') \rangle \langle N(d\mathbf{R}'') \rangle}{d\mathbf{R}' \, d\mathbf{R}''}$$

$$= \langle \varrho(\mathbf{R}')\varrho(\mathbf{R}'') \rangle - \langle \varrho(\mathbf{R}') \rangle \langle \varrho(\mathbf{R}'') \rangle \quad (3.153)$$

where $\varrho(\mathbf{R})$ stands for the local density at $\mathbf{R}$ for a given configuration $\mathbf{R}^N$ of the system. [Note the difference between $\varrho(\mathbf{R})$ and $\varrho^{(1)}(\mathbf{R})$.] Combining

(3.153) with (3.152), we get

$$\langle \varrho(\mathbf{R}')\varrho(\mathbf{R}'')\rangle - \langle \varrho(\mathbf{R}')\rangle\langle \varrho(\mathbf{R}'')\rangle = \varrho^2[g(R) - 1] \qquad (3.154)$$

or

$$\langle \varrho(\mathbf{R}')\varrho(\mathbf{R}'')\rangle = \varrho^2 g(R), \quad R = |\mathbf{R}'' - \mathbf{R}'| \text{ and } \mathbf{R}' \neq \mathbf{R}'' \qquad (3.155)$$

The last result also can be viewed as a definition of $g(R)$, i.e., this function conveys the *correlation* in the local densities at two points $\mathbf{R}'$ and $\mathbf{R}''$ in the fluid. Note that if we also allow the case $\mathbf{R}'' = \mathbf{R}'$, then (3.155) should be modified to read

$$\langle \varrho(\mathbf{R}')\varrho(\mathbf{R}'')\rangle = \varrho\, \delta(\mathbf{R}' - \mathbf{R}'') + \varrho^2 g(R) \qquad (3.156)$$

Computations of local density fluctuations have been reported by Fisher and Adamovich (1963) and by Chay and Frank (1972).

## 3.11. THE WORK REQUIRED TO FORM A CAVITY IN A FLUID

A quantity of considerable interest in the study of fluids and, in particular, aqueous solutions, is the work required to create a cavity of radius $\sigma$ at some fixed position $\mathbf{R}_0$. In addition to its importance as a property of the fluid, the work of cavity formation may also be viewed as a first step in the process of solubility. The latter aspect will be dealt with in Chapter 7.

A cavity of *radius* $\sigma$ at a point $\mathbf{R}_0$ in the fluid is defined as the spherical region of radius $\sigma$ centered at $\mathbf{R}_0$ from which the centers of all other particles are excluded.[8] Figure 3.2 depicts such a cavity.

It is important to realize that a cavity may seem to be "filled" in the conventional sense, yet be empty according to the definition given above. Two extreme cases are illustrated in Fig. 3.3. In (a) we have a cavity of radius $\sigma$ which is completely "filled" by a molecule of radius $\sigma_a/2$, yet, since no *center* of a molecule falls in this cavity, it is empty according to our definition. In (b) we have a cavity which is filled, since one center of a particle occupies it, yet it looks almost empty in the conventional sense.

The work required to create a cavity of radius $\sigma$ (keeping $T$, $V$, and $N$ constant) is

$$\Delta A_{\text{cav}}(\mathbf{R}_0, \sigma) = A(T, V, N; \mathbf{R}_0, \sigma) - A(T, V, N) \qquad (3.157)$$

---

[8] One can easily generalize the concept of a cavity to a region $S$ of any geometric form. For most of our applications, however, we shall be interested only in spherical cavities.

# Molecular Distribution Functions and Thermodynamics

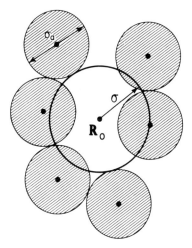

Fig. 3.2. A cavity of radius $\sigma$ at $\mathbf{R}_0$. The diameter of the particles is $\sigma_a$.

Here, $A(T, V, N; \mathbf{R}_0, \sigma)$ stands for the Helmholtz free energy of a system in the $T$, $V$, $N$ ensemble having a cavity of radius $\sigma$ centered at $\mathbf{R}_0$. Clearly, in a homogeneous fluid, all points in the vessel are considered to be equivalent (except for a negligible region near the boundaries of the system), and hence $\Delta A_{\text{cav}}(\mathbf{R}_0, \sigma)$ does not depend on $\mathbf{R}_0$. Nevertheless, we shall keep $\mathbf{R}_0$ in our notation to stress the fact that we are concerned with a cavity at a *fixed* position in the system.

From the definition of the concept of a cavity, it follows that all centers of the $N$ particles are excluded from the spherical region of radius $\sigma$ cen-

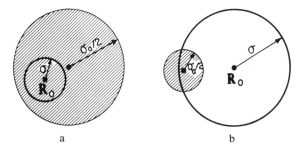

Fig. 3.3. (a) A cavity of radius $\sigma$ "filled" by a particle of diameter $\sigma_a$ yet empty according to the definition of a cavity. (b) A cavity of radius $\sigma$ filled according to the definition of a cavity.

tered at $\mathbf{R}_0$. Denoting this region by $V(\mathbf{R}_0, \sigma)$, we get

$$\exp[-\beta \Delta A_{\text{cav}}(\mathbf{R}_0, \sigma)] = \left\{ (q^N/\Lambda^{3N} N!) \int \cdots \int_{V-V(\mathbf{R}_0,\sigma)} d\mathbf{R}^N \exp[-\beta U_N(\mathbf{R}^N)] \right\}$$
$$\times \left\{ (q^N/\Lambda^{3N} N!) \int \cdots \int_V d\mathbf{R}^N \exp[-\beta U_N(\mathbf{R}^N)] \right\}^{-1}$$
(3.158)

Note that the only difference in the two integrals is in the limits of the integration. For simplicity, we have considered a system of spherical particles; most of our conclusions hold for nonspherical particles as well.

The rhs of (3.158) now can be interpreted as the probability of observing a cavity of radius $\sigma$ at $\mathbf{R}_0$. We recall that the probability density of observing any specific configuration $\mathbf{R}^N$ is given by

$$P(\mathbf{R}^N) = \frac{\exp[-\beta U_N(\mathbf{R}^N)]}{\int \cdots \int d\mathbf{R}^N \exp[-\beta U_N(\mathbf{R}^N)]}$$

Suppose we choose a region $S$ in our system. The probability of finding the centers of *all* the particles in $S$ is obtained by summing over all the events (configurations) conforming to the requirement that all $\mathbf{R}_i$ ($i = 1, 2, \ldots, N$) are within $S$. Hence,

$$P(S) = \int \cdots \int_S d\mathbf{R}^N \, P(\mathbf{R}^N) \tag{3.159}$$

As a particular case, if $S$ coincides with the entire volume of the system, then $P(V) = 1$. Similarly, if all the centers are to be excluded from $V(\mathbf{R}_0, \sigma)$, we have the case in which $S = V - V(\mathbf{R}_0, \sigma)$ and therefore

$$P_{\text{cav}}(\mathbf{R}_0, \sigma) = \int \cdots \int_{V-V(\mathbf{R}_0,\sigma)} d\mathbf{R}^N \, P(\mathbf{R}^N) \tag{3.160}$$

As a very simple example for which we can compute $P_{\text{cav}}$ explicitly, consider a small region of radius $\sigma = \sigma_a/2$ (where $\sigma_a$ is the effective diameter of the particles). This region can be occupied by at most one particle at any given time. Hence, the probability that this region is occupied is $\varrho 4\pi\sigma^3/3$, whereas the probability that it is empty is $1 - \varrho 4\pi\sigma^3/3$.

Comparing (3.158) with (3.160), we arrive at the important relation

$$P_{\text{cav}}(\mathbf{R}_0, \sigma) = \exp[-\beta \Delta A_{\text{cav}}(\mathbf{R}_0, \sigma)] \tag{3.161}$$

**Molecular Distribution Functions and Thermodynamics**     117

On the rhs of (3.161), we have the work required to form a cavity of radius $\sigma$ at $\mathbf{R}_0$, keeping $T$, $V$, and $N$ constant. On the lhs, we have the probability of observing such a cavity in a system in the $T$, $V$, $N$ ensemble. A similar relation between work and probability was obtained in Section 2.6. Note also that in all of our considerations, we have required that all centers of particles be excluded from a spherical region of radius $\sigma$. We have not required that the radius of the cavity be exactly $\sigma$. The region of integration in (3.160) is $V - V(\mathbf{R}_0, \sigma)$, which, of course, does not preclude the possibility that some of the allowed configurations form a cavity having a radius larger than $\sigma$. In other words, $P_{\text{cav}}(\mathbf{R}_0, \sigma)$ is the probability of finding at $\mathbf{R}_0$ a cavity, the radius of which is at least $\sigma$.

We now turn to another useful relation, between the quantity $\Delta A_{\text{cav}}(\mathbf{R}_0, \sigma)$ and the pseudo-chemical potential of a hard-sphere solute.

Consider a system of $N$ particles with an effective hard-core diameter $\sigma_a$. By "effective," we mean that two molecules at a distance of $\sigma_a$ from each other feel almost infinite repulsive forces. Certainly there is no unique method of determining the values of $\sigma_a$ for a real molecule. Nevertheless, for, say, Lennard-Jones particles the parameter $\sigma$ is a reasonable choice for an effective hard-core diameter.

In Section 3.6, we obtained the pseudo-chemical potential for a one-component system. We repeat the same process here, but instead of adding the $(N + 1)$th particle, we add a hard-sphere particle of diameter $\sigma_{\text{HS}}$ to a *fixed* position $\mathbf{R}_0$ in a system of particles having an effective hard-core diameter $\sigma_a$. The work associated with this process, keeping $T$, $V$, $N$ constant, is given by

$$\exp(-\beta\bar{\mu}_{\text{HS}}) = \exp\{-\beta[A(T, V, N; \mathbf{R}_0, \sigma_{\text{HS}}) - A(T, V, N)]\}$$

$$= \left\{(q^N/\Lambda^{3N}N!) \int_V \cdots \int d\mathbf{R}^N \exp[-\beta U_N(\mathbf{R}^N) - \beta U(\mathbf{R}^N/\mathbf{R}_0)]\right\}$$

$$\times \left\{(q^N/\Lambda^{3N}N!) \int_V \cdots \int d\mathbf{R}^N \exp[-\beta U_N(\mathbf{R}^N)]\right\}^{-1} \quad (3.162)$$

Note that since the hard-sphere solute is presumed to have no internal structure, its addition to the system does not introduce an internal partition function [compare with Eq. (3.89)]. The quantity $U(\mathbf{R}^N/\mathbf{R}_0)$ stands for the total interaction energy between the $N$ molecules in the configuration $\mathbf{R}^N$ and the hard-sphere particle at $\mathbf{R}_0$. More specifically,

$$U(\mathbf{R}^N/\mathbf{R}_0) = \sum_{i=1}^{N} U(\mathbf{R}_i, \mathbf{R}_0) \quad (3.163)$$

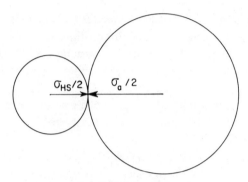

Fig. 3.4. A hard-sphere solute with diameter $\sigma_{HS}$ and a particle with an effective hard-core diameter $\sigma_a$. The distance of "closest" approach is $\sigma = (\sigma_{HS} + \sigma_a)/2$.

and, according to our assumption, $\sigma_a$ is the hard-core diameter of the particle. This statement means that

$$U(\mathbf{R}_i, \mathbf{R}_j) = \infty \quad \text{for} \quad R_{ij} = |\mathbf{R}_j - \mathbf{R}_i| < \sigma_a; \quad i,j = 1, 2, \ldots, N \quad (3.164)$$

For $R_{ij} \geq \sigma_a$, the potential function can have any form. However, the interaction between a particle of the system and the hard-sphere solute is presumed to have the form

$$U(\mathbf{R}_i, \mathbf{R}_0) = \begin{cases} \infty & \text{for} \quad R_{i0} < \sigma = \tfrac{1}{2}(\sigma_{HS} + \sigma_a), \\ 0 & \text{for} \quad R_{i0} \geq \sigma = \tfrac{1}{2}(\sigma_{HS} + \sigma_a), \end{cases} \quad i = 1, 2, \ldots, N \quad (3.165)$$

That is, a particle is not permitted to come closer than $\sigma$ to the center of the hard-sphere solute (see Fig. 3.4). The condition (3.165) can be rewritten in an equivalent form as

$$\exp[-\beta U(\mathbf{R}_i, \mathbf{R}_0)] = \begin{cases} 0 & \text{for} \quad R_{i0} < \sigma, \\ 1 & \text{for} \quad R_{i0} \geq \sigma, \end{cases} \quad i = 1, 2, \ldots, N \quad (3.166)$$

Using (3.163) and (3.166) in (3.162), we get

$$\begin{aligned}
\exp(-\beta \bar{\mu}_{HS}) &= \int \cdots \int_V d\mathbf{R}^N \, P(\mathbf{R}^N) \exp[-\beta U(\mathbf{R}^N/\mathbf{R}_0)] \\
&= \int \cdots \int_V d\mathbf{R}^N \, P(\mathbf{R}^N) \prod_{i=1}^{N} \exp[-\beta U(\mathbf{R}_i, \mathbf{R}_0)] \\
&= \int \cdots \int_{V-V(\mathbf{R}_0,\sigma)} d\mathbf{R}^N \, P(\mathbf{R}^N) \quad (3.167)
\end{aligned}$$

Molecular Distribution Functions and Thermodynamics 119

The last equality in (3.167) follows from the property of the unit step function (3.166), which nullifies the integrand whenever $R_{i0} < \sigma$ (for $i = 1, 2, \ldots, N$). Since we have a product of $N$ such factors (each of which operates on one vector $\mathbf{R}_i$) in the integrand of (3.167), their effect is to reduce the region of integration, for each $\mathbf{R}_i$, from $V$ to $V - V(\mathbf{R}_0, \sigma)$. Comparing with (3.160) and (3.167), we arrive at

$$\exp(-\beta \bar{\mu}_{\mathrm{HS}}) = P_{\mathrm{cav}}(\mathbf{R}_0, \sigma) = \exp[-\beta \, \Delta A_{\mathrm{cav}}(\mathbf{R}_0, \sigma)] \qquad (3.168)$$

with the condition

$$\sigma = (\sigma_{\mathrm{HS}} + \sigma_a)/2 \qquad (3.169)$$

Thus, the work required to produce a cavity of *radius* $\sigma$ at $\mathbf{R}_0$ (which may be chosen at any point in the fluid) in a system of particles with an effective hard-core diameter $\sigma_a$ is equal to the work required to introduce a hard-sphere particle of diameter $\sigma_{\mathrm{HS}}$ [given by $\sigma_{\mathrm{HS}} = 2\sigma - \sigma_a$, as in (3.169)] into a fixed position $\mathbf{R}_0$.

It is very important to realize that the last statement is valid only when we refer to a *fixed* position for both processes. The equivalence of the two processes is quite clear on intuitive grounds. Creation of a cavity means imposing a restriction on the centers of all particles, keeping them out of a certain region. This constraint is achieved simply by putting a hard-sphere solute at the position $\mathbf{R}_0$, provided that we have properly chosen its diameter by relation (3.169). In other words, as far as the solvent (i.e., the particles of the system) is concerned, there is no difference if we create a cavity at $\mathbf{R}_0$ or put a hard-sphere solute there, provided that (3.169) is fulfilled.

All of the above reasoning breaks down if we remove the condition of *fixed* position for both of the processes. We recall that the total work required to add a hard-sphere solute to the system (at $T$, $V$, $N$ constant) is equal to the chemical potential of the solute, which can be written as (see Section 3.6)

$$\mu_{\mathrm{HS}} = \bar{\mu}_{\mathrm{HS}} + kT \ln(\varrho_{\mathrm{HS}} \Lambda_{\mathrm{HS}}^3) \qquad (3.170)$$

where $\varrho_{\mathrm{HS}} = 1/V$ is the (number) density of the solute and $\Lambda_{\mathrm{HS}}^3$ its momentum partition function. We also recall (see Section 3.6 for more details) that the second term on the rhs of (3.170) is the contribution to $\mu_{\mathrm{HS}}$ that is associated with the release of the constraint of the fixed position for the solute. The analog of this contribution in the case of a cavity is not evident. In the first place, a "free cavity" has no momentum partition function akin to the quantity $\Lambda_{HS}^3$ of a real solute. Second, there is no clear-cut definition of the (number) density of "free cavities of diameter $\sigma$." For these reasons, the equivalence of the process of introducing a hard-sphere solute and the creation of a cavity should be stressed only when reference is made to a *fixed* position in the liquid. We return to this topic in Chapter 7 when dealing with aqueous solutions of simple solutes.

## 3.12. PERTURBATION THEORIES OF LIQUIDS

In this section, we present another demonstration of the application of the general theorem of Section 3.2 to obtain a first-order term in a perturbation expansion of the free energy.

Consider a system in the $T, V, N$ ensemble obeying the pairwise additivity assumption for the total potential energy:

$$U(\mathbf{X}^N) = \tfrac{1}{2} \sum_{i \neq j} U(\mathbf{X}_i, \mathbf{X}_j) \tag{3.171}$$

Now, suppose we can separate the pair potential into two parts

$$U(\mathbf{X}_i, \mathbf{X}_j) = U^0(\mathbf{X}_i, \mathbf{X}_j) + U^1(\mathbf{X}_i, \mathbf{X}_j) \tag{3.172}$$

Hence,

$$U(\mathbf{X}^N) = U^0(\mathbf{X}^N) + U^1(\mathbf{X}^N) \tag{3.173}$$

$U^0(\mathbf{X}^N)$ is referred to as the total potential energy of the unperturbed system, whereas $U^1(\mathbf{X}^N)$ is considered to be the perturbation energy. The basic distribution function in the unperturbed system is

$$P^0(\mathbf{X}^N) = \frac{\exp[-\beta U_N^0(\mathbf{X}^N)]}{\int \cdots \int d\mathbf{X}^N \exp[-\beta U_N^0(\mathbf{X}^N)]} \tag{3.174}$$

and the Helmholtz free energy of the perturbed and unperturbed systems is given by

$$\exp[-\beta A(T, V, N)] = (q^N/\Lambda^{3N} N!) \int \cdots \int d\mathbf{X}^N \exp[-\beta U_N(\mathbf{X}^N)] \tag{3.175}$$

$$\exp[-\beta A^0(T, V, N)] = (q^N/\Lambda^{3N} N!) \int \cdots \int d\mathbf{X}^N \exp[-\beta U_N^0(\mathbf{X}^N)] \tag{3.176}$$

Defining the difference

$$A^1(T, V, N) = A(T, V, N) - A^0(T, V, N) \tag{3.177}$$

## Molecular Distribution Functions and Thermodynamics

and using relations (3.173)–(3.176), we get

$$\exp[-\beta A^1(T, V, N)] = \frac{\int \cdots \int d\mathbf{X}^N \exp[-\beta U_N^0(\mathbf{X}^N) - \beta U_N^1(\mathbf{X}^N)]}{\int \cdots \int d\mathbf{X}^N \exp[-\beta U_N^0(\mathbf{X}^N)]}$$

$$= \int \cdots \int d\mathbf{X}^N P^0(\mathbf{X}^N) \exp[-\beta U_N^1(\mathbf{X}^N)]$$

$$= \langle \exp[-\beta U_N^1(\mathbf{X}^N)] \rangle_0 \qquad (3.178)$$

The symbol $\langle \ \rangle_0$ stands for an average over the unperturbed system (here, in the $T$, $V$, $N$ ensemble). Clearly, even when we assume the pairwise additivity (3.171), the average in (3.178) is not an average of a pairwise quantity. However, if the perturbation energy $U_N^1(\mathbf{X}^N)$ is considered to be "small" compared with $kT$, we can expand the exponential function on the rhs of (3.178) to obtain

$$\exp[-\beta U_N^1(\mathbf{X}^N)] = 1 - \beta U_N^1(\mathbf{X}^N) + \cdots \qquad (3.179)$$

For the present purposes, we assume that $|\beta U_N^1(\mathbf{X}^N)| \ll 1$ for all possible configurations $\mathbf{X}^N$, so that expansion up to first order as in (3.179) is justified. We now have

$$\langle \exp[-\beta U_N^1(\mathbf{X}^N)] \rangle_0 = 1 - \beta \langle U_N^1(\mathbf{X}_N) \rangle_0 + \cdots \qquad (3.180)$$

where the average on the rhs of (3.180) is over a pairwise quantity. Hence, using (3.171) and the general theorem (3.4), we get

$$\langle \exp[-\beta U_N^1(\mathbf{X}^N)] \rangle = 1 - \tfrac{1}{2}\beta \int d\mathbf{X}_1 \int d\mathbf{X}_2 \, U^1(\mathbf{X}_1, \mathbf{X}_2)\varrho^{0(2)}(\mathbf{X}_1, \mathbf{X}_2) \quad (3.181)$$

where $\varrho^{0(2)}(\mathbf{X}_1, \mathbf{X}_2)$ is the pair distribution function for the unperturbed system. Thus, for the Helmholtz free energy of the system, we have

$$A(T, V, N) = A^0(T, V, N) + A^1(T, V, N)$$

$$= A^0(T, V, N) - kT \ln \langle \exp[-\beta U_N^1(\mathbf{X}^N)] \rangle_0$$

$$= A^0(T, V, N) - kT \ln[1 - \beta \langle U_N^1(\mathbf{X}^N) \rangle_0 + \cdots]$$

$$= A^0(T, V, N) + \langle U_N^1(X^N) \rangle_0 + \cdots$$

$$\approx A^0(T, V, N) + \tfrac{1}{2} \int d\mathbf{X}_1 \int d\mathbf{X}_2 \, U^1(\mathbf{X}_1, \mathbf{X}_2)\varrho^{0(2)}(\mathbf{X}_1, \mathbf{X}_2) \quad (3.182)$$

The last relation can be simplified for spherical particles as

$$A(T, V, N) \approx A^0(T, V, N) + \tfrac{1}{2}N\varrho \int_0^\infty U^1(R)g^0(R)4\pi R^2 \, dR \qquad (3.183)$$

where $g^0(R)$ is the pair correlation function for the unperturbed system. Relation (3.183) is useful whenever we know the free energy of the unperturbed system and when the perturbation energy is small compared with $kT$. It is clear that if we take more terms in the expansion (3.179), we end up with integrals involving higher-order molecular distribution functions. Therefore, such an expansion is useful only for the cases discussed in this section. For a recent review on the application of perturbation theories to liquids see Barker and Henderson (1972).

*Chapter 4*
# Theory of Solutions

## 4.1. INTRODUCTION

This chapter is concerned with a few aspects of the theory of solutions which are either of fundamental character or are believed to be particularly useful in the study of aqueous solutions. With these limitations in mind, we have eliminated, for instance, discussion of lattice theories of solutions,[1] which have played an important role in the development of the thermodynamics of mixing. We also confine ourselves to first-order expansion theories, which, for all practical purposes, are the most useful parts of the complete and general schemes.

The bulk of the subject matter of Section 4.2 is concerned with concepts and relationships obtainable by direct generalization of the corresponding topics in one-component systems. We therefore omit detailed discussion whenever the generalization procedure is obvious.

One of the most fruitful and ubiquitous concepts which appears in the theory of solutions is the concept of "ideality." In fact, there exist three fundamentally different kinds of ideal mixtures. We shall elaborate on these throughout the chapter.

In Section 4.5, we develop the Kirkwood–Buff theory of solutions as our central tool, from which we derive most of our subsequent relations and conclusions. In fact, because of its extreme generality, the Kirkwood–Buff theory is also the most adequate for handling aqueous solutions. The various advantages of this theory are summarized in the beginning of Sec. 4.5.

---

[1] In Chapters 6 and 7, we discuss some aspects of the interstitial models for water. These may be classified formally as lattice models. However, in our application, we shall stress the "mixture-model" aspect, rather than the lattice aspect of the model.

In addition, because of its relative simplicity, we have used the Kirkwood–Buff theory for characterization of the various ideal solutions, as well as for the study of first-order deviations from ideal solutions. Although formal theories, such as that of McMillan and Mayer (1945), exist which provide expressions for higher-order deviations from ideality, their practical usefulness is limited to first-order terms only. Higher-order terms usually involve higher-order molecular distribution functions, about which little is known.

## 4.2. MOLECULAR DISTRIBUTION FUNCTIONS IN MIXTURES; DEFINITIONS

Before embarking on the concepts of molecular distribution functions in mixtures, it is appropriate to digress to a brief discussion of the intermolecular potential function for two particles of different species. We denote by $U_{AB}(\mathbf{X}', \mathbf{X}'')$ the work required to bring two molecules of species $A$ and $B$ from infinite separation to the final configuration $\mathbf{X}', \mathbf{X}''$. (We adopt the convention that the first vector in the parentheses, $\mathbf{X}'$, describes the configuration of the first species $A$; similarly, $\mathbf{X}''$ describes the configuration of the second species $B$. This convention will be applied for any other pairwise function as well.)

For spherical molecules, $U_{AB}(\mathbf{R}', \mathbf{R}'')$ is a function of the separation $R = |\mathbf{R}'' - \mathbf{R}'|$ only. Hence, it is clear that

$$U_{AB}(R) = U_{BA}(R) \tag{4.1}$$

In principle, the function $U_{AB}(R)$ does not have to bear any resemblence to the corresponding functions $U_{AA}(R)$ and $U_{BB}(R)$ of the pure substances $A$ and $B$, respectively. However, it is convenient, when working with, say, Lennard-Jones particles, to adopt the following "combination rule" [for more details, see Prigogine (1957)]:

$$\sigma_{AB} = \tfrac{1}{2}(\sigma_{AA} + \sigma_{BB}) = \sigma_{BA} \tag{4.2}$$

$$\varepsilon_{AB} = (\varepsilon_{AA}\varepsilon_{BB})^{1/2} = \varepsilon_{BA} \tag{4.3}$$

Relation (4.2) is expected to be "exact" for hard spheres.[2] We adopt

---

[2] The word "exact" used here does not imply that we can prove this statement exactly. Given $U_{AA}(R)$ and $U_{BB}(R)$ for hard spheres with diameters $\sigma_{AA}$ and $\sigma_{BB}$, respectively, the function $U_{AB}(R)$ is still undefined. The most plausible definition of $U_{AB}(R)$ would be

$$U_{AB}(R) = \begin{cases} \infty & \text{for } R < \tfrac{1}{2}(\sigma_{AA} + \sigma_{BB}) \\ 0 & \text{for } R \geq \tfrac{1}{2}(\sigma_{AA} + \sigma_{BB}) \end{cases}$$

it also for cases where $\sigma_{AA}$ and $\sigma_{BB}$ represent only effective hard-core diameters. Relation (4.3) is also adopted basically on grounds of convenience, although it has some theoretical justifications which will not concern us here.

The situation becomes more complex when dealing with nonspherical particles, in which case there is no simple way of relating $U_{AB}$ to $U_{AA}$ and $U_{BB}$. (As always in this book, we assume that the molecules are *rigid*, so that the configuration of each molecule is completely specified by six coordinates, say $\mathbf{X}' = x', y', z', \phi', \theta', \psi'$. The configuration of the pair requires 12 coordinates $\mathbf{X}'$, $\mathbf{X}''$. However, six of these are redundant since we can always translate and rotate the pair, keeping the relative configuration fixed.)

A system of two components $A$ and $B$ with composition $N_A$ and $N_B$, respectively, in a specified configuration $\mathbf{X}^{N_A}$, $\mathbf{X}^{N_B}$ has a total interaction energy

$$U_{N_A,N_B}(\mathbf{X}^{N_A}, \mathbf{X}^{N_B}) = \tfrac{1}{2} \sum_{i \neq j} U_{AA}(\mathbf{X}_i, \mathbf{X}_j) + \tfrac{1}{2} \sum_{i \neq j} U_{BB}(\mathbf{X}_i, \mathbf{X}_j)$$

$$+ \sum_{i=1}^{N_A} \sum_{j=1}^{N_B} U_{AB}(\mathbf{X}_i, \mathbf{X}_j) \qquad (4.4)$$

Here, we have assumed pairwise additivity of the total potential energy, and adopted the convention that the order of arguments in the parentheses corresponds to the order of species as indicated by the subscript of $U$. Thus, $\mathbf{X}_j$ in the first sum on the rhs of (4.4) is the configuration of the $j$th molecule ($j = 1, 2, \ldots, N_A$) of species $A$, whereas $\mathbf{X}_j$ in the last term on the rhs of (4.4) stands for the $j$th molecule ($j = 1, 2, \ldots, N_B$) of species $B$.

Some care should be exercised when specifying the configuration of the whole system, which is symbolized in (4.4) by $\mathbf{X}^{N_A}$, $\mathbf{X}^{N_B}$. To make this more explicit, we can choose different conventions, such as $(\mathbf{X}_1^A, \mathbf{X}_2^A, \ldots, \mathbf{X}_{N_A}^A, \mathbf{X}_1^B, \mathbf{X}_2^B, \ldots, \mathbf{X}_{N_B}^B)$, $(\mathbf{X}_1, \mathbf{X}_2, \ldots, \mathbf{X}_{N_A}, \mathbf{Y}_1, \mathbf{Y}_2, \ldots, \mathbf{Y}_{N_B})$, or $(\mathbf{X}_1, \mathbf{X}_2, \ldots, \mathbf{X}_{N_A}, \mathbf{X}_{N_A+1}, \mathbf{X}_{N_A+2}, \ldots, \mathbf{X}_{N_A+N_B})$. In the first choice, we use different superscripts for the two species; in the second, we use different symbols for the vectors, i.e., $\mathbf{X}$ for $A$ and $\mathbf{Y}$ for $B$; and in the last, we use consecutive indices to designate the various species, i.e., $1, 2, \ldots, N_A$ for $A$, and $N_A + 1, N_A + 2, \ldots, N_A + N_B$ for $B$.

In the following sections, we shall not adhere to any particular choice of convention, but shall adopt the most convenient one for the particular case under consideration.

The basic probability density in the canonical ensemble is

$$P(\mathbf{X}^{N_A+N_B}) = P(\mathbf{X}^{N_A}, \mathbf{X}^{N_B})$$

$$= \frac{\exp[-\beta U_{N_A,N_B}(\mathbf{X}^{N_A}, \mathbf{X}^{N_B})]}{\int \cdots \int d\mathbf{X}^{N_A} d\mathbf{X}^{N_B} \exp[-\beta U_{N_A,N_B}(\mathbf{X}^{N_A}, \mathbf{X}^{N_B})]} \quad (4.5)$$

where obvious shorthand notation such as $\mathbf{X}^{N_A+N_B}$ and $d\mathbf{X}^{N_B}$ has been applied. The singlet distribution function for the $A$ species is defined, in complete analogy with the definition in Section 2.2, by

$$\varrho_A^{(1)}(\mathbf{X}') = \int \cdots \int d\mathbf{X}^{N_A+N_B} P(\mathbf{X}^{N_A+N_B}) \sum_{i=1}^{N_A} \delta(\mathbf{X}_i^A - \mathbf{X}')$$

$$= N_A \int \cdots \int d\mathbf{X}^{N_A+N_B} P(\mathbf{X}^{N_A+N_B}) \delta(\mathbf{X}_1^A - \mathbf{X}') \quad (4.6)$$

and similarly

$$\varrho_B^{(1)}(\mathbf{X}') = N_B \int \cdots \int d\mathbf{X}^{N_A+N_B} P(\mathbf{X}^{N_A+N_B}) \delta(\mathbf{X}_1^B - \mathbf{X}') \quad (4.7)$$

As in the case of a one-component system, $\varrho_A^{(1)}(\mathbf{X}')$ is the average density of $A$ molecules in the configuration $\mathbf{X}'$. In a homogeneous and isotropic fluid, we have (see Section 2.2 for more details)

$$\varrho_A^{(1)}(\mathbf{X}') = N_A/V 8\pi^2 \quad (4.8)$$

$$\varrho_B^{(1)}(\mathbf{X}') = N_B/V 8\pi^2 \quad (4.9)$$

The average local density of $A$ molecules at $\mathbf{R}'$ is defined by[3]

$$\varrho_A^{(1)}(\mathbf{R}') = \int d\mathbf{\Omega}' \, \varrho_A^{(1)}(\mathbf{X}') = N_A/V \quad (4.10)$$

and a similar definition applies to $\varrho_B^{(1)}(\mathbf{R}')$.

In a similar fashion, one defines the pair distribution functions for the four different pairs $AA$, $AB$, $BA$, and $BB$. For instance,

$$\varrho_{AA}^{(2)}(\mathbf{X}', \mathbf{X}'')$$

$$= \int \cdots \int d\mathbf{X}^{N_A+N_B} P(\mathbf{X}^{N_A+N_B}) \sum_{i \neq j} \delta(\mathbf{X}_i^A - \mathbf{X}') \delta(\mathbf{X}_j^A - \mathbf{X}'')$$

$$= N_A(N_A - 1) \int \cdots \int d\mathbf{X}^{N_A+N_B} P(\mathbf{X}^{N_A+N_B}) \delta(\mathbf{X}_1^A - \mathbf{X}') \delta(\mathbf{X}_2^A - \mathbf{X}'') \quad (4.11)$$

---

[3] As in Section 2.2, we use $\varrho_A^{(1)}$ to denote two different functions, $\varrho_A^{(1)}(\mathbf{R}')$ and $\varrho_A^{(1)}(\mathbf{X}')$.

# Theory of Solutions

and, for different species,

$$\varrho_{AB}^{(2)}(\mathbf{X}', \mathbf{X}'')$$

$$= \int \cdots \int d\mathbf{X}^{N_A+N_B} P(\mathbf{X}^{N_A+N_B}) \sum_{i=1}^{N_A} \sum_{j=1}^{N_B} \delta(\mathbf{X}_i^A - \mathbf{X}') \delta(\mathbf{X}_j^B - \mathbf{X}'')$$

$$= N_A N_B \int \cdots \int d\mathbf{X}^{N_A+N_B} P(\mathbf{X}^{N_A+N_B}) \delta(\mathbf{X}_1^A - \mathbf{X}') \delta(\mathbf{X}_1^B - \mathbf{X}'') \quad (4.12)$$

The pair correlation functions $g_{\alpha\beta}(\mathbf{X}', \mathbf{X}'')$ (where $\alpha$ and $\beta$ can be either $A$ or $B$) are defined by

$$\varrho_{\alpha\beta}^{(2)}(\mathbf{X}', \mathbf{X}'') = \varrho_{\alpha}^{(1)}(\mathbf{X}')\varrho_{\beta}^{(1)}(\mathbf{X}'')g_{\alpha\beta}(\mathbf{X}', \mathbf{X}'') \quad (4.13)$$

and the spatial pair correlation functions by

$$g_{\alpha\beta}(\mathbf{R}', \mathbf{R}'') = [1/(8\pi^2)^2] \int d\mathbf{\Omega}' \int d\mathbf{\Omega}'' g_{\alpha\beta}(\mathbf{X}', \mathbf{X}'') \quad (4.14)$$

As in a one-component system, the functions $g_{\alpha\beta}(\mathbf{R}', \mathbf{R}'')$ depend only on the scalar distance $R = |\mathbf{R}'' - \mathbf{R}'|$. Hence, for the spatial pair correlation function, we must have

$$g_{AB}(R) = g_{BA}(R) \quad (4.15)$$

Next, the conditional distribution functions are introduced. For instance,

$$\varrho_{AB}(\mathbf{X}'/\mathbf{X}'') = \varrho_{AB}^{(2)}(\mathbf{X}', \mathbf{X}'')/\varrho_B^{(1)}(\mathbf{X}'') = \varrho_A^{(1)}(\mathbf{X}')g_{AB}(\mathbf{X}', \mathbf{X}'') \quad (4.16)$$

As in the one-component case, $\varrho_{AB}(\mathbf{X}'/\mathbf{X}'')$ may be interpreted as the density of $A$ particles in configuration $\mathbf{X}'$, given a $B$ particle in a fixed configuration $\mathbf{X}''$. (For more details, see Sections 2.3 and 2.4.) In a two-component system, we have *four* conditional distribution functions corresponding to the four pairs of species $AA$, $AB$, $BA$, and $BB$.

Higher-order molecular distribution functions can be easily defined by a simple generalization of the corresponding definitions in a one-component system.

## 4.3. MOLECULAR DISTRIBUTION FUNCTIONS IN MIXTURES; PROPERTIES

Most of the properties of the molecular distribution functions discussed in Chapter 2 hold for mixtures as well. In this section, we dwell upon some new features that are specific to multicomponent systems. We shall be mainly interested in the properties of the various pair correlation functions in a mixture of two components of spherical molecules.

Let $A$ and $B$ be two simple spherical molecules interacting through pair potentials which we designate by $U_{AA}(R)$, $U_{AB}(R)$, and $U_{BB}(R)$. For simplicity, we may think of Lennard-Jones particles obeying the following relations:

$$U_{AA}(R) = 4\varepsilon_{AA}[(\sigma_{AA}/R)^{12} - (\sigma_{AA}/R)^6] \qquad (4.17)$$

$$U_{BB}(R) = 4\varepsilon_{BB}[(\sigma_{BB}/R)^{12} - (\sigma_{BB}/R)^6] \qquad (4.18)$$

$$U_{AB}(R) = U_{BA}(R) = 4\varepsilon_{AB}[(\sigma_{AB}/R)^{12} - (\sigma_{AB}/R)^6] \qquad (4.19)$$

with the additional requirements

$$\sigma_{AB} = \sigma_{BA} = \tfrac{1}{2}(\sigma_{AA} + \sigma_{BB}) \qquad (4.20)$$

$$\varepsilon_{AB} = \varepsilon_{BA} = (\varepsilon_{AA}\varepsilon_{BB})^{1/2} \qquad (4.21)$$

This is a convenient scheme of potential functions, by the use of which we shall illustrate some of the features that are novel to mixtures. Most of the arguments that will be given are more general, however, and do not necessarily depend on the choice of this particular scheme. Some features of the various pair correlation functions are similar to those in the one-component system; for instance,

$$g_{AB}(R \leq \sigma_{AB}) \approx 0 \qquad (4.22)$$

$$g_{AB}(R \to \infty) = 1 \qquad (4.23)$$

$$g_{AB}(R) \xrightarrow{\varrho \to 0} \exp[-\beta U_{AB}(R)] \qquad (4.24)$$

In the last relation, we require that the *total* density $\varrho = \varrho_A + \varrho_B$ tend to zero to validate the limiting behavior (4.24).

Before proceeding to mixtures at high densities, it is instructive to recall the density dependence of $g(R)$ for a one-component system (see Section 2.5). We have noticed that the second, third, etc. peaks of $g(R)$ develop as the density increases. The illustrations in Section 2.5 were given for Lennard-Jones particles with $\sigma = 1.0$, and increasing (number) density $\varrho$. It is clear, however, that the important parameter determining the form of $g(R)$ is the dimensionless quantity $\varrho\sigma^3$ (assuming for the moment that $\varepsilon/kT$ is fixed). This can be illustrated schematically with the help of Fig. 4.1. In the two boxes, we have the same number density, whereas the volume density, defined below, is quite different. Clearly, the behavior of these two

# Theory of Solutions

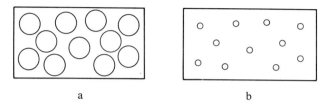

Fig. 4.1. Schematic illustration of two systems with the same *number* density but with different volume density. Clearly the system in (a) will behave like a "liquid," whereas (b) will show gaslike behavior.

systems will differ markedly even when $A$ and $B$ are hard spheres differing only by their diameters. Hence, the form of $g(R)$ will be quite different for these two systems. The reason is that in (a) the particles fill the volume of the vessel to a larger extent compared with (b). In other words, the average separation between the particles in (b) is larger than in (a) when measured in units of $\sigma$, although they are almost the same in absolute units.

Now, suppose we consider mixtures of $A$ and $B$ (in which $\sigma_{AA} \gg \sigma_{BB}$) at different compositions. If we study the dependence of, say, $g_{AB}(R)$ on the mole fraction $x_A$, we find that at $x_A \approx 1$, $g_{AB}(R)$ behaves as in the case of a high-density fluid, whereas at $x_A \approx 0$, we observe the behavior of the low-density fluid. In order to stress those effects specific to the properties of the mixtures, it is advisable to examine the variations of the pair correlation function with mole fraction when the total "volume density" is constant. The latter is defined as follows. In a one-component system of particles with effective diameter $\sigma$, the ratio of the volume occupied by the particles to the total volume of the system is[4]

$$\eta = \frac{N}{V}\frac{4\pi(\sigma/2)^3}{3} = \frac{\varrho\pi\sigma^3}{6} \qquad (4.25)$$

The total volume density of a mixture of two components $A$ and $B$ is similarly defined by

$$\eta = \tfrac{1}{6}\pi(\varrho_A\sigma_{AA}^3 + \varrho_B\sigma_{BB}^3) = \tfrac{1}{6}\pi\varrho(x_A\sigma_{AA}^3 + x_B\sigma_{BB}^3) \qquad (4.26)$$

In the second equation on the rhs of (4.26), we have expressed $\eta$ in terms of the total (number) density and the mole fractions. An extensive examina-

---

[4] The meaning of $\eta$ as a "volume density" is clear for hard spheres of diameter $\sigma$. For real molecules, this quantity measures the *effective* volume density of the system.

tion of the dependence of the various pair correlation functions on $\eta$ and on $x_A$ has been carried out by Throop and Bearman (1965, 1966) [further computations are reported by Grundke (1972) and Grundke et al. (1973)]. We shall illustrate some of the most salient features of this behavior for a system of Lennard-Jones particles obeying relations (4.17)–(4.21), with the parameters[5]

$$\sigma_{AA} = 1.0, \quad \sigma_{BB} = 1.5$$
$$\varepsilon_{AA}/kT = \varepsilon_{BB}/kT = 0.5, \quad \eta = 0.45$$
(4.27)

Figure 4.2 shows a set of pair correlation functions for this system at four mole fractions, $x_A = 0.99, 0.8, 0.4, 0.01$. Before we discuss some special features of these curves, we recall that the positions of the maxima of $g(R)$ for a one-component system occur roughly at integral multiples of $\sigma$. (This is not exactly so, as discussed in Section 2.5, first because of the random character of the packing of the spheres in the liquid, and second because the position of the minimum of the Lennard-Jones potential is at $R = 2^{1/6}\sigma$, which is slightly larger than $\sigma$ itself.) In Fig. 4.2, we have indicated the positions of all the maxima of the curves, but in the rest of our discussion, we refer only to the rounded figures which are integral multiples of $\sigma_{AA}$, $\sigma_{BB}$, or combinations of these.

First consider the set of functions $g_{AA}(R)$, $g_{AB}(R)$, $g_{BA}(R)$, and $g_{BB}(R)$ at $x_A = 0.99$. Here, $g_{AA}(R)$ is almost identical with the pair correlation function for pure $A$. The peaks occur at about $\sigma_{AA}$, $2\sigma_{AA}$, $3\sigma_{AA}$, and $4\sigma_{AA}$. Since $\eta = 0.45$ in (4.27) corresponds to quite a high density, we have four well-pronounced peaks. The function $g_{AB}(R)$ has the first peak at $\sigma_{AB}$. [The exact value of $\sigma_{AB}$ is $\frac{1}{2}(\sigma_{AA} + \sigma_{BB}) = 1.25$, but due to errors in the numerical computation and the fact that the minimum of $U_{AB}$ is at $2^{1/6}\sigma_{AB}$, we actually obtain a maximum at about $R = 1.3$.] The second, third, and fourth peaks are determined *not* by multiples of $\sigma_{AB}$, but by the addition of $\sigma_{AA}$.[6] That is, the maxima are at $R \approx \sigma_{AB}, \sigma_{AB} + \sigma_{AA}, \sigma_{AB} + 2\sigma_{AA}$, etc. This is an important feature of a dilute solution of $B$ in $A$, where the spacing

---

[5] All the computations for these illustrations were performed by a numerical solution of the Percus–Yevick equations. For more details on the numerical procedure, see Appendix 9-E.

[6] The second peak of $g_{AB}(R)$ is clearly related to $\sigma_{AB} + \sigma_{AA}$ and the third to $\sigma_{AB} + 2\sigma_{AA}$. If we had chosen $\sigma_{AA} = 1.0$ and $\sigma_{BB} = 2.0$, then we could not have distinguished between $\sigma_{AB} + 2\sigma_{AA}$ and $\sigma_{AB} + \sigma_{BB}$. It is for this reason that we have chosen the values of $\sigma_{AA} = 1.0$ and $\sigma_{BB} = 1.5$, which lead to less ambiguity in the interpretation of the first few peaks.

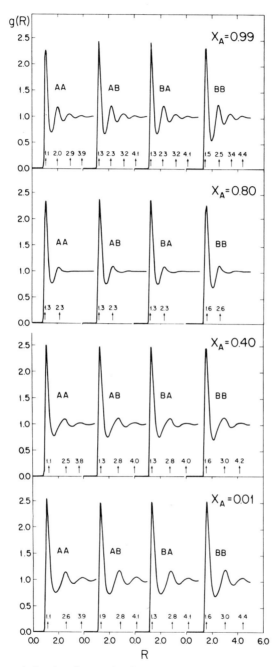

Fig. 4.2. Pair correlation functions $g_{\alpha\beta}(R)$ for two-component systems of Lennard-Jones particles with the parameters given in (4.27). The total volume density $\eta$ is the same for all the curves. The mole fraction of the component $A$ (with $\sigma_{AA} = 1.0$) is indicated in each row, where all the four $g_{\alpha\beta}(R)$ are shown. The pair of species is indicated next to each curve. The locations of the maxima are shown on the abscissa.

between the maxima is determined by $\sigma_{AA}$ (the diameter of the dominating species). The molecular reason for this is very simple. The spacing between, say, the first and the second peaks is determined by the size of the molecule that will most probably fill the space between the two molecules under observation. Because of the prevalence of $A$ molecules in this case, they are the most likely to fill the space between $A$ and $B$. The situation is depicted schematically in Fig. 4.3. In each row, we show the most likely filling of space between a pair of molecules for the case of $x_A \approx 1$, i.e., for a very dilute solution of $B$ in $A$. The first row shows the approximate locations of the first three peaks of $g_{AA}(R)$; other rows correspond successively to $g_{AB}(R)$, $g_{BA}(R)$, and $g_{BB}(R)$.

In Fig. 4.3, the dark circles correspond to the "solvent" molecules (i.e., the prevalent component $A$ with $\sigma_{AA} = 1.0$), whereas the open circles denote the molecules for which $g(R)$ is observed. The function $g_{BA}(R)$ in

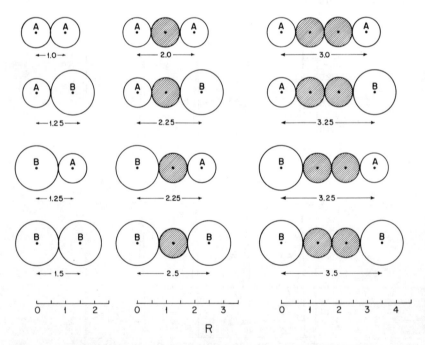

Fig. 4.3. Configurations corresponding to the first three peaks of $g_{\alpha\beta}(R)$ for a system of $B$ diluted in $A$ (e.g., $x_A = 0.99$ of Fig. 4.2). The two unshaded particles are the ones under "observation," i.e., these are the particles for which $g_{\alpha\beta}(R)$ is considered. The shaded particles, which here are invariably of species $A$, are the ones that fill the space between the observed particles. The locations of the expected peaks of $g_{\alpha\beta}(R)$ can be estimated with the help of the scale at the bottom of the figure.

**Theory of Solutions**

the present case must be exactly equal to $g_{AB}(R)$, as is quite evident in Fig. 4.2. [Because of the approximate nature of the computations (see Appendix 9-E), the curves $g_{AB}(R)$ and $g_{BA}(R)$ may come out a little differently; however, for most of our purposes, we shall view these two curves as identical, as is implied by the theory.]

The second row in Fig. 4.2 corresponds to $x_A = 0.80$. The most remarkable change is the almost complete disappearance of the third and fourth peaks. The second peak is less pronounced than in the case of $x_A = 0.99$. This is a very interesting feature and we return to a further elaboration of it later. We note in connection with this case that the separation between the peaks is still dominated by $\sigma_{AA}$, which is in accordance with the expected behavior at composition $x_A = 0.80$. In the third row, we have the curves for $x_A = 0.40$. Here, the separation between the peaks is determined by $\sigma_{BB}$ (though it is not very clear for the separation between the second and the third peaks). The last row corresponds to $x_A = 0.01$, i.e., $A$ diluted in $B$. Clearly, the separation between the peaks is determined by $\sigma_{BB}$, since $B$ is now the prevalent component.

Figure 4.4 shows a more detailed composition dependence of $g_{AA}(R)$ in the region $1.2 \leq R \leq 3.0$. The important point to be noted is the way the location of the second peak changes from about $\sigma_{AA} + \sigma_{AA}$ at $x_A = 0.99$ to about $\sigma_{AA} + \sigma_{BB}$ at $x_A = 0.01$. The height of the second peak is maximum for $x_A = 0.99$; it gradually decreases when composition changes, until at about $x_A = 0.65$, the curve is almost flat in the region between $\sigma_{AA} + \sigma_{AA}$ and $\sigma_{AA} + \sigma_{BB}$. When $x_A$ decreases further, a *new* peak starts to develop at $\sigma_{AA} + \sigma_{BB}$ which reaches a maximal value at $x_A = 0.01$.

It is important to stress the fact that the decay of the second peak of $g_{AA}(R)$ as a function of composition is *not* a result of a decrease in the density or in $\varepsilon/kT$ (as we witnessed in Section 2.5 for a one-component system). As a matter of fact, this is precisely the reason why we have chosen to keep the volume density, rather than the number density, fixed. With a fixed value of $\eta$, we get a pronounced structure for $g_{AA}(R)$ at both limiting cases of almost pure $A$ or $B$. Therefore, the peculiar behavior of $g_{AA}(R)$ as a function of $x_A$ must be attributed to a new feature that is specific to mixtures. To obtain further insight, we shall elaborate on the physical reason for this behavior. We recall that the location of the *second* peak is determined principally by the size of the particles that fill the space between the two "observed" particles. If $x_A = 0.99$, it is most likely that the filling of the space will be done by $A$ molecules. Similarly, for $x_A = 0.01$, it is most probable that $B$ molecules will be filling the space. The strong peak at $2\sigma_{AA}$ in the first case, and at $\sigma_{AA} + \sigma_{BB}$ in the second case, reflects the high

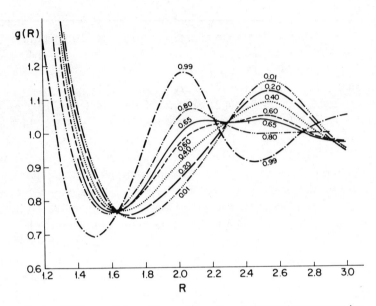

Fig. 4.4. A "closeup" view of the variation of $g_{AA}(R)$ in the important region $2\sigma_{AA} \lesssim R \lesssim \sigma_{AA} + \sigma_{BB}$ with the composition $x_A$ (as indicated next to each curve). Note that as $x_A$ decreases from 0.99 to 0.01, the second peak at $R \approx 2\sigma_{AA} = 2.0$ gradually decays. The function is almost flat at about $x_A \approx 0.65$. At lower mole fractions, a new peak evolves, at $R \approx \sigma_{AA} + \sigma_{BB} = 2.5$, and becomes more pronounced as $x_A \to 0$.

degree of certainty with which the system chooses the species for filling the space between any pair of observed particles. As the mole fraction of $A$ decreases, the $B$ molecules become competitive with $A$ for the "privilege" of filling the space. At about $x_A \approx 0.65$, $B$ is in a state of emulating $A$. (The fact that this occurs at $x_A \approx 0.65$ and not, say, at $x_A \approx 0.5$, is a result of the difference in $\sigma$ of the two components. Since $B$ is "larger" than $A$, its prevalence as volume occupant is effective at $x_B \approx 0.35 < 0.5$.) The fading of the second peak manifests the inability of the system to make a decision as to which kind of particle should be filling the space.

This whole argument holds for any other pair correlation function and also to third, fourth, etc., peaks. The implication of this behavior extends beyond mere competition between various species to fill the space between particles. We recall that the pair correlation function is related to the potential of average force (see Section 2.6). For the particular pair of species $AA$, we write

$$g_{AA}(R) = \exp[-\beta W_{AA}(R)] \tag{4.28}$$

Theory of Solutions 135

By extending the same arguments as in the case of a one-component system, we arrive at the conclusion that the gradient of $W_{AA}(R)$ is related to the average force between the two particles at distance $R$ from each other. The fact that $g_{AA}(R)$ is almost flat in the region $2\sigma_{AA} \leq R \leq \sigma_{AA} + \sigma_{BB}$ at $x_A \approx 0.65$ implies that the average *force* operating between the two particles is almost zero in this region, whereas strong attractive and repulsive forces occur in the same region for $x_A = 0.01$ and $x_A = 0.99$, respectively.

## 4.4. MIXTURES OF VERY SIMILAR COMPONENTS

In this section, we consider a system of two components in the $T$, $P$, $N_A$, $N_B$ ensemble. Similar arguments and results apply to multicomponent systems. We have chosen the $T$, $P$, $N_A$, $N_B$ ensemble because the isothermal–isobaric systems are the most common ones in actual experiments.

By *very similar* components we mean, in the present context, that the potential energy of interaction among a group of $n$ molecules in a configuration $\mathbf{X}^n$ is independent of the species we assign to each configuration $\mathbf{X}_i$. For example, the pair potential $U_{AA}(\mathbf{X}', \mathbf{X}'')$ is the same as the pair potential $U_{AB}(\mathbf{X}', \mathbf{X}'')$ or $U_{BB}(\mathbf{X}', \mathbf{X}'')$, provided that the configuration of the pair is the same in each case.[7] Clearly, we do not expect that this property will be fulfilled exactly for any pair of different real molecules. However, for molecules differing in, say isotopic constitution, this may hold to a good approximation.

The chemical potential of $A$ is defined by

$$\mu_A = \left(\frac{\partial G}{\partial N_A}\right)_{T,P,N_B} = G(T, P, N_A + 1, N_B) - G(T, P, N_A, N_B) \quad (4.29)$$

where G is the Gibbs free energy, and the last equality is valid by virtue of the reasoning given in Section 3.5.

---

[7] Although the meaning of this statement is obvious, some care is necessary for a more precise definition. It is implicitly understood that the centers and the convention of the orientational angles have been chosen in the *same way* for the two molecules $A$ and $B$. For instance, for $H_2O$ and $D_2O$, we adopt the same convention for $\mathbf{R}$ and $\mathbf{\Omega}$ for the two molecules, in wich case the pair potential will be almost the same for $H_2O$–$H_2O$ and $D_2O$–$D_2O$. It will differ markedly if we choose different conventions for the configurations of $H_2O$ and $D_2O$. If the two components $A$ and $B$ are very different, say $H_2O$ and $CH_4$, then the choice of a convention for $\mathbf{X}$ cannot be made in the "same way" for the two particles. In such cases, the concept of "very similar" is not expected to hold either, and the whole problem becomes irrelevant.

The connection between the chemical potential and statistical mechanics is (see Section 3.5 and Appendix 9-F for more details)

$$\exp(-\beta\mu_A) = \frac{q_A \int dV \int d\mathbf{X}^{N_A+1} d\mathbf{X}^{N_B} \exp[-\beta PV - \beta U_{N_A+1,N_B}(\mathbf{X}^{N_A+1}, \mathbf{X}^{N_B})]}{\Lambda_A^3 (N_A + 1) \int dV \int d\mathbf{X}^{N_A} d\mathbf{X}^{N_B} \exp[-\beta PV - \beta U_{N_A,N_B}(\mathbf{X}^{N_A}, \mathbf{X}^{N_B})]} \tag{4.30}$$

where $\Lambda_A^3$ and $q_A$ are the momentum and the internal partition function of an $A$ molecule, respectively. An obvious shorthand notation has been applied for the total potential energy of the system. The configuraton $\mathbf{X}^{N_A}$, $\mathbf{X}^{N_B}$ denotes the total configuration of $N_A$ molecules of type $A$ and $N_B$ molecules of type $B$.

Next, consider a system of $N$ particles of type $A$ *only*. The chemical potential for such a system (with the same $P$ and $T$ as before) is

$$\exp(-\beta\mu_A^p) = \frac{q_A \int dV \int d\mathbf{X}^{N+1} \exp[-\beta PV - \beta U_{N+1}(\mathbf{X}^{N+1})]}{\Lambda_A^3 (N + 1) \int dV \int d\mathbf{X}^N \exp[-\beta PV - \beta U_N(\mathbf{X}^N)]} \tag{4.31}$$

where we have denoted by $\mu_A^p$ the chemical potential of *pure A* at the same $P$ and $T$ as above.

Now let us choose $N$ in (4.31) equal to $N_A + N_B$ in (4.30). The assumption of *very similar* implies, according to its definition, the two equalities

$$U_{N+1}(\mathbf{X}^{N+1}) = U_{N_A+1,N_B}(\mathbf{X}^{N_A+1}, \mathbf{X}^{N_B}) \tag{4.32}$$

$$U_N(\mathbf{X}^N) = U_{N_A,N_B}(\mathbf{X}^{N_A}, \mathbf{X}^{N_B}) \tag{4.33}$$

Using (4.32) and (4.33) in (4.30) and (4.31), we get for the ratio of the latter pair of equations

$$\exp(-\beta\mu_A + \beta\mu_A^p) \approx (N + 1)/(N_A + 1) \tag{4.34}$$

Rearranging (4.34) and noting that for macroscopic systems

$$x_A = N_A/N \approx (N_A + 1)/(N + 1)$$

we get

$$\mu_A(T, P, x_A) = \mu_A^p(T, P) + kT \ln x_A \tag{4.35}$$

Here, we have expressed the chemical potential in terms of only intensive parameters $T$, $P$, and $x_A$.

Relation (4.35) is important since it gives an explicit dependence of the chemical potential on the composition, the fruitfulness of which was recognized long ago. We stress, however, that this relation has been obtained at the expense of the strong assumption that the two components are *very similar*. We know from experiment that a relation such as (4.35) holds also under much weaker conditions. We shall see in Section 4.6 that the assumption of "very similar" can be relaxed considerably and still yield relation (4.35). In fact, relation (4.35) could never have been so useful had it been restricted to the extreme case of very similar components.

The entire argument can be generalized easily to multicomponent systems of very similar molecules. The result would be the same as (4.35).

## 4.5. THE KIRKWOOD-BUFF THEORY OF SOLUTIONS

The Kirkwood–Buff theory of solutions (Kirkwood and Buff, 1951) provides new relations between thermodynamic quantities and molecular distribution functions. Moreover, these relations are very general and indeed enjoy all of the advantages that we listed in connection with the compressibility equation (Section 3.9). Because of its great importance, we shall recapitulate the main features of these relations that make them powerful. (1) The theory is valid for any kind of particle, not necessarily spherical. (2) Only the *spatial* pair correlation functions appear in the relations, even when the particles are not spherical. (3) No assumption of additivity of the total potential energy is invoked at any stage of the derivation, hence its more universal validity over theories that explicitly depend on the assumption of pairwise additivity. In particular, this tool is to be preferred in the treatment of complex fluids such as water and aqueous solutions. We demonstrate such applications in the following chapters. Here, we derive the formal relations. The arguments we use throughout are basically a generalization of those used in deriving the compressibility equation in Section 3.9.

Consider a grand canonical ensemble characterized by the variables $T$, $V$, and $\boldsymbol{\mu}$, where $\boldsymbol{\mu} = (\mu_1, \mu_2, \ldots, \mu_c)$ stands for the vector comprising the chemical potentials of all the $c$ components in the system.

The normalization conditions for the singlet and the pair distribution functions follow directly from their definitions. Here, we use the indices $\alpha$

and $\beta$ to denote the species, i.e., $\alpha, \beta = 1, 2, \ldots, c$. Hence, the two normalization conditions are

$$\int \varrho_\alpha^{(1)}(\mathbf{X}') \, d\mathbf{X}' = \langle N_\alpha \rangle \tag{4.36}$$

$$\int \varrho_{\alpha\beta}^{(2)}(\mathbf{X}', \mathbf{X}'') \, d\mathbf{X}' \, d\mathbf{X}'' = \begin{cases} \langle N_\alpha N_\beta \rangle & \text{if } \alpha \neq \beta \\ \langle N_\alpha(N_\alpha - 1) \rangle & \text{if } \alpha = \beta \end{cases}$$

$$= \langle N_\alpha N_\beta \rangle - \langle N_\alpha \rangle \delta_{\alpha\beta} \tag{4.37}$$

where the symbol $\langle \; \rangle$ stands for an average in the grand canonical ensemble. In (4.37), we must make a distinction between two cases: $\alpha \neq \beta$ and $\alpha = \beta$. The two cases can be combined into a single equation by using the Kronecker delta function $\delta_{\alpha\beta}$. For homogeneous and isotropic fluids, we also have the relations

$$\varrho_\alpha^{(1)}(\mathbf{X}') = \varrho_\alpha/8\pi^2 \tag{4.38}$$

$$\varrho_{\alpha\beta}^{(2)}(\mathbf{X}', \mathbf{X}'') = \varrho_\alpha \varrho_\beta g_{\alpha\beta}(\mathbf{X}', \mathbf{X}'')/(8\pi^2)^2 \tag{4.39}$$

where $\varrho_\alpha$ is the average number density of molecules of species $\alpha$, i.e., $\varrho_\alpha = \langle N_\alpha \rangle/V$. We also recall the definition of the *spatial* pair correlation function

$$g_{\alpha\beta}(\mathbf{R}', \mathbf{R}'') = (8\pi^2)^{-2} \int d\mathbf{\Omega}' \, d\mathbf{\Omega}'' \, g_{\alpha\beta}(\mathbf{X}', \mathbf{X}'') \tag{4.40}$$

which, as usual, is a function of the scalar distance $R = |\mathbf{R}'' - \mathbf{R}'|$.

From (4.36) and (4.37), we get

$$\int \varrho_{\alpha\beta}^{(2)}(\mathbf{X}', \mathbf{X}'') \, d\mathbf{X}' \, d\mathbf{X}'' - \int \varrho_\alpha^{(1)}(\mathbf{X}') \, d\mathbf{X}' \int \varrho_\beta^{(1)}(\mathbf{X}'') \, d\mathbf{X}''$$

$$= \int [\varrho_{\alpha\beta}^{(2)}(\mathbf{X}', \mathbf{X}'') - \varrho_\alpha^{(1)}(\mathbf{X}')\varrho_\beta^{(1)}(\mathbf{X}'')] \, d\mathbf{X}' \, d\mathbf{X}''$$

$$= \langle N_\alpha N_\beta \rangle - \langle N_\alpha \rangle \delta_{\alpha\beta} - \langle N_\alpha \rangle \langle N_\beta \rangle \tag{4.41}$$

Using relations (4.38)–(4.40), we can simplify (4.41) to

$$\varrho_\alpha \varrho_\beta \int [g_{\alpha\beta}(\mathbf{R}', \mathbf{R}'') - 1] \, d\mathbf{R}' \, d\mathbf{R}'' = \langle N_\alpha N_\beta \rangle - \langle N_\alpha \rangle \delta_{\alpha\beta} - \langle N_\alpha \rangle \langle N_\beta \rangle \tag{4.42}$$

We now define the quantity

$$G_{\alpha\beta} = \int_0^\infty [g_{\alpha\beta}(R) - 1] 4\pi R^2 \, dR \tag{4.43}$$

# Theory of Solutions

Combining (4.42) and (4.43), we get

$$G_{\alpha\beta} = V\left(\frac{\langle N_\alpha N_\beta\rangle - \langle N_\alpha\rangle\langle N_\beta\rangle}{\langle N_\alpha\rangle\langle N_\beta\rangle} - \frac{\delta_{\alpha\beta}}{\langle N_\alpha\rangle}\right) \quad (4.44)$$

This is a generalization of a similar relation for a one-component system derived in Section 3.9. Here, we have a connection between the cross fluctuations in the number of particles of various species and an integral involving only the *spatial* pair correlation functions for the corresponding pair of species $\alpha$ and $\beta$.

Next, we establish a connection between the fluctuations in the number of molecules and thermodynamic quantities. We start with the grand canonical partition function for a $c$-component system[8]

$$\Xi(T, V, \boldsymbol{\mu}) = \sum_{\mathbf{N}} Q(T, V, \mathbf{N}) \exp(\beta\boldsymbol{\mu} \cdot \mathbf{N}) \quad (4.45)$$

where $\mathbf{N} = N_1, N_2, \ldots, N_c$ and the summation is over each of the $N_i$ from zero to infinity. The exponential function includes the scalar product

$$\boldsymbol{\mu} \cdot \mathbf{N} = \sum_{i=1}^{c} \mu_i N_i \quad (4.46)$$

The average number of, say, $\alpha$ molecules in this ensemble is

$$\langle N_\alpha\rangle = \Xi^{-1} \sum_{\mathbf{N}} N_\alpha Q(T, V, \mathbf{N}) \exp(\beta\boldsymbol{\mu} \cdot \mathbf{N})$$

$$= kT\left[\frac{\partial \ln \Xi(T, V, \boldsymbol{\mu})}{\partial \mu_\alpha}\right]_{T,V,\boldsymbol{\mu}'_\alpha} \quad (4.47)$$

where $\boldsymbol{\mu}_\alpha'$ stands for the set $\mu_1, \mu_2, \ldots, \mu_c$ excluding $\mu_\alpha$.
Differentiating (4.47) with respect to $\mu_\beta$, we get

$$kT\left(\frac{\partial\langle N_\alpha\rangle}{\partial \mu_\beta}\right)_{T,V,\boldsymbol{\mu}'_\beta} = \Xi^{-1} \sum_{\mathbf{N}} N_\alpha N_\beta Q(T, V, \mathbf{N}) \exp(\beta\boldsymbol{\mu}\cdot\mathbf{N}) - \langle N_\alpha\rangle\langle N_\beta\rangle$$

$$= \langle N_\alpha N_\beta\rangle - \langle N_\alpha\rangle\langle N_\beta\rangle \quad (4.48)$$

By symmetry of the arguments with respect to interchanging the indices $\alpha$ and $\beta$, we get

$$kT\left(\frac{\partial\langle N_\alpha\rangle}{\partial \mu_\beta}\right)_{T,V,\boldsymbol{\mu}'_\beta} = kT\left(\frac{\partial\langle N_\beta\rangle}{\partial \mu_\alpha}\right)_{T,V,\boldsymbol{\mu}'_\alpha} = \langle N_\alpha N_\beta\rangle - \langle N_\alpha\rangle\langle N_\beta\rangle \quad (4.49)$$

---

[8] Note that $\beta$ in (4.45) stands for $(kT)^{-1}$. It is also used as a species index. There is no reason for confusion since in the latter case, $\beta$ always appears as a subscript.

Combining relations (4.49) with (4.44), we get

$$B_{\alpha\beta} \equiv \frac{kT}{V}\left(\frac{\partial\langle N_\alpha\rangle}{\partial\mu_\beta}\right)_{T,V,\mu'_\beta} = kT\left(\frac{\partial\varrho_\alpha}{\partial\mu_\beta}\right)_{T,\mu'_\beta} = \varrho_\alpha\varrho_\beta G_{\alpha\beta} + \varrho_\alpha\delta_{\alpha\beta} \quad (4.50)$$

Note that $G_{\alpha\beta} = G_{\beta\alpha}$, by virtue of the symmetry with respect to interchanging the $\alpha$ and $\beta$ indices in either (4.44) or (4.50). In fact, one can provide a direct argument based on the definition (4.43) of $G_{\alpha\beta}$, which employs only the spatial pair correlation function $g_{\alpha\beta}(R)$.

Relation (4.50) is already a connection between thermodynamics and molecular distribution functions. However, since the derivatives in (4.50) are taken at constant chemical potentials, these relations are of importance mainly in osmotic experiments. We are interested in derivatives at constant temperature and pressure. To obtain these requires some simple manipulations in partial derivatives. We define the elements of the matrix **A** by

$$A_{\alpha\beta} \equiv \frac{V}{kT}\left(\frac{\partial\mu_\alpha}{\partial\langle N_\beta\rangle}\right)_{T,V,N'_\beta} = \frac{1}{kT}\left(\frac{\partial\mu_\alpha}{\partial\varrho_\beta}\right)_{T,\varrho'_\beta} \quad (4.51)$$

where again we use $\mathbf{N}'_\beta$ and $\mathbf{\varrho}'_\beta$ to denote vectors, from which we have excluded the components $N_\beta$ and $\varrho_\beta$, respectively. Using the chain rule of differentiation, we get the obvious identities

$$\delta_{\alpha\gamma} = \left(\frac{\partial\mu_\alpha}{\partial\mu_\gamma}\right)_{T,\mu'_\gamma} = \sum_{\beta=1}^{c}\left(\frac{\partial\mu_\alpha}{\partial\varrho_\beta}\right)_{T,\varrho'_\beta}\left(\frac{\partial\varrho_\beta}{\partial\mu_\gamma}\right)_{T,\mu'_\gamma} = \sum_{\beta=1}^{c} A_{\alpha\beta}B_{\beta\gamma} \quad (4.52)$$

where the elements $B_{\alpha\beta}$ are defined in (4.50). Equation (4.52) can be written in matrix notation as

$$\mathbf{A} \cdot \mathbf{B} = \mathbf{I} \quad (4.53)$$

with **I** the unit matrix. From (4.53), we can solve for **A** if we know **B**. Taking the inverse of the matrix **B**, we get for the elements of **A** the relation

$$A_{\alpha\beta} = B^{\alpha\beta}/|\mathbf{B}| \quad (4.54)$$

where $B^{\alpha\beta}$ stands for the cofactor of the element $B_{\alpha\beta}$ in the determinant $|\mathbf{B}|$. [The cofactor of $B_{\alpha\beta}$ is obtained by eliminating the row and the column containing $B_{\alpha\beta}$ in the determinant $|\mathbf{B}|$, and multiplying[9] the result by $(-1)^{\alpha+\beta}$.] The existence of the inverse of the matrix **B** is equivalent to a

---

[9] Here $\alpha$ and $\beta$ must take numerical values, otherwise $(-1)^{\alpha+\beta}$ is meaningless. In the following applications, we shall take $\alpha$ and $\beta$ to stand for, say, components $A$ and $B$, respectively. In this case, we may assign the number 1, say, to $A$, and the number 2 to $B$,

# Theory of Solutions

condition of stability of the system. We shall not elaborate on this question any further, but merely point out that for a one-component system, this reduces to the statement that the derivative $(\partial \mu/\partial \varrho)_T$ is nonzero.

Since the $B_{\alpha\beta}$ are already expressible in terms of the $G_{\alpha\beta}$ through (4.50), relation (4.54) also connects $A_{\alpha\beta}$ with the molecular quantities $G_{\alpha\beta}$.

Next, we transform the volume, as an independent variable, into the pressure. This can be achieved via the thermodynamic identity[10]

$$\left(\frac{\partial \mu_\alpha}{\partial N_\beta}\right)_{T,V,N'_\beta} = \left(\frac{\partial \mu_\alpha}{\partial N_\beta}\right)_{T,P,N'_\beta} + \left(\frac{\partial \mu_\alpha}{\partial P}\right)_{T,N}\left(\frac{\partial P}{\partial N_\beta}\right)_{T,V,N'_\beta} \quad (4.55)$$

Using the identity

$$\left(\frac{\partial P}{\partial N_\beta}\right)_{T,V,N'_\beta}\left(\frac{\partial N_\beta}{\partial V}\right)_{T,P,N'_\beta}\left(\frac{\partial V}{\partial P}\right)_{T,N} = -1 \quad (4.56)$$

and the definition of the partial molar volume

$$\bar{V}_\alpha = \left(\frac{\partial V}{\partial N_\alpha}\right)_{T,P,N'_\alpha} = \left(\frac{\partial \mu_\alpha}{\partial P}\right)_{T,N} \quad (4.57)$$

we get from (4.55) the relation

$$\mu_{\alpha\beta} \equiv \left(\frac{\partial \mu_\alpha}{\partial N_\beta}\right)_{T,P,N'_\beta} = \left(\frac{\partial \mu_\alpha}{\partial N_\beta}\right)_{T,V,N'_\beta} - \frac{\bar{V}_\alpha \bar{V}_\beta}{V \varkappa_T} \quad (4.58)$$

where $\varkappa_T$ is the isothermal compressibility of the system

$$\varkappa_T = -\frac{1}{V}\left(\frac{\partial V}{\partial P}\right)_{T,N} \quad (4.59)$$

We now have all the necessary relations to express the thermodynamic quantities $\mu_{\alpha\beta}$, $\bar{V}_\alpha$, and $\varkappa_T$ in terms of the $G_{\alpha\beta}$. The general solution is quite involved. Therefore, we specialize to the case of two components $A$ and $B$.

Let us collect all the relations for this special case. From (4.58) and (4.54),

$$\mu_{\alpha\beta} = \frac{kT}{V}\frac{B^{\alpha\beta}}{|\mathbf{B}|} - \frac{\bar{V}_\alpha \bar{V}_\beta}{V \varkappa_T}, \quad \alpha = A, B;\ \beta = A, B \quad (4.60)$$

---

[10] From here on, for convenience of notation, we use $N_\alpha$ for $\langle N_\alpha \rangle$.

We have also the two Gibbs–Duhem relations

$$\varrho_A \mu_{AA} + \varrho_B \mu_{AB} = 0 \tag{4.61}$$

$$\varrho_A \mu_{AB} + \varrho_B \mu_{BB} = 0 \tag{4.62}$$

and the identity

$$\varrho_A \bar{V}_A + \varrho_B \bar{V}_B = 1 \tag{4.63}$$

Relations (4.60) to (4.63) constitute seven equations (the first one comprises four equations), from which we can solve for the seven thermodynamic quantities $\mu_{\alpha\beta}$ ($\alpha = A, B$; $\beta = A, B$), $\bar{V}_\alpha$ ($\alpha = A, B$), and $\varkappa_T$.

Before proceeding to solve these equations, we write the explicit form of the determinant

$$|\mathbf{B}| = \begin{vmatrix} \varrho_A + \varrho_A{}^2 G_{AA} & \varrho_A \varrho_B G_{AB} \\ \varrho_A \varrho_B G_{AB} & \varrho_B + \varrho_B{}^2 G_{BB} \end{vmatrix}$$

$$= \varrho_A \varrho_B [1 + \varrho_A G_{AA} + \varrho_B G_{BB} + \varrho_A \varrho_B (G_{AA} G_{BB} - G_{AB}^2)] \tag{4.64}$$

and the various cofactors

$$B^{AA} = \varrho_B + \varrho_B{}^2 G_{BB}, \quad B^{AB} = B^{BA} = -\varrho_A \varrho_B G_{AB}, \quad B^{BB} = \varrho_A + \varrho_A{}^2 G_{AA} \tag{4.65}$$

[Note the different meanings assigned to $B$ in (4.65).]

It will be convenient to write[11]

$$\eta = \varrho_A + \varrho_B + \varrho_A \varrho_B (G_{AA} + G_{BB} - 2G_{AB}) \tag{4.66}$$

$$\zeta = 1 + \varrho_A G_{AA} + \varrho_B G_{BB} + \varrho_A \varrho_B (G_{AA} G_{BB} - G_{AB}^2) \tag{4.67}$$

By straightforward algebra, we can solve Eqs. (4.60)–(4.63) and express all the thermodynamic quantities $\mu_{\alpha\beta}$, $\bar{V}_\alpha$ and $\varkappa_T$ in terms of the $G_{\alpha\beta}$:

$$\varkappa_T = \zeta/kT\eta \tag{4.68}$$

$$\bar{V}_A = [1 + \varrho_B(G_{BB} - G_{AB})]/\eta \tag{4.69}$$

$$\bar{V}_B = [1 + \varrho_A(G_{AA} - G_{AB})]/\eta \tag{4.70}$$

$$\mu_{AA} = \varrho_B kT/\varrho_A V\eta, \quad \mu_{BB} = \varrho_A kT/\varrho_B V\eta, \quad \mu_{AB} = \mu_{BA} = -kT/V\eta \tag{4.71}$$

---

[11] From the stability conditions of the system (Prigogine and Defay, 1954), it can be proven that $\eta > 0$ and $\zeta > 0$ always. The first follows from the stability condition applied to the chemical potential, say, in (4.71). We must have $\mu_{AB} < 0$, hence $\eta > 0$. Furthermore, since $\varkappa_T > 0$, it follows from (4.68) that $\zeta > 0$ also.

# Theory of Solutions

This completes the process of expressing the thermodynamic quantities in terms of the molecular quantities. Let us examine a few limiting cases. In the limit $\varrho_B \to 0$, we have

$$\lim_{\varrho_B \to 0} \eta = \varrho_A, \quad \lim_{\varrho_B \to 0} \zeta = 1 + \varrho_A G_{AA}^\circ \qquad (4.72)$$

Hence, (4.68) reduces to

$$\lim_{\varrho_B \to 0} \varkappa_T = (1 + \varrho_A G_{AA}^\circ)/kT\varrho_A \qquad (4.73)$$

which is just the compressibility equation for a one-component system ($G_{AA}^\circ$ being the limiting value of $G_{AA}$ as $\varrho_B \to 0$). Note, however, that relation (4.68) is not an "obvious" generalization of the one-component compressibility equation, and could not have been guessed intuitively. Similarly, from (4.69) and (4.70), we get

$$\lim_{\varrho_B \to 0} \bar{V}_A = \frac{1}{\varrho_A}, \quad \lim_{\varrho_B \to 0} \bar{V}_B = \frac{1 + \varrho_A^\circ(G_{AA}^\circ - G_{AB}^\circ)}{\varrho_A^\circ} \qquad (4.74)$$

Thus, for $A$, we simply get the molar (strictly molecular) volume of pure $A$, whereas for $B$, we get the partial molar volume at infinite dilution. The limiting behavior of the derivatives of the chemical potentials will be discussed in detail in Sections 4.8 and 4.9 of this chapter.

We now derive some relations which will prove useful in later applications of the theory. All of the following relations are obtainable by the application of simple identities between partial derivatives, such as

$$\varrho_A \left(\frac{\partial \mu_A}{\partial \varrho_B}\right)_{T,P} + \varrho_B \left(\frac{\partial \mu_B}{\partial \varrho_B}\right)_{T,P} = 0 \qquad (4.75)$$

$$\left(\frac{\partial \mu_A}{\partial \varrho_B}\right)_{T,\mu_B} \left(\frac{\partial \varrho_B}{\partial \mu_B}\right)_{T,\mu_A} \left(\frac{\partial \mu_B}{\partial \mu_A}\right)_{T,\varrho_B} = -1 \qquad (4.76)$$

$$\left(\frac{\partial \mu_B}{\partial \varrho_B}\right)_{T,P} = \left(\frac{\partial \mu_B}{\partial \varrho_B}\right)_{T,\mu_A} + \left(\frac{\partial \mu_B}{\partial \mu_A}\right)_{T,\varrho_B} \left(\frac{\partial \mu_A}{\partial \varrho_B}\right)_{T,P} \qquad (4.77)$$

From (4.75)–(4.77), we eliminate the required derivative at constant $P$ and $T$, namely

$$\left(\frac{\partial \mu_B}{\partial \varrho_B}\right)_{T,P} = \frac{\varrho_A(\partial \mu_B/\partial \varrho_B)_{T,\mu_A}(\partial \mu_A/\partial \varrho_B)_{T,\mu_B}}{\varrho_A(\partial \mu_A/\partial \varrho_B)_{T,\mu_B} - \varrho_B(\partial \mu_B/\partial \varrho_B)_{T,\mu_A}} \qquad (4.78)$$

On the rhs of (4.78), we have only quantities that are expressible in terms of the $G_{\alpha\beta}$ through (4.50). Carrying out this substitution yields

$$\left(\frac{\partial \mu_B}{\partial \varrho_B}\right)_{T,P} = \frac{kT}{\varrho_B(1 + \varrho_B G_{BB} - \varrho_B G_{AB})} = kT\left(\frac{1}{\varrho_B} - \frac{G_{BB} - G_{AB}}{1 + \varrho_B G_{BB} - \varrho_B G_{AB}}\right) \quad (4.79)$$

The second form on the rhs of (4.79) will turn out to be particularly useful for the study of very dilute solutions of $B$ in $A$.

Using (4.75), we also get

$$\left(\frac{\partial \mu_A}{\partial \varrho_B}\right)_{T,P} = -\frac{\varrho_B}{\varrho_A}\left(\frac{\partial \mu_B}{\partial \varrho_B}\right)_{T,P} = \frac{-kT}{\varrho_A(1 + \varrho_B G_{BB} - \varrho_B G_{AB})} \quad (4.80)$$

Similarly, if we interchange the roles of $A$ and $B$, we obtain

$$\left(\frac{\partial \mu_A}{\partial \varrho_A}\right)_{T,P} = \frac{kT}{\varrho_A(1 + \varrho_A G_{AA} - \varrho_A G_{AB})} \quad (4.81)$$

$$\left(\frac{\partial \mu_B}{\partial \varrho_A}\right)_{T,P} = \frac{-kT}{\varrho_B(1 + \varrho_A G_{AA} - \varrho_A G_{AB})} \quad (4.82)$$

Note that

$$\left(\frac{\partial \mu_B}{\partial \varrho_A}\right)_{T,P} \neq \left(\frac{\partial \mu_A}{\partial \varrho_B}\right)_{T,P} \quad (4.83)$$

The relation between these two derivatives can be obtained by taking the ratio of (4.80) and (4.82):

$$\left(\frac{\partial \mu_A}{\partial \varrho_B}\right)_{T,P} = \left(\frac{\partial \mu_B}{\partial \varrho_A}\right)_{T,P} \frac{\varrho_B(1 + \varrho_A G_{AA} - \varrho_A G_{AB})}{\varrho_A(1 + \varrho_B G_{BB} - \varrho_B G_{AB})} = \left(\frac{\partial \mu_B}{\partial \varrho_A}\right)_{T,P} \frac{\varrho_B}{\varrho_A} \frac{\bar{V}_B}{\bar{V}_A} \quad (4.84)$$

Another useful relation is

$$\left(\frac{\partial \varrho_A}{\partial \varrho_B}\right)_{T,P} = \frac{(\partial \varrho_A / \partial \mu_A)_{T,P}}{(\partial \varrho_B / \partial \mu_A)_{T,P}} = -\frac{1 + \varrho_A G_{AA} - \varrho_A G_{AB}}{1 + \varrho_B G_{BB} - \varrho_B G_{AB}} = -\frac{\bar{V}_B}{\bar{V}_A} \quad (4.85)$$

Finally, we obtain the derivative of the chemical potential with respect to the mole fraction via

$$\left(\frac{\partial \mu_A}{\partial x_A}\right)_{T,P} = \left(\frac{\partial \mu_A}{\partial \varrho_A}\right)_{T,P}\left(\frac{\partial \varrho_A}{\partial x_A}\right)_{T,P} = \left(\frac{\partial \mu_A}{\partial \varrho_A}\right)_{T,P}(\varrho_A + \varrho_B)\varrho_B \bar{V}_B \quad (4.86)$$

# Theory of Solutions

In the last form on the rhs of (4.86), we have used the identities

$$\left(\frac{\partial \varrho_A}{\partial x_A}\right)_{T,P} = N\left(\frac{\partial \varrho_A}{\partial N_A}\right)_{T,P,N} = N\frac{1 - \varrho_A \bar{V}_A}{V} = \varrho \varrho_B \bar{V}_B \quad (4.87)$$

where $N = N_A + N_B$ and $\varrho = \varrho_A + \varrho_B$. From (4.86), (4.70), and (4.81), we get

$$\left(\frac{\partial \mu_A}{\partial x_A}\right)_{T,P} = \frac{kT\varrho_B\varrho}{\varrho_A \eta} = kT\left(\frac{1}{x_A} - \frac{\varrho_B \varDelta_{AB}}{1 + \varrho_B x_A \varDelta_{AB}}\right) \quad (4.88)$$

where we used the notation

$$\varDelta_{AB} = G_{AA} + G_{BB} - 2G_{AB} \quad (4.89)$$

Relation (4.88) will be most useful for the study of symmetric ideal solutions carried out in the next section.

## 4.6. SYMMETRIC IDEAL SOLUTIONS; NECESSARY AND SUFFICIENT CONDITIONS

In Section 4.4 we showed that in a mixture of "very similar" components, the chemical potential of each component has the form

$$\mu_\alpha = \mu_\alpha^p + kT \ln x_\alpha \quad (4.90)$$

where $\mu_\alpha^p$ is the chemical potential of pure[12] species $\alpha$ (at the same temperature and pressure of the solution under observation) and $x_\alpha$ is its mole fraction. The same result holds for multicomponent systems and for mixtures at constant volume.

In this section, we confine ourselves to treatment of mixtures of two components $A$ and $B$, characterized by the thermodynamic variables $T$, $P$, $N_A$, $N_B$.

A symmetric ideal (SI) solution is defined as a solution for which the chemical potential of each component obeys relation (4.90) in the entire range of compositions, keeping $T$, $P$ constant, i.e.,

$$\mu_\alpha(T, P, x_\alpha) = \mu_\alpha^p(T, P) + kT \ln x_\alpha, \quad 0 \leq x_\alpha \leq 1; \ T, P \text{ constant} \quad (4.91)$$

Clearly, from the requirement that (4.91) hold in the entire range of com-

---

[12] The superscript p on $\mu_\alpha^p$ stands for *pure* species. We reserve the symbol $\mu_\alpha^\circ$ to denote standard chemical potential. See Section 4.8.

position, it follows that, in particular, we may put $x_\alpha = 1$; hence, the meaning of $\mu_\alpha^p$ as the chemical potential of the pure $\alpha$ is established.

For most practical applications, it is also useful to require that (4.91) hold also in a small neighborhood of $T$ and $P$ so that we may differentiate (4.91) with respect to these variables. There is no need to require the validity of (4.91) for *any* $T, P$, since no real mixture is expected to fulfill such an exaggerated requirement. In the context of the present section, we confine ourselves to the definition (4.91) for *fixed* temperature and pressure. This may not conform with a common practice in thermodynamics, where ideality is defined in terms of entropy and enthalpy of mixing. We shall comment further on this matter at the end of this section.

Since (4.91) holds for any $x_A$, we can differentiate it to obtain, say, for component $A$,

$$\left(\frac{\partial \mu_A}{\partial x_A}\right)_{T,P} = \frac{kT}{x_A}, \quad 0 \leq x_A \leq 1; \ T, P \text{ constant} \quad (4.92)$$

If (4.92) holds, then by integration we get

$$\mu_A = kT \ln x_A + c(T, P), \quad 0 \leq x_A \leq 1$$

The constant of integration $c$ is identified by substituting $x_A = 1$. Hence, we conclude that (4.91) and (4.92) are equivalent definitions of SI solutions.

We now recall the result (4.88) of the previous section:

$$\left(\frac{\partial \mu_A}{\partial x_A}\right)_{T,P} = kT\left(\frac{1}{x_A} - \frac{x_B \varrho \Delta_{AB}}{1 + \varrho x_A x_B \Delta_{AB}}\right) \quad (4.93)$$

where $\varrho = \varrho_A + \varrho_B$ is the total density and

$$\Delta_{AB} = G_{AA} + G_{BB} - 2G_{AB} \quad (4.94)$$

A comparison of (4.92) with (4.93) shows that at any finite density[13] $\varrho$, a necessary and a sufficient condition for a SI solution (in a binary system at $T, P$ constant) is

$$\Delta_{AB} = 0 \quad \text{for} \quad 0 \leq x_A \leq 1 \quad (4.95)$$

---

[13] The case of very low densities, $\varrho \to 0$, will be discussed separately in Section 4.10. Here, we are interested in solutions at liquid densities. (Note, however, that here $\varrho$ is not an independent variable; it is determined by $T, P, x_A$.) Also, we shall assume throughout that all the $G_{\alpha\beta}$ are finite quantities.

## Theory of Solutions

In the context of the present section, we emphasize the requirement $0 \leq x_A \leq 1$. A different kind of ideality is obtained in the limit of, say, $x_A \to 0$.

The condition (4.95) is clearly a sufficient condition for SI solutions, since by substitution in (4.93), we get (4.92). Conversely, in order that (4.92) and (4.93) be equivalent, we must have

$$x_B \varrho \, \Delta_{AB} = 0 \quad \text{for} \quad 0 \leq x_A \leq 1 \tag{4.96}$$

Since $\varrho$ is presumed to be nonzero, (4.96) implies (4.95), and hence (4.95) is also a necessary condition for SI solutions.

The condition (4.95) is a very general one for SI solutions. It should be recognized that this condition does not depend on any modelistic assumption for the solution. For instance, in simple lattice models of solutions[14] [see, for example, Guggenheim (1952)], one gets a *sufficient* condition for SI solutions of the form

$$W = W_{AA} + W_{BB} - 2W_{AB} = 0 \tag{4.97}$$

where $W_{\alpha\beta}$ are the interaction energies between the species $\alpha$ and $\beta$ situated on adjacent lattice points. We stress here that (4.97) is meaningful only within the context of a simple *lattice* model, and, furthermore, even then it is a *sufficient* but not a necessary condition.

We now define the concept of "similarity" between two components $A$ and $B$ whenever they fulfill the condition (4.95). We shall soon see that the concept of "similarity" defined here implies a far less stringent requirement on the two components than the concept of "very similar" defined in Section 4.4.

For simplicity, consider a system with pairwise additive potential (the conclusions at which we shall arrive are independent of this assumption). The condition of "very similar" implies that[15]

$$U_{AA} = U_{AB} = U_{BA} = U_{BB} \tag{4.98}$$

We have seen in Section 4.4 that (4.98) is a sufficient condition for a SI solution. Here, we shall arrive at the same conclusion by an indirect argument.

---

[14] The particles are presumed to have the "same" size, i.e., they may exchange lattice sites without disturbing the geometry of the lattice.
[15] Here, we do not include in the notation the configuration of the corresponding pair of species; the molecules are not necessarily spherical.

Clearly, condition (4.98) is expected to hold for almost identical components. If this were the only case in which ideality had been observed, the whole concept of a SI solution would have had little importance. We shall now see that condition (4.95) is a considerably weaker condition than (4.98). Let us examine the following series of conditions:

$$
\begin{aligned}
&\text{(a)} \ U_{AA} = U_{AB} = U_{BA} = U_{BB} \\
&\text{(b)} \ g_{AA} = g_{AB} = g_{BA} = g_{BB} \\
&\text{(c)} \ G_{AA} = G_{AB} = G_{BB} \\
&\text{(d)} \ G_{AA} + G_{BB} - 2G_{AB} = 0
\end{aligned}
\quad \text{(for } 0 \leq x_A \leq 1\text{)} \quad (4.99)
$$

Each of the conditions in (4.99) follows from its predecessor. Symbolically, we can write

$$\text{(a)} \Rightarrow \text{(b)} \Rightarrow \text{(c)} \Rightarrow \text{(d)} \qquad (4.100)$$

The first relation, (a) $\Rightarrow$ (b), follows directly from the formal definition of the pair correlation function. A qualitative argument is the following: $g_{\alpha\beta}$ governs the average density of molecules of species $\alpha$ at a certain relative configuration with respect to a molecule of species $\beta$. If the field of force produced by a $\beta$ molecule is the same as that produced by an $\alpha$ molecule, then the local density of molecules at any point is independent of the species of the molecule that produces the field of force.

The second relation, (b) $\Rightarrow$ (c), follows from the definition of $G_{\alpha\beta}$ [see Eq. (4.43)], and the third relation (c) $\Rightarrow$ (d) is obvious.

Since we have shown that the condition (d) is a *sufficient* condition for SI solutions, any condition that precedes (d) will also be a sufficient condition. In particular, (a) is a sufficient condition for SI solutions (a conclusion that was derived directly in Section 4.4).

It is very important to stress that, in general, the arrows in (4.100) may not be reversed. For instance, condition (c) implies an equality of the *integrals*, which is a far weaker requirement than equality of the *integrands* $g_{\alpha\beta}$ in (b) (there are an infinite number of integrands that produce the same value of the integral $G_{\alpha\beta}$). It is also obvious that (d) is much weaker than (c), i.e., the $G_{\alpha\beta}$ may be quite different and yet fulfill (d).

Furthermore, since the condition (d) was shown to be also a *necessary* condition for SI solutions, or symbolically

$$\text{SI} \Rightarrow \text{(d)} \qquad (4.101)$$

it must also be the weakest *sufficient* condition. This can be proven as follows: Suppose we are given a new sufficient condition (e) which is claimed to be weaker than (d), i.e.,

$$(d) \Rightarrow (e) \quad [(e) \text{ weaker than } (d)] \quad (4.102)$$

$$(e) \Rightarrow SI \quad [(e) \text{ is a sufficient condition for SI}] \quad (4.103)$$

It follows from (4.101) and (4.103) that

$$(e) \Rightarrow SI \Rightarrow (d) \quad (4.104)$$

Hence, comparing (4.102) and (4.104), we conclude that (e) and (d) must be equivalent.

We now elaborate on the meaning of the condition (4.95) for similarity between two components on a molecular level. First, we note that the concept of "very similar" defined in Section 4.4 is independent of temperature or pressure. This is not the case, however, for the concept of "similarity." To illustrate the situation, we consider a two-component system of Lennard-Jones particles with the following parameters:

$$\sigma_{AA} = 1.0, \quad \sigma_{BB} = 0.5, \quad \sigma_{AB} = \tfrac{1}{2}(\sigma_{AA} + \sigma_{BB})$$
$$\varepsilon_{AA}/kT = 0.5, \quad \varepsilon_{BB}/kT = 0.5, \quad \varepsilon_{AB} = (\varepsilon_{AA}\varepsilon_{BB})^{1/2} \quad (4.105)$$

For simplicity of the computation, we assume also that the total density is sufficiently low, so that

$$G_{\alpha\beta} = \int_0^\infty \{\exp[-\beta U_{\alpha\beta}(R)] - 1\} 4\pi R^2 \, dR \quad (4.106)$$

Clearly, if $\sigma_{BB}$ were chosen to be equal to $\sigma_{AA}$ in (4.105), we would have two "very similar" components. We have made the two components different by choosing different diameters in (4.105). The value of $\Delta_{AB}$ computed for this case is about 0.76; hence, the two components are not similar. We can now change the value of $\varepsilon_{BB}/kT$ and follow the dependence of $\Delta_{AB}$ on this parameter. Figure 4.5 shows the dependence of $\Delta_{AB}$ on $\varepsilon_{BB}/kT$, where all other parameters are as in (4.105). Note that at about $\varepsilon_{BB}/kT = 0.70$, we get $\Delta_{AB} = 0$. Thus, the two components, which may seem to be quite different in the conventional sense, are considered to be "similar" according to our definition. This illustrates the idea that the "energy" and "diameter" parameters in the Lennard-Jones potential can be adjusted in such a way that the resulting value of $\Delta_{AB}$ will be zero. In other words, the dissimilarity in the $\sigma_{\alpha\beta}$ can be compensated by a dissimi-

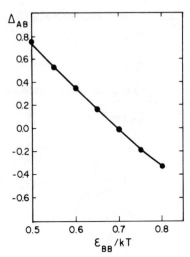

Fig. 4.5. Dependence of $\Delta_{AB}$ on $\varepsilon_{BB}/kT$ for a system of Lennard-Jones particles with parameters given in Eqs. (4.105). The computations were carried out for a system at very low density ($\varrho = \varrho_A + \varrho_B \approx 0.01$) for which the limiting form (4.106) is valid.

larity in the $\varepsilon_{\alpha\beta}$. Furthermore, if we take the final set of parameters for which we obtained $\Delta_{AB} = 0$ and change either $P$ or $T$, the "similarity" between the two particles will, in general, be destroyed. Note that although $\Delta_{AB} = 0$ at some specific $T$ and $P$, its temperature dependence may be quite different from zero. Such a system may not conform with the requirements of SI solutions in the thermodynamic sense (see below).

For more complex molecules, there can be more possibilities for compensations leading to SI behavior of the solution. For example, ethylene bromide and propylene bromide are certainly not "very similar." The very fact that their "size" is different excludes the possibility that their pair potentials will be identical, yet, experimentally, they exhibit the SI behavior [see, for instance, Guggenheim (1952)].

A deeper insight into the molecular significance of the quantity $\Delta_{AB}$ can be gained by looking at the meaning of the quantities $G_{\alpha\beta}$. We recall the definition of $G_{\alpha\beta}$ in (4.43):

$$G_{\alpha\beta} = \int_0^\infty [g_{\alpha\beta}(R) - 1] 4\pi R^2 \, dR \qquad (4.107)$$

Suppose that we are situated on an $A$ molecule, and observe the local

Theory of Solutions 151

densities in spherical shells around the center of this molecule. The local density of, say, $B$ molecules at a distance $R$ is $\varrho_B g_{BA}(R)$; hence, the average number of $B$ particles in a spherical shell of width $dR$ at distance $R$ from an $A$ particle is $\varrho_B g_{BA}(R) 4\pi R^2 \, dR$. On the other hand, $\varrho_B 4\pi R^2 \, dR$ is the average number of $B$ particles in the same spherical shell, the origin of which has been chosen at random (see also discussion in Section 2.5). Therefore, the quantity $\varrho_B [g_{BA}(R) - 1] 4\pi R^2 \, dR$ measures the excess (or deficiency) in the number of $B$ particles in a spherical shell of volume $4\pi R^2 \, dR$ around an $A$ molecule, relative to the number that would have been measured there using the bulk density $\varrho_B$. Hence, the quantity $\varrho_B G_{BA}$ is a sort of average excess of $B$ particles around $A$. Similarly, $\varrho_A G_{AB}$ is the average excess of $A$ particles around $B$. Thus, $G_{AB}$ is the average excess of $A$ (or $B$) particles around $B$ (or $A$) per unit density of $A$ (or $B$). Henceforth, we refer to $G_{AB}$ as a sort of "affinity" of $A$ toward $B$ (and vice versa).[16] A similar meaning is ascribed to $G_{AA}$ and $G_{BB}$.

Two comments are now in order. First, the quantities $G_{\alpha\beta}$ depend on the molecular properties of $\alpha$ and $\beta$ as well as on the macroscopic variables, say $T$ and $P$. Second, the integral usually gets its major contribution from within a short distance of a few molecular diameters [we exclude systems near the critical point, in which $g(R)$ may be of long range]. Hence, the concept of "affinity," as defined here, is concerned with the local environments of the molecules.

The conditions (4.95) for SI solutions can be stated as follows: The affinity of $A$ toward $B$ is the arithmetic average of the affinities of $A$ toward $A$ and $B$ toward $B$. This is true for all compositions $0 \leq x_A \leq 1$ and at a given $T$, $P$. In Table 4.1, we show the variation of all $G_{\alpha\beta}$ for the example cited above. Note that as we increase $\varepsilon_{BB}/kT$, the difference in the affinities $G_{AA} - G_{AB}$ is positive and decreasing, whereas the difference $G_{BB} - G_{AB}$ is negative and increasing (in absolute magnitude), so that at about $\varepsilon_{BB}/kT \approx 0.70$, we get $\Delta_{AB} = (G_{AA} - G_{AB}) + (G_{BB} - G_{AB}) = 0$. Note that in this particular example, $\Delta_{AB}$ is independent of composition. This follows from the particularly simple case of (4.106), in which $g_{\alpha\beta}(R) = \exp[-\beta U_{\alpha\beta}(R)]$. In general, $g_{\alpha\beta}$ and, hence, $G_{\alpha\beta}$ would be composition dependent.

We end this section by considering the phenomenological characterization of SI solutions by their partial molar entropies and enthalpies.

---

[16] The term "affinity" in this context is used as a short name for a long sentence, and should not be taken too literally. Certainly it should not be confused with the term affinity used in chemical reactions.

## Table 4.1

Variation of $G_{AA}$, $G_{BB}$, and $G_{AB}$ with $\varepsilon_{BB}/kT$ for Lennard-Jones Particles with Parameters Given in Eqs. (4.105)[a]

| $\varepsilon_{BB}/k$ | $G_{AA}$ | $G_{AB}$ | $G_{BB}$ | $G_{AA} - G_{AB}$ | $G_{BB} - G_{AB}$ | $\Delta_{AB}$ |
|---|---|---|---|---|---|---|
| 0.50 | 2.57 | 1.08 | 0.35 | 1.49 | −0.73 | 0.76 |
| 0.55 | 2.57 | 1.23 | 0.43 | 1.34 | −0.80 | 0.54 |
| 0.60 | 2.57 | 1.36 | 0.52 | 1.21 | −0.84 | 0.37 |
| 0.65 | 2.57 | 1.50 | 0.61 | 1.07 | −0.89 | 0.18 |
| 0.70 | 2.57 | 1.64 | 0.70 | 0.93 | −0.94 | −0.01 |
| 0.75 | 2.57 | 1.77 | 0.79 | 0.80 | −0.98 | −0.18 |
| 0.80 | 2.57 | 1.89 | 0.89 | 0.68 | −1.00 | −0.32 |

[a] At sufficiently low total density so that (4.106) is valid.

From the definition (4.90), we get by differentiation

$$\bar{S}_\alpha = -(\partial \mu_\alpha / \partial T)_P = S_\alpha^P - k \ln x_\alpha \qquad (4.108)$$

$$\bar{H}_\alpha = \mu_\alpha + T\bar{S}_\alpha = H_\alpha^P \qquad (4.109)$$

$$\bar{V}_\alpha = (\partial \mu_\alpha / \partial P)_T = V_\alpha^P \qquad (4.110)$$

where $S_\alpha^P$, $H_\alpha^P$, and $V_\alpha^P$ are the molar quantities of pure $\alpha$. It is easily verified that relations (4.108) and (4.109), if obeyed in the entire range of compositions, form an equivalent definition of SI solutions.

An alternative, although equivalent, way of describing ideal solutions phenomenologically is by introducing excess thermodynamic functions, defined by

$$G^E = G - G^{ideal} = G - \left[ \sum_{\alpha=1}^{c} N_\alpha(\mu_\alpha^P + kT \ln x_\alpha) \right] \qquad (4.111)$$

$$S^E = S - S^{ideal} = S - \left[ \sum_{\alpha=1}^{c} N_\alpha(S_\alpha^P - kT \ln x_\alpha) \right] \qquad (4.112)$$

$$V^E = V - V^{ideal} = V - \sum_{\alpha=1}^{c} N_\alpha V_\alpha^P \qquad (4.113)$$

$$H^E = H - H^{ideal} = H - \sum_{\alpha=1}^{c} N_\alpha H_\alpha^P \qquad (4.114)$$

# Theory of Solutions

The ideal solutions are characterized by zero excess thermodynamic functions. In thermodynamics, ideal solutions are often characterized by their excess entropy and enthalpy. Note that to obtain these, we need to assume differentiability of (4.91) with respect to temperature. This assumption is still quite weaker than the requirement that (4.91) be valid at *all* $T$ and $P$.

## 4.7. SMALL DEVIATIONS FROM SYMMETRIC IDEAL (SI) SOLUTIONS

Any solution of two components can be viewed as a SI solution, with a correction term which takes into account the extent of dissimilarity between the two components. More precisely, we can integrate relation (4.93) to obtain

$$\mu_A(T, P, x_A) = \mu_A^p(T, P) + kT \ln x_A + kT \int_0^{x_B} \frac{\varrho x_B' \Delta_{AB}}{1 + \varrho x_A' x_B' \Delta_{AB}} dx_B' \quad (4.115)$$

where we put[17] $dx_B' = -dx_A'$, and we note that in general $\varrho$ and $\Delta_{AB}$ are composition dependent. The constant of integration has been obtained by putting $x_B = 0$; hence, $\mu_A^p(T, P)$ is identified as the chemical potential of pure $A$ at the given $T$ and $P$. We can define an activity coefficient by

$$kT \ln \gamma_A^S = kT \int_0^{x_B} \frac{\varrho x_B' \Delta_{AB}}{1 + \varrho x_B' x_A' \Delta_{AB}} dx_B' \quad (4.116)$$

and rewrite (4.115) as

$$\mu_A(T, P, x_A) = \mu_A^p(T, P) + kT \ln(x_A \gamma_A^S) \quad (4.117)$$

The superscript S on $\gamma_A^S$ serves to indicate that this particular activity coefficient is for *symmetric ideal* behavior, and should be distinguished from other activity coefficients introduced later which have different meanings.

In the general case, (4.116) is not expected to be useful, since we know nothing about the analytical dependence of $\varrho$ or $\Delta_{AB}$ on composition. We therefore confine ourselves to a discussion of "first-order" deviations from SI solutions.

We recall that SI solutions are characterized by the condition $\Delta_{AB} = 0$. It is therefore natural to measure deviations from "similarity" by the parameter $\Delta_{AB}$. By first-order deviations from SI solutions, we refer to the

---

[17] The prime on $x_A$ and $x_B$ is used here only to distinguish these from the variables that are not subject to integration.

cases in which

$$\varrho x_A x_B \Delta_{AB} \ll 1 \quad \text{for} \quad 0 \leq x_A \leq 1 \quad (4.118)$$

[Note again that we exclude here the case of ideal gas mixtures attainable as $\varrho \to 0$ (see also Section 4.10). Also, since condition (4.118) is required to hold in the entire range of composition, it is, in fact, a condition on $\Delta_{AB}$ only.]

In practical cases, we have two components which are quite similar (in the usual sense) but not "similar" (in the sense of the definition of the previous section) and therefore we expect (4.118) to be a valid approximation. In fact, first-order theories of regular and athermal solutions are special cases of (4.118), although their phenomenological characterization is more general [for more details, see Guggenheim (1952)].

Expanding the integrand of (4.115) to first order in the parameter $\Delta_{AB}$, we get

$$\mu_A(T, P, x_A) = \mu_A{}^\text{p}(T, P) + kT \ln x_A + kT \int_0^{x_B} \varrho^\circ x_B{}' \Delta_{AB} \, dx_B{}' \quad (4.119)$$

where $\varrho^\circ$ is defined by

$$\varrho^\circ = \lim_{\Delta_{AB} \to 0} \varrho, \quad T, P, x_A \text{ constant} \quad (4.120)$$

and is the total density of the system at $\Delta_{AB} = 0$, keeping $T, P, x_A$ constant. Clearly, we cannot realize the limiting procedure of (4.120) in practice. We can, of course, do this for a theoretical model, or if we have a hypothetical series of mixtures at the same $T, P, x_A$ but having different values of $\Delta_{AB}$ that tend to zero.

A special and simple case occurs when $\varrho^\circ \Delta_{AB}$ is independent of composition (in the theory of lattice models of mixtures, a similar condition is assumed); hence, the integral in (4.119) can be evaluated immediately and we get

$$\mu_A(T, P, x_A) = \mu_A{}^\text{p}(T, P) + kT \ln x_A + \tfrac{1}{2} kT \varrho^\circ x_B{}^2 \Delta_{AB}$$
$$= \mu_A{}^\text{p}(T, P) + kT \ln(x_A \gamma_A{}^\text{S}) \quad (4.121)$$

where here $\gamma_A{}^\text{S}$ includes only the first-order deviations from SI solutions (for the special case presumed in (4.121)].

The limiting behavior of $\gamma_A{}^\text{S}$ in (4.121) is

$$\lim_{\Delta_{AB} \to 0} \gamma_A{}^\text{S} = 1, \quad T, P, x_A \text{ constant} \quad (4.122)$$

The meaning of the limiting process in (4.122) is the same as in (4.120).

Theory of Solutions

A mixture for which (4.121) is valid may be referred to as a simple mixture.[18] Within the realm of simple solutions, one may further distinguish between "regular" and "athermal" solutions as two special cases.

## 4.8. DILUTE IDEAL (DI) SOLUTIONS

In this section, we discuss a different class of ideal solutions which have been of central importance in the study of solution thermodynamics. We shall refer to a dilute ideal (DI) solution whenever one of the components is very dilute in the solvent. The term "very dilute" depends on the system under consideration, and we shall define it more precisely in what follows. The solvent may have a single component or be a mixture of several components. Here, however, we confine ourselves to two-component systems. The solute, say $A$, is the component diluted in the solvent $B$.

The fact that we make a distinction between the solute and the solvent requires separate discussion of the behavior of each of them. We shall be mainly concerned with the solute, and only briefly discuss the relevant relations for the solvent.

The characterization of a DI solution can be carried out along different but equivalent routes. In fact, we mentioned one example already in Section 3.6, where we introduced the chemical potential of a single particle in a solvent. Here, we have chosen the Kirkwood–Buff theory to provide the basic relations from which we derive the limiting behavior of DI solutions. The appropriate relations needed from Section 4.5 are (4.50), (4.79), and (4.54) which, when specialized to a two-component system, can be rewritten as

$$\left(\frac{\partial \mu_A}{\partial \varrho_A}\right)_{T,\mu_B} = \frac{kT}{\varrho_A^2 G_{AA} + \varrho_A} = kT\left(\frac{1}{\varrho_A} - \frac{G_{AA}}{1 + \varrho_A G_{AA}}\right) \tag{4.123}$$

$$\left(\frac{\partial \mu_A}{\partial \varrho_A}\right)_{T,P} = kT\left(\frac{1}{\varrho_A} - \frac{G_{AA} - G_{AB}}{1 + \varrho_A G_{AA} - \varrho_A G_{AB}}\right) \tag{4.124}$$

$$\left(\frac{\partial \mu_A}{\partial \varrho_A}\right)_{T,\varrho_B} = kT\left(\frac{1}{\varrho_A} - \frac{G_{AA} + \varrho_B(G_{AA}G_{BB} - G_{AB}^2)}{1 + \varrho_A G_{AA} + \varrho_B G_{BB} + \varrho_A \varrho_B(G_{AA}G_{BB} - G_{AB}^2)}\right) \tag{4.125}$$

[18] The term "simple mixture" was introduced by Guggenheim (1959). Originally, it was defined in terms of the excess Gibbs free energy. From (4.121) and (4.111), we get

$$G^{\mathrm{E}} = \tfrac{1}{2}kT\varrho^\circ(\varDelta_{AB})(N_A x_B^2 + N_B x_A^2) = w(T, P)x_A x_B$$

where $w$ is a function of $T$ and $P$ only. The last form for $G^{\mathrm{E}}$ has been employed by Guggenheim to define "simple" solutions.

where in each case we have separated the singular part, $\varrho_A^{-1}$, as a first term on the rhs. [This is shown explicitly in (4.123); the others can be obtained in a simple fashion from (4.79) and (4.54), respectively.]

Note that the response of the chemical potential to variations in the density $\varrho_A$ is different for each set of thermodynamic variables. The three derivatives in (4.123)–(4.125) correspond to three different processes. The first corresponds to a process in which the chemical potential of the solvent is kept constant (the temperature being constant in all three cases) and therefore is useful in the study of osmotic experiments. This is the simplest expression of the three, and it should be noted that if we simply drop the condition of $\mu_B$ constant, we get the appropriate derivative for the pure $A$ component system. This is not an accidental result; in fact, this is the case where strong resemblance exists between the behavior of the solute $A$ in a solvent $B$ under constant $\mu_B$ and a system $A$ in a vacuum which replaces the solvent. We return to this analogy later and compare the virial expansion of the pressure with the corresponding virial expansion of the osmotic pressure.

The second derivative, (4.124), is the most important one from the practical point of view, since it is concerned with a system under constant pressure. The third relation, (4.125), is concerned with a system under constant volume, which is rarely useful in practice.

A common feature of all the derivatives (4.123)–(4.125) is the $\varrho_A^{-1}$ divergence as $\varrho_A \to 0$ (which is the reason for the particular and convenient form in which we write them; note also that we always assume here that all the $G_{\alpha\beta}$ are finite quantities).

For sufficiently low solute density, $\varrho_A \to 0$, the first term on the rhs of each of Eqs. (4.123)–(4.125) is the dominant one; hence, we get the limiting form of these equations:

$$\left(\frac{\partial \mu_A}{\partial \varrho_A}\right)_{T,\mu_B} = \left(\frac{\partial \mu_A}{\partial \varrho_A}\right)_{T,P} = \left(\frac{\partial \mu_A}{\partial \varrho_A}\right)_{T,\varrho_B} = \frac{kT}{\varrho_A}, \qquad \varrho_A \to 0 \quad (4.126)$$

which, upon integration, yields

$$\mu_A(T, \mu_B, \varrho_A) = \mu_A^\circ(T, \mu_B) + kT \ln \varrho_A$$
$$\mu_A(T, P, \varrho_A) = \mu_A^\circ(T, P) + kT \ln \varrho_A \qquad (\varrho_A \to 0) \quad (4.127)$$
$$\mu_A(T, \varrho_B, \varrho_A) = \mu_A^\circ(T, \varrho_B) + kT \ln \varrho_A$$

A few comments regarding Eqs. (4.127) are now in order.

1. The precise condition $\varrho_A$ must satisfy for the approximations (4.126) to hold clearly depends on the independent variables we have chosen to describe the system. For instance, if $\varrho_A G_{AA} \ll 1$ in (4.123), then the term $G_{AA}/(1 + \varrho_A G_{AA})$ is negligible compared with $\varrho_A^{-1}$, and we may assume the validity of (4.126). The corresponding requirement for (4.124) is that $\varrho_A(G_{AA} - G_{AB}) \ll 1$, which is clearly different from the previous condition, if only because the latter depends on $G_{AA}$ as well as on $G_{AB}$. Similarly, the precise condition under which the limiting behavior of (4.125) is obtained involves all three $G_{\alpha\beta}$. We can define a DI solution, for each case, whenever $\varrho_A$ is sufficiently small that the limiting behavior of either (4.126) or (4.127) is valid.

2. Once the limiting behavior (4.127) has been attained, we see that all three equations have the same formal form, i.e., a constant of integration, independent of $\varrho_A$, and a term of the form $kT \ln \varrho_A$. This is quite a remarkable observation which holds only in this limiting case. This uniform appearance of the chemical potential already disappears in the first-order deviation from a DI solution, a topic which is discussed in the next section.

3. The quantities $\mu_A^\circ$ that appear in (4.127) are constants of integration, and, as such, depend on the thermodynamic variables, which are different in each case. We may refer to these quantities as the "standard chemical potentials" of $A$ in the corresponding set of thermodynamic variables. It is important to stress that these quantities, in contrast to $\mu_A^p$ of the previous section, are *not* chemical potentials of $A$ in any real system. We shall see in Section 4.11 that the difference between two standard chemical potentials may be taken as the difference between two chemical potentials. However, such a meaning may not be ascribed to each standard chemical potential separately. Therefore, in the meantime, we refer to $\mu_A^\circ$ merely as a constant of integration.

4. Instead of starting with relations (4.123)–(4.125), we could have started from relation (4.88) of the Kirkwood–Buff theory, namely

$$\left(\frac{\partial \mu_A}{\partial x_A}\right)_{T,P} = kT\left(\frac{1}{x_A} - \frac{\varrho_B \Delta_{AB}}{1 + \varrho_B x_A \Delta_{AB}}\right) \qquad (4.128)$$

(similar derivatives at constant $T, \mu_B$ or $T, \varrho_B$ can be obtained; we omit these for the present discussion).

A limiting behavior of (4.128) is obtained for $x_A \to 0$, in which case the first term on the rhs of (4.128) becomes dominant, i.e.,

$$\left(\frac{\partial \mu_A}{\partial x_A}\right)_{T,P} = \frac{kT}{x_A}, \qquad x_A \to 0 \qquad (4.129)$$

which, upon integration [in the region for which (4.129) is valid], yields

$$\mu_A(T, P, x_A) = \mu_A^*(T, P) + kT \ln x_A, \qquad x_A \to 0 \qquad (4.130)$$

Again, $\mu_A^*(T, P)$ is merely a constant of integration. It is different from $\mu_A^\circ(T, P)$ in (4.127) and therefore it is wise to use a different superscript to stress this difference. The exact relation between $\mu_A^\circ$ and $\mu_A^*$ can be obtained by noting that $x_A = \varrho_A/\varrho$, where $\varrho$ is the total density of the solution. Hence, from (4.130), we get

$$\begin{aligned}\mu_A &= \mu_A^*(T, P) + kT \ln \varrho_A - kT \ln \varrho \\ &= \mu_A^\circ(T, P) + kT \ln \varrho_A \end{aligned} \qquad (4.131)$$

Hence,

$$\mu_A^\circ(T, P) = \mu_A^*(T, P) - kT \ln \varrho_B^\circ \qquad (4.132)$$

where we put $\varrho_B^\circ$ in place of $\varrho$, which is permissible since as $\varrho_A \to 0$, $\varrho \to \varrho_B^\circ$. Hence, the total density is essentially the density of pure $B$ at the given $T$ and $P$. In a similar fashion, we can take other sets of thermodynamic variables, say $T$ and $\mu_B$, and derive relations analogous to (4.130)–(4.132). In addition, instead of $\varrho_A$ or $x_A$ as a concentration variable, we can use the molality of $A$, which is related to $x_A$ (for dilute solutions) by

$$m_A = 1000 x_A / M_B \qquad (4.133)$$

with $M_B$ the molecular weight of $B$. From (4.133) and (4.130), we get

$$\begin{aligned}\mu_A &= \mu_A^*(T, P) + kT \ln(M_B m_A / 1000) \\ &= [\mu_A^*(T, P) + kT \ln(M_B/1000)] + kT \ln m_A \\ &= \mu_A^{**}(T, P) + kT \ln m_A \end{aligned} \qquad (4.134)$$

where we have introduced a new standard chemical potential $\mu_A^{**}$, which is different from both $\mu_A^\circ$ and $\mu_A^*$.

Finally, we note that it is often tempting to identify $\mu_A^*$ of (4.130) with $\mu_A^p$ of Section 4.6 (at least as far as using the same symbol for the two quantities is concerned). The reason is that (4.130) is quite similar to, say, relation (4.91). However, the important difference is that (4.91) is presumed to hold in the entire range of compositions $0 \leq x_A \leq 1$, whereas (4.130) is valid only in the limiting case of $x_A \to 0$. Thus, (4.130) cannot be used to lend meaning to $\mu_A^*(T, P)$ by substituting $x_A = 1$. This is one

# Theory of Solutions

reason why relations (4.127) are preferable for characterizing DI solutions. A more profound reason for this is given in Section 4.11.

We now discuss briefly the behavior of the solvent in a DI solution of a two-component system. The simplest way of obtaining the chemical potential of the solvent $B$ is to apply the Gibbs–Duhem relation, which, in combination with (4.129), yields

$$x_B\left(\frac{\partial \mu_B}{\partial x_A}\right)_{T,P} + x_A\left(\frac{\partial \mu_A}{\partial x_A}\right)_{T,P} = -x_B\left(\frac{\partial \mu_B}{\partial x_B}\right)_{T,P} + kT = 0, \qquad x_A \to 0 \quad (4.135)$$

Hence, upon integration in the region for which (4.135) is valid, we get

$$\mu_B(T, P, x_B) = C(T, P) + kT \ln x_B, \qquad x_A \to 0 \qquad (4.136)$$

Since the condition $x_A \to 0$ is the same as the condition $x_B \to 1$, we can substitute $x_B = 1$ in (4.136) to identify the constant of integration as the chemical potential of pure $B$ at the given $T$ and $P$, i.e.,

$$\mu_B(T, P, x_B) = \mu_B^p(T, P) + kT \ln x_B, \qquad x_B \to 1 \qquad (4.137)$$

Note that (4.137) has the same form as, say, (4.91), except for the restriction $x_B \to 1$ in the former.

## 4.9. SMALL DEVIATIONS FROM DILUTE IDEAL SOLUTIONS

In Section 4.8, we considered the limiting behavior of the chemical potential as $\varrho_A \to 0$. We have seen that the formal appearance of the chemical potential is independent of the thermodynamic variables used to describe the system. In this section, we discuss first-order deviations from DI solutions. In fact, these nonideal cases are of foremost importance in practical applications. There exist formal statistical mechanical expressions for the higher-order deviations of DI behavior; however, their practical value is questionable since they usually involve higher-order molecular distribution functions. As in the previous section, we derive all the necessary relations from the Kirkwood–Buff theory, and we will be mainly concerned with the behavior of the solute $A$.

Consider again the general relations (4.123)–(4.125) of the previous section. In each of these, we can expand the nonsingular term on the rhs in power series of $\varrho_A$. The leading term in each case is

$$\left(\frac{\partial \mu_A}{\partial \varrho_A}\right)_{T,\mu_B} = kT\left(\frac{1}{\varrho_A} - G^\circ_{AA} + \cdots\right) \tag{4.138}$$

$$\left(\frac{\partial \mu_A}{\partial \varrho_A}\right)_{T,P} = kT\left[\frac{1}{\varrho_A} - (G^\circ_{AA} - G^\circ_{AB}) + \cdots\right] \tag{4.139}$$

$$\left(\frac{\partial \mu_A}{\partial \varrho_A}\right)_{T,\varrho_B} = kT\left\{\frac{1}{\varrho_A} - \frac{G^\circ_{AA} + \varrho_B{}^\circ[G^\circ_{AA}G^\circ_{BB} - (G^\circ_{AB})^2]}{1 + \varrho_B{}^\circ G^\circ_{BB}} + \cdots\right\} \tag{4.140}$$

The superscript degree in (4.138)–(4.140) stands for the limiting value of the corresponding quantity as $\varrho_A \to 0$. Note that the limit $\varrho_A \to 0$ is taken under different conditions in each case, i.e., $T$ and $\mu_A$ are constants in the first, $T$ and $P$ are constants in the second, etc.

These relations can be integrated, in the region of $\varrho_A$ for which the first-order expansion is valid, to obtain

$$\mu_A(T, \mu_B, \varrho_A) = \mu_A{}^\circ(T, \mu_B) + kT \ln \varrho_A - kTG^\circ_{AA}\varrho_A + \cdots \tag{4.141}$$

$$\mu_A(T, P, \varrho_A) = \mu_A{}^\circ(T, P) + kT \ln \varrho_A - kT(G^\circ_{AA} - G^\circ_{AB})\varrho_A + \cdots \tag{4.142}$$

$$\mu_A(T, \varrho_B, \varrho_A) = \mu_A{}^\circ(T, \varrho_B) + kT \ln \varrho_A - kT\left[G^\circ_{AA} - \frac{\varrho_B{}^\circ(G^\circ_{AB})^2}{1 + \varrho_B{}^\circ G^\circ_{BB}}\right]\varrho_A + \cdots \tag{4.143}$$

A comparison of these relations with (4.127) clearly shows that the uniformity shown in (4.127) breaks down once we consider deviations from DI behavior. The first-order terms in (4.141)–(4.143) depend on the thermodynamic variables we choose to describe our system.

In a very formal fashion, we can introduce activity coefficients to account for the first-order deviations from the DI behavior. These are defined by

$$kT \ln \gamma_A{}^D(T, \mu_B, \varrho_A) = -kTG^\circ_{AA}\varrho_A \tag{4.144}$$

$$kT \ln \gamma_A{}^D(T, P, \varrho_A) = -kT(G^\circ_{AA} - G^\circ_{AB})\varrho_A \tag{4.145}$$

$$kT \ln \gamma_A{}^D(T, \varrho_B, \varrho_A) = -kT\left(G^\circ_{AA} - \frac{\varrho_B{}^\circ(G^\circ_{AB})^2}{1 + \varrho_B{}^\circ G^\circ_{BB}}\right)\varrho_A \tag{4.146}$$

# Theory of Solutions

so that Eqs. (4.141)–(4.143) can be rewritten as

$$\mu_A(T, \mu_B, \varrho_A) = \mu_A^\circ(T, \mu_B) + kT \ln[\varrho_A \gamma_A^D(T, \mu_B, \varrho_A)] \quad (4.147)$$

$$\mu_A(T, P, \varrho_A) = \mu_A^\circ(T, P) + kT \ln[\varrho_A \gamma_A^D(T, P, \varrho_A)] \quad (4.148)$$

$$\mu_A(T, \varrho_B, \varrho_A) = \mu_A^\circ(T, P) + kT \ln[\varrho_A \gamma_A^D(T, \varrho_B, \varrho_A)] \quad (4.149)$$

We note, first, that the activity coefficients in (4.144)–(4.146) differ fundamentally from the activity coefficient introduced in Section 4.7. To stress this difference, we have used the superscript D to denote deviations from DI behavior, whereas in Section 4.7, we used the superscript S to denote deviations from SI behavior. (In the next section, we shall elaborate on an example for which three kinds of ideality can be distinguished in a very simple and explicit manner.) Furthermore, each of the activity coefficients defined in (4.144)–(4.146) depends on the thermodynamic variables, say $T$ and $\mu_B$ or $T$ and $P$, etc. This has been stressed in the notation. In practical applications, however, one usually knows which variables have been chosen, in which case one can drop the arguments in the notation for $\gamma_A^D$.

The limiting behavior of the activity coefficients defined in (4.144)–(4.146) is, for example,

$$\lim_{\varrho_A \to 0} \gamma_A^D(T, P, \varrho_A) = 1, \quad T, P \text{ constant} \quad (4.150)$$

Note the difference between (4.150) and (4.122). Here, the limiting behavior can be attained *experimentally*, simply by letting $\varrho_A \to 0$, keeping $T$ and $P$ constant.

Clearly, we can, in each case, transform to other concentration variables such as mole fraction or molality and obtain the appropriate activity coefficient. We shall not elaborate on this since it requires a relatively simple transformation of variables. We stress, however, that the number density (or the molar concentration) is the more "natural" choice of a concentration scale, and the corresponding standard chemical potentials enjoy some advantages which are not shared by standard chemical potentials based on either the mole fraction or the molality. More details are given in Section 4.11.

Consider next the "content" of the first-order contribution to the activity coefficients in (4.144)–(4.146). Note that all of these include a "new" quantity $G_{AA}^\circ$. It is "new" in the following sense: We recall that $G_{\alpha\beta}$ was assigned the meaning of an "affinity" of species $\alpha$ toward $\beta$ (see Section 4.6). The DI solutions are characterized by the complete absence of a contribution from "solute–solute" affinity. It is quite clear on qualitative grounds that the standard chemical potential is determined by the "solvent–solvent" and the "solvent–solute" affinities (this will be shown more explicitly in the next section). Thus, the effect of solute–solute affinity becomes operative only when we increase the solute concentration, so that the solute molecules "see" each other, which is the reason for the appearance of $G_{AA}^\circ$ in (4.144)–(4.146).

In addition to $G_{AA}^\circ$, relation (4.145) also includes $G_{AB}^\circ$, and relation (4.146) also includes $G_{BB}^\circ$. The last two quantities, as mentioned above, do not convey new properties of the system, since their basic content is already carried by the standard chemical potential.

The quantity $G_{AA}$, for example, is often referred to as containing the solute–solute interaction. In this book, we reserve the term "interaction" for the direct intermolecular interaction operating between two particles. Thus, two hard-sphere solutes of diameter $\sigma$ do not *interact* with each other at a distance $R > \sigma$, yet the solute–solute affinity conveyed by $G_{AA}$ may be different from zero. Furthermore, the use of the term solute–solute interaction for $G_{AA}$ may lead to some misinterpretations. Consider, for instance, an imperfect gas, where deviations from the ideal gas law are noticeable. Here, one can imagine a process of "switching off" all the interactions between the particles. Such a process will immediately turn our system into an ideal gas. Care must be employed when extending the analogy to solutions. Suppose we have a dilute solution of $A$ in $B$ in which deviations from DI behavior are significant. Here, if we "switch off" the *interactions* between the solute particles, we do *not* obtain a DI solution. It is difficult to point out the precise quantity that should be "switched off" in this case. As an example, in (4.144), we should "switch off" the correlation between the solutes, i.e., put $g_{AA}(R) - 1 \equiv 0$ in order to produce a zero solute–solute affinity, and hence DI solutions. The situation is clearly more complex with relations (4.145) and (4.146). Therefore, care must be exercised in identifying DI solutions as arising from the absence of solute–solute interactions.

From the practical point of view, the most important set of thermodynamic variables is, of course, $T$, $P$, $\varrho_A$, employed in (4.145). However, relation (4.144) is also useful, and has enjoyed considerable attention in the study of osmotic experiments, where $\mu_B$ is kept constant. We already noted in Section 4.8 that this set of variables provides relations that bear a remarkable analogy to the virial expansion of various quantities of real gases. We demonstrate this point by extracting the first-order expansion of the osmotic pressure $\pi$ in the solute density $\varrho_A$. This can be obtained

## Theory of Solutions

by the use of the thermodynamic relation

$$\left(\frac{\partial \pi}{\partial \varrho_A}\right)_{T,\mu_B} = \varrho_A \left(\frac{\partial \mu_A}{\partial \varrho_A}\right)_{T,\mu_B} \qquad (4.151)$$

Using either (4.138) or the more general expression (4.123) in (4.151), we get

$$\left(\frac{\partial \pi}{\partial \varrho_A}\right)_{T,\mu_B} = \frac{kT}{\varrho_A G_{AA} + 1} \xrightarrow{\varrho_A \to 0} kT(1 - G_{AA}^{\circ}\varrho_A + \cdots) \qquad (4.152)$$

Upon integration, we get

$$\pi/kT = \varrho_A - \tfrac{1}{2} G_{AA}^{\circ} \varrho_A^2 + \cdots \qquad (4.153)$$

which is better known in the form

$$\pi/kT = \varrho_A + B_2^* \varrho_A^2 + \cdots \qquad (4.154)$$

where $B_2^*$ is the analog of the second virial coefficient in the density expansion of the pressure

$$P/kT = \varrho + B_2 \varrho^2 + \cdots \qquad (4.155)$$

We now turn to a brief treatment of the chemical potential of the solvent $B$ for a system deviating slightly from the DI behavior. The simplest way of doing this is to use relation (4.88) from the Kirkwood–Buff theory, which, when written for the $B$ component and expanded to first order in $x_A$, yields

$$\left(\frac{\partial \mu_B}{\partial x_B}\right)_{T,P} = kT\left(\frac{1}{x_B} - \varrho^{\circ} \Delta_{AB}^{\circ} x_A + \cdots\right), \qquad x_A \to 0 \qquad (4.156)$$

where $\varrho^{\circ}$ and $\Delta_{AB}^{\circ}$ are the limiting values of $\varrho$ and $\Delta_{AB}$ as $x_A \to 0$. Integrating (4.156) yields

$$\mu_B(T, P, x_B) = \mu_B^{\text{p}}(T, P) + kT \ln x_B + \int_0^{x_A} \varrho^{\circ} \Delta_{AB}^{\circ} x_A' \, dx_A', \qquad x_A \to 0 \qquad (4.157)$$

Here, $\varrho^{\circ} \Delta_{AB}^{\circ}$ is independent of composition; hence we can integrate to obtain

$$\mu_B(T, P, x_A) = \mu_B^{\text{p}}(T, P) + kT \ln x_B + \tfrac{1}{2} \varrho_B^{\circ} \Delta_{AB}^{\circ} x_A^2, \qquad x_A \to 0 \qquad (4.158)$$

Note that in the limit $\varrho_A \to 0$, we can replace $\varrho^{\circ}$ by $\varrho_B^{\circ}$, the density of the pure solvent $B$ at the given $T$, $P$. It is important to stress the difference between (4.158) and relation (4.121), which looks very similar. Here, the expansion is valid for small $x_A$, but otherwise the value of $\varrho_B{}^\circ \varDelta_{AB}^\circ$ is unrestricted. In (4.121), on the other hand, we have an expansion in $\varDelta_{AB}$ which is required to hold for all compositions $0 \leq x_A \leq 1$.

## 4.10. A COMPLETELY SOLVABLE EXAMPLE

This section is devoted to illustrating the existence of three fundamentally different types of ideal mixtures. As we have demonstrated in the previous sections, one can further distinguish between concepts of ideality by the choice of different sets of thermodynamic variables. These, however, are not considered to be fundamentally different.

The first and simplest class is that of the ideal gas (IG) mixtures, which, as in the case of an ideal gas, are characterized by the complete absence (or neglect) of all intermolecular forces. This class is of least importance in the study of solution chemistry.

The second class, referred to as symmetric ideal (SI) solutions, emerges whenever the various components are "similar" to each other. There are no restrictions on the magnitude of the intermolecular forces or on the densities. The third class, dilute ideal (DI) solutions, consists of those solutions for which at least one component is very diluted in the remaining solvent, which may be a one-component or a multicomponent system. Again, there is no restriction on the strength of the intermolecular forces, the total density, or the degree of similarity of the various components.

The occurrence of all three types of ideality can be demonstrated by the use of Eq. (4.88) from the Kirkwood–Buff theory of solutions. We limit the discussion to a two-component system:

$$\left(\frac{\partial \mu_A}{\partial x_A}\right)_{T,P} = kT\left(\frac{1}{x_A} - \frac{\varrho x_B \varDelta_{AB}}{1 + \varrho x_A x_B \varDelta_{AB}}\right) \qquad (4.159)$$

We distinguish between the following three limiting cases of this equation.

1. In an ideal gas mixture, all the pair correlation functions are identically unity, $g_{\alpha\beta}(R) \equiv 1$; hence, $G_{\alpha\beta} = 0$ and also $\varDelta_{AB} = 0$. This is the case for a complete absence of interaction between the particles, a situation which is never realized for real systems. However, for systems of very low total density, $\varrho \to 0$, the particles are, on the average, very far from each other; hence, the intermolecular forces have a negligible effect on the properties of the mixture. This case is also referred to as an ideal gas mixture,

# Theory of Solutions

although it is quite different from the case in which intermolecular forces are presumed to be absent. Clearly, putting either $\Delta_{AB} = 0$ or $\varrho \to 0$ in (4.159), we get the typical ideal behavior

$$(\partial \mu_A / \partial x_A)_{T,P} = kT/x_A \qquad (4.160)$$

2. For a real system with any type of intermolecular force and at a finite density $\varrho$, the second term on the rhs of (4.159) may vanish if the two components are "similar" in the sense of Section 4.6; i.e., if $\Delta_{AB} = 0$ over the entire range of compositions, we get the SI solutions. (Note the essential difference between this and the previous case.)

3. For a real system of two components which are not similar but in which $x_B \to 0$, we get from (4.159) the DI solutions.

Any arbitrary mixture of two components can be viewed as deviating from one of the ideal reference cases. This can be written symbolically as

$$\begin{aligned} \mu_A &= c_1 + kT \ln(x_A \gamma_A^{\mathrm{I}}) \\ &= c_2 + kT \ln(x_A \gamma_A^{\mathrm{S}}) \\ &= c_3 + kT \ln(x_A \gamma_A^{\mathrm{D}}) \end{aligned} \qquad (4.161)$$

where the constants $c_i$ are independent of $x_A$. Here, $\gamma_A^{\mathrm{I}}$, $\gamma_A^{\mathrm{S}}$, and $\gamma_A^{\mathrm{D}}$ are activity coefficients that incorporate the correction due to nonideality[19] (I for IG, S for SI, and D for DI).

We now consider a particular example of a system which, on the one hand, is not trivial, since interactions between particles are taken into account, yet is sufficiently simple so that all three activity coefficients can be written in an explicit form.

We choose a two-component system for which the pressure (or the total density) is sufficiently low that the pair correlation function for each pair of species has the form

$$g_{\alpha\beta}(R) = \exp[-\beta U_{\alpha\beta}(R)] \qquad (4.162)$$

where, for simplicity, we assume that all the pair potentials are spherically symmetric. The general expression for the chemical potential of, say, $A$ in this system is obtained by a simple extension of the one-component expression given in Section 3.5 (for the purpose of this section, we assume that

---

[19] For the purpose of demonstration, we have chosen $T$, $P$, $x_A$ as the thermodynamic variables. A parallel treatment can be carried out for any other set of thermodynamic variables.

the internal partition function has been included in the factor $\Lambda_A{}^3$):

$$\mu_A = kT\ln(\varrho_A\Lambda_A{}^3) + \varrho_A \int_0^1 d\xi \int_0^\infty U_{AA}(R)g_{AA}(R,\xi)4\pi R^2\,dR$$

$$+ \varrho_B \int_0^1 d\xi \int_0^\infty U_{AB}(R)g_{AB}(R,\xi)4\pi R^2\,dR$$

$$= kT\ln(\varrho_A\Lambda_A{}^3) + 2kTB_{AA}\varrho_A + 2kTB_{AB}\varrho_B \tag{4.163}$$

where we have used (4.162) [with $\xi U_{\alpha\beta}(R)$ replacing $U_{\alpha\beta}(R)$] so that integration over $\xi$ becomes immediate. Also, we have used the more familiar notation

$$B_{\alpha\beta} = -\tfrac{1}{2}\int_0^\infty \{\exp[-\beta U_{\alpha\beta}(R)] - 1\}4\pi R^2\,dR \tag{4.164}$$

It is possible to analyze Eq. (4.163) for the various kinds of ideality, but for the purpose of this section, we prefer to first transform it so that $\mu_A$ is expressed as a function of $T$, $P$, and $x_A$. To do this, we use the analog of the virial expansion for mixtures, which reads (Section 1.8)

$$P/kT = (\varrho_A + \varrho_B) + [x_A{}^2 B_{AA} + 2x_A x_B B_{AB} + x_B{}^2 B_{BB}](\varrho_A + \varrho_B)^2 + \cdots \tag{4.165}$$

The factor in the square brackets can be viewed as the second virial coefficient for the mixture of two components. We now invert this relation by assuming an expansion of the form

$$\varrho = \varrho_A + \varrho_B = (P/kT) + CP^2 + \cdots \tag{4.166}$$

This is substituted on the rhs of (4.165), and on equating coefficients of equal powers of $P$, we get

$$\varrho = \varrho_A + \varrho_B = (P/kT) - (x_A{}^2 B_{AA} + 2x_A x_B B_{AB} + x_B{}^2 B_{BB})(P/kT)^2 + \cdots \tag{4.167}$$

Substituting $\varrho_A = x_A\varrho$ and $\varrho_B = x_B\varrho$ in (4.163) and using the expansion (4.167) for $\varrho$, we get the final form of the chemical potential:

$$\mu_A(T,P,x_A) = kT\ln(x_A\Lambda_A{}^3) + kT\ln(P/kT)$$
$$+ kT\ln[1 - (x_A{}^2 B_{AA} + 2x_A x_B B_{AB} + x_B{}^2 B_{BB})P/kT]$$
$$+ (2kTB_{AA}x_A + 2kTB_{AB}x_B)$$
$$\times [(P/kT) - (x_A{}^2 B_{AA} + 2x_A x_B B_{AB} + x_B{}^2 B_{BB})(P/kT)^2]$$
$$= kT\ln(x_A\Lambda_A{}^3) + kT\ln(P/kT)$$
$$+ PB_{AA} - Px_B{}^2(B_{AA} + B_{BB} - 2B_{AB}) \tag{4.168}$$

where, in the last form of (4.168), we have retained only first-order terms in the pressure (except for the logarithmic term). We now view expression (4.168) in various ways, according to the choice of reference ideal state.

### 4.10.1. Ideal Gas Mixture as a Reference System

If $P \to 0$ (or if no interactions exist, so that *all* $B_{\alpha\beta} = 0$), we get from (4.168)

$$\mu_A^{\text{IG}} = kT \ln(x_A \Lambda_A^3) + kT \ln(P/kT)$$
$$= \mu_A^{\circ g}(T, P) + kT \ln x_A \qquad (4.169)$$

Here, $\mu_A^{\circ g}(T, P)$ is defined as the IG standard chemical potential (note its dependence on both $T$ and $P$).

Comparing (4.168) with (4.169), we see that the correction due to deviations from the IG mixture can be included in the activity coefficient

$$kT \ln \gamma_A^{\text{I}} = PB_{AA} - Px_B^2(B_{AA} + B_{BB} - 2B_{AB}) \qquad (4.170)$$

and hence

$$\mu_A(T, P, x_A) = \mu_A^{\circ g}(T, P) + kT \ln(x_A \gamma_A^{\text{I}}) \qquad (4.171)$$

which is a particular case of the first form on the rhs of (4.161).

### 4.10.2. Symmetric Ideal Solution as a Reference System

Suppose that the two components $A$ and $B$ are "similar" in the sense of Section 4.6, which, in this case, means [see Eq. (4.95)]

$$B_{AA} + B_{BB} - 2B_{AB} = 0 \qquad (4.172)$$

Hence, in such a case, the solution will be SI; i.e., from (4.168), we have

$$\mu_A^{\text{SI}} = kT \ln(x_A \Lambda_A^3) + kT \ln(P/kT) + PB_{AA}$$
$$= \mu_A^{\text{p}}(T, P) + kT \ln x_A \qquad (4.173)$$

Note that $\mu_A^{\text{p}}$ is the chemical potential of pure $A$ at this particular $T$ and $P$ and therefore it includes the effect of $A$–$A$ interactions through $B_{AA}$. If, on the other hand, the system is not SI, then we define the activity coefficient as

$$kT \ln \gamma_A^{\text{S}} = -Px_B^2(B_{AA} + B_{BB} - 2B_{AB}) \qquad (4.174)$$

and (4.168) is now rewritten as

$$\mu_A(T, P, x_A) = \mu_A^{\text{p}} + kT \ln(x_A \gamma_A^{\text{S}}) \qquad (4.175)$$

where $\gamma_A^S$ incorporates the deviations due to the dissimilarity between the components. Equation (4.175) is a particular case of the second form on the rhs of (4.161).

### 4.10.3. Dilute Ideal Solution as a Reference System

If $A$ is diluted in $B$, i.e., when $x_A \to 0$, we also have $x_B \to 1$. Hence, in this case, we have a DI solution. Equation (4.168) reduces to

$$\mu_A^{DI} = kT \ln(x_A \Lambda_A^3) + kT \ln(P/kT) + P(2B_{AB} - B_{BB})$$
$$= \mu_A^* + kT \ln x_A \qquad (4.176)$$

where $\mu_A^*$ is defined by (4.176) for this particular example. Since dilute solutions are preferably expressed in terms of $\varrho_A$ rather than $x_A$, we can transform (4.176) by putting $x_A = \varrho_A/\varrho$ and using (4.167) to get

$$\mu_A^{DI} = kT \ln(\varrho_A \Lambda_A^3) + 2PB_{AB} = \mu_A^\circ + kT \ln \varrho_A \qquad (4.177)$$

Note that $\mu_A^*$ and $\mu_A^\circ$ contain $B_{AB}$ and $B_{BB}$, i.e., they include solute–solvent and solvent–solvent interactions, but not solute–solute interactions. [The word "interaction" is appropriate in the present context since, in the present limiting case, we know that $g_{\alpha\beta}(R)$ depends only on the *direct* interaction between the pair of species $\alpha$ and $\beta$, as we have assumed in (4.162).]

The activity coefficients defined for the two representations (4.176) and (4.177) are obtained from comparison with (4.168) as

$$kT \ln \gamma_A^{D,x} = 2x_A P(B_{AA} + B_{BB} - 2B_{AB}) \qquad (4.178)$$

$$kT \ln \gamma_A^{D,\varrho} = 2kT\varrho_A(B_{AA} - B_{AB}) \qquad (4.179)$$

Note that in (4.178) and (4.179), we retained only the first-order terms in $x_A$ and in $\varrho_A$, respectively. Using these activity coefficients, the chemical potential in (4.168) can be rewritten in two alternative forms:

$$\mu_A(T, P, x_A) = \mu_A^* + kT \ln(x_A \gamma_A^{D,x}) \qquad (4.180)$$

$$\mu_A(T, P, \varrho_A) = \mu_A^\circ + kT \ln(\varrho_A \gamma_A^{D,\varrho}) \qquad (4.181)$$

The notation $\gamma_A^{D,x}$ and $\gamma_A^{D,\varrho}$ has been introduced to distinguish between the two cases. Table 4.2 summarizes the three different ways of splitting the chemical potential according to the various reference systems.

## Table 4.2
## Various Ways of Splitting the Chemical Potential in (4.168)[a]

| Reference ideal system | Standard chemical potential | Activity coefficient |
|---|---|---|
| 1. Ideal gas mixture, Eq. (4.171) | $\mu_A^{og} = kT\ln(\Lambda_A^3 P/kT)$ | $kT\ln\gamma_A^{\mathrm{I}} = pB_{AA} - Px_B^2(B_{AA} + B_{BB} - 2B_{AB})$ |
| 2. Symmetric ideal solution, Eq. (4.175) | $\mu_A^{p} = kT\ln(\Lambda_A^3 P/kT) + PB_{AA}$ | $kT\ln\gamma_A^{\mathrm{S}} = -Px_B^2(B_{AA} + B_{BB} - 2B_{AB})$ |
| 3. Dilute ideal solution, Eq. (4.180) (mole fraction) | $\mu_A^{*} = kT\ln(\Lambda_A^3 P/kT) + P(2B_{AB} - B_{BB})$ | $kT\ln\gamma_A^{\mathrm{D},x} = 2Px_A(B_{AA} + B_{BB} - 2B_{AB})$ |
| Eq. (4.181) (density) | $\mu_A^{\circ} = kT\ln\Lambda_A^3 + 2PB_{AB}$ | $kT\ln\gamma_A^{\mathrm{D},\varrho} = 2kT\varrho_A(B_{AA} - B_{AB})$ |

[a] Note that in the activity coefficients for the dilute solutions, we have retained first-order terms in $x_A$ or $\varrho_A$.

## 4.11. STANDARD THERMODYNAMIC QUANTITIES OF TRANSFER

This section is devoted to a detailed elaboration of thermodynamic quantities employed in the literature to deal with dilute solutions. These are the so-called "standard" free energies, entropies, enthalpies, etc. associated with the transfer of a solute from one phase to another. There are actually many quantities which are referred to as standard quantities. We deal essentially with one of these, the one which is most directly amenable to a molecular interpretation.

Because of our ultimate intention to interpret experimental quantities, we shall be concerned with systems at constant temperature and pressure, which are the most prevalent in experimental work in this field.

The general expressions for the chemical potential and the pseudo-chemical potential in the $T, V, N$ ensemble have been developed in Sections 3.5 and 3.6. The analogous relations in the $T, P, N$ ensemble are derived in Appendix 9-F. Here, we summarize the basic quantities that will be required. The chemical potential and the pseudo-chemical potential of a component $A$ are defined in the $T, P, N_A, N_B$ ensemble by

$$\mu_A = G(T, P, N_A + 1, N_B) - G(T, P, N_A, N_B) \quad (4.182)$$

$$\bar{\mu}_A = G(T, P, N_A + 1, N_B; \mathbf{R}_0) - G(T, P, N_A, N_B) \quad (4.183)$$

In (4.183), the added $A$ molecule is assumed to be placed at a fixed position $\mathbf{R}_0$. For nonspherical molecules, we may require also that the orientation be fixed. Here we deal, for simplicity, with spherical molecules only.

The basic statistical mechanical expressions for the chemical potential are

$$\mu_A = kT \ln(\varrho_A \Lambda_A^3 q_A^{-1}) - kT \ln\langle \exp[-\beta B_A(\mathbf{R}_0)] \rangle$$
$$= \bar{\mu}_A(T, P, \varrho_A) + kT \ln(\varrho_A \Lambda_A^3)$$
$$= \mu_A^\circ(T, P, \varrho_A) + kT \ln \varrho_A \quad (4.184)$$

where $B_A(\mathbf{R}_0)$ stands for the total interaction energy of the $A$ molecule at $\mathbf{R}_0$ with the rest of the system. The symbol $\langle \ \rangle$ stands for an average in the $T, P, N_A, N_B$ ensemble. $\bar{\mu}_A$ is the pseudo-chemical potential, defined in (4.183), and $\mu_A^\circ(T, P, \varrho_A)$ is a "nonconventional" standard chemical potential of $A$. The last statement needs further elaboration; the (conven-

**Theory of Solutions** 171

tional) standard chemical potential is defined by

$$\begin{aligned}
\mu_A^\circ(T, P) &= \lim_{\varrho_A \to 0} [\mu_A(T, P, \varrho_A) - kT \ln \varrho_A] \\
&= kT \ln(\Lambda_A^3 q_A^{-1}) - \lim_{\varrho_A \to 0} \{kT \ln \langle \exp[-\beta B_A(\mathbf{R}_0)] \rangle\} \\
&= kT \ln \Lambda_A^3 + G(T, P, N_A = 1, N_B; \mathbf{R}_0) - G(T, P, N_B) \\
&= kT \ln \Lambda_A^3 + \bar{\mu}_A(T, P) \quad (4.185)
\end{aligned}$$

where, in the last form on the rhs of (4.185), we produce the pseudo-chemical potential by adding *one* $A$ molecule to a fixed position $\mathbf{R}_0$ in pure $B$. The most important feature of the (conventional) standard chemical potential is its independence of the concentration $\varrho_A$. The nonconventional standard chemical potential, defined in (4.184), is generally dependent on $\varrho_A$, and clearly, from (4.184) and (4.185), we have

$$\mu_A^\circ(T, P) = \lim_{\varrho_A \to 0} \mu_A^\circ(T, P, \varrho_A) \quad (4.186)$$

Let us now focus our attention on DI solutions of $A$, in two solvents $\alpha$ and $\beta$, in which the chemical potentials have the form

$$\mu_A^\alpha(T, P, \varrho_A^\alpha) = \mu_A^{\circ\alpha}(T, P) + kT \ln \varrho_A^\alpha, \quad \varrho_A^\alpha \to 0 \quad (4.187)$$

$$\mu_A^\beta(T, P, \varrho_A^\beta) = \mu_A^{\circ\beta}(T, P) + kT \ln \varrho_A^\beta, \quad \varrho_A^\beta \to 0 \quad (4.188)$$

Note that the solvents $\alpha$ and $\beta$ may be either a pure liquid or a mixture of several components.

In Section 4.8, we stressed the fact that the standard chemical potential $\mu_A^\circ$ is not a chemical potential of $A$ in any real system, and should be viewed merely as a constant of integration (it is sometimes referred to as a chemical potential of $A$ in a hypothetical state; however, we refrain from such an interpretation since it does not lead to any useful meaning on a molecular level). In most practical applications, we encounter *differences* in standard chemical potentials rather than a single standard chemical potential. We shall now see that differences in standard chemical potentials may be assigned the meaning of differences in chemical potentials. This is obtained directly from (4.187) and (4.188), which, upon subtraction, yield

$$\mu_A^\alpha(T, P, \varrho_A^\alpha) - \mu_A^\beta(T, P, \varrho_A^\beta) = \mu_A^{\circ\alpha}(T, P) - \mu_A^{\circ\beta}(T, P) + kT \ln(\varrho_A^\alpha/\varrho_A^\beta) \quad (4.189)$$

The quantity on the lhs of (4.189) is the free energy change upon transferring an $A$ molecule from $\beta$ (at $T, P, \varrho_A^\beta$) to $\alpha$ (at $T, P, \varrho_A^\alpha$).

Consider now a special case of (4.189) in which we have chosen the condition $\varrho_A^\alpha = \varrho_A^\beta = \varrho_A$. In this particular case, we get

$$\mu_A^\alpha(T, P, \varrho_A) - \mu_A^\beta(T, P, \varrho_A) = \mu_A^{\circ\alpha}(T, P) - \mu_A^{\circ\beta}(T, P) \equiv \Delta\mu_A^\circ(\beta \to \alpha) \tag{4.190}$$

Hence, $\Delta\mu_A^\circ(\beta \to \alpha)$ is the change in the Gibbs free energy for the transfer of an $A$ molecule from $\beta$ to $\alpha$ at constant $T$ and $P$, provided that the density of $A$ is the same in the two phases and that it is small enough to ensure the validity of (4.187) and (4.188). In the following, we refer to the above statement as the *experimental interpretation* of $\Delta\mu_A^\circ(\beta \to \alpha)$, since the above process can be achieved in an actual experiment.

A second interpretation of $\Delta\mu_A^\circ(\beta \to \alpha)$, which does not correspond to a realizable experiment, is referred to as a *molecular interpretation*, and follows directly from (4.185):

$$\Delta\mu_A^\circ(\beta \to \alpha) = \bar{\mu}_A^\alpha - \bar{\mu}_A^\beta \tag{4.191}$$

i.e., $\Delta\mu_A^\circ(\beta \to \alpha)$ is the change of the Gibbs free energy for the transfer of a single $A$ molecule from a fixed position in $\beta$ to a fixed position in $\alpha$, the process being carried out at constant $T$ and $P$. Note that within the context of the present discussion, the above statement strictly refers to the limiting case of DI solutions. However, since (4.191) can also be obtained from (4.184) by the application of the nonconventional standard chemical potentials, the restriction to a DI solution may be relaxed.

The most important aspect of the molecular interpretation of the quantity $\Delta\mu_A^\circ(\beta \to \alpha)$ is its local character. We have seen in Sections (3.5) and (3.6) that the pseudo-chemical potential of simple solutes depends on the local environment of the molecule fixed at $\mathbf{R}_0$. Hence, $\Delta\mu_A^\circ(\beta \to \alpha)$ in (4.191) can be used as a probe for characterizing the difference of the solvation properties of $\alpha$ and $\beta$ with respect to the solute $A$. This interpretation will dominate our discussion of aqueous solutions in Chapter 7.

Thus far, we have dwelled upon the interpretation of $\Delta\mu_A^\circ(\beta \to \alpha)$. Let us consider a method for its experimental determination. From (4.189) we see that if $A$ is in *equilibrium* between the two phases, then

$$\mu_A^\alpha = \mu_A^\beta \tag{4.192}$$

and hence

$$0 = \Delta\mu_A^\circ(\beta \to \alpha) + kT \ln(\varrho_A^\alpha/\varrho_A^\beta)_{\text{eq}} \tag{4.193}$$

where the subscript eq has been appended to stress the fact that we are considering the specific case in which the ratio $\varrho_A^\alpha/\varrho_A^\beta$ is measured at

Theory of Solutions 173

equilibrium. Relation (4.193) provides a simple means of determining $\Delta\mu_A{}^\circ(\beta \to \alpha)$ through a measurement of the ratio $(\varrho_A{}^\alpha/\varrho_A{}^\beta)_{eq}$. Using relation (4.184) instead of (4.189), we can obtain the equation analogous to (4.193) which is not restricted to DI solutions. The result is exactly the same as (4.193) with the understanding that $\Delta\mu_A{}^\circ(\beta \to \alpha)$ is the difference in the *nonconventional* standard chemical potentials of $A$ in the two phases.

Before considering an example of the application of the nonconventional standard chemical potential, we note that the above experimental interpretation of $\Delta\mu_A{}^\circ(\beta \to \alpha)$ has been derived for the specific set of variables $T$, $P$, $\varrho_A$. Similar considerations apply to other variables such as $T$, $P$, $x_A$ or $T$, $P$, $m_A$. However, the molecular interpretation given above applies only to the standard states based on the density $\varrho_A$ as a solute concentration. Furthermore, application of the nonconventional standard chemical potential, as demonstrated below, may not be extended to the set of variables $T$, $P$, $x_A$ or $T$, $P$, $m_A$.

Consider a solution of benzene ($A$) in water. The two phases $\alpha$ and $\beta$ will be (almost) pure benzene and water, respectively. Suppose, for simplicity, that $A$ forms a DI solution in $\beta$ (this is not a necessary assumption for the present discussion). Then, the chemical potential of $A$ in the two phases can be written as

$$\mu_A{}^\alpha(T, P, \varrho_A{}^\alpha) = \mu_A^{\circ\alpha}(T, P, \varrho_A{}^\alpha) + kT \ln \varrho_A{}^\alpha \qquad (4.194)$$

$$\mu_A{}^\beta(T, P, \varrho_A{}^\beta) = \mu_A^{\circ\beta}(T, P) + kT \ln \varrho_A{}^\beta \qquad (4.195)$$

Note that in (4.194), we use the nonconventional standard chemical potential. At equilibrium between the two phases, we have

$$\mu_A^{\circ\alpha}(T, P, \varrho_A{}^\alpha) - \mu_A^{\circ\beta}(T, P) = -kT \ln(\varrho_A{}^\alpha/\varrho_A{}^\beta)_{eq} \qquad (4.196)$$

Here, $\varrho_A{}^\beta$ is the concentration (or solubility) of benzene in *water*, whereas $\varrho_A{}^\alpha$ is the concentration of benzene in (almost) pure benzene. Clearly, the rhs of (4.196) is a measurable quantity. The lhs of (4.196) is, according to the molecular interpretation, the change in Gibbs free energy for the transfer of a benzene molecule from a fixed position in pure water to a fixed position in (almost) pure benzene. Due to its local character, this quantity measures the difference in the "solvation" properties of benzene and water with respect to a single benzene molecule.[20]

---

[20] A different quantity $\mu_A^{\circ\beta}(T, P) - \mu_A{}^p(T, P)$ is often used in the literature. This, however, does *not* correspond to a transfer of $A$ from pure benzene to a dilute aqueous solution of benzene, nor to any other real process of transfer between the two phases.

Some care must be exercised when using the nonconventional standard chemical potential. The difficulty arises from their dependence on the density $\varrho_A$. For instance, in Eq. (4.193), when it is applied to DI solutions, $\Delta\mu_A{}^\circ(\beta \to \alpha)$ is *independent* of $\varrho_A$. Hence, for its interpretation, it is sufficient to consider the transfer of single $A$ from a fixed position in *pure $\beta$* to a fixed position in *pure $\alpha$*. If, on the other hand, we use the nonconventional standard chemical potential, as in the example (4.196), then the lhs of the equation is dependent on $\varrho_A{}^\alpha$ (and, in general, on $\varrho_A{}^\beta$, too). Hence, the corresponding process is a transfer of $A$ from a fixed position in one phase to that of another, where the density of $A$ in the two phases is determined by the equilibrium condition.

We now consider some other thermodynamic quantities of transfer for the processes mentioned above. First, we introduce a notation that will be useful later. We refer to process I as the process of transfer described immediately following relation (4.190). Likewise, the process of transfer described immediately following relation (4.191) will be referred to as process II. The free energy changes corresponding to these two processes are denoted by $\Delta\mu_A{}^\circ(\mathrm{I})$ and $\Delta\mu_A{}^\circ(\mathrm{II})$, respectively. From (4.190) and (4.191), we have

$$\Delta\mu_A{}^\circ(\mathrm{I}) = \Delta\mu_A{}^\circ(\mathrm{II}) \qquad (4.197)$$

which follows from the equivalence of the two interpretations of $\Delta\mu_A{}^\circ(\beta \to \alpha)$ in (4.190) and (4.191). In other words, the free energy changes for the two processes I and II are equal. This equality is *unique* to the free energy and is not shared by other thermodynamic quantities of transfer, as we shall demonstrate below.

### 4.11.1. Entropy

The standard entropy of transfer for process I is computed by differentiating (4.187) and (4.188) with respect to temperature at constant $P$, $N_A$, $N_B$, i.e.,

$$-\bar{S}_A{}^\alpha = \left(\frac{\partial \mu_A{}^\alpha}{\partial T}\right)_{P,N_A,N_B} = \left(\frac{\partial \mu_A{}^{\circ\alpha}}{\partial T}\right)_P + \left[\frac{\partial}{\partial T}(kT \ln \varrho_A{}^\alpha)\right]_{P,N_A,N_B} \qquad (4.198)$$

$$-\bar{S}_A{}^\beta = \left(\frac{\partial \mu_A{}^\beta}{\partial T}\right)_{P,N_A,N_B} = \left(\frac{\partial \mu_A{}^{\circ\beta}}{\partial T}\right)_P + \left[\frac{\partial}{\partial T}(kT \ln \varrho_A{}^\beta)\right]_{P,N_A,N_B} \qquad (4.199)$$

Theory of Solutions 175

Note that, in general, $\varrho_A{}^\alpha$ and $\varrho_A{}^\beta$ depend on temperature, and we have

$$\left[\frac{\partial}{\partial T}(\ln \varrho_A{}^\alpha)\right]_{P,N_A,N_B} = \left[\frac{\partial}{\partial T}\left(\ln \frac{N_A{}^\alpha}{V^\alpha}\right)\right]_{P,N_A,N_B} = -\left[\frac{\partial}{\partial T}(\ln V^\alpha)\right]_{P,N_A,N_B}$$
(4.200)

where $V^\alpha$ is the volume of phase $\alpha$. For pure solvents, the above derivative is simply the coefficient of the thermal expansion of $\alpha$.

Subtracting (4.199) from (4.198) gives

$$-(\bar{S}_A{}^\alpha - \bar{S}_A{}^\beta) = \frac{\partial}{\partial T}[\Delta\mu_A{}^\circ(\beta \to \alpha)] + k \ln\left(\frac{\varrho_A{}^\alpha}{\varrho_A{}^\beta}\right) - kT\frac{\partial}{\partial T}\left(\ln \frac{V^\alpha}{V^\beta}\right)$$
(4.201)

As in the passage from (4.189) to (4.190), we now consider the special case $\varrho_A{}^\alpha = \varrho_A{}^\beta = \varrho_A$ in (4.201), to obtain the standard entropy change corresponding to process I, i.e.,

$$\Delta S_A{}^\circ(\mathrm{I}) = -\frac{\partial}{\partial T}[\Delta\mu_A{}^\circ(\mathrm{I})] + kT\frac{\partial}{\partial T}\left(\ln \frac{V^\alpha}{V^\beta}\right) \qquad (4.202)$$

Note that the substitution of $\varrho_A{}^\alpha = \varrho_A{}^\beta = \varrho_A$ was made *after* and not before the differentiation with respect to temperature. We see that in order to obtain the standard entropy change for process I, one needs, in addition to the temperature derivative of $\Delta\mu_A{}^\circ(\mathrm{I})$, the coefficients of thermal expansion of the two phases.

Next, we compute the standard entropy change for process II. This can be obtained most directly by differentiating (4.191) with respect to temperature [note the meaning of $\bar{\mu}_A{}^\alpha$ and $\bar{\mu}_A{}^\beta$ as defined in (4.183)]:

$$\Delta S_A{}^\circ(\mathrm{II}) = -\frac{\partial}{\partial T}(\bar{\mu}_A{}^\alpha - \bar{\mu}_A{}^\beta) = -\frac{\partial}{\partial T}[\Delta\mu_A{}^\circ(\mathrm{I})] \qquad (4.203)$$

Comparing (4.202) with (4.203), we get

$$\Delta S_A{}^\circ(\mathrm{I}) = \Delta S_A{}^\circ(\mathrm{II}) + kT\frac{\partial}{\partial T}\left(\ln \frac{V^\alpha}{V^\beta}\right) \qquad (4.204)$$

Relation (4.204) shows that, in general, the standard entropy changes for processes I and II are different. It is also clear that $\Delta S_A{}^\circ(\mathrm{II})$ is obtained more easily from experimental quantities than is $\Delta S_A{}^\circ(\mathrm{I})$. The latter requires knowledge of the coefficients of thermal expansion of the two phases.

### 4.11.2. Enthalpy

The standard enthalpy changes associated with the two processes I and II can be obtained most directly by using the thermodynamic relations

$$\Delta H_A^\circ(\text{I}) = \Delta \mu_A^\circ + T \Delta S_A^\circ(\text{I}) \qquad (4.205)$$

$$\Delta H_A^\circ(\text{II}) = \Delta \mu_A^\circ + T \Delta S_A^\circ(\text{II}) \qquad (4.206)$$

Note that the difference between $\Delta H_A^\circ(\text{I})$ and $\Delta H_A^\circ(\text{II})$ arises only from the difference between $\Delta S_A^\circ(\text{I})$ and $\Delta S_A^\circ(\text{II})$.

### 4.11.3. Volume

Using a procedure similar to the one taken to compute the standard entropy changes, but differentiating with respect to pressure instead of temperature, we get the relations

$$\Delta V_A^\circ(\text{I}) = \frac{\partial}{\partial P}(\Delta \mu_A^\circ) + kT \frac{\partial}{\partial P}\left(\ln \frac{V^\alpha}{V^\beta}\right) \qquad (4.207)$$

$$\Delta V_A^\circ(\text{II}) = \frac{\partial}{\partial P}(\Delta \mu_A^\circ) \qquad (4.208)$$

Again we see that the evaluation of $\Delta V_A^\circ(\text{II})$ is simpler than that of $\Delta V_A^\circ(\text{I})$, the latter requiring knowledge of the isothermal compressibilities of the two solvents $\alpha$ and $\beta$.

In a similar fashion, one can derive other standard thermodynamic quantities of transfer for processes I and II. The expressions for process II will always be simpler than the corresponding expressions for process I.

*Chapter 5*
# Generalized Molecular Distribution Functions and the Mixture-Model Approach to Liquids

## 5.1. INTRODUCTION

This chapter provides a bridge between the formal theories of liquids and various modelistic approaches devised for the study of aqueous fluids. Such a bridge is needed for the comfortable accommodation, in a single book, of both a fundamental theory, based on first principles of statistical mechanics, and various, basically heuristic, approaches.

The general strategy of constructing such a bridge is the following: First we start with the fundamental notions of ordinary molecular distribution functions (MDF). These are generalized in a purely formal fashion to obtain what we shall call "generalized molecular distribution functions" (GMDF). This step establishes a sound basis in the fundamentals of statistical mechanics. Some aspects of these GMDF's and their merits are discussed in the following three sections.

In the next step, we focus our attention on one of the GMDF's, the *singlet* GMDF, which by a reinterpretation, can be used to view a one-component system as a mixture of various quasicomponents. This step provides a firm and rigorous basis for the so-called mixture-model (MM) approach to liquids. It now requires only a small step to reach the various modelistic approaches to the theory of water and aqueous solutions. These are viewed as approximate versions of the exact MM approach, and will be discussed in detail in the next chapter. Historically, these approximate theories have been suggested on intuitive grounds and for many years have played a central role in our comprehension of the unusual, and important, properties of water and aqueous solutions.

In the previous chapters, we derived some relations between thermodynamic quantities and *ordinary* MDF's. We recall, however, that most of these relations are strongly dependent on the assumption of pairwise additivity of the total potential energy. Such an assumption is believed to be a good approximation for simple fluids, but it is almost universally recognized to be a very poor one for aqueous fluids. This difficulty calls for new methods of investigating complex liquids which do not depend on any assumption on the potential energy of the system. One such possibility is the Kirkwood–Buff theory of solutions, which, although very general, is mainly useful for solutions rather than for one-component systems. Indeed, as we shall see, we can incorporate some of the results of the Kirkwood–Buff theory into the MM approach to liquids and hence obtain new results for one-component systems.

Furthermore, even within the assumption of pairwise additivity, there are thermodynamic quantities which, although very important, are not expressible in terms of the pair distribution functions. For example, the heat capacity of a system can hardly be investigated within the framework of ordinary MDF's; its investigation is rendered somewhat more feasible within the framework of the GMDF's.

In Chapter 2, we outlined the information on the local mode of packing conveyed by the pair correlation function. Similarly, new and complementary information is furnished via the GMDF's, which seem to be of particular value in the study of complex fluids such as water.

Although we shall be using mainly the first- and second-order GMDF's, it is instructive to discuss the general procedure for defining any GMDF.

We recall the general definition of the $n$th-order MDF, say in the $T$, $V$, $N$ ensemble, which we write in the following two equivalent forms:

$$\varrho^{(n)}(\mathbf{S}_1, \ldots, \mathbf{S}_n) = [N!/(N-n)!] \int \cdots \int d\mathbf{R}_{n+1} \cdots d\mathbf{R}_N$$
$$\times P(\mathbf{S}_1, \ldots, \mathbf{S}_n; \mathbf{R}_{n+1}, \ldots, \mathbf{R}_N)$$
$$= \sum_{\substack{i_1=1 \\ i_1 \neq i_2 \cdots \neq i_n}}^{N} \cdots \sum_{i_n=1}^{N} \int \cdots \int d\mathbf{R}^N$$
$$\times P(\mathbf{R}^N)[\delta(\mathbf{R}_{i_1} - \mathbf{S}_1) \cdots \delta(\mathbf{R}_{i_n} - \mathbf{S}_n)] \quad (5.1)$$

Generalized Molecular Distribution Functions 179

Here, $P(\mathbf{R}^N)$ is the basic probability density in the $T$, $V$, $N$ ensemble. In the first form on the rhs of (5.1), we have made the distinction between *fixed* variables $\mathbf{S}_1, \ldots, \mathbf{S}_n$ and dummy variables $\mathbf{R}_{n+1}, \ldots, \mathbf{R}_N$ which undergo integration. The second form on the rhs of (5.1) is more amenable for pursuing the required generalization. We first recognize that the brackets in the integrand comprise a stipulation on the range of integration, i.e., they serve to extract from the entire configurational space only those configurations (or regions) for which the vector $\mathbf{R}_{i_1}$ attains the value $\mathbf{S}_1, \ldots$, the vector $\mathbf{R}_{i_n}$ attains the value $\mathbf{S}_n$. Alternatively, we can break up the above statement into its following logical ingredients. We choose a *property*; here, the property is, "the location of, say, particle $i$." We then impose a *condition* on this property, which in this case reads, "the location of particle $i$ is $\mathbf{S}_i$." Next, we combine several of these conditions applied to $n$ particles and arrive at the content of the square brackets in the integral in (5.1).

Although the above analysis is futile for the special case of ordinary MDF's, it is quite useful for obtaining insight into the structure of our generalization process. The generalization can be carried out on both the *property* and the *condition* imposed on it. The distinction between property and condition is arbitrary and is made for convenience only. As a matter of fact, the generalization procedure involves only one concept. This will be demonstrated, along with a few examples, in the next section.

## 5.2. THE SINGLET GENERALIZED MOLECULAR DISTRIBUTION FUNCTION

In this section, we present a special case of the generalization procedure outlined in the previous section. We also establish new notation that will be useful for later applications. Consider the ordinary singlet MDF:

$$N_L^{(1)}(\mathbf{S}_1)\,d\mathbf{S}_1 = d\mathbf{S}_1 \int \cdots \int d\mathbf{R}^N\, P(\mathbf{R}^N) \sum_{i=1}^N \delta(\mathbf{R}_i - \mathbf{S}_1)$$

$$= N\,d\mathbf{S}_1 \int \cdots \int d\mathbf{R}^N\, P(\mathbf{R}^N)\, \delta(\mathbf{R}_1 - \mathbf{S}_1) \qquad (5.2)$$

Here, $N_L^{(1)}(\mathbf{S}_1)\,d\mathbf{S}_1$ is the average number[1] of particles occupying the element of volume $d\mathbf{S}_1$. For the present treatment, we specialize our discussion to

---

[1] We use the letter $N$ rather than $\varrho$ for density of particles. This is done in order to unify the system of notation for the continuous as well as discrete cases that are treated in this section.

spherical molecules only. As we already stressed in Section 2.2, the quantity defined in (5.2) can be assigned two different meanings. The first follows from the first form on the rhs of (5.2), which is identified as an average quantity in the $T, V, N$ ensemble. The second form on the rhs of (5.2) provides the probability of finding particle 1 in the element of volume $d\mathbf{S}_1$. Clearly, this probability is given by $N_L^{(1)}(\mathbf{S}_1)\, d\mathbf{S}_1/N$.

In order to systematize the procedure of generalization, let us rewrite (5.2), yet in a somewhat more complicated way. For each configuration $\mathbf{R}^N$, we define the *property* of the particle $i$ as

$$L_i(\mathbf{R}^N) = \mathbf{R}_i \tag{5.3}$$

The property of particle $i$ defined in (5.3) is simply its location $\mathbf{R}_i$. This is the reason for using the letter $L$ in the definition of the function $L_i(R^N)$ and as a subscript in (5.2).

Next, we define the "counting function" of the property $L$ by

$$N_L^{(1)}(\mathbf{R}^N, \mathbf{S}_1)\, d\mathbf{S}_1 = \sum_{i=1}^{N} \delta[L_i(\mathbf{R}^N) - \mathbf{S}_1]\, d\mathbf{S}_1 \tag{5.4}$$

which is the number of particles whose property $L$ attains a value within $d\mathbf{S}_1$ at $\mathbf{S}_1$, given the specific configuration $\mathbf{R}^N$. The average number (here in the $T, V, N$ ensemble) of such particles is

$$N_L^{(1)}(\mathbf{S}_1)\, d\mathbf{S}_1 = \langle N_L^{(1)}(\mathbf{R}^N, \mathbf{S}_1) \rangle\, d\mathbf{S}_1$$

$$= d\mathbf{S}_1 \int \cdots \int d\mathbf{R}^N\, P(\mathbf{R}^N) \sum_{i=1}^{N} \delta[L_i(\mathbf{R}^N) - \mathbf{S}_1] \tag{5.5}$$

which is the same as (5.2). For the same property $L$, we can distinguish between various *conditions*. For instance, we may count all the particles whose property $L$ has the value within a given region $D$, instead of $d\mathbf{S}_1$. However, the latter can be obtained by a simple integration (or summation, in discrete cases) over the appropriate region. Therefore, the generalization procedure involves essentially the property of the particles under consideration.

We now illustrate a few examples of properties that may replace $L$ in (5.4) and (5.5).

### 5.2.1. Coordination Number (CN)

A simple property which has been the subject of many investigations [see Bernal and King (1968)] is the coordination number (CN). We recall that the *average* coordination number can be obtained from the pair distribution function (Section 2.6). Here, we are interested in more detailed information on the distribution of CN's.

Let $R_C$ be a fixed number, to serve as the radius of the first coordination shell. If $\sigma$ is the effective diameter of the particles of the system, a reasonable choice of $R_C$ for our purposes is $\sigma \lesssim R_C \lesssim 1.5\sigma$. There exists a certain latitude in the choice of $R_C$. The range given above seems to be in conformity with the current concept of the radius of the *first* coordination sphere around a given particle. In what follows, we assume that $R_C$ has been fixed and we omit it from the notation.

The *property* to be considered here is the CN of the particle $i$ at a given configuration $\mathbf{R}^N$. This is defined by

$$C_i(\mathbf{R}^N) = \sum_{j=1, j \neq i}^{N} H(|\mathbf{R}_j - \mathbf{R}_i| - R_C) \tag{5.6}$$

where $H(x)$ is a unit step function

$$H(x) = \begin{cases} 0 & \text{if } x > 0 \\ 1 & \text{if } x \leq 0 \end{cases} \tag{5.7}$$

Each summand in (5.6) contributes unity whenever $|\mathbf{R}_j - \mathbf{R}_i| < R_C$, i.e., whenever the center of particle $j$ falls within the coordination sphere of particle $i$. Hence, $C_i(\mathbf{R}^N)$ is the number of particles ($j \neq i$) that falls in the coordination sphere of particle $i$ for a given configuration $\mathbf{R}^N$. Next, we define the counting function for this property by

$$N_C^{(1)}(\mathbf{R}^N, K) = \sum_{i=1}^{N} \delta[C_i(\mathbf{R}^N) - K] \tag{5.8}$$

Here, we have used the Kronecker delta function $\delta(x - K)$ instead of the more common notation $\delta_{x,K}$ for the sake of unity of notation. The meaning of $\delta$ as a Dirac or Kronecker delta should be clear from the context. In the sum of (5.8), we scan all the particles ($i = 1, 2, \ldots, N$) of the system in a given configuration $\mathbf{R}^N$. Each particle whose CN is exactly $K$ contributes unity to the sum (5.8), and zero otherwise. Hence, the sum in (5.8) counts all particles whose CN is $K$ for the particular configuration $\mathbf{R}^N$. A schematic illustration of this counting procedure is shown in Fig. 5.1. The average number of such particles is

$$N_C^{(1)}(K) = \langle N_C^{(1)}(\mathbf{R}^N, K) \rangle = N \int \cdots \int d\mathbf{R}^N \, P(\mathbf{R}^N) \, \delta[C_1(\mathbf{R}^N) - K] \tag{5.9}$$

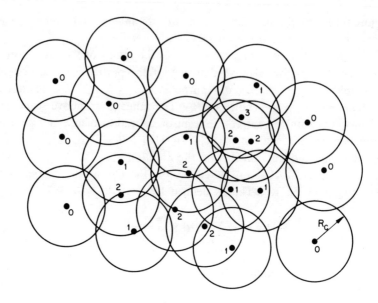

Fig. 5.1. A two-dimensional example of counting the coordination number of each molecule for a specific configuration of the whole system. A sphere of radius $R_C$ is drawn about the center of each molecule, indicated by the dark circles. The number written near each center is the coordination number corresponding to this particular example. There are altogether 23 particles distributed in the following manner: $N(0) = 9$, $N(1) = 7$, $N(2) = 6$, $N(3) = 1$ (for the particular configuration shown here).

We define the following quantity:

$$x_C(K) = N_C^{(1)}(K)/N = \int \cdots \int d\mathbf{R}^N P(\mathbf{R}^N) \, \delta[C_1(\mathbf{R}^N) - K] \quad (5.10)$$

From the definition of $N_C^{(1)}(K)$ in (5.9), it follows that $x_C(K)$ is the mole fraction of particles whose coordination number is equal to $K$. On the other hand, the second form on the rhs of (5.10) provides the probabilistic meaning of $x_C(K)$, i.e., this is the probability that a specific particle, say 1, will be found with CN equal to $K$.

The quantity $x_C(K)$ can be viewed as a component of a vector

$$\mathbf{x}_C = (x_C(0), x_C(1), \ldots) \quad (5.11)$$

which gives the "composition" of the system with respect to the classification according to the CN's, i.e., each component gives the mole fraction of particles with a given CN. The average CN of particles in the

# Generalized Molecular Distribution Functions

system is given by[2]

$$\langle K \rangle = \sum_{K=0}^{\infty} K x_C(K) \tag{5.12}$$

It is important to note that the distribution vector (5.11) contains more information on the system than the average quantity $\langle K \rangle$.

We also use this particular example to demonstrate that changes in the *condition* can be achieved easily. For instance, with the same property (CN), we may ask for the average number of particles whose CN is less than or equal to, say, five. This is obtained from (5.9):

$$N_C^{(1)}(K \leq 5) = \sum_{K=0}^{5} N_C^{(1)}(K) \tag{5.13}$$

Therefore, there is no need to treat separately the generalization procedure for the conditions imposed on a given property.

The CN, as defined above, may be viewed as a property conveying the local density around the particles. By a simple transformation, we can make this statement more precise. The local density around particle $i$ for a given configuration $\mathbf{R}^N$ is defined by

$$D_i(\mathbf{R}^N) = C_i(\mathbf{R}^N)(\tfrac{4}{3}\pi R_C^3)^{-1} \tag{5.14}$$

Hence, the average number of particles having local density equal to $\eta$ is

$$N_D^{(1)}(\eta) = \left\langle \sum_{i=1}^{N} \delta[D_i(\mathbf{R}^N) - \eta] \right\rangle \tag{5.15}$$

Another quantity conveying a similar meaning will be introduced later in this section.

## 5.2.2. Binding Energy (BE)

The next property is referred to as binding energy (BE) and is defined for particle $i$ for the configuration $\mathbf{R}^N$ as follows:

$$B_i(\mathbf{R}^N) = U_N(\mathbf{R}_1, \ldots, \mathbf{R}_{i-1}, \mathbf{R}_i, \mathbf{R}_{i+1}, \ldots, \mathbf{R}_N)$$
$$- U_{N-1}(\mathbf{R}_1, \ldots, \mathbf{R}_{i-1}, \mathbf{R}_{i+1}, \ldots, \mathbf{R}_N) \tag{5.16}$$

---

[2] Note that $\langle K \rangle$, as defined in (5.12), coincides with the definition of the average CN given in Eq. (2.76), provided that we choose $R_C$ of this section to coincide with $R_M$ of Chapter 2.

This is the work required to bring a particle from infinite distance, with respect to the other particles, to the position $\mathbf{R}_i$. For a system of pairwise additive potentials, (5.16) reduces to

$$B_i(\mathbf{R}^N) = \sum_{j=1, j\neq i}^{N} U(\mathbf{R}_i, \mathbf{R}_j) \tag{5.17}$$

The counting function for this property is

$$N_B^{(1)}(\mathbf{R}^N, \nu) \, d\nu = d\nu \sum_{i=1}^{N} \delta[B_i(\mathbf{R}^N) - \nu] \tag{5.18}$$

which is the number of particles having BE between $\nu$ and $\nu + d\nu$, for the specified configuration $\mathbf{R}^N$. Note that since $\nu$ is a continuous variable, the $\delta$-function in (5.18) is the Dirac delta function. The average number of particles having BE between $\nu$ and $\nu + d\nu$ is

$$N_B^{(1)}(\nu) \, d\nu = d\nu \left\langle \sum_{i=1}^{N} \delta[B_i(\mathbf{R}^N) - \nu] \right\rangle \tag{5.19}$$

The corresponding mole fraction is

$$x_B(\nu) \, d\nu = N_B^{(1)}(\nu) \, d\nu / N \tag{5.20}$$

with the normalization condition

$$\int_{-\infty}^{\infty} x_B(\nu) \, d\nu = 1 \tag{5.21}$$

The function $x_B(\nu)$ is referred to as the distribution of BE. In analogy with the vector (5.11), which has discrete components, we often write $\mathbf{x}_B$ for the whole distribution function, the components of which are $x_B(\nu)$. This function will play an important role in later applications.

### 5.2.3. Volume of the Voronoi Polyhedron (VP)

Another local property of interest in the study of liquids is the Voronoi polyhedron (VP), or the Dirichlet region, defined as follows.

Consider a specific configuration $\mathbf{R}^N$ and a particular particle $i$. Let us draw all the segments $l_{ij}$ ($j = 1, \ldots, N$, $j \neq i$) connecting the centers of particles $i$ and $j$. Let $P_{ij}$ be the plane perpendicular to and bisecting the line $l_{ij}$. Each plane $P_{ij}$ divides the entire space into two parts. Denote by $V_{ij}$ the part of space that includes the point $\mathbf{R}_i$. The VP of particle $i$ for the

# Generalized Molecular Distribution Functions 185

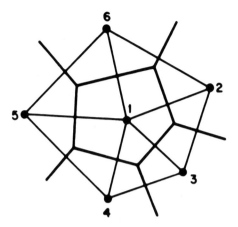

Fig. 5.2. Construction of the Voronoi polygon of particle 1 in a two-dimensional system of particles.

configuration $\mathbf{R}^N$ is defined as the intersection of all the $V_{ij}$ ($j = 1, \ldots, N$, $j \neq i$):

$$(\text{VP})_i = \bigcap_{j=1, j\neq i}^{N} V_{ij}(\mathbf{R}_i, \mathbf{R}_j) \qquad (5.22)$$

A two-dimensional illustration of a construction of a VP is shown in Fig. 5.2. It is clear from the definition that the region $(\text{VP})_i$ includes all the points in space that are nearer to $\mathbf{R}_i$ than to any $\mathbf{R}_j$ ($j \neq i$). Furthermore, each VP contains the center of one and only one particle. A two-dimensional illustration is given in Fig. 5.3.

The concept of VP can be used to generate a few local properties[3]; the one we shall be using is the volume of the VP, which we denote by

$$\psi_i(\mathbf{R}^N) = \text{volume of } (\text{VP})_i \qquad (5.23)$$

The counting function for this property is

$$N_\psi^{(1)}(\mathbf{R}^N, \phi) \, d\phi = d\phi \sum_{i=1}^{N} \delta[\psi_i(\mathbf{R}^N) - \phi] \qquad (5.24)$$

---

[3] Note that the *form* of the VP is also a *property* which can be considered in the context of this section. Other properties of interest are the number of faces of the VP, the surface area of the VP, etc. The distribution functions defined in this section involve random variables whose values are real numbers. If we choose the *form* of the VP as a random variable, then its range of variation is the space of geometric figures and not real numbers.

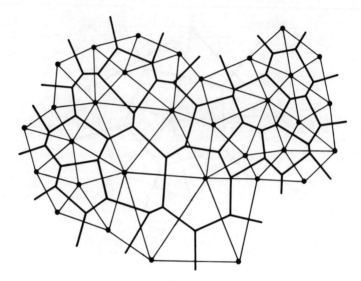

Fig. 5.3. A two-dimensional system with its network of Voronoi polygons. Heavy lines: boundaries of the Voronoi polygons. Light lines: the dual network of polygons, constructed by lines $l_{ij}$ connecting pairs of particles which contribute a boundary for the Voronoi polygons of $i$ and $j$.

and its average is

$$N_\psi^{(1)}(\phi)\, d\phi = d\phi \Big\langle \sum_{i=1}^{N} \delta[\psi_i(\mathbf{R}^N) - \phi] \Big\rangle \tag{5.25}$$

Clearly, $N_\psi^{(1)}(\phi)\, d\phi$ is the average number of particles whose VP has a volume between $\phi$ and $\phi + d\phi$.

### 5.2.4. Combination of Properties

One way of generating new properties from previous ones is by combination. For instance, the counting function for BE *and* the volume of the VP is

$$N_{B,\psi}^{(1)}(\mathbf{R}^N, \nu, \phi)\, d\nu\, d\phi = d\nu\, d\phi \sum_{i=1}^{N} \delta[B_i(\mathbf{R}^N) - \nu]\, \delta[\psi_i(\mathbf{R}^N) - \phi] \tag{5.26}$$

which counts the number of particles having BE between $\nu$ and $\nu + d\nu$ *and* the volume of the VP between $\phi$ and $\phi + d\phi$. The average number of

### Generalized Molecular Distribution Functions

such particles is

$$N_{B,\psi}^{(1)}(v, \phi) \, dv \, d\phi = dv \, d\phi \left\langle \sum_{i=1}^{N} \delta[B_i(\mathbf{R}^N) - v] \, \delta[\psi_i(\mathbf{R}^N) - \phi] \right\rangle \quad (5.27)$$

Note that although we have combined two *properties*, we still have a *singlet* GMDF. This should be distinguished from the *pair* GMDF introduced in Section 5.4 [see, for instance, Eq. (5.41)]. The singlet GMDF introduced in (5.27) is of potential importance in the study of liquid water. A related singlet GMDF, which conveys similar information to the one in (5.27), but is simpler for computational purposes, is constructed by the combination of BE and CN, i.e.,

$$N_{B,C}^{(1)}(v, K) \, dv = dv \left\langle \sum_{i=1}^{N} \delta[B_i(\mathbf{R}^N) - v] \, \delta[C_i(\mathbf{R}^N) - K] \right\rangle \quad (5.28)$$

Note that the first $\delta$ on the rhs of (5.28) is a Dirac delta function, whereas the second is a Kronecker delta function.

The general procedure of defining GMDF's is now clear. We first define a *property* which is a function definable on the configurational space, say $G_i(\mathbf{R}^N)$, and then introduce its distribution function in the appropriate ensemble. Later, we shall find the $T$, $P$, $N$ more useful than the $T$, $V$, $N$ ensemble for actual applications.

## 5.3. ILLUSTRATIVE EXAMPLES OF GMDF's

In this section, we present some examples of GMDF's. We confine ourselves to spherical particles in two dimensions. (In Chapter 6, we present some further examples for particles interacting via noncentral forces.) All the illustrations given in this section were obtained by a Monte Carlo computation on a two-dimensional system, consisting of 36 Lennard-Jones particles, for which the pair potential is presumed to have the form [for more details, see Ben-Naim (1973b)]

$$U(R) = 4\varepsilon[(\sigma/R)^{12} - (\sigma/R)^6] \quad (5.29)$$

Figure 5.4 shows the distribution of CN, $x_C(K)$, and its density dependence. The Lennard-Jones parameters for these computations are[4]

$$\sigma = 1.0, \quad \varepsilon/kT = 0.5 \quad (5.30)$$

---

[4] We choose here a dimensionless length parameter $\sigma$. The number density $\varrho$ is therefore the number of particles per unit area $\sigma^2$.

and the coordination radius is chosen as

$$R_C = 1.5\sigma \tag{5.31}$$

Also included in the figure are the radial distribution functions for these systems. The number density is indicated next to each curve. The behavior

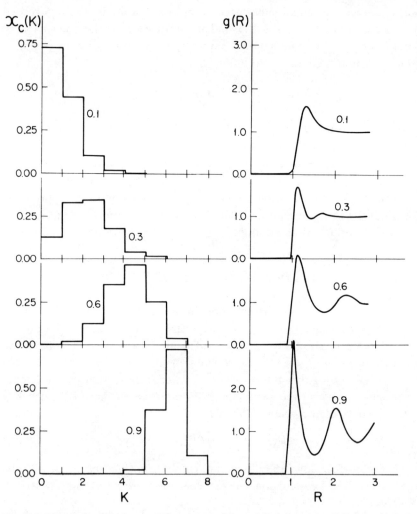

Fig. 5.4. The singlet functions $x_C(K)$ and the pair correlation functions $g(R)$ for spherical particles interacting via a Lennard-Jones potential ($\sigma = 1.0$ and $\varepsilon/kT = 0.5$). The number density is indicated next to each curve. These curves were obtained by Monte Carlo simulation with 36 particles and for about $3$–$4 \times 10^5$ configurations.

# Generalized Molecular Distribution Functions

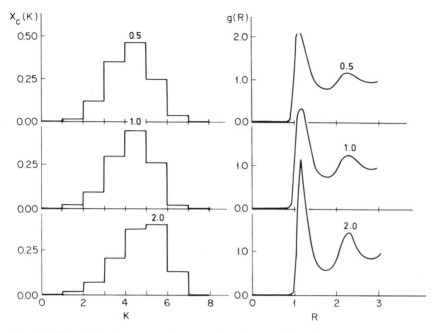

Fig. 5.5. The singlet functions $x_C(K)$ and the pair correlation functions $g(R)$ for the same system as in Fig. 5.4 but with a fixed density $\varrho = 0.6$ and various energy parameters $\varepsilon/kT$ (which are indicated next to each curve).

of these functions is according to our intuitive expectations. As the density increases, the mole fraction of particles with relatively higher CN increases.

Figure 5.5 shows the dependence of $x_C(K)$ and $g(R)$ on $\varepsilon/kT$ for the same system at a fixed number density $\varrho = 0.6$.

Figures 5.6 and 5.7 show the dependence of $x_B(v)$ on the density and on $\varepsilon/kT$, respectively. The most pronounced feature of all these curves is the essentially single peak.[5] In order to understand this phenomenon, it is useful to consider a limiting case, namely a system of hard-sphere particles. Clearly, the BE of such particles can attain one of two values: zero or infinity. It is zero whenever the particle under observation does not "penetrate" into any other particle, and it is infinity whenever it comes closer than $\sigma$ to any other particle. However, since the latter

---

[5] This does not exclude the possibility that some small wiggles will occur in these functions. However, within the approximate scheme of the present computation, the functions $x_B(v)$ seem to be quite smooth and exhibit one maximum.

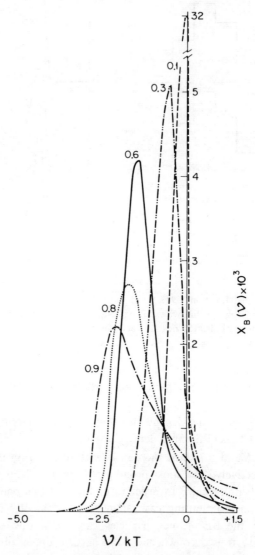

Fig. 5.6. The singlet functions $x_B(v)$ as a function of density (indicated next to each curve) for spherical particles with parameters $\sigma = 1.0$, $\varepsilon/kT = 0.5$.

# Generalized Molecular Distribution Functions

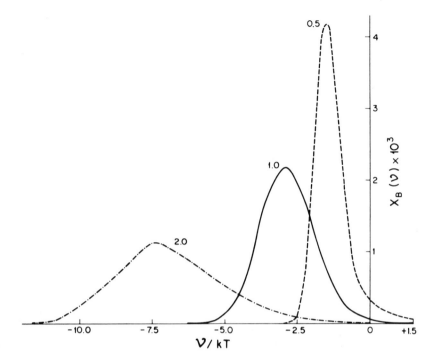

Fig. 5.7. The singlet function $x_B(\nu)$ for spherical particles with different energy parameters $\varepsilon/kT$ (indicated next to each curve). The density for each curve is $\varrho = 0.6$.

event has zero probability [because $P(\mathbf{R}^N)$ vanishes whenever two hard spheres penetrate into each other], it follows that $x_B(\nu)$ must be a Dirac delta function

$$x_B(\nu) = \delta(\nu) \quad \text{(hard spheres)} \quad (5.32)$$

Next, suppose that the intermolecular potential for simple particles can be represented by

$$U(R) = U^{\mathrm{HS}}(R) + U^{\mathrm{SI}}(R) \quad (5.33)$$

where the "soft interaction" (SI) is viewed as a perturbation to the hard-sphere potential function. Clearly, if the perturbation is small, the properties of the system will be dominated by the hard-sphere potential. In particular, we expect that $x_B(\nu)$ will have essentially a single sharp peak. The average value of $\nu$ must be given by

$$\langle \nu \rangle = 2\langle U_N(\mathbf{R}^N)\rangle/N \quad (5.34)$$

This equality can be shown to follow directly from the definitions (5.19) and (5.20); indeed,

$$\langle \nu \rangle = \int_{-\infty}^{\infty} \nu x_B(\nu) \, d\nu$$

$$= \frac{1}{N} \int_{-\infty}^{\infty} \nu \, d\nu \int \cdots \int d\mathbf{R}^N \, P(\mathbf{R}^N) \sum_{i=1}^{N} \delta[B_i(\mathbf{R}^N) - \nu]$$

$$= \frac{1}{N} \int \cdots \int d\mathbf{R}^N \, P(\mathbf{R}^N) \sum_{i=1}^{N} \int_{-\infty}^{\infty} \nu \, d\nu \, \delta[B_i(\mathbf{R}^N) - \nu]$$

$$= \frac{1}{N} \int \cdots \int d\mathbf{R}^N \, P(\mathbf{R}^N) \sum_{i=1}^{N} B_i(\mathbf{R}^N)$$

$$= \frac{2}{N} \int \cdots \int d\mathbf{R}^N \, P(\mathbf{R}^N) U_N(\mathbf{R}^N) = \frac{2\langle U_N(\mathbf{R}^N) \rangle}{N} \quad (5.35)$$

In Fig. 5.6, it is observed that as the density decreases, the peak of $x_B(\nu)$ shifts to the right and becomes sharper. Similar behavior is exhibited in Fig. 5.7, in which $\varepsilon/kT$ is varied.

As an example of a singlet GMDF of a combined property, Fig. 5.8 shows the distribution function defined by

$$x_{B,C}(\nu, K) = N_{B,C}^{(1)}(\nu, K)/N \quad (5.36)$$

which is related to the probability[6] of finding a particle with binding energy between $\nu$ and $\nu + d\nu$ *and* coordination number $K$. The most important feature of this distribution function is that the average BE defined for each CN is a monotonic function of the CN. More precisely, we define the average value of $\nu$ for each value of $K$ by

$$\langle \nu(K) \rangle = \int_{-\infty}^{\infty} \nu x_{B,C}(\nu, K) \, d\nu \bigg/ \int_{-\infty}^{\infty} x_{B,C}(\nu, K) \, d\nu \quad (5.37)$$

[Note that the symbol $\langle \nu(K) \rangle$ stands for an average over $\nu$ with a *fixed* value of $K$; a better notation could be $\langle \nu \rangle(K)$.]

In (5.37), we have constructed the *conditional* average of the BE, given a fixed value for the CN. The average is computed with the corresponding conditional probability (see next section for more details)

$$x_{B/C}(\nu/K) \, d\nu = x_{B,C}(\nu, K) \, d\nu \bigg/ \int_{-\infty}^{\infty} x_{B,C}(\nu, K) \, d\nu \quad (5.38)$$

---

[6] Here, $x_{B,C}(\nu, K)$ is a density function with respect to $\nu$ and a distribution function with respect to $K$; i.e., $x_{B,C}(\nu, K) \, d\nu$ has the meaning of probability. Figure 5.8 shows the values of the probability for the events $K$ and $\Delta\nu$ as explained in the legend.

# Generalized Molecular Distribution Functions

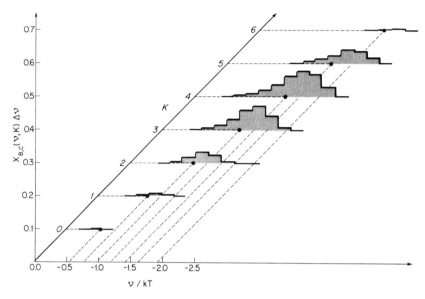

Fig. 5.8. The mole fractions (or probabilities) $x_{B,C}(\nu, K)\,\Delta\nu$ of particles with a given coordination number $K$ and with binding energy between $\nu/kT$ and $(\nu + \Delta\nu)/kT$. The dark circles indicate the average binding energy for each coordination number (the exact location of the average is within the interval $\nu$ and $\nu + \Delta\nu$). Each curve is the conditional distribution of binding energy for a given coordination number. These functions pertain to a system of spherical particles with parameters $\sigma = 1.0$, $\varepsilon/kT = 0.5$, and $\varrho = 0.6$.

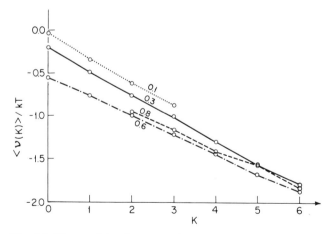

Fig. 5.9. The conditional average binding energy as a function of the coordination number [defined in Eq. (5.37)] for a system of spherical particles with parameters $\sigma = 1.0$, $\varepsilon/kT = 0.5$. The density is indicated next to each curve.

which is the probability of finding a specific particle with BE between $v$ and $v + dv$, given that the particle has CN equal to $K$. The function $\langle v(K) \rangle$ is shown in Fig. 5.9. It is clearly seen, and intuitively obvious, that $\langle v(K) \rangle$ decreases as $K$ increases. A fundamentally different behavior is expected for complex fluids such as water, as will be discussed in the next chapter.

## 5.4. PAIR AND HIGHER-ORDER GMDF's

The extension of the concepts of GMDF's to higher orders in the number of particles is quite straightforward. We elaborate only on the pair distribution function, which will be of foremost importance in later applications. The ordinary pair distribution function is written in the notation of Section 5.2 as

$$N^{(2)}_{L;L}(\mathbf{S}_1, \mathbf{S}_2)\, d\mathbf{S}_1\, d\mathbf{S}_2$$
$$= d\mathbf{S}_1\, d\mathbf{S}_2 \Big\langle \sum_{\substack{i=1 \\ i \neq j}}^{N} \sum_{j=1}^{N} \delta[L_i(\mathbf{R}^N) - \mathbf{S}_1]\, \delta[L_j(\mathbf{R}^N) - \mathbf{S}_2] \Big\rangle \quad (5.39)$$

which is the average number of *pairs* of particles; for one of them, the value of the property $L$ is $\mathbf{S}_1$, and the second has the value $\mathbf{S}_2$ for the same property. The structure of (5.39) is similar to that of (5.27), with the exception that here we are concerned with *properties* assigned to two *different* particles.

The generalization of (5.39) can be carried out in various ways. For instance, we may assign the same property to the two particles, different properties, or even combined properties. Examples are

$$N^{(2)}_{B;B}(v, v')\, dv\, dv' = dv\, dv' \Big\langle \sum_{\substack{i=1 \\ i \neq j}}^{N} \sum_{j=1}^{N} \delta[B_i(\mathbf{R}^N) - v]\, \delta[B_j(\mathbf{R}^N) - v'] \Big\rangle \quad (5.40)$$

$$N^{(2)}_{B;C}(v, K)\, dv = dv \Big\langle \sum_{\substack{i=1 \\ i \neq j}}^{N} \sum_{j=1}^{N} \delta[B_i(\mathbf{R}^N) - v]\, \delta[C_j(\mathbf{R}^N) - K] \Big\rangle \quad (5.41)$$

$$N^{(2)}_{L,B;L,B}(\mathbf{S}_1, v_1, \mathbf{S}_2, v_2) = \Big\langle \sum_{\substack{i=1 \\ i \neq j}}^{N} \sum_{j=1}^{N} \delta[L_i - \mathbf{S}_1]\, \delta[B_i - v_1]$$
$$\times\, \delta[L_j - \mathbf{S}_2]\, \delta[B_j - v_2] \Big\rangle \quad (5.42)$$

The convention used for the notation is the following. The superscript to $N$ stands for the order of the GMDF. Here, it is two, whereas in Sec-

## Generalized Molecular Distribution Functions

tion 2, it was one (singlet). The subscript indicates the set of properties. Properties assigned to the same particle are separated by a comma, and those assigned to different particles by a semicolon. For instance, in (5.40), we have the same property, BE, assigned to the two particles. In (5.41), we have BE for one particle and CN for the second. In (5.42), we assign the combined properties $L$ and BE to the two particles [an obvious shorthand notation for the properties in (5.42) has been employed].

Note that in Section 5.2, we did not use the location ($L$) for constructing combined properties. For instance, the combination of the properties $L$ and BE for the singlet GMDF is identical with the singlet GMDF with respect to BE alone. The specification of the location of the particle is of no importance in this case as long as we are concerned with homogeneous fluids. New and useful pair GMDF's are obtained by incorporating $L$ in combination with other properties.

Various conditional distribution functions can be introduced. For instance,

$$N_{L,B/L,B}(\mathbf{S}_1, \nu_1/\mathbf{S}_2, \nu_2) = \frac{N^{(2)}_{L,B;L,B}(\mathbf{S}_1, \nu_1, \mathbf{S}_2, \nu_2)}{N^{(1)}_{L,B}(\mathbf{S}_2, \nu_2)} \qquad (5.43)$$

is the average density of particles at $\mathbf{S}_1$ and $\nu_1$ given a particle at $\mathbf{S}_2$ and $\nu_2$. Similarly, we can define correlation functions such as

$$g^{(2)}_{L,B;L,B}(\mathbf{S}_1, \nu_1, \mathbf{S}_2, \nu_2) = \frac{N^{(2)}_{L,B;L,B}(\mathbf{S}_1, \nu_1, \mathbf{S}_2, \nu_2)}{N^{(1)}_{L,B}(\mathbf{S}_1, \nu_1) N^{(1)}_{L,B}(\mathbf{S}_2, \nu_2)} \qquad (5.44)$$

In a homogeneous system, the correlation function defined in (5.44) is a function of $\nu_1$, $\nu_2$, and the scalar separation between the two locations $|\mathbf{S}_2 - \mathbf{S}_1|$. The generalized pair correlation function in (5.44) can also be viewed as an ordinary pair correlation function between two "species" $\nu_1$ and $\nu_2$. This point of view will be discussed further in Section 5.7.

We will not elaborate further on higher-order GMDF's, since these will not be needed for the purposes of this book.

## 5.5. RELATIONS BETWEEN THERMODYNAMIC QUANTITIES AND GMDF's

In this section, we present some formal relations between thermodynamic quantities and GMDF's. We elaborate on some simple examples here, and defer to Section 5.8 the discussion of more general relations.

Two quite obvious connections follow almost directly from the definitions:

$$E = N\varepsilon^K + \tfrac{1}{2} \int_{-\infty}^{\infty} \nu N_B^{(1)}(\nu)\, d\nu \qquad (5.45)$$

$$V = \int_0^{\infty} \phi N_\nu^{(1)}(\phi)\, d\phi \qquad (5.46)$$

where $\varepsilon^K$ is the average kinetic energy per particle (including any internal energy that may exist). Relation (5.45) can be verified by substituting $N_B^{(1)}(\nu)$ from its definition (5.19):

$$E = N\varepsilon^K + \tfrac{1}{2} \int_{-\infty}^{\infty} \nu\, d\nu \int \cdots \int d\mathbf{R}^N\, P(\mathbf{R}^N) \sum_{i=1}^{N} \delta[B_i(\mathbf{R}^N) - \nu]$$

$$= N\varepsilon^K + \int \cdots \int d\mathbf{R}^N\, P(\mathbf{R}^N) \left[ \tfrac{1}{2} \sum_{i=1}^{N} B_i(\mathbf{R}^N) \right] = N\varepsilon^K + \langle U_N(\mathbf{R}^N) \rangle \qquad (5.47)$$

where we have interchanged the order of integration and followed similar steps as in (5.35). The result is the familiar expression for the internal energy as derived in Section 3.3. The point that should be stressed is that in (5.45), $E$ is expressed in terms of a *singlet* GMDF, whereas in Section 3.3, we derived an expression for $E$ involving the ordinary *pair* distribution function, which, in the notation of this chapter, is written as

$$E = N\varepsilon^K + \tfrac{1}{2} \int \int d\mathbf{R}_1\, d\mathbf{R}_2\, U(\mathbf{R}_1, \mathbf{R}_2) N_{L;L}^{(2)}(\mathbf{R}_1, \mathbf{R}_2) \qquad (5.48)$$

This example seems to be representative of a general result: When a given quantity is expressed in terms of GMDF's, it requires in general a lower-order distribution function compared with the corresponding expression in terms of ordinary MDF's. We shall encounter other examples later. It should be realized that although the GMDF's may seem to be more complex than ordinary MDF's, this does not necessarily imply that their computation, either analytically or numerically, should be more difficult.

The second relation (5.46) is obvious and follows from the fact that the volumes of the VP of all particles add up to give the total volume of the system. [In the $T, V, N$ ensemble, (5.46) gives the *exact* volume $V$ of the system. On the other hand, in the $T, P, N$ ensemble, $V$ stands for the *average* volume of the system.]

We now discuss various derivatives of $E$ and $V$ that will be useful in later applications.

### 5.5.1. Heat Capacity at Constant Volume

In Section 3.8, we saw that the heat capacity (at constant volume) is expressible in terms of ordinary MDF's of up to order *four*. The order is reduced to two by using GMDF's. This is demonstrated below.

Differentiating (5.45) with respect to temperature gives

$$C_V = NC^K + \frac{1}{2} \int_{-\infty}^{\infty} \nu \frac{\partial N_B^{(1)}(\nu)}{\partial T} d\nu = NC^K + \Delta C_V \qquad (5.49)$$

In this section, we assume that $N_B^{(1)}(\nu)$ has been defined in the $T, V, N$ ensemble, and all the derivatives that follow are at constant $N$ and $V$. $C^K$ is the contribution to the heat capacity per particle due to the kinetic (and internal) energies. We shall be mainly concerned with the second term on the rhs of (5.49), denoted by $\Delta C_V$, which arises from the existence of interactions among the particles.

By direct differentiation of $N_B^{(1)}(\nu)$ in (5.19), using the basic distribution function in the $T, V, N$ ensemble,

$$P(\mathbf{R}^N) = \exp[-\beta U_N(\mathbf{R}^N)]/Z_N \qquad (5.50)$$

we get

$$\Delta C_V = \frac{1}{2kT^2} \int_{-\infty}^{\infty} \nu \, d\nu \Big\{ \int \cdots \int d\mathbf{R}^N P(\mathbf{R}^N) \sum_{i=1}^{N} \delta[B_i(\mathbf{R}^N) - \nu] U_N(\mathbf{R}^N)$$

$$- \int \cdots \int d\mathbf{R}^N P(\mathbf{R}^N) \sum_{i=1}^{N} \delta[B_i(\mathbf{R}^N) - \nu] \int \cdots \int d\mathbf{R}^N P(\mathbf{R}^N) U_N(\mathbf{R}^N) \Big\}$$

$$= \frac{N}{2kT^2} \int_{-\infty}^{\infty} \nu \, d\nu \{\langle \delta[B_1(\mathbf{R}^N) - \nu] U_N(\mathbf{R}^N) \rangle$$

$$- \langle \delta[B_1(\mathbf{R}^N) - \nu] \rangle \langle U_N(\mathbf{R}^N) \rangle \} \qquad (5.51)$$

We now recall that

$$U_N(\mathbf{R}^N) = \tfrac{1}{2} \sum_{i=1}^{N} B_i(\mathbf{R}^N) \qquad (5.52)$$

Putting (5.52) in (5.51) and interchanging the order of integration, we find

$$\Delta C_V = \frac{N}{2kT^2} [\langle U_N(\mathbf{R}^N) B_1(\mathbf{R}^N) \rangle - \langle U_N(\mathbf{R}^N) \rangle \langle B_1(\mathbf{R}^N) \rangle]$$

$$= \frac{N}{2kT^2} \left( \frac{1}{2} \Big\langle \sum_{i=1}^{N} B_i B_1 \Big\rangle - \frac{1}{2} \Big\langle \sum_{i=1}^{N} B_i \Big\rangle \langle B_1 \rangle \right)$$

$$= \frac{N}{2kT^2} \left[ \frac{1}{2} \langle B_1^2 \rangle + \frac{1}{2}(N-1) \langle B_1 B_2 \rangle - \frac{1}{2} N \langle B_1 \rangle \langle B_1 \rangle \right]$$

$$= \frac{N}{4kT^2} [(\langle B_1^2 \rangle - \langle B_1 \rangle^2) + (N-1)(\langle B_1 B_2 \rangle - \langle B_1 \rangle \langle B_2 \rangle)] \qquad (5.53)$$

In (5.53), we used an obvious shorthand notation for $B_i(\mathbf{R}^N)$. In the last form on the rhs of (5.53), we see that $\Delta C_V$ depends on the fluctuation of the BE of a single particle and on the cross fluctuation of BE between two different particles. It should be noted that in (5.53) we are concerned with fluctuations of *molecular* quantities and not of macroscopic quantities (compare with Section 1.4, for example).

All of the averages in (5.53) can be now rewritten as integrals over the singlet and the pair GMDF's based on BE. To do this, we use the relations

$$\int_{-\infty}^{\infty} v N_B^{(1)}(v)\, dv = N\langle B_1 \rangle \tag{5.54}$$

$$\int_{-\infty}^{\infty} v^2 N_B^{(1)}(v)\, dv = N\langle B_1^2 \rangle \tag{5.55}$$

$$\int_{-\infty}^{\infty}\int_{-\infty}^{\infty} vv' N_{B;B}^{(2)}(v, v')\, dv\, dv' = N(N-1)\langle B_1 B_2 \rangle \tag{5.56}$$

Relations (5.54)–(5.56) can be easily verified using arguments similar to those employed in (5.35). Substituting these in (5.53), we obtain the final expression for the heat capacity in the $T, V, N$ ensemble:

$$C_V = NC^K + (1/4kT^2)\left\{ \int v^2 N_B^{(1)}(v)\, dv + \int\int vv' N_{B;B}^{(2)}(v, v')\, dv\, dv' \right.$$
$$\left. - \left[ \int v N_B^{(1)}(v)\, dv \right]^2 \right\} \tag{5.57}$$

Here, we have expressed $C_V$ explicitly in terms of the singlet and pair GMDF's. We recall that the heat capacity can also be expressed in terms of ordinary MDF's up to order four (see Section 3.8). In this respect, relation (5.57) for $C_V$ is somewhat simpler than the previous one. A different interpretation of the term $\Delta C_V$, using the mixture-model approach, will be developed in Section 5.9.

### 5.5.2. Heat Capacity at Constant Pressure

In this and in the following examples, we assume that the GMDF's have been defined in the $T, P, N$ ensemble, using the basic distribution function

$$P(\mathbf{R}^N, V) = \frac{\exp[-\beta U_N(\mathbf{R}^N) - \beta PV]}{\int dV \int \cdots \int d\mathbf{R}^N \exp[-\beta U_N(\mathbf{R}^N) - \beta PV]} \tag{5.58}$$

## Generalized Molecular Distribution Functions

The heat capacity at constant pressure is obtained from the temperature derivative of the enthalpy

$$C_P = (\partial H/\partial T)_{N,P} \tag{5.59}$$

where $H$ stands for the average enthalpy in the $T, P, N$ ensemble, defined by

$$H = \int_0^\infty dV \int \cdots \int d\mathbf{R}^N P(\mathbf{R}^N, V)[N\varepsilon^K + U_N(\mathbf{R}^N) + PV] \tag{5.60}$$

The term in the square brackets under the integral sign can be viewed as the "enthalpy" of the system for a given volume $V$ and configuration $\mathbf{R}^N$.

As in (5.45), the enthalpy $H$ can be expressed in terms of the singlet GMDF's $N_B^{(1)}(\nu)$ and $N_\psi^{(1)}(\phi)$, which have been defined in the $T, P, N$ ensemble. An argument similar to the one employed in (5.47) yields

$$H = N\varepsilon^K + \tfrac{1}{2} \int_{-\infty}^\infty \nu N_B^{(1)}(\nu)\, d\nu + P \int_0^\infty \phi N_\psi^{(1)}(\phi)\, d\phi \tag{5.61}$$

By direct differentiation of (5.61) with respect to temperature, and by using similar arguments (we omit the details) to those employed in the derivation of (5.53) and (5.57), we get for $C_P$

$$C_P = NC^K + \frac{N}{4kT^2}[(\langle B_1^2 \rangle - \langle B_1 \rangle^2) + (N-1)(\langle B_1 B_2 \rangle - \langle B_1 \rangle \langle B_2 \rangle)]$$

$$+ \frac{NP}{2kT^2}[(\langle B_1 \psi_1 \rangle - \langle B_1 \rangle \langle \psi_1 \rangle) + (N-1)(\langle B_1 \psi_2 \rangle - \langle B_1 \rangle \langle \psi_2 \rangle)] \tag{5.62}$$

$$C_P = NC^K + \frac{1}{2kT^2}\left\{ \frac{1}{2} \int \nu^2 N_B^{(1)}(\nu)\, d\nu + \frac{1}{2} \int\int \nu\nu' N_{B;B}^{(2)}(\nu, \nu')\, d\nu\, d\nu' \right.$$

$$+ P \int\int \nu\phi N_{B,\psi}^{(1)}(\nu, \phi)\, d\nu\, d\phi + P \int\int \nu\phi N_{B;\psi}^{(2)}(\nu, \phi)\, d\nu\, d\phi$$

$$\left. - \frac{1}{2}\left[\int \nu N_B^{(1)}(\nu)\, d\nu\right]^2 - P \int \nu N_B^{(1)}(\nu)\, d\nu \int \phi N_\psi^{(1)}(\phi)\, d\phi \right\} \tag{5.63}$$

Relations (5.62) and (5.63) should be compared with (5.53) and (5.57), respectively. Here, the expression is more complicated due to the appearance of cross fluctuation between BE and the volume of the VP for a single particle, as well as between a pair of particles. Nevertheless, we note that the order of the GMDF's needed does not exceed two. Again, we recall the expression for $C_P$ cited in Section 1.4. There, we were concerned with fluctuations in macroscopic quantities, whereas in (5.62), we are concerned with fluctuations of molecular quantities, i.e., quantities such as BE or the volume of VP assigned to single particles.

### 5.5.3. Coefficient of Thermal Expansion

The coefficient of thermal expansion is defined as

$$\alpha = \frac{1}{V}\left(\frac{\partial V}{\partial T}\right)_{N,P} \tag{5.64}$$

where $V$ is the average volume in the $T$, $P$, $N$ ensemble. Using (5.46) with $N_\psi^{(1)}(\phi)$ defined in the $T$, $P$, $N$ ensemble, we get by direct differentiation of $V$, using arguments similar to those above,

$$\left(\frac{\partial V}{\partial T}\right)_{N,P} = \frac{N}{2kT^2}[(\langle B_1\psi_1\rangle - \langle B_1\rangle\langle\psi_1\rangle) + (N-1)(\langle B_1\psi_2\rangle - \langle B_1\rangle\langle\psi_2\rangle)]$$

$$+ \frac{PN}{kT^2}[(\langle\psi_1^2\rangle - \langle\psi_1\rangle^2) + (N-1)(\langle\psi_1\psi_2\rangle - \langle\psi_1\rangle\langle\psi_2\rangle)] \tag{5.65}$$

The corresponding expression in terms of GMDF's is

$$\left(\frac{\partial V}{\partial T}\right)_{N,P} = \frac{1}{kT^2}\left\{\frac{1}{2}\int\int \phi v N_{B,\psi}^{(1)}(v,\phi)\,d\phi\,dv + \frac{1}{2}\int\int \phi v N_{B;\psi}^{(2)}(v,\phi)\,d\phi\,dv\right.$$

$$+ P\int \phi^2 N_\psi^{(1)}(\phi)\,d\phi + P\int\int \phi\phi' N_{\psi;\psi}^{(2)}(\phi,\phi')\,d\phi\,d\phi'$$

$$\left. - \frac{1}{2}\int \phi N_\psi^{(1)}(\phi)\,d\phi \int v N_B^{(1)}(v)\,dv - P\left[\int \phi N_\psi^{(1)}(\phi)\,d\phi\right]^2\right\} \tag{5.66}$$

We recall that the coefficient of thermal expansion is expressible in terms of cross fluctuations between the (macroscopic) volume and the enthalpy in the $T$, $P$, $N$ ensemble (Section 1.4). Here again, we are concerned with fluctuations in molecular quantities only.

### 5.5.4. Isothermal Compressibility

The isothermal compressibility is defined by

$$\varkappa_T = -\frac{1}{V}\left(\frac{\partial V}{\partial P}\right)_{N,T} \tag{5.67}$$

# Generalized Molecular Distribution Functions

Using the expression (5.46) for the volume, assuming that $N_\psi^{(1)}(\phi)$ has been defined in the $T$, $P$, $N$ ensemble, we get, by direct differentiation with respect to pressure,

$$\left(\frac{\partial V}{\partial P}\right)_{N,T} = \frac{-N}{kT}[(\langle \psi_1^2 \rangle - \langle \psi_1 \rangle^2) + (N-1)(\langle \psi_1 \psi_2 \rangle - \langle \psi_1 \rangle \langle \psi_2 \rangle)] \quad (5.68)$$

$$\left(\frac{\partial V}{\partial P}\right)_{N,T} = \frac{-1}{kT}\left\{\int \phi^2 N_\psi^{(1)}(\phi)\, d\phi + \int\int \phi\phi' N_{\psi;\psi}^{(2)}(\phi, \phi')\, d\phi\, d\phi' \right.$$

$$\left. - \left[\int \phi N_\psi^{(1)}(\phi)\, d\phi\right]^2 \right\} \quad (5.69)$$

Note that here we encounter only fluctuations in the volume of the VP of singles and pairs of particles. The compressibility relation derived in Section 3.9 is admittedly simpler than (5.69), yet, as in (5.69), it also employed MDF's of order two.

In concluding this section, we recall that at present, the most reliable source of information on the ordinary MDF's are the direct computational procedures using either the Monte Carlo or the molecular dynamic method. It is expected that these methods will also provide the appropriate information on the GMDF's. Computation of the latter should not pose any additional difficulties to those already encountered in the computation of ordinary MDF's. Once we get such information on the singlet and pair GMDF's, all of the quantities discussed in this section can be computed easily using one- and two-dimensional integrals.

## 5.6. THE MIXTURE-MODEL (MM) APPROACH; GENERAL CONSIDERATIONS

The general idea of the mixture-model (MM) approach is very simple and quite old. In particular, it has been applied extensively in the theory of water and aqueous solutions. In this section, we outline the basic procedure employed in the MM approach. We choose two examples in which the idea of a *classification* procedure, a fundamental step in the theory, is very simple and vivid. Then, with a little imagination, we can generalize the classification procedure so that contact with the notion of GMDF's is achieved. This is deferred to the next section.

Consider a system of $N$ molecules in a volume $V$ and temperature $T$. Furthermore, we assume that the density is very low, so that the partition function can be written as

$$Q(T, V, N) = q^N/N! \qquad (5.70)$$

Here, $q$ is the full partition function of a single molecule, which can be written as

$$q = \sum_{i=1}^{\infty} \exp(-\beta E_i) \qquad (5.71)$$

where the summation in (5.71) is carried over all the states of a single molecule, and the $E_i$ are the corresponding energy levels. Note that the independent variables are $T$, $V$, $N$ and that in (5.70) there exists no trace of any "mixture" in our system.

Suppose we now wish to split the sum in (5.71) into two (or more) sums. This can be achieved in many ways, for instance,

$$\begin{aligned}
\sum_{i=1}^{\infty} \exp(-\beta E_i) &= \sum_{i=1}^{5} \exp(-\beta E_i) + \sum_{i=6}^{\infty} \exp(-\beta E_i) \\
&= \sum_{i=\text{odd}} \exp(-\beta E_i) + \sum_{i=\text{even}} \exp(-\beta E_i) \\
&= \exp(-\beta E_{10}) + \sum_{\substack{i=1 \\ i \neq 10}}^{\infty} \exp(-\beta E_i)
\end{aligned} \qquad (5.72)$$

The splittings in (5.72) can be arbitrary or motivated by theoretical or experimental reasoning. The important thing is that once we have been given a rule by which we can classify all the states into two groups, we can split the partition function into a sum of two partition functions

$$q = q_A + q_B \qquad (5.73)$$

where $q_A$ and $q_B$ are the "partition functions" defined by the first and second sums, respectively. We can also refer to a molecule in any one of the states of the first group as an $A$-cule and any molecule in the remaining group of states as a $B$-cule. In this way, we have completed our *classification procedure*, and from now on, we can view our system as a mixture of two components, $A$ and $B$. It should be stressed that by adopting the above point of view we have changed nothing in the physics of the system. Of course, the classification procedure must be exhaustive and unique. That is, each molecule, at any given moment, is classified as belonging to one and to only one group.

## Generalized Molecular Distribution Functions

Substituting (5.73) into (5.70) and applying the binomial theorem, we get

$$Q(T, V, N) = \frac{(q_A + q_B)^N}{N!} = \sum_{N_A + N_B = N} \frac{q_A^{N_A} q_B^{N_B}}{N_A! \, N_B!} \qquad (5.74)$$

where the sum on the rhs of (5.74) extends over all $N_A$ and $N_B$ for which $N_A + N_B = N$. We write

$$Q(T, V, N_A, N_B) = q_A^{N_A} q_B^{N_B}/(N_A! \, N_B!) \qquad (5.75)$$

and rewrite (5.74) as

$$Q(T, V, N) = \sum_{N_A + N_B = N} Q(T, V, N_A, N_B) \qquad (5.76)$$

Thus far, we have taken only formal steps of rewriting the same quantity $Q$ in a new form. The latter suggests, however, a new point of view. We notice that (5.75) has the *form* of a partition function of an ideal gas mixture of two components $A$ and $B$, where $N_A$ and $N_B$ are the number of $A$ and $B$ molecules, respectively.

Of course, $N_A$ and $N_B$ in (5.75) are not independent variables, i.e., we cannot prepare a system with any chosen values of $N_A$ and $N_B$, as in the case of a real mixture of two components. Hence, we refer to the $A$-cules and the $B$-cules as quasicomponents. One may envisage a device which prevents the conversion of molecules between $A$ and $B$. Such a device may be called an inhibitor (or an anticatalyst) for the conversion "reaction" $A \rightleftarrows B$. A system in the presence of this inhibitor is referred to as being "frozen in" with respect to the conversion $A \rightleftarrows B$. Clearly, the partition function of our system in the "frozen-in" state is (5.75) and not (5.74).

We next pose the following question: Can we find a "frozen-in" system, i.e., fixed numbers $N_A$ and $N_B$, for which the partition function (5.75) is practically equal to (5.74)? It turns out that this can, indeed, be done. Since the arguments that follow are of fundamental importance in many problems in statistical mechanics, we elaborate on them in considerable detail. In what follows, we always assume that $N_A$ and $N_B$ are sufficiently large that we can safely use the Stirling approximation in the form[7]

$$\ln n! = n \ln n - n \qquad (5.77)$$

---

[7] A better approximation is

$$\ln n! = \tfrac{1}{2} \ln(2\pi n) + n \ln n - n$$

However, for $n$ of the order of $10^{23}$, the term of order $\ln n$ is negligible with respect to $n$. Therefore, (5.77) is an excellent approximation valid for all our purposes.

We wish to find the largest term in the sum of (5.74). To do this, we view $Q(T, V, N_A, N_B)$ as a function of $N_A$ ($N_B$ being determined by $N_B = N - N_A$) and seek the condition for an extremum of this function, i.e.,

$$0 = \left[\frac{\partial \ln Q(T, V, N_A, N_B)}{\partial N_A}\right]_{T,N,V} = \ln q_A - \ln q_B - \ln N_A^* + \ln(N-N_A^*) \quad (5.78)$$

where we have used the Stirling approximation before taking the derivative. We have also denoted by $N_A^*$ the value of $N_A$ for which the function $Q(T, V, N_A, N - N_A)$ has a maximum.[8] Writing $N_B^* = N - N_A^*$, the condition (5.78) can be rewritten in a more familiar form as

$$K \equiv q_A/q_B = N_A^*/N_B^* \quad (5.79)$$

where $K$ is interpreted as the "equilibrium constant" for the conversion $A \rightleftarrows B$.

Now comes the crucial step. We replace the whole sum in (5.74) by a single term, namely

$$Q(T, V, N) = \sum_{N_A+N_B=N} Q(T, V, N_A, N_B) \approx Q(T, V, N_A^*, N_B^*) \quad (5.80)$$

The reader may wonder about the validity of (5.80), for two reasons. First, we started with a one-component system described by *three* variables $T$, $V$, and $N$, and suddenly we have on the rhs of (5.80) *four* variables $T$, $V$, $N_A^*$, and $N_B^*$. Second, the sum over all possible values of $N_A$ in the complete partition function (5.74) is over an extremely large number of terms; how could all of that be replaced by just a single, albeit the maximal, term?

The answer to the first question is relatively simple. The variables $N_A^*$ and $N_B^*$ are not really independent. Relation (5.79) imposes a dependence between the two; in fact, each of them is a function of $T$, $V$, and $N$, as can be seen explicitly by solving (5.79) to get

$$N_A^* = KN/(1 + K), \qquad N_B^* = N/(1 + K)$$

---

[8] It is easily verified that the condition (5.78) leads to a maximum. Indeed,

$$\left[\frac{\partial^2 \ln Q(T, V, N_A, N_B)}{\partial N_A^2}\right]_{T,V,N} = -\frac{1}{N_A} - \frac{1}{N-N_A} = -\frac{N}{N_A N_B} < 0$$

This is true for any $N_A$ and, in particular, for $N_A^*$.

## Generalized Molecular Distribution Functions

The elucidation of the second question is more difficult, and is of central importance for many arguments that will be employed in later applications. We now show that within the realm of the Stirling approximation (5.77), the approximate relation (5.80) is actually an *equality*, provided that we choose $N_A^*$ (and $N_B^*$) to fulfill (5.79), and, indeed,

$$Q(T, V, N_A^*, N_B^*)$$

$$\stackrel{(5.75)}{=} \frac{q_A^{N_A^*} q_B^{N_B^*}}{N_A^*! \, N_B^*!} \stackrel{(5.77)}{=} \left(\frac{q_A e}{N_A^*}\right)^{N_A^*} \left(\frac{q_B e}{N_B^*}\right)^{N_B^*}$$

$$= e^{N_A^* + N_B^*} \left(\frac{q_A}{N_A^*}\right)^{N_A^*} \left(\frac{q_B}{N_B^*}\right)^{N_B^*} \stackrel{(5.79)}{=} e^{N_A^* + N_B^*} \left(\frac{q_A}{N_A^*}\right)^{N_A^*} \left(\frac{q_A}{N_A^*}\right)^{N_B^*}$$

$$= e^{N_A^* + N_B^*} \left(\frac{q_A}{N_A^*}\right)^{N_A^* + N_B^*} \stackrel{(5.79)}{=} e^{N_A^* + N_B^*} \left(\frac{q_A + q_B}{N_A^* + N_B^*}\right)^{N_A^* + N_B^*}$$

$$= e^N \left(\frac{q_A + q_B}{N}\right)^N \stackrel{(5.77)}{=} \frac{(q_A + q_B)^N}{N!} \stackrel{(5.74)}{=} Q(T, V, N) \qquad (5.81)$$

The argument used in each step is indicated above the equality sign of that step.[9] The final result of (5.81) is

$$Q(T, V, N) = Q(T, V, N_A^*, N_B^*) \qquad (5.82)$$

In deriving (5.82), we have used the Stirling approximation in the form $n! = (n/e)^n$. In fact, the more useful approximation is the logarithmic form (5.77) which, when applied to our case, leads to the relation between the free energies

$$A(T, V, N) = A(T, V, N_A^*, N_B^*) \qquad (5.83)$$

It is the last result, rather than (5.82), that will be considered as an equality, provided that we understand this equality in the same sense as we understood (5.77).

The equality (5.83) is equivalent to the statement that the free energy of the system at equilibrium [i.e., $A(T, V, N)$] is equal to the free energy of the "frozen-in" system [$A(T, V, N_A^*, N_B^*)$], provided that $N_A^*$ and $N_B^*$ are chosen to fulfill (5.79) and that $N_A^* + N_B^* = N$. In other words, if we introduce an inhibitor (for the conversion $A \rightleftarrows B$) to a system in

[9] In the second application of (5.79) in (5.81), we use the identity

$$\frac{q_A}{N_A^*} = \frac{q_B}{N_B^*} = \frac{q_A + q_B}{N_A^* + N_B^*}$$

which follows directly from (5.79).

which $N_A = N_A^*$ and $N_B = N_B^*$, then the free energy of the system does not change. This is not true for any values of $N_A$ and $N_B$, and, in general, it is not true for other thermodynamic quantities. Special examples will be discussed in detail in the following chapters.

The condition (5.79) is equivalent to the statement that the chemical potentials of $A$ and $B$ are equal; i.e., for any given $N_A$ and $N_B$, we can define the chemical potentials by

$$\mu_A = -kT\left[\frac{\partial \ln Q(T, V, N_A, N_B)}{\partial N_A}\right]_{T,V,N_B} = -kT \ln q_A + kT \ln N_A \quad (5.84)$$

$$\mu_B = -kT\left[\frac{\partial \ln Q(T, V, N_A, N_B)}{\partial N_B}\right]_{T,V,N_A} = -kT \ln q_B + kT \ln N_B \quad (5.85)$$

Note that $\mu_A$ and $\mu_B$ are definable by keeping $T$, $V$, $N_B$ and $T$, $V$, $N_A$ constant, respectively. This means that either we employ an inhibitor to "freeze in" the system, so that we can actually add, say, $dN_A$, keeping $N_B$ constant, or we have an analytical expression for the partition function in terms of $N_A$ and $N_B$, such as (5.75), which can be differentiated analytically. We henceforth assume that, in principle, we can express $Q$ in terms of the variables $T$, $V$, $N_A$, $N_B$, so that differentiations of the form (5.84) and (5.85) are meaningful.

Substituting $N_A = N_A^*$ and $N_B = N_B^*$ in (5.84) and (5.85), and using (5.79), we get the result

$$\mu_A = \mu_B \quad (5.86)$$

Conversely, if (5.86) is valid, then (5.79) follows. Hence, the two conditions are equivalent.

We now consider a different case of the MM approach. As we have stressed, the essential step is to provide a classification procedure. The following is a very simple example. Consider a system of $N$ particles at a given $T$ and $V$, the partition function of which is

$$Q(T, V, N) = (q^N/N!) \int \cdots \int d\mathbf{R}^N \exp[-\beta U_N(\mathbf{R}^N)] \quad (5.87)$$

[Note that in (5.87), $q$ is the internal partition function for a single molecule. In this respect, it differs from $q$ in (5.70), which also includes the volume of the system. For ideal gases, (5.87) reduces to $(qV)^N/N!$, to be compared with $q^N/N!$ of (5.70).]

We now envisage an imaginary partition that divides the vessel containing our system into two parts, as in Fig. 5.10. Each particle whose

# Generalized Molecular Distribution Functions

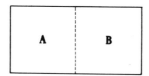

Fig. 5.10. A one-component system viewed as a mixture of two components. Each molecule whose center falls in the left-hand side is referred to as an $A$-cule. Similarly, a $B$-cule is a molecule whose center falls in the right-hand side.

center falls in one part will be referred to as an $A$-cule, and those whose center falls in the second part as $B$-cules. This is a classification procedure that is unique and exhaustive, since at any given time, each particle must be in one and only one part of the vessel. At any given moment, we may have $N_A$ $A$-cules and $N_B$ $B$-cules such that $N_A + N_B = N$. The partition function (5.87) can be rewritten [as in (5.74)] in the form

$$Q(T, V, N) = \frac{q^N}{N!} \sum_{N_A=0}^{N} \frac{N!}{N_A! N_B!} Z_{N_A, N_B} = \sum_{N_A=0}^{N} \frac{q_A^{N_A} q_B^{N_B}}{N_A! N_B!} Z_{N_A, N_B} \qquad (5.88)$$

where $Z_{N_A, N_B}$ is the configurational partition function for a system of $N$ particles, of which $N_A$ *specific* particles are $A$-cules and the remaining are $B$-cules. The argument leading to (5.88) is somewhat more complex than the previous example given in (5.74). The configurational integral in (5.87) extends over *all* possible configurations of $N$ particles in the entire volume of the system. These configurations are now collected in a different way. First, we collect all configurations for which particles $1, \ldots, N_A$ are in one part and the remaining particles are in the second part. The configurational partition function for such a system is exactly $Z_{N_A, N_B}$. We also realize that since all particles in the system are equivalent, there are precisely $N!/(N_A! N_B!)$ ways of choosing the $N_A$ *specific* particles to be placed in one part and the remaining $N_B$ particles in the other. Each of these choices produces the same configurational partition function $Z_{N_A, N_B}$. Finally, we must sum over all possible pairs of $N_A$ and $N_B$ in order to exhaust the entire configurational space of the system, which is the expression written explicitly in (5.88).

The analogy with the previous example is quite clear. In the present case, the "inhibitor" to the conversion $A \rightleftarrows B$ may be taken to be a *real* partition that divides the vessel into two compartments. Once we place such a partition in the system, $N_A$ and $N_B$ become independent variables and $Z_{N_A, N_B}$ becomes the relevant configurational partition function. (Of course, in the present example, $q_A = q_B = q$.) The value of $N_A$ for which the summand in (5.88) is maximum is denoted by $N_A^*$ (and $N_B^* = N - N_A^*$). Using the same arguments as in the previous example, we can write the approximate equality

$$Q(T, V, N) = \frac{q_A^{N_A^*} q_B^{N_B^*}}{N_A^*! N_B^*!} Z_{N_A^*, N_B^*} \quad (5.89)$$

That is, we replace the whole sum in (5.88) by the maximum term, and assume that the equality in (5.89) holds in the same sense as in (5.82).

The procedure employed in this section will be generalized in the next section to obtain the basis of the so-called mixture-model approach to liquids. In fact, the treatment of a large number of problems in physical chemistry and especially in biophysics rests on arguments similar to those given above.

## 5.7. THE MIXTURE-MODEL APPROACH TO LIQUIDS; CLASSIFICATIONS BASED ON LOCAL PROPERTIES OF THE MOLECULES

In the previous section, we presented two very simple classification procedures for splitting a one-component system into a mixture of, say two, quasicomponents. Here, we extend that idea to include more subtle procedures which will be particularly useful in the study of aqueous fluids. Most of the ground work for this extension has been carried out in Section 5.2, where various singlet GMDF's were introduced.

Consider, for instance, the singlet GMDF, based on the coordination number (CN). The quantity [Eq. (5.10)]

$$x_C(K) = N_C^{(1)}(K)/N = \int \cdots \int d\mathbf{R}^N P(\mathbf{R}^N) \, \delta[C_1(\mathbf{R}^N) - K] \quad (5.90)$$

has been interpreted in two ways. First, this is the mole fraction of particles having CN equal to $K$. Second, from the form of the integral on the rhs of (5.90), it follows that $x_C(K)$ is the probability that a specific particle, say number one, will be found with a CN equal to $K$. As a matter of fact, this is a particular example of a very general statement: The mole fraction of molecules possessing a specific property is equal to the probability that a specific particle will acquire that property. It is because of this dual

## Generalized Molecular Distribution Functions

meaning that we may refer to the vector

$$\mathbf{x}_C = (x_C(0), x_C(1), \ldots) \quad (5.91)$$

as a quasicomponent distribution function (QCDF). It gives the "composition" of the system when viewed as a mixture of quasicomponents, and it also gives the probability distribution for finding a specific particle in various CN's. The normalization condition for $\mathbf{x}_C$ is

$$\sum_{K=0}^{\infty} x_C(K) = 1 \quad (5.92)$$

Note, however, that since the particles exert strong repulsive forces when brought to a very close distance from each other, the occurrence of a large CN for a given particle must be a very improbable event. Therefore, the sum over $K$ in (5.92) effectively extends from $K = 0$ to, say, $K \approx 12$ (depending, of course, on the choice of $R_C$); i.e., the mole fractions $x_C(K)$, with $K > 12$, will in general be negligible.

A mixture of quasicomponents must be distinguished from a mixture of real components in essentially two respects. First, the quasicomponents do not differ in their chemical composition or in their structure; they are characterized by the nature of their *local environment*, which, in the above example, is the CN. A more important difference is that a system of quasicomponents cannot be "prepared" in any desired composition, i.e., the components of the vector $\mathbf{x}_C$ cannot be chosen at will [even when they satisfy the normalization condition (5.92)]. One consequence of this restriction is that quasicomponents have no existence in the pure state.[10]

There exists a certain analogy between a mixture of quasicomponents and a mixture of chemically reacting species. For the sake of simplicity, consider a dimerization reaction

$$2A \rightleftarrows A_2 \quad (5.93)$$

In general, the average numbers of $A$ and $A_2$ molecules are not independent. The condition of chemical equilibrium imposes a relation between $N_A$ and $N_{A_2}$, although this relation can rarely be written explicitly. In the same manner, there exist equilibrium conditions on the $x_C(K)$, a simple example of which was discussed in detail in the previous section. We have also noted

---

[10] Of course, for some special cases, one may envisage a system of pure quasicomponents. For instance, a solid in the regular close packing state is composed of particles having CN equal to $K = 12$. This, in general, cannot be realized in a liquid.

that if we find an inhibitor to the reaction (5.93), the numbers $N_A$ and $N_{A_2}$ become virtually independent, and therefore we can prepare a system with an arbitrary composition. This is not possible, however, in a mixture of quasicomponents. Suppose we could find a hypothetical inhibitor such that the transformation among the various species becomes "frozen in," in which case the equilibrium condition does not impose dependence among the $x_C(K)$. However, we still face another kind of dependence which arises from the very definition of our species, i.e., we are still unable to prepare a mixture with any arbitrary composition $\mathbf{x}_C$. The reason is very simple. Our species are distinguishable according to their local environment. This environment is also built up of molecules, each of which belongs to some species. Therefore, a molecule forming a particular environment for another molecule cannot, at the same time, possess an arbitrary environment of its own. A simple example may clarify the situation. Let $N$ be the total number of molecules and let us assume a choice of $R_C \approx 1.5\sigma$ for the radius of the first coordination sphere. At a given configuration $\mathbf{R}^N$ of the system, the following composition is impossible:

$$N_C^{(1)}(0, \mathbf{R}^N) = N - 1, \; N_C^{(1)}(12, \mathbf{R}^N) = 1, \; \text{all } N_C^{(1)}(K, \mathbf{R}^N) = 0 \text{ for } K \neq 0, 12$$

Clearly, it is impossible to have one particle with CN equal to 12, and at the same time all other particles with CN equal to zero. This situation contradicts the meaning of the concept of CN. It should be noted, however, that in the present example, we have taken a fixed configuration $\mathbf{R}^N$. It is possible, in principle, that average vectors $[N_C^{(1)}(0), N_C^{(1)}(1), \ldots]$ will not be subject to the same restrictions that are operative for vectors for a specific configuration, such as the one given above.

In the above example, we elaborated on a classification procedure based on a *discrete* parameter, the value of the CN. It can be easily generalized to any other *property* used in Section 5.2 for constructing various singlet GMDF's. For instance, the quantity $x_B(\nu) \, d\nu$ defined in (5.20) is the mole fraction of molecules having BE between $\nu$ and $\nu + d\nu$. This is an example of a QCDF function based on a continuous parameter, $\nu$. Another example is the function $x_\psi(\phi) \, d\phi = N_\psi^{(1)}(\phi) \, d\phi / N$ based on the volume of the VP of the particles.

It should be noted that in all of the above-mentioned examples, the adoption of the MM approach depends solely on the rule of classification of molecules into various groups. In each case, the rule must be unique and exhaustive, i.e., for each configuration, each particle belongs to one and only one group.

**Generalized Molecular Distribution Functions** 211

In most of the applications introduced in the next chapter, we specialize to the simplest case of MM, which we refer to as the two-structure model. It must be remembered, however, that although we are applying the term "mixture model," we have in fact invoked no modelistic assumptions thus far in our treatment. This will be done in the next chapter when we examine various ad hoc models for water.

## 5.8. GENERAL RELATIONS BETWEEN THERMODYNAMICS AND QUASICOMPONENT DISTRIBUTION FUNCTIONS (QCDF)

In this section, we shall be working in the $T$, $P$, $N$ ensemble, and all the distribution functions are presumed to be defined in this ensemble. We denote by $\mathbf{x}$ either the vector or the function which serves as a QCDF. An appropriate subscript will be used to indicate the property employed in the classification procedure. For instance, using the CN as a property, the components of $\mathbf{x}_C$ are the quantities $x_C(K)$. Similarly, using the BE as a property, the components of $\mathbf{x}_B$ are the quantities $x_B(\nu)$. When reference is made to a general QCDF, we simply write $\mathbf{x}$ without a subscript. Once a QCDF is given, we can obtain the average number of each quasicomponent directly from the components of the vector $\mathbf{N} = N\mathbf{x}$.

Let $E$ be any extensive thermodynamic quantity expressed as a function of the variables $T$, $P$, and $N$ (where $N$ is the total number of molecules in the system). Viewing the same system as a mixture of quasicomponents, we can express $E$ as a function of the new set of variables $T$, $P$, and $\mathbf{N}$. For concreteness, consider the QCDF based on the concept of CN. The two possible functions mentioned above are then[11]

$$E(T, P, N) = E(T, P, N_C^{(1)}(0), N_C^{(1)}(1), \ldots) \qquad (5.94)$$

In writing (5.94), it is assumed that we can, in principle, express the quantity $E$ in terms of the variables $T$, $P$, and $\mathbf{N}_C^{(1)}$. A simple case in which this can be done explicitly was presented in Section 5.6.

For the sake of simplicity, we henceforth use $N(K)$ in place of $N_C^{(1)}(K)$, so that the treatment will be valid for any discrete QCDF. Since $E$ is an

---

[11] Note that in (5.94), we use the same symbol $E$ to denote two different functions; on the lhs, we have a function of three variables $T$, $P$, $N$, whereas on the rhs, we have a function of the variables $T$, $P$, $N_C^{(1)}(0)$, $N_C^{(1)}(1)$, ....

extensive quantity, it has the property

$$E(T, P, \alpha N(0), \alpha N(1), \ldots) = \alpha E(T, P, N(0), N(1), \ldots) \quad (5.95)$$

for any real $\alpha \geq 0$; i.e., $E$ is a homogeneous function of order one with respect to the variables $N(0), N(1), \ldots$, keeping $T$, $P$ constant. For such a function, the Euler theorem states that

$$E(T, P, \mathbf{N}) = \sum_{i=0}^{\infty} \bar{E}_i(T, P, \mathbf{N}) N(i) \quad (5.96)$$

where $\bar{E}_i(T, P, \mathbf{N})$ is the partial molar (or molecular) quantity, defined by

$$\bar{E}_i(T, P, \mathbf{N}) = [\partial E/\partial N(i)]_{T,P,N(j)}, \quad j \neq i \quad (5.97)$$

In (5.96) and (5.97), we have stressed the fact that the partial molar quantities depend on the whole vector $\mathbf{N}$.

At this point, it is important to digress to a discussion of the meaning of the partial derivatives introduced in (5.97). We recall that the variables $N(i)$ are *not* independent; therefore, it is impossible to take the derivatives of (5.97) "experimentally." One cannot, in general, add $dN(i)$ of the $i$-cules, while keeping all the $N(j)$, $j \neq i$, constant, a process which can certainly be achieved in a mixture of independent components. However, if we assume that, in principle, $E$ can be expressed in terms of the variables $T$, $P$, and $\mathbf{N}$, then $\bar{E}_i$ is the component of the gradient of $E$ along the $i$th axis. Here, we must assume that in the neighborhood of the equilibrium vector $\mathbf{N}$, there is a sufficiently dense set of vectors (which describe various "frozen-in" systems) so that the gradient of $E$ exists along each axis. We note, however, that the existence of such a dense set of vectors $\mathbf{N}$ may not always be guaranteed. For example, we have seen that the function $x_B(v)$ for a system of hard spheres is a delta function

$$x_B(v) = \delta(v) \quad (5.98)$$

and there are no other vectors in the "neighborhood" of $x_B(v)$ which belong to the "frozen-in" systems.

The generalization of (5.96) and (5.97) to the case of a continuous QCDF requires the application of the technique of functional differentiation. We introduce the generalized Euler theorem by way of analogy [for more details, see Appendix 9-B and Ben-Naim (1972f,g).]

The generalization can be easily visualized if we rewrite (5.96) in the form

$$E(T, P, \mathbf{N}) = \sum_{i=0}^{\infty} \bar{E}(T, P, \mathbf{N}; i) N(i) \quad (5.99)$$

## Generalized Molecular Distribution Functions

where we have introduced the (discrete) variable $i$ as one of the arguments of the function $\bar{E}$. If $\mathbf{N}$ is a vector derived from a QCDF based on a continuous variable, say $v$, then the generalization of (5.99) is simply

$$E(T, P, \mathbf{N}) = \int_{-\infty}^{\infty} \bar{E}(T, P, \mathbf{N}; v) N(v) \, dv \tag{5.100}$$

where $\bar{E}(T, P, \mathbf{N}; v)$ is the functional derivative of $E(T, P, \mathbf{N})$ with respect to $N(v)$, symbolized as

$$\bar{E}(T, P, \mathbf{N}; v) = \delta E(T, P, \mathbf{N})/\delta N(v) \tag{5.101}$$

By analogy with the discrete case, we may assign to $\bar{E}(T, P, \mathbf{N}; v)$ the meaning of a partial molar quantity of the appropriate $v$-cule. The functional derivative in (5.101) is viewed here as a limiting case of (5.97) when the index $i$ refers to a continuous variable.

Thus far, we have used the symbol $E$ to designate any extensive thermodynamic quantity. In this respect, relations (5.96) and (5.100) provide general formal connections between thermodynamics and QCDF's (or singlet GMDF's). Note that for each thermodynamic quantity, we can have different representations of the form (5.96) or (5.99), depending on the choice of QCDF. For example, the average volume can be written as

$$V(T, P, \mathbf{N}_C^{(1)}) = \sum_{K=0}^{\infty} \bar{V}_K(T, P, \mathbf{N}_C^{(1)}) N_C^{(1)}(K) \tag{5.102}$$

$$V(T, P, \mathbf{N}_B^{(1)}) = \int_{-\infty}^{\infty} \bar{V}(T, P, \mathbf{N}_B^{(1)}; v) N_B^{(1)}(v) \, dv \tag{5.103}$$

$$V(T, P, \mathbf{N}_\psi^{(1)}) = \int_{0}^{\infty} \bar{V}(T, P, \mathbf{N}_\psi^{(1)}; \phi) N_\psi^{(1)}(\phi) \, d\phi \tag{5.104}$$

where $\bar{V}_K(T, P, \mathbf{N}_C^{(1)})$, $\bar{V}(T, P, \mathbf{N}_B^{(1)}; v)$, and $\bar{V}(T, P, \mathbf{N}_\psi^{(1)}; \phi)$ are the partial molar volumes of the $K$-cule, the $v$-cule, and the $\phi$-cule, respectively. In general, the explicit dependence of $V$ on the various singlet GMDF's is not known and therefore we cannot evaluate the corresponding partial molar volumes explicitly. Fortunately, there are two examples which are of considerable importance and for which such an evaluation is possible.

Consider relation (5.46) for the average volume

$$V(T, P, \mathbf{N}_\psi^{(1)}) = \int_{0}^{\infty} \phi N_\psi^{(1)}(\phi) \, d\phi \tag{5.105}$$

Note that this relation is based on the fact that the volumes of the VP of all the particles add up to build the total volume of the system. Here we have an example of an explicit dependence between $V$ and $\mathbf{N}_\psi^{(1)}$ that could have been "guessed." Therefore, the partial molar volume of the $\phi$-cule can be obtained by taking the functional derivative of $V$ with respect to $N_\psi^{(1)}(\phi')$ (for details, see Appendix 9-B), i.e.,

$$\bar{V}(T, P, \mathbf{N}_\psi^{(1)}; \phi') = \delta V(T, P, \mathbf{N}_\psi^{(1)})/\delta N_\psi^{(1)}(\phi') = \phi' \quad (5.106)$$

This is a remarkable result. It states that the partial molar volume of the $\phi'$-cule is exactly equal to the volume of its VP. We note that, in general, the partial molar volume of a species is not related in a simple manner to the actual volume which it contributes to the total volume of the system. We note also that, in this particular example, the partial molar volume $\bar{V}(T, P, \mathbf{N}_\psi^{(1)}; \phi')$ is independent of $T$, $P$, $\mathbf{N}_\psi^{(1)}$. This fact will be further exploited in the next section.

A second example is based on relation (5.45) for the average internal energy $E$, which, in the $T$, $P$, $N$ ensemble, is given by

$$E(T, P, \mathbf{N}_B^{(1)}) = N\varepsilon^K + \tfrac{1}{2} \int_{-\infty}^{\infty} \nu N_B^{(1)}(\nu)\, d\nu \quad (5.107)$$

Note that in (5.107), $E$ stands for the energy, whereas in previous expressions in this section, we have used $E$ for any extensive thermodynamic quantity. We also recall that this relation was derived in Section 5.5 on the basis of direct arguments which do not depend on the mixture-model approach.

Since the normalization condition for $\mathbf{N}_B^{(1)}$ is

$$\int_{-\infty}^{\infty} N_B^{(1)}(\nu)\, d\nu = N \quad (5.108)$$

we can rewrite (5.107) as

$$E(T, P, \mathbf{N}_B^{(1)}) = \int_{-\infty}^{\infty} (\varepsilon^K + \tfrac{1}{2}\nu) N_B^{(1)}(\nu)\, d\nu \quad (5.109)$$

This is again an explicit relation between the energy and the singlet GMDF $\mathbf{N}_B^{(1)}$. By direct functional differentiation, we obtain (for details, see Appendix 9-B)

$$\bar{E}(T, P, \mathbf{N}_B^{(1)}; \nu') = \delta E(T, P, \mathbf{N}_B^{(1)})/\delta N_B^{(1)}(\nu') = \varepsilon^K + \tfrac{1}{2}\nu' \quad (5.110)$$

Thus, the partial molar energy of the $v'$-cule is equal to its average kinetic energy and half of its BE. Here again, the partial molar energy does not depend on composition, although it still depends on $T$ through $\varepsilon^K$.

## 5.9. REINTERPRETATION OF SOME THERMODYNAMIC QUANTITIES USING THE MIXTURE-MODEL APPROACH

In the previous section, we reinterpreted relations (5.45) and (5.46) as special cases of the generalized Euler theorem, i.e.,

$$V(T, P, \mathbf{N}_\psi^{(1)}) = \int_0^\infty \phi N_\psi^{(1)}(\phi) \, d\phi \qquad (5.111)$$

$$E(T, P, \mathbf{N}_B^{(1)}) = \int_{-\infty}^\infty (\varepsilon^K + \tfrac{1}{2}v) N_B^{(1)}(v) \, dv \qquad (5.112)$$

Here, by adoption of the MM approach, the quantities $\phi$ and $(\varepsilon^K + \tfrac{1}{2}v)$ are assigned the meaning of partial molar volume and energy, respectively. We now treat some other thermodynamic quantities which are of importance in the study of aqueous fluids. We recall the relations for $C_V$, $C_P$, $(\partial V/\partial T)_P$, and $(\partial V/\partial P)_T$ expressed in Section 5.5 in terms of the GMDF's. Here, a new interpretation is given to the same quantities within the realm of the MM approach.

Before turning to specific examples, we consider a general expression, say, for the volume in terms of any chosen singlet GMDF, $\mathbf{N}^{(1)}$ [see, for instance, (5.102)–(5.104)]:

$$V(T, P, \mathbf{N}^{(1)}) = \int \bar{V}(T, P, \mathbf{N}^{(1)}; \eta) N^{(1)}(\eta) \, d\eta \qquad (5.113)$$

where $\eta$ is any parameter used in the particular classification procedure. The temperature derivative of (5.113) will, in general, produce two terms:

$$\left(\frac{\partial V}{\partial T}\right)_{P,N} = \int \frac{\partial \bar{V}}{\partial T} N^{(1)}(\eta) \, d\eta + \int \bar{V} \frac{\partial N^{(1)}(\eta)}{\partial T} \, d\eta \qquad (5.114)$$

The significance of the two terms in (5.114) is the following. The first includes the contributions from the changes in the partial molar volumes of the various species at a fixed distribution $N^{(1)}(\eta)$. The second term includes the temperature dependence of the distribution of particles among the various species; i.e., if we "freeze in" the equilibrium for conversion among the quasicomponents, then $N^{(1)}(\eta)$ is fixed and no contribution arises from this term. Hence, we may refer to this term as the relaxation term. In the general case, (5.114) can hardly be of any use since we know nothing about the partial molar volumes. However, using (5.111) and (5.112), we get some simple relations which are quite useful.

Consider first the temperature derivatives of (5.111) and (5.112):

$$\left(\frac{\partial V}{\partial T}\right)_{P,N} = \int_0^\infty \phi \, \frac{\partial N_v^{(1)}(\phi)}{\partial T} \, d\phi \tag{5.115}$$

$$\left(\frac{\partial E}{\partial T}\right)_{P,N} = NC^K + \frac{1}{2} \int_{-\infty}^\infty v \, \frac{\partial N_B^{(1)}(v)}{\partial T} \, dv \tag{5.116}$$

Since $\phi$ is independent of temperature, we get in (5.115) only the contribution due to the "structural changes" in the system, i.e., the redistribution of particles among the various species caused by the change in temperature. As a very simple example of the distribution $N_v^{(1)}(\phi)$, suppose that the volume of the VP of each particle may have only one of two values, say $\phi_A$ and $\phi_B$, which are independent of temperature. In this case, we have

$$N_v^{(1)}(\phi) = N_A \, \delta(\phi - \phi_A) + N_B \, \delta(\phi - \phi_B) \tag{5.117}$$

and hence

$$\left(\frac{\partial V}{\partial T}\right)_{P,N} = \frac{\partial}{\partial T} \left[\int_0^\infty \phi N_v^{(1)}(\phi) \, d\phi\right] = \frac{\partial}{\partial T}(N_A \phi_A + N_B \phi_B)$$

$$= (\phi_A - \phi_B)\left(\frac{\partial N_A}{\partial T}\right)_{P,N} \tag{5.118}$$

which means that the temperature dependence of the volume results only from the "excitation" between the two states $A$ and $B$. In (5.115), we have more complex structural changes taking place among the infinite number of species.

Similarly, the heat capacity[12] in (5.116) receives a contribution from the properties of the single particle (kinetic and internal energies), and a second contribution due to the existence of interactions among the particles (see Section 5.5 for more details). The latter contribution is viewed within the realm of the MM approach as a relaxation term, i.e., a redistri-

---

[12] If $E$ is expressed as a function of $T$, $V$, $N_B^{(1)}$, we get the heat capacity at constant volume. To get the heat capacity at constant pressure, we must take the enthalpy rather than the energy in (5.116). However, for the purpose of this section, we ignore the differences between energy and enthalpy.

bution of particles among the various $\nu$ species arising from the change in temperature. As a very simple example of a distribution $N_B^{(1)}(\nu)$, we suppose that the BE may attain only one of two values, say $\nu_1$ and $\nu_2$, which are independent of temperature. In such a case, relation (5.116) reduces to

$$\left(\frac{\partial E}{\partial T}\right)_{P,N} = NC^K + \frac{1}{2}\frac{\partial}{\partial T}\left\{\int_{-\infty}^{\infty}[N_1\,\delta(\nu-\nu_1)+N_2\,\delta(\nu-\nu_2)]\,d\nu\right\}$$

$$= NC^K + \frac{1}{2}(\nu_1 - \nu_2)\left(\frac{\partial N_1}{\partial T}\right)_{P,N} \qquad (5.119)$$

where the second term arises from the thermal excitation from one state to the other. In the next chapter, we encounter a more detailed application of a relation similar to (5.119).

In a similar fashion, one may consider the pressure derivatives of the volume and the energy in (5.111) and (5.112) to get

$$\left(\frac{\partial V}{\partial P}\right)_{T,N} = \int_0^{\infty} \phi\,\frac{\partial N_\nu^{(1)}(\phi)}{\partial P}\,d\phi \qquad (5.120)$$

$$\left(\frac{\partial E}{\partial P}\right)_{T,N} = \frac{1}{2}\int_{-\infty}^{\infty} \nu\,\frac{\partial N_B^{(1)}(\nu)}{\partial P}\,d\nu \qquad (5.121)$$

In this case, all of the pressure dependences of the volume and the energy are viewed as relaxation terms.

The main thesis of this section is that by adoption of the MM approach, a new interpretation may be assigned to some thermodynamic quantities. This reinterpretation is purely formal; it may, however, be useful if we have some information on the structure of the singlet GMDF's involved. Two simple examples have been given in (5.118) and (5.119) and more will be presented in subsequent chapters.

## 5.10. SOME THERMODYNAMIC IDENTITIES IN THE MIXTURE-MODEL APPROACH

In this section, we consider classification procedures that provide a QCDF that has only two components. A very simple example was treated in detail in Section 5.6. For the purpose of this section, as well as for most of the later applications, we can construct such a two-component system from any of the singlet GMDF's which have been previously introduced. For example, using $x_C(K)$ in (5.10), we can regroup the particles with different CN into two classes. First, we select an integer $K^*$ (say $K^* = 5$) and define the following two mole fractions:

$$x_L = \sum_{K=0}^{K^*} x_C(K) \qquad (5.122)$$

$$x_H = \sum_{K=K^*+1}^{\infty} x_C(K) \qquad (5.123)$$

Clearly, $x_L$ is the mole fraction of particles having a CN smaller than or equal to $K^*$. These may be referred to as particles with "low local density" (for more details, see Section 5.2). Similarly, $x_H$ is referred to as the mole fraction of "high-local-density" particles (i.e., particles for which $K > K^*$). In this way, the system is viewed as a mixture of two quasi-components, $L$ and $H$. This point of view is called a two-structure model (TSM). In a similar fashion, one can construct TSM's from any other discrete or continuous QCDF. Therefore, the following treatment applies for any TSM, not necessarily the one defined in (5.122) and (5.123).

Within the realm of TSM's, the following natural question may be asked. Suppose that $x_L$ and $x_H$ are the mole fractions of the two components; how are they expected to respond to variation of temperature or pressure or addition of solute? This question has been the subject of many investigations. Here, we shall derive a few identities which will be found useful for later applications.

Given a two-component system of, say, $L$ and $H$ at chemical equilibrium, we have the condition

$$\Delta\mu = \mu_L - \mu_H = 0 \qquad (5.124)$$

Let $N_L = Nx_L$ and $N_H = Nx_H$ be the average number of $L$-cules and $H$-cules, respectively, and $N = N_L + N_H$ the total number of particles in the system. Viewing $\Delta\mu$ as a function of the variables $T$, $P$, $N_L$, and $N_H$, we can write its total differential as

$$0 = d(\Delta\mu)_{\mathrm{eq}} = \left(\frac{\partial \Delta\mu}{\partial T}\right)_{P,N_L,N_H} dT + \left(\frac{\partial \Delta\mu}{\partial P}\right)_{T,N_L,N_H} dP$$
$$+ \left(\frac{\partial \Delta\mu}{\partial N_L}\right)_{T,P,N_H} dN_L + \left(\frac{\partial \Delta\mu}{\partial N_H}\right)_{T,P,N_L} dN_H \qquad (5.125)$$

Clearly, the total differential of $\Delta\mu$ along the *equilibrium* line (eq) is zero. However, if we "freeze in" the conversion reaction $L \rightleftarrows H$, then $N_L$

## Generalized Molecular Distribution Functions

and $N_H$ become virtually independent. Denote by

$$\Delta S = \bar{S}_L - \bar{S}_H, \quad \Delta H = \bar{H}_L - \bar{H}_H, \quad \Delta V = \bar{V}_L - \bar{V}_H, \quad \mu_{\alpha\beta} = \partial^2 G/\partial N_\alpha \, \partial N_\beta \tag{5.126}$$

where $\alpha$ and $\beta$ stand for either $L$ or $H$. We get from (5.125), after some rearrangement,

$$\left(\frac{\partial N_L}{\partial T}\right)_{P,N,\text{eq}} = (\mu_{LL} - 2\mu_{LH} + \mu_{HH})^{-1}(\Delta H/T) \tag{5.127}$$

$$\left(\frac{\partial N_L}{\partial P}\right)_{T,N,\text{eq}} = -(\mu_{LL} - 2\mu_{LH} + \mu_{HH})^{-1}\Delta V \tag{5.128}$$

$$\left(\frac{\partial N_L}{\partial N_S}\right)_{T,P,N,\text{eq}} = -(\mu_{LL} - 2\mu_{LH} + \mu_{HH})^{-1}\left[\frac{\partial(\Delta\mu)}{\partial N_S}\right]_{T,P,N_L,N_H} \tag{5.129}$$

On the lhs of (5.127)–(5.129), we have a derivative at equilibrium, i.e., the change of $N_L$ with $T$, $P$, or $N_S$ along the equilibrium line ($N_S$ is the number of solute molecules added to the system). Note that all the quantities on the rhs of (5.127)–(5.129) contain partial derivatives pertaining to a system in which the equilibrium has been "frozen in."

The quantity $(\mu_{LL} - 2\mu_{LH} + \mu_{HH})$ appearing in the above relations is always positive. We present a direct proof of this contention. Consider a system at equilibrium with composition $N_L$ and $N_H$. Now, suppose that, as a result of a fluctuation (at $T$, $P$, $N$ constant), $N_L$ has changed into $N_L + dN_L$ and $N_H$ into $N_H + dN_H$. The condition of stability of the system requires that if we allow the system to relax back to its equilibrium position, the Gibbs free energy must decrease, i.e.,

$$G(N_L + dN_L, N_H + dN_H) \geq G(N_L, N_H) \tag{5.130}$$

Expanding $G$ to second order in $dN_L$ and $dN_H$ around the equilibrium state $N_L$, $N_H$ (holding $T$, $P$, $N$ fixed), we get

$$G(N_L + dN_L, N_H + dN_H)$$
$$= G(N_L, N_H) + \left(\frac{\partial G}{\partial N_L}\right) dN_L + \left(\frac{\partial G}{\partial N_H}\right) dN_H$$
$$+ \frac{1}{2}\left(\frac{\partial^2 G}{\partial N_L^2} dN_L^2 + 2\frac{\partial^2 G}{\partial N_L \partial N_H} dN_L \, dN_H + \frac{\partial^2 G}{\partial N_H^2} dN_H^2\right) + \cdots \tag{5.131}$$

Using the equilibrium condition (5.124), the notation in (5.126), the relation $dN_L + dN_H = 0$, and the inequality (5.130), we get from (5.131)

$$\mu_{LL} - 2\mu_{LH} + \mu_{HH} \geq 0 \tag{5.132}$$

which is the condition for a minimum of $G$.

Using relation (4.71), we can also write this quantity as

$$\mu_{LL} - 2\mu_{LH} + \mu_{HH} = \frac{kT}{x_L x_H V[\varrho_L + \varrho_H + \varrho_L \varrho_H (G_{LL} + G_{HH} - 2G_{LH})]} \tag{5.133}$$

a relation which will be used in Chapter 7.

Finally, we derive some general relations between the pair correlation functions of the various quasicomponents. We begin with the simplest case of a TSM, and denote by $L$ and $H$ the two species, and by $W$ any molecule in the system. We denote by $g_{\alpha\beta}(R)$ the pair correlation function for the pair of species $\alpha$ and $\beta$.[13] Then, $\varrho_\alpha g_{\alpha\beta}(R)$ is the local density of $\alpha$-cules at a distance $R$ from a $\beta$-cule. Conservation of the total number of $W$ molecules around an $L$-cule gives

$$\varrho_L g_{LL}(R) + \varrho_H g_{HL}(R) = \varrho_W g_{WL}(R) \tag{5.134}$$

where we denote by $g_{WL}(R)$ the pair correlation function between an $L$-cule and any molecule of the system. Similarly, considering the total density of molecules around an $H$-cule, we get the equality

$$\varrho_H g_{HH}(R) + \varrho_L g_{LH}(R) = \varrho_W g_{WH}(R) \tag{5.135}$$

Multiplying (5.134) by $\varrho_L$ and (5.135) by $\varrho_H$ and summing the two equations, we get

$$\varrho_L^2 g_{LL}(R) + 2\varrho_L \varrho_H g_{HL}(R) + \varrho_H^2 g_{HH}(R) = \varrho_W[\varrho_L g_{WL}(R) + \varrho_H g_{WH}(R)]$$
$$= \varrho_W^2 g_{WW}(R) \tag{5.136}$$

where $g_{WW}(R)$ is the ordinary pair correlation function in the system when viewed as a one-component system. Dividing through by $\varrho_W^2$, we get

$$x_L^2 g_{LL}(R) + 2x_L x_H g_{LH}(R) + x_H^2 g_{HH}(R) = g_{WW}(R) \tag{5.137}$$

---

[13] Note that by the adoption of the MM approach, $g_{\alpha\beta}(R)$ is an ordinary pair correlation function for the two species $\alpha$, $\beta$. However, viewing the same system as a one-component system, $g_{\alpha\beta}(R)$ is considered as a generalized pair correlation function for the two "properties" $\alpha$ and $\beta$, assigned to the two particles.

## Generalized Molecular Distribution Functions

Relations similar to (5.137) can be generalized to any number of quasi-components. In the case of a discrete QCDF, we get the relation

$$\sum_{K'=0}^{\infty} \sum_{K=0}^{\infty} x(K)x(K')g(K, K', R) = g(R) \qquad (5.138)$$

where $g(R)$ is the ordinary pair correlation function of the system, and $g(K, K', R)$ is the pair correlation function between the quasicomponents $K$ and $K'$. Similarly, for a continuous QCDF, we get

$$\int dv \int dv' \, x(v)x(v')g(v, v', R) = g(R) \qquad (5.139)$$

All of the above relations can be verified directly from the definitions. For instance, using Eqs. (5.42) and (5.44), we prove (5.139) as follows [note that in (5.42) and (5.44), $N_{L,B}^{(1)}$ is a density]

$$\int dv_1 \int dv_2 \, x_B(v_1)x_B(v_2) g_{L,B;L,B}^{(2)}(\mathbf{S}_1, v_1, \mathbf{S}_2, v_2)$$

$$= \varrho^{-2} \sum_{\substack{i=1 \\ i \neq j}}^{N} \sum_{j=1}^{N} \int d\mathbf{R}^N P(\mathbf{R}^N) \int dv \int dv'$$

$$\times \delta[L_i - \mathbf{S}_1] \, \delta[B_i - v_1] \, \delta[L_j - \mathbf{S}_2] \, \delta[B_j - v_2]$$

$$= \varrho^{-2} \sum_{\substack{i=1 \\ i \neq j}}^{N} \sum_{j=1}^{N} \int d\mathbf{R}^N P(\mathbf{R}^N) \, \delta[L_i - \mathbf{S}_1] \, \delta[L_j - \mathbf{S}_2]$$

$$= g(\mathbf{S}_1, \mathbf{S}_2) = g(R) \qquad (5.140)$$

where $R = |\mathbf{S}_2 - \mathbf{S}_1|$.

## Chapter 6
# Liquid Water

## 6.1. INTRODUCTION

In this chapter, we begin our excursion into the field of liquid water and aqueous solutions. From the very outset, and along the whole trip, it is important to bear in mind that our systems are not "simple," and that some fundamental difficulties are encountered over and above the usual difficulties met in the theory of simple fluids. Therefore, any theoretical approach that we pursue includes many approximations, some of which, admittedly, are very poor. Nevertheless, remarkable progress has been achieved in the last decade toward understanding the properties of aqueous systems, progress which is mainly due to the study of very simplified models.

There are essentially two routes along which the study of water has been carried out. Both employ models, though they differ in the level at which the model is introduced. In the first method, one proposes a model on the molecular level, i.e., one devises a model for the molecules themselves. The second method employs a model for the whole system. We describe both methods in the following sections. Here, we outline some of the basic assumptions that are involved in most theories, without which it is doubtful that any progress could have been achieved in this area.

Our most fundamental starting point should be the quantum mechanical partition function for the whole system. However, since at present we cannot solve the Schrödinger equation for such a complex system, we resort to the partially classical partition function, written as

$$Q(T, V, N) = \frac{q^N}{(8\pi^2)^N \Lambda^{3N} N!} \int \cdots \int d\mathbf{X}^N \exp[-\beta U_N(\mathbf{X}^N)] \quad (6.1)$$

where we have separated the internal partition function $q$ from the configurational partition function. The former can be treated by classical or quantum mechanical methods. In the following discussion, we assume the validity of the partition function in this form. This does not imply that quantum mechanical effects are necessarily negligible for water at room temperature. Any attempt to include even first-order quantum mechanical corrections to the partition functions (6.1) would lead to insurmountable difficulties [see Friedmann (1962, 1964)].

The quantity $q$ can be further factorized into rotational, vibrational, and electronic partition functions. We will not need explicit knowledge of this quantity. It should be recognized, however, that a molecule experiencing strong interaction with its neighboring molecules may be perturbed to a significant extent. This, in turn, affects the internal partition function $q$. The very writing of (6.1) implies that we disregard these effects. This is the meaning of the separation of $q^N$ from the configurational partition function.

The next problem is concerned with the content of the quantity $U_N(\mathbf{X}^N)$. It is well known that major progress in the theory of simple fluids could not have been achieved without the assumption of pairwise additivity for the total potential energy, written as

$$U_N(\mathbf{X}^N) = \sum_{i<j} U(\mathbf{X}_i, \mathbf{X}_j) \qquad (6.2)$$

It is widely believed that such an assumption is a good approximation for simple nonpolar fluids such as liquid argon. In a formal manner, one may write the total potential energy $U_N(\mathbf{X}^N)$ as a sum of contributions due to pairs, triplets, quadruplets, etc., of particles

$$U_N(\mathbf{X}^N) = \sum_{i<j} U(\mathbf{X}_i, \mathbf{X}_j) + \sum_{i<j<k} U^{(3)}(\mathbf{X}_i, \mathbf{X}_j, \mathbf{X}_k) + \cdots + U^{(N)}(\mathbf{X}_1, \ldots, \mathbf{X}_N) \qquad (6.3)$$

where $U^{(n)}(\mathbf{X}_1, \mathbf{X}_2, \ldots, \mathbf{X}_n)$ is referred to as the $n$th-order potential.

Consider as a simple example a system of three particles. The total potential energy of the system at a given configuration can be written as

$$U_3(\mathbf{X}_1, \mathbf{X}_2, \mathbf{X}_3) = U(\mathbf{X}_1, \mathbf{X}_2) + U(\mathbf{X}_1, \mathbf{X}_3) + U(\mathbf{X}_2, \mathbf{X}_3) + U^{(3)}(\mathbf{X}_1, \mathbf{X}_2, \mathbf{X}_3) \qquad (6.4)$$

Recent quantum mechanical computations for a triplet of water molecules have shown that deviations from the additivity assumption for such a system are quite appreciable [for recent discussions, see Rao (1972) and Lentz and Scheraga (1973)]. Similarly, one expects that higher-order potentials may contribute to a significant extent to the total potential energy. In spite of the intensive study of this problem, we have no knowledge as yet of the analytical form of any of the potential functions appearing in (6.3) or (6.4). As a kind of practical procedure, one can incorporate all nonadditivity effects in a formal way by introducing the idea of an *effective* pair potential, for which we assume the resolution (Sinanoglu, 1967; Stillinger, 1970)

$$U_N(\mathbf{X}^N) = \sum_{i<j} U_{\text{eff}}(\mathbf{X}_i, \mathbf{X}_j) \tag{6.5}$$

The choice of the effective pair potential to be used in (6.5) is based on quantum mechanical calculations, on intuitive arguments, and on some knowledge of the mode of packing of water molecules in the condensed phases. Some examples are described in Section 6.4. Once we have chosen such an effective pair potential, we have, in fact, committed ourselves to a model of particles. That is, we define the particles of our system in terms of their (effective) pair potential. This is exactly the same procedure that one employs when studying Lennard-Jones or hard-sphere particles. In all cases, one must bear in mind that the model particles may represent the real particles only to a limited extent. Therefore, any result obtained from the theory strictly pertains to the system of these model particles, and inference of the behavior of the real system should be made with care.

The next difficulty is technical in essence. It involves the execution of an actual computation of average quantities by one of the numerical methods of statistical mechanics. Use of an angle-dependent pair potential function introduces a minor modification in the formal appearance of the theory, but vastly increases the amount of computational time required to accomplish a project which would normally take a relatively short time in the angle-independent case.

Because of the above difficulties, it is no wonder that many scientists have incessantly searched for other routes to studying liquid water and its solutions. The most successful approach has been the devising of various ad hoc models for water. In subsequent sections, we describe some of these theories and view them as approximate versions of the general mixture-model approach treated in Chapter 5.

## 6.2. SURVEY OF PROPERTIES OF WATER

In this section, we present a brief survey of the properties of water. We refrain from duplicating material that has been discussed fully by Eisenberg and Kauzmann (1969). The reader is presumed to be familiar with the basic properties of water in the gas, liquid, and solid states. The most important facts that we will need as background for the theoretical development are included in the following three subsections.

### 6.2.1. The Single Water Molecule

It is well established that in the water molecule, the nuclei of oxygen and two hydrogens form an isosceles triangle, the geometry of which is depicted in Fig. 6.1. The equilibrium O–H bond length is 0.957 Å and the H–O–H angle is 104.52°. Further information on the geometry of HDO and $D_2O$ may be found in the book by Eisenberg and Kauzmann (1969). Of course, the molecular dimensions may depend upon the particular quantum mechanical state of the molecules. However, it has been established that deviations from the equilibrium values cited above for water at room temperature are very small. Hence, we will assume that a single water molecule has a fixed and rigid geometry. We also assume that the basic geometry of the water molecule is maintained in the liquid and solid states [this topic has been reviewed by Fletcher (1971)].

The center of the oxygen nucleus is chosen as the center of the molecule (which is slightly different from the center of mass of the molecule). Figure 6.1 depicts the water molecule according to the van der Waals radii assigned to the oxygen (1.4 Å) and the hydrogen (1.2 Å) [values taken

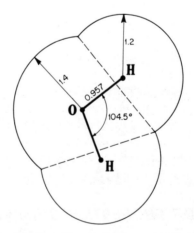

Fig. 6.1. Geometry of a single water molecule. The O–H distance and the H–O–H angle are indicated, as are the van der Waals radii of the hydrogen and the oxygen.

from Pauling (1960)]. It is sometimes convenient to view a water molecule as a sphere of radius 1.41 Å (this is about half of the average distance between two closest water molecules, averaged over all orientations, as manifested by the first peak of the radial distribution function of water; see below.)

An important characteristic feature of the charge distribution in a water molecule follows from the hybridization of the $2s$ and the $2p$ orbitals of the oxygen atom. As a result of this hybridization, two lobes of charge are created by the unshared electron pairs (or lone-pair electrons) which are symmetrically located above and below the molecular plane. These two directions, together with the two O–H directions, give rise to the tetrahedral geometry which is one of the most prominent structural features of the water molecule, and is probably the one responsible for the particular mode of packing of the molecules in the liquid and solid states.

### 6.2.2. The Structure of Ice and Ice Polymorphs

Nine forms of ice are known. Ordinary ice (referred to as hexagonal ice, denoted as $I_h$) has a structure isomorphous to the wurtzite form of ZnS. A detailed description of the geometry and properties of the various forms of ice can be found in Eisenberg and Kauzmann (1969), Fletcher (1970, 1971), and Franks (1972).

The most important feature of the structure of ice that is relevant to most of our considerations of the structure of water is the local tetrahedral geometry around each oxygen. That is, each oxygen is surrounded by four oxygens situated at the vertices of a regular tetrahedron, at a distance of 2.76 Å from the central oxygen. This basic unit of geometry is shown in Fig. 6.2.

It is convenient to introduce four unit vectors originating from the center of the oxygen and pointing toward the four corners of the regular tetrahedron. Let $\mathbf{h}_{ik}$ ($k = 1, 2$) be the two unit vectors belonging to the $i$th molecule and pointing approximately along the two O–H directions. The remaining two vectors $\mathbf{l}_{ik}$ ($k = 1, 2$) are along the directions of the lone pairs of electrons. Using the terminology of hydrogen-bond formation, we may identify the $\mathbf{h}_{ik}$ as the directions along which molecule $i$ forms a hydrogen bond as a donor molecule, whereas $\mathbf{l}_{ik}$ is the direction along which the same molecule participates as an acceptor in a hydrogen bond.

It is important to realize that the H–O–H angle of 104.54° is slightly smaller than the characteristic tetrahedral angle of 109.46° (see Fig. 6.2). There exists sound evidence that the basic geometry of a water molecule

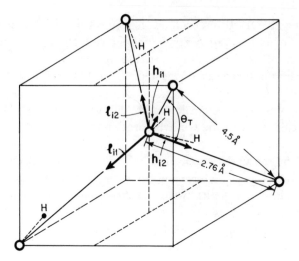

Fig. 6.2. Basic geometry around a water molecule in ice. Two unit vectors $\mathbf{h}_{i1}$ and $\mathbf{h}_{i2}$ are directed along the O–H bonds, and two unit vectors $\mathbf{l}_{i1}$ and $\mathbf{l}_{i2}$ are along the lone-pair directions. Together the four unit vectors point to the four vertices of a regular tetrahedron.

in ice is almost the same as in the vapor [Eisenberg and Kauzmann (1969), and Fletcher (1971)].

One fascinating problem is the configurational degeneracy that arises from the myriad arrangements of the hydrogens along the O–O lines, subject to the two so-called Bernal and Fowler (1933) conditions: (1) Each O–O line accommodates one and only one hydrogen, situated at a distance of about 1 Å from one of the oxygens. (2) Each oxygen has two hydrogens that are at a distance of 1 Å and two at a distance of 1.76 Å from its center. (We disregard the possibility of ionization in ice, i.e., we assume that the concentration of ions such as OH$^-$ and H$_3$O$^+$ is negligible.) Pauling (1935, 1960) was the first to compute the number of such configurations. His computation is now a classical example of a successful prediction based on an elementary statistical argument. We outline Pauling's solution of the problem.

Consider a perfect ice, containing $N$ water molecules. There are $N$ oxygens, $2N$ hydrogens, and $2N$ O–O bonds. The question is, in how many ways can we distribute the $2N$ hydrogens, subject to the two restrictions cited above?

Suppose that first we disregard condition 2, i.e., each hydrogen may be placed at one of the two "sides" of an O–O bond. Altogether there will

# Liquid Water

be $2^{2N}$ possible arrangements. Of course, many of these are inconsistent with the second condition. To account for the latter, let us consider a particular oxygen and count all the possible arrangements of hydrogens in its immediate environment. There are altogether 16 such arrangements distributed in the following manner: (1) One arrangement in which all of the hydrogens are close to the oxygen. (2) Four arrangements in which three hydrogens are close and one is far. (3) Six arrangements in which two hydrogens are close and two are far. (4) Four arrangements in which one is close and three are far. (5) One arrangement in which all four hydrogens are far from the oxygen. A schematic discription of these possibilities is given in Fig. 6.3.

The Pauling approximate solution is the following. Since there are six out of 16 acceptable arrangements about each oxygen (i.e., arrangement consistent with the second condition), we assume that 6/16 is the probability that a given oxygen has the correct arrangement around it. (This involves the assumption that all 16 arrangements counted above are equally probable, or "elementary events." Hence, the probability of the acceptable event is the ratio of the number of acceptable to the total number of events.) Furthermore, assuming that all of the oxygens are independent, we get $(6/16)^N$ as the probability of finding simultaneously all of the oxygens in an arrangement consistent with the second condition. Hence, the total number of accepted configurations is

$$\Omega = (6/16)^N 2^{2N} = (3/2)^N = (1.5)^N \tag{6.6}$$

This number has been used to estimate the residual entropy of ice,

$$S_0 = k \ln \Omega = 0.805 \quad \text{e.u.}$$

which is in remarkable agreement with the corresponding experimental value of 0.82 e.u.

Agreement between the theoretical and experimental values of $S_0$ has helped to establish the fact that the hydrogens are distributed in such a

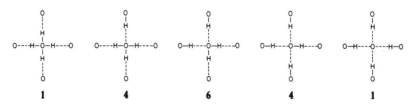

Fig. 6.3. Five possible arrangements of the hydrogens around a given oxygen. The numbers indicate the degeneracy of each configuration.

way that a pair of hydrogens is always close to an oxygen, which lends further support to the contention that the water molecules preserve their integrity in the ice structure. [Direct determination of the O–H distance in ice has been reported by Peterson and Levy (1957), using neutron diffraction data from heavy ice.]

In 1967, an exact solution of the analogous two-dimensional ice problem was presented by Lieb (1967). The result is

$$\Omega = (\tfrac{4}{3})^{3N/2} = (1.539\ldots)^N \qquad (6.7)$$

which is surprisingly close to the approximate value [Eq. (6.6)] computed by Pauling. For a review of further developments of this topic, see Lebowitz (1968).

A variety of interesting properties are manifested by the high-pressure polymorphs of ice. For the purpose of the study of liquid water, it is useful to remember that a large number of structures can be formed around a water molecule in the solid state. In particular, we draw attention to the fact that both open and close-packed structures are possible. There is no doubt that the open structure of ice, $I_h$, is maintained because of strong directional forces (hydrogen bonds) operating along the directions of the four unit vectors, as depicted in Fig. 6.2. The high-pressure polymorphs of ice are characterized by relatively higher densities. That is, each water molecule experiences a higher local density than in ice $I_h$. In spite of the fact that the number of nearest neighbors is larger in these structures, their internal energies are higher than those of ice $I_h$, which indicates that the average binding energy of a water molecule in an open structure may be stronger than the binding energy of the same molecule in a more closely packed structure. Although we will not make any direct inferences from this fact, it will be useful to remember it when dealing with some of the anomalous properties of water and aqueous solutions. In fact, we will see later that the relation between local density and binding energy is a very important aspect of the mode of packing of water molecules in the liquid state.

### 6.2.3. Liquid Water

The anomalous properties of water have been extensively reviewed in the literature (Dorsey, 1940; Pauling, 1960; Kavanau, 1964; Berendsen, 1967; Ives and Lemon, 1968; Eisenberg and Kauzmann, 1969; Frank, 1970; Horne, 1972; and Franks, 1972). In this section, we present a very

# Liquid Water

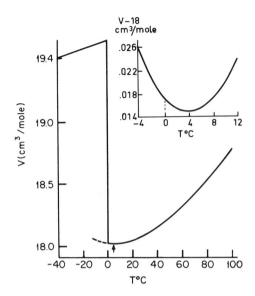

Fig. 6.4. The temperature dependence of the molar volume of water. The scale of the figure in the upper right is magnified to show the minimum at about 4°C. [Redrawn with changes from Eisenberg and Kauzmann (1969).]

brief survey of some of the properties to which we shall refer in subsequent sections.

The most unusual and prominent property of water is the temperature dependence of the density between 0 and 4°C. The fact that ice contracts upon melting, though a rare phenomenon, is not a unique property of water. Other solids with similar open structures, such as germanium, also contract upon melting. The outstanding property of water is the continual *increase* in density upon increase of temperature between 0 and 4°C. Relevant experimental data are presented in Fig. 6.4. Note that the relative decrease in volume is quite small compared with the molar volume at these temperatures. It is also worthwhile noting that this anomalous phenomenon of water disappears at high pressures, and the temperature dependence of the volume becomes "normal."

The coefficient of isothermal compressibility $\varkappa_T = [\partial(\ln \varrho)/\partial P]_T$ decreases between 0 and 46°C (at 1 atm), whereas the compressibility of most liquids increases monotonically as temperature increases. The experimental values of $\varkappa_T$ for water are shown in Fig. 6.5. For recent results see Fine and Millero (1973).

The value of the heat capacity of water (both $C_V$ and $C_P$) is much higher than the value expected from the classical contributions due to the various degrees of freedom of a water molecule. Neglecting the contribution from vibrational excitation, one expects that a mole of water molecules will contribute $3R/2$ due to translational degrees of freedom, and $3R/2$ due to the rotational degrees of freedom ($R = 1.98$ cal/mole deg is the gas constant). Thus, the estimated heat capacity of water due to its kinetic degrees of freedom is about $3R$, i.e., 5.96 cal/mole deg, which is indeed quite close to the value of the heat capacity of water vapor. Figure 6.6 shows that the actual value of $C_V$ of water in the liquid region is about 18 cal/mole deg, which is quite large compared with the corresponding values of $C_V$ for the solid and gaseous phases. It is also worthwhile noting that the $C_P$ of liquid water passes through a minimum at about 35°C, whereas $C_V$ decreases monotically from about 18.1 at 0°C to about 16.2 cal/mole deg at 100°C. The variation in $C_P$ of liquid water is shown schematically in Fig. 6.6.

The high value of the heat capacity is not a unique property of water. Other liquids, such as ammonia, have high heat capacities when compared to the values expected from classical considerations of the contributions from various degrees of freedom.

Other well-known outstanding properties of water are the high melting and boiling temperatures (compared, for instance, with the isoelectronic sequence of hydride molecules such as $NH_3$, HF, and $CH_4$), the high value of the heat and entropy of vaporization, and the high value of the dielectric constant.

All of these properties reflect the role of the strong and highly directional forces, which dictate a unique mode of packing of water molecules

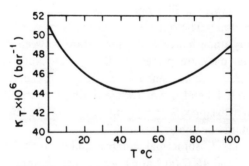

Fig. 6.5. The isothermal compressibility of water as a function of temperature. [Redrawn with changes from Eisenberg and Kauzmann (1969).]

# Liquid Water

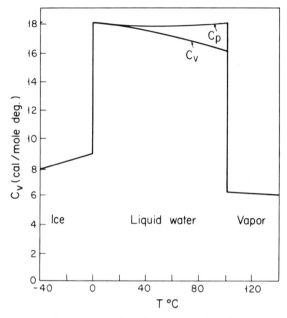

Fig. 6.6. The heat capacity of water as a function of temperature. [Redrawn with changes from Eisenberg and Kauzmann (1969).]

in the liquid state. In the following sections, we elaborate on the various theoretical tools that can be employed to illuminate the origin of these properties from the molecular point of view.

## 6.3. THE RADIAL DISTRIBUTION FUNCTION OF WATER

Perhaps some of the most important information on the mode of molecular packing of water in the liquid state is contained in the radial distribution function, which, in principle, can be obtained by processing X-ray or neutron scattering data. There are, however, several difficulties in extracting the proper information from the experimental data. First, it should be kept in mind that the full orientation-dependent pair correlation function cannot be obtained from such an experiment. Instead, only information on the spatial pair correlation function is accessible. We recall the definition of this function,

$$g(\mathbf{R}_1, \mathbf{R}_2) = [1/(8\pi^2)^2] \int \int d\mathbf{\Omega}_1 \, d\mathbf{\Omega}_2 \, g(\mathbf{R}_1, \mathbf{\Omega}_1, \mathbf{R}_2, \mathbf{\Omega}_2) \qquad (6.8)$$

and remind the reader that the function $g(\mathbf{R}_1, \mathbf{R}_2)$ is, in fact, a function of the scalar distance $R = |\mathbf{R}_2 - \mathbf{R}_1|$.

Furthermore, water, as a heteroatomic liquid, produces a diffraction pattern that reflects the combined effects of O–O, O–H, and H–H correlations. Thus, in principle, we have three distinct atom pair-correlation functions: $g_{OO}(R)$, $g_{OH}(R)$, and $g_{HH}(R)$. Experimental data cannot, at present, be resolved to obtain these three functions separately. Therefore, the only information obtained is a weighted average of these three functions. It has been estimated (Narten, 1970; Narten and Levy, 1972) that the relative contribution to the total scattering from the O–H and H–H pairs is about 12% and 2%, respectively. Figure 6.7 shows the resultant function, denoted by $G(R)$, and the pair correlation function $g_{OO}(R)$, as obtained by Narten and Levy (1971). The essential difference between the two functions is the prominent peak of $G(R)$ at $R \sim 1$ Å, which is due to the intramolecular distance of the O–H pair. Other features of the function are mainly due to the O–O pair distribution. We henceforth refer only to the function $g_{OO}(R)$ or, simply, to $g(R)$ of water.

Figure 6.8 shows the radial distribution function of water and argon as a function of the reduced distance $R^* = R/\sigma$, where $\sigma$ has been taken as 2.82 Å for water and 3.4 Å for argon. [These "effective" diameters are roughly the locations of the first maximum of the function $g(R)$ for water and argon, respectively.] Note that the location of the first peak of $g(R)$ of water at 2.82 Å is very close to 2.76 Å, which is the value of the nearest-neighbor distance of O–O in ice $I_h$. This indicates a high probability of obtaining nearest neighbors at the correct position, as in ice.

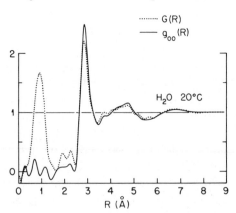

Fig. 6.7. The functions $G(R)$ and $g_{oo}(R)$ for water at 20°C. [Redrawn with changes from Narten and Levy (1971).]

# Liquid Water

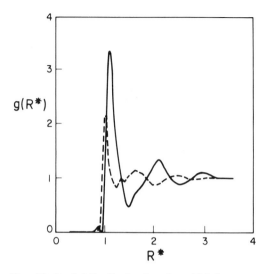

Fig. 6.8. Radial distribution function $g(R^*)$ for water (dashed line) at 4°C and 1 atm, and for argon (solid line) at 84.25°K and 0.71 atm, as a function of the reduced distance $R^* = R/\sigma$. [Data for argon was provided by N. S. Gingrich, and for water by A. H. Narten, for which the author is very grateful.]

The most important differences between the two curves in Fig. 6.8 are the following: (1) The first coordination number, defined by

$$n_{\mathrm{CN}} = \varrho \int_0^{R_M} g(R) 4\pi R^2 \, dR \quad (6.9)$$

where $R_M$ may be taken, conveniently, as the location of the first minimum (following the first maximum at $R^* \approx 1$), measures the average number of centers of molecules in a sphere of radius $R_M$ drawn about the center of a given molecule (the latter is not included in the counting). The value of $n_{\mathrm{CN}}$ for water at 4°C is about 4.4, whereas the value for argon (at 84.25°K and 0.71 bar) is about 10. The value of 4.4 for the coordination number is strongly reminiscent of the exact coordination number, 4, in the ice $I_h$ structure. This is one indication that some of the "structure" is preserved upon the melting of ice. (2) The location of the second peak of $g(R)$ of argon is at about $R^* = 2$. In Chapter 2, we showed that this phenomenon reflects the tendency of the spherical molecules to pack in roughly concentric and equidistant spheres around a given particle. In mixtures of two kinds of spherical molecules, as we saw in Chapter 4, the locations of the peaks of $g_{\alpha\beta}(R)$ beyond the first one are determined by the diameters of the particles that fill the space between the two particles under consideration. In contrast to the spherical case, the spatial correlation between water molecules is mainly determined by the strong directional forces acting between the molecules. This is reflected in the location of the second peak of $g(R)$, which for water is $R^* = 1.6$ or $R \approx 4.5$ Å. Incidentally, this distance coincides almost exactly with the distance of the second nearest neighbors in the ice $I_h$ lattice. Indeed, a simple computation (see Fig. 6.2) shows that the ideal second neighbors in ice are found to be at a distance of

$$2 \times 2.76 \times \sin(\theta_T/2) = 4.5 \qquad (6.10)$$

where $\theta_T$ is the tetrahedral angle ($\theta_T = 109.46°$). The difference in the nature of the mode of packing of molecules in water and that of spherical particles is shown schematically in Fig. 6.9. It is thus seen that the strong directional forces (hydrogen bonds) dictate a particular pattern of packing of the water molecules.

These two features of the radial distribution function lead to the following conclusion: The basic geometry around a single molecule in

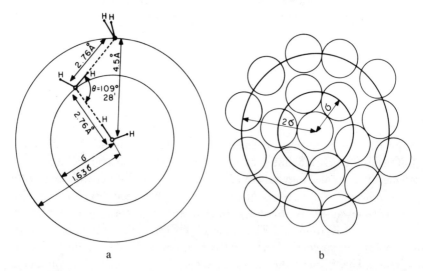

Fig. 6.9. Schematic description of the distribution of first- and second-nearest neighbors (a) in water and (b) in a simple fluid. The tetrahedral orientation of the hydrogen bond induces a radial distribution of first- and second-nearest neighbors at $\sigma$ and $1.63\sigma$, respectively, $\sigma = 2.76$ Å being the O–O distance in ice $I_h$. The almost equidistant and concentric nature of the packing of particles in a simple fluid produces corresponding neighbors at $\sigma$ and $2\sigma$.

**Liquid Water** 237

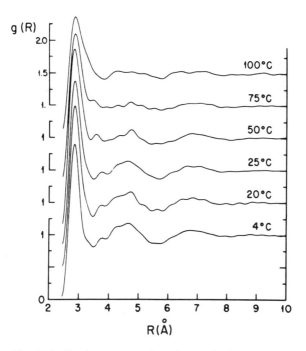

Fig. 6.10. The temperature dependence of $g(R)$ of water. [Redrawn with changes from Narten and Levy (1971).]

water is, to a large extent, similar to that of ice. This is to say that, on the average, each molecule has a coordination number of about four, and, furthermore, there is a high probability that triplets of molecules will be found with nearly the same geometry as triplets of molecules in successive lattice points in ice. It is important to emphasize that this conclusion pertains only to the *local* environment of a water molecule, and that no information whatsoever is furnished by $g(R)$ on the "structure" of the extended layer of molecules. In other words, if one were to sit at the center of a water molecule and observe the local geometry in a sphere of radius, say, 5 Å, one would see most of the time a picture very similar to what he would see if he sat on an ice molecule, with frequent distortions caused by thermal agitations typical to the liquid state. The distortions become so large at distances greater than, say, 5 Å that one can hardly recognize a characteristic pattern of ice beyond that distance.

The temperature dependence of $g(R)$ of water is shown in Fig. 6.10. It has been noted by Narten and Levy (1972) that there is a gradual shift in the location of the first peak, from 2.84 Å at 4°C to about 2.94 Å at

200°C. A more characteristic feature is the rapid decay of the second peak, which is almost unrecognizable at 100°C.

The temperature dependence of the coordination number of water is also in sharp contrast to that of a simple fluid. According to computations made by Samoilov (1957), the coordination number of water increases slightly from 4.4 at 4°C to about 4.9 at 83°C. Whereas the coordination number of argon decreases substantially with rising temperature, Narten and Levy (1972), using more recent data on water, concluded that the coordination number of water is almost constant with temperature. [Note that differences in the definition of the concept of coordination number may exist. For a survey of the various definitions, the reader is referred to Pings (1968).]

We conclude this section by noting that although experimental data provide only the spatial correlation function, theoretical calculations are, in principle, capable of furnishing the more information-rich function $g(\mathbf{R}, \mathbf{\Omega})$. We discuss some aspects of these functions in subsequent sections of this chapter.

## 6.4. EFFECTIVE PAIR POTENTIAL FOR WATER

In Section 1.7, we stressed the fact that any pair potential ever used in the theory of simple fluids has actually been a *model pair potential* for real particles. Alternatively, we may adopt the point of view that we are developing a theory for *model* particles, interacting via an exact pair potential. This is the way we introduced the idea of "Lennard-Jones particles," which, of course, are not real particles.

The same situation occurs in the theory of water, a far more complex fluid than the common simple ones dealt with by most theorists of the liquid state.

In this section, we present some of the characteristics of an effective pair potential that may be used for simulating the properties of water. Because of the rather crude and preliminary stage of this subject, it is more appropriate to speak of *waterlike* particles that are presumed to interact according to some specified pair potential.

Let us consider the various ingredients that are expected to contribute to the interaction energy between two real water molecules.

1. At very short distances, say $R < 2$ Å, the two molecules exert strong repulsive forces on each other, thereby preventing excessive interpenetration. A reasonable description of the potential energy in this region can be

# Liquid Water

achieved with a Lennard-Jones function of the form

$$U_{LJ}(R) = 4\varepsilon[(\sigma/R)^{12} - (\sigma/R)^6] \qquad (6.11)$$

Since water and neon are isoelectronic molecules, it is appropriate to take for $\varepsilon$ and $\sigma$ in (6.11) the corresponding values for neon (Corner, 1948), namely

$$\varepsilon = 5.01 \times 10^{-15} \text{ erg} = 7.21 \times 10^{-2} \text{ kcal/mole}, \quad \sigma = 2.82 \text{ Å} \qquad (6.12)$$

2. At a large distance, say a few molecular diameters, the interaction between the two molecules can be described as resulting from a few ideal electric multipoles. The most important term is obviously the dipole–dipole interaction, namely,

$$U_{DD}(\mathbf{X}_1, \mathbf{X}_2) = R_{12}^{-3}[\boldsymbol{\mu}_1 \cdot \boldsymbol{\mu}_2 - 3(\boldsymbol{\mu}_1 \cdot \mathbf{u}_{12})(\boldsymbol{\mu}_2 \cdot \mathbf{u}_{12})] \qquad (6.13)$$

where $\boldsymbol{\mu}_i$ is the dipole moment vector of the $i$th particle and $\mathbf{u}_{12}$ is a unit vector along the direction $\mathbf{R}_2 - \mathbf{R}_1$. The dipole moment for a single water molecule is $1.84 \text{ D} = 1.84 \times 10^{-18}$ esu·cm.

In principle, one can add to (6.13) interactions between higher multipoles, and thus arrive at a more precise description of the interaction between the two molecules. This procedure is not useful, however, since our knowledge of higher moments of a water molecule ends effectively with the dipole moment. Furthermore, even if we knew the exact values of a few multipole moments, it is unlikely that these could be used to describe the interaction between two water molecules at a short separation.

3. The intermediate range of distances, say $2 \text{ Å} \lesssim R \lesssim 5 \text{ Å}$, is the most difficult one to describe analytically. We know that two water molecules can engage in hydrogen bonding (HB), a phenomenon that must be dealt with in quantum mechanical language. It is clear that the HB energy is stronger than typical interaction energies operating between simple, nonpolar molecules, yet it is much weaker than typical energies of chemical bonds. It is also highly orientational dependent; hence, any function that is supposed to describe the interaction energy in this range must be a function of at least six coordinates, consisting of the separation and the relative orientation of the two molecules. We use the term HB to refer to the potential function operative in this intermediate range of distances.

A rough estimate of the order of magnitude of the HB energy can be obtained from the heat of sublimation of ice, which is about 11.65 kcal/mole. Since one mole of ice contains two "moles" of hydrogen bonds, then, assuming that all the interaction energy comes from the nearest-neighbor bonding, we arrive at the estimate of 5.82 kcal/mole of bonds. Indeed, various estimates of the HB energies range between 2 and 7 kcal/mole of bonds. (The meaning of the term hydrogen bond varies according to the context in which it is used. There is not as yet a precise definition of this concept.)

The question of the nature of the hydrogen bond has been the subject of numerous investigations [recent work is reported by Kollman (1972), Kollman and Allen (1969, 1970, 1972), Morokuma and Pederson (1968), Morokuma and Winick (1970), Hankins *et al.* (1970); and Lentz and Scheraga (1973); for reviews, the reader is referred to Pauling (1960), Pimentel and McClellan (1960), Eisenberg and Kauzmann (1969), Franks (1972), and Kollman and Allen (1972)]. In spite of considerable effort expended on this question, we are still far from knowing the analytical behavior of the potential function bearing the HB character of the interaction. Furthermore, even if we knew the exact potential function giving the interaction energy for a pair of water molecules at any specified orientation, it is doubtful that such a function would be suitable for the study of water in the liquid state. Strong nonadditive effects (sometimes referred to as cooperative effects; see Frank, 1958), which are known to play an important role in the condensed phase, are bound to render a knowledge of the exact pair potential for two water molecules almost irrelevant to the study of liquid water.[1]

At present, there seems to be no other way but to resort to our intuition for constructing an analytical form of a pair potential for water which best describes the prominent features of the HB interaction. Of course, as time passes, more accurate data, from quantum mechanical calculations, will be accumulated. Such information may be utilized to perfect the analytical description of the pair, as well as higher-order, potentials for water.

We describe two possible analytical forms of a potential function which reflect some of our knowledge as well as our expectations of such an effective potential function.

---

[1] One of the most vivid manifestations of such a nonadditivity effect has been demonstrated by Coulson and Eisenberg (1966). They showed that the average dipole moment of a water molecule in ice is about 2.60 D compared to the value of 1.84 D for a single molecule. A full study of this phenomenon must be carried out by direct quantum mechanical computations on a small number of water molecules. For instance, Weissmann and Cohan (1965) have shown that the dipole moment of water increases to 2.4 D upon participation in a hydrogen bond. For a recent discussion of this topic see Crowe and Santry (1973).

# Liquid Water

## 6.4.1. Construction of a HB Potential by Gaussian Functions

Recognizing the fact that the hydrogen bond occurs only along four directions pointing to the vertices of a regular tetrahedron, we can use the four unit vectors introduced in Section 6.2 (see Fig. 6.2) to construct a HB potential function. This we denote by $U_{HB}(\mathbf{X}_1, \mathbf{X}_2)$, so that the full effective pair potential is a superposition of three terms

$$U(\mathbf{X}_1, \mathbf{X}_2) = U_{LJ}(R_{12}) + U_{DD}(\mathbf{X}_1, \mathbf{X}_2) + U_{HB}(\mathbf{X}_1, \mathbf{X}_2) \quad (6.14)$$

Figure 6.11 depicts two waterlike molecules in a "favorable" orientation for the formation of a hydrogen bond. This means that the O–O distance is about 2.76 Å, that one molecule, serving as a donor, has its O–H direction at about $\mathbf{R}_{12} = \mathbf{R}_2 - \mathbf{R}_1$, and that the second molecule, serving as an acceptor, has the lone-pair direction along $-\mathbf{R}_{12}$. Mathematically, we can transcribe the above statement in terms of a product of three Gaussian functions.

Let $G_\sigma(x)$ be an unnormalized Gaussian function

$$G_\sigma(x) = \exp(-x^2/2\sigma^2) \quad (6.15)$$

which is symmetric about $x = 0$ and has a maximum height of one. Its qualitative form is shown in Fig. 6.12. Now, consider the following combination of such Gaussian functions:

$$U_{HB}(\mathbf{X}_i, \mathbf{X}_j) = \varepsilon_H G_{\sigma'}(R_{ij} - R_H) \Big\{ \sum_{\alpha,\beta=1}^{2} G_\sigma[(\mathbf{h}_{i\alpha} \cdot \mathbf{u}_{ij}) - 1] G_\sigma[(\mathbf{l}_{j\beta} \cdot \mathbf{u}_{ij}) + 1]$$

$$+ \sum_{\alpha,\beta=1}^{2} G_\sigma[(\mathbf{l}_{i\alpha} \cdot \mathbf{u}_{ij}) - 1] G_\sigma[(\mathbf{h}_{j\beta} \cdot \mathbf{u}_{ij}) + 1] \Big\} \quad (6.16)$$

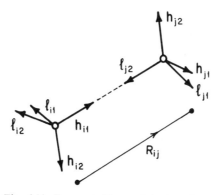

Fig. 6.11. Two waterlike particles in a favorable orientation to form a hydrogen bond.

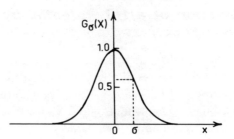

Fig. 6.12. Gaussian function $G_\sigma(x)$. The value of the function at $x = 0$ is unity. The value drops to $\exp(-0.5) = 0.605\ldots$ at $x = \sigma$.

The above function, though cumbersome in appearance, is quite simple in content. Consider first the function $G_{\sigma'}(R_{ij} - R_H)$, where $R_H$ is the intermolecular distance at which we expect a hydrogen bond to be formed (pending fulfillment of the other conditions listed below). A reasonable choice is $R_H = 2.76$ Å. This Gaussian function attains its maximum value whenever $R_{ij} = R_H$, and the rate at which the interaction energy drops as $|R_{ij} - R_H|$ increases is controlled by the parameter $\sigma'$. Next, we must stipulate the relative orientation of the pair of molecules. The factor $G_\sigma[(\mathbf{h}_{i\alpha} \cdot \mathbf{u}_{ij}) - 1]$ attains its maximum value whenever the unit vector $\mathbf{h}_{i\alpha}$ (i.e., one of the O–H directions, $\alpha = 1, 2$, of the $i$th molecule) is in the direction of the unit vector $\mathbf{u}_{ij} = \mathbf{R}_{ij}/R_{ij}$. Similarly, $G_\sigma[(\mathbf{l}_{j\beta} \cdot \mathbf{u}_{ij}) + 1]$ attains its maximum value whenever the direction of the lone pair of the $j$th molecule is in the direction $-\mathbf{u}_{ij}$. Thus, the product of these three Gaussian functions attains a value close to unity only if, simultaneously, $R_{ij}$ is about $R_H$, the direction of $\mathbf{h}_{i\alpha}$ is about that of $\mathbf{u}_{ij}$, and the direction of $\mathbf{l}_{j\beta}$ is about that of $-\mathbf{u}_{ij}$. Such a configuration is said to be "favorable" for HB formation. Clearly, if all of the above three conditions are fulfilled, then the interaction energy is about $\varepsilon_H$, a parameter that reflects the strength of the hydrogen bond. The sum of the various terms in the curly brackets of (6.16) arises from the total of eight possible favorable directions for HB formation (four when $i$ is a donor and four when $i$ is an acceptor). The variances $\sigma$ and $\sigma'$ are considered as parameters that, in principle, can be determined once detailed knowledge of the variation of the HB energy with distance and orientation is available. Note that of the eight terms in the curly brackets, only one may be appreciably different from zero at any given configuration $\mathbf{X}_i$, $\mathbf{X}_j$. This can always be achieved by choosing a sufficiently small value for $\sigma$.

## 6.4.2. Construction of a HB Potential Function Based on the Bjerrum Model for Water

The second example of waterlike particles is essentially an extension of Bjerrum's four-point-charge model for the water molecule (Bjerrum, 1951; Ben-Naim and Stillinger, 1972). Figure 6.13 shows this model. It consists of four point charges situated at the vertices of a regular tetrahedron, the center of which is assumed to coincide with the center of the oxygen atom. The point charges are located 1 Å from the center of the tetrahedron. Two charges are positive, $+\eta e$ (where $e$ is the magnitude of the electron charge and $\eta$ is a parameter that is chosen to produce the dipole moment of a water molecule, but which may be regarded essentially as an adjustable parameter), and two charges are negative, $-\eta e$. Together, they simulate the two partially shielded protons and the two pairs of unshared electrons, respectively.

The "hydrogen bond" potential in this case consists of the 16 Coulombic interactions between the pairs of point charges situated on different molecules, which can be written as

$$U_{\mathrm{HB}}(\mathbf{X}_1, \mathbf{X}_2) = (\eta e)^2 \sum_{\alpha_1, \alpha_2=1}^{4} (-1)^{\alpha_1+\alpha_2}/d_{\alpha_1\alpha_2}(\mathbf{X}_1, \mathbf{X}_2) \quad . (6.17)$$

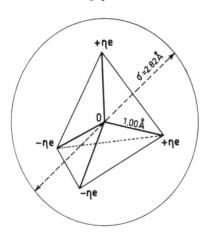

Fig. 6.13. Bjerrum's four-point-charge model for a water molecule. The oxygen atom coincides with the center of a regular tetrahedron. The fractions of charge $\pm \eta e$ are placed at the vertices of the tetrahedron at a distance of 1.0 Å from the center. The van der Waals diameter of the molecule is 2.82 Å.

where $\alpha_1$ and $\alpha_2$ run over the four point charges of molecules 1 and 2, respectively, in such a way that even and odd values of $\alpha_i$ correspond to positive and negative charges. $d_{\alpha_1\alpha_2}(\mathbf{X}_1, \mathbf{X}_2)$ is the distance between the charges $\alpha_1$ and $\alpha_2$ at the specified configuration of the pair $\mathbf{X}_1, \mathbf{X}_2$.

In the construction of the total pair potential, we need not take an additional term for the dipole–dipole interaction, as in (6.14). However, in order to avoid divergence of the potential function when two charges of opposite signs come very close to each other, we use a "switching function" $S(R_{12})$, which serves to suppress this possibility by switching off the HB potential at close distances. The specific function $S(R_{12})$ chosen by Ben-Naim and Stillinger (1972) is

$$S(R_{12}) = \begin{cases} 0 & \text{for } 0 \leq R_{12} \leq R_1 \\ \dfrac{(R_{12} - R_1)^2(3R_2 - R_1 - 2R_{12})}{(R_2 - R_1)^3} & \text{for } R_1 \leq R_{12} \leq R_2 \\ 1 & \text{for } R_2 \leq R_{12} \leq \infty \end{cases} \quad (6.18)$$

where $R_1$ and $R_2$, as well as $\eta$, are adjusted so that the minimum of the potential functions is attained at $R_{12} = 2.76$ Å (at the specific orientation of the pair of molecules referred to below as the symmetric eclipsed one) and that the experimental second virial coefficient $B_2(T)$ is reproduced (see next section). The parameters chosen for this particular model are

$$\eta = 0.19, \quad R_1 = 2.0379 \text{ Å}, \quad R_2 = 3.1877 \text{ Å} \quad (6.19)$$

Note that the value of $\eta = 0.19$ is slightly larger than the value of $\eta = 0.17$ chosen by Bjerrum to reproduce the dipole moment of a free water molecule.[2] The combined potential function is thus

$$U(\mathbf{X}_1, \mathbf{X}_2) = U_{\text{LJ}}(R_{12}) + S(R_{12})U_{\text{HB}}(\mathbf{X}_1, \mathbf{X}_2) \quad (6.20)$$

The various ingredients of this potential function are shown schematically in Fig. 6.14 for one particular line of approach for the pair of molecules. Note that the LJ part of the potential serves essentially as the repulsive part at short distances; the contribution of the attractive part is negligible compared to the attraction produced by the electrostatic interaction. For more details on this particular function, and the determination of the various parameters, the reader is referred to Ben-Naim and Stillinger (1972).

---

[2] The Bjerrum value of $\eta = 0.17$ corresponds to a dipole moment of $\mu = 1.87$ D, whereas the value of $\eta = 0.19$ corresponds to $\mu = 2.1$ D. Such an increase in the dipole moment is indeed expected for a water molecule in the condensed state (see also Section 6.3).

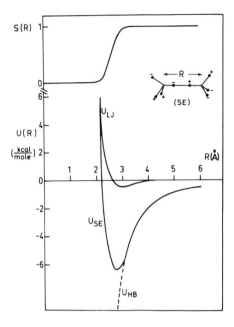

Fig. 6.14. Schematic description of the various ingredients of the potential function based on the Bjerrum model for water; $U_{SE}(R)$ is the full potential function for the symmetric eclipse (SE) approach of two water molecules (drawn at the upper right part of the figure). The three components of $U_{SE}$ for the same line of approach are also included. [See Eq. (6.20).]

## 6.5. VIRIAL COEFFICIENTS OF WATER

At low densities, the pressure can be expanded in a power series in the density $\varrho$,

$$P/kT = \varrho + B_2(T)\varrho^2 + B_3(T)\varrho^3 + \cdots \quad (6.21)$$

where $B_k(T)$ is called the $k$th virial coefficient. The importance of this expansion is in providing systematic corrections for deviations from ideal gas behavior. One interesting aspect of the virial coefficients is the following: The $k$th coefficient is determined by the behavior of a system of $k$ particles (Mayer and Mayer, 1940). For instance, the statistical mechanical expressions for the second and third virial coefficients are given by (see Section 1.8)

$$B_2(T) = -(1/16\pi^2) \int d\mathbf{X}_2 \{\exp[-\beta U(\mathbf{X}_1, \mathbf{X}_2)] - 1\} \qquad (6.22)$$

$$\begin{aligned}B_3(T) = -[1/3(8\pi^2)^2] \int d\mathbf{X}_2 \int d\mathbf{X}_3 \, \{&\exp[-\beta U_3(\mathbf{X}_1, \mathbf{X}_2, \mathbf{X}_3)] \\
-\, &\exp[-\beta U(\mathbf{X}_1, \mathbf{X}_2) - \beta U(\mathbf{X}_2, \mathbf{X}_3)] \\
-\, &\exp[-\beta U(\mathbf{X}_1, \mathbf{X}_2) - \beta U(\mathbf{X}_1, \mathbf{X}_3)] \\
-\, &\exp[-\beta U(\mathbf{X}_1, \mathbf{X}_3) - \beta U(\mathbf{X}_2, \mathbf{X}_3)] \\
+\, &\exp[-\beta U(\mathbf{X}_1, \mathbf{X}_2)] + \exp[-\beta U(\mathbf{X}_1, \mathbf{X}_3)] \\
+\, &\exp[-\beta U(\mathbf{X}_2, \mathbf{X}_3)] - 1\}\end{aligned} \qquad (6.23)$$

Note that (6.23) differs from the expression given in Section 1.8. The latter is obtained from the former if the total potential energy $U_3(\mathbf{X}_1, \mathbf{X}_2, \mathbf{X}_3)$ is pairwise additive. Equations (6.22) and (6.23) are special cases of a more general scheme which provides relations between virial coefficients and integrals involving interactions among a set of a small number of particles. This is known as the Mayer cluster theory [see, for example, Mayer and Mayer (1940), Hill (1956), and Münster (1969)].

From (6.22), we see that the second virial coefficient is a sixfold integral involving the pair potential for two water molecules. The numerical evaluation of such integrals is much more difficult than that of the corresponding one-dimensional integral for a system of spherical particles. Furthermore, the integral in (6.22), even when evaluated exactly, cannot be expected to be a stringent test of the potential function. Indeed, there can be an infinite number of functions $U(\mathbf{X}_1, \mathbf{X}_2)$ which will lead to the same virial coefficient $B_2(T)$. A somewhat stronger condition imposed on the potential is to require that it reproduce the temperature dependence of the second virial coefficient. [Note, however, that even for spherical particles, the temperature dependence of the second virial coefficient does not uniquely determine the pair potential $U(R)$ (Hirschfelder et al., 1954; Frisch and Helfand, 1960).]

Table 6.1 shows some values of $B_2(T)$ computed by using three different pair potentials, and their corresponding experimental values. The third column presents the values computed by Rowlinson (1949), using the Stockmayer potential function (Stockmayer, 1941), which is a superposition of a LJ and a dipole–dipole potential:

$$U(\mathbf{X}_1, \mathbf{X}_2) = U_{\mathrm{LJ}}(R_{12}) + U_{\mathrm{DD}}(\mathbf{X}_1, \mathbf{X}_2) \qquad (6.24)$$

## Table 6.1
### Second Virial Coefficient for Water (cm$^3$/mole)

| Temperature, °C | Experimental[a] | Stockmayer potential[b] | Modified Stockmayer potential[c] | Potential function Eq. (6.20)[d] |
|---|---|---|---|---|
| 100 | −450 | −460 | −450 | −466 |
| 200 | −197 | −202 | −205 | −190 |
| 300 | −112 | −115 | −123 | −107 |
| 400 | −72 | −74 | −80 | −65 |

[a] Data from Hirschfelder et al. (1954). More recent data are available (Kell et al., 1968) which are in qualitative agreement with these values.
[b] Calculated by Rowlinson (1949).
[c] Calculated by Rowlinson (1951).
[d] Calculated by Ben-Naim and Stillinger (1972).

The fourth column contains values computed by Rowlinson (1951), using a modified Stockmayer potential, which includes, in addition to the dipole–dipole term, a dipole–quadrupole term. The fifth column contains values of $B_2(T)$ computed by Ben-Naim and Stillinger (1972), using the potential function given in (6.20) with the parameters in (6.19). Note that although the potential functions are quite different, the computed values of $B_2(T)$ can be made to fit, to a reasonable extent, the experimental values. It is most important to remember that such a fit gives only a rough indication that the parameters used in the potential function have been chosen within a "reasonable" range of values; it does not provide sound evidence that these parameters are the "correct" ones, nor does it lend any support to the functional *form* of the particular function that has been chosen.

One further comment is worthwhile. Suppose we could have found the exact pair potential operating between a pair of water molecules. Using such a function in (6.22) should give a perfect fit to the experimental values of the second virial coefficient. However, as discussed in Section 6.1, a knowledge of such a pair potential is far from sufficient for the study of liquid water, since the presence of other molecules in the system modifies to a large extent the interaction between a given pair of molecules. This argument was the basic justification for constructing an *effective* pair potential, by the use of which we hope to simulate the properties of *liquid* water. This effective pair potential, therefore, does not necessarily have to be consistent with the second virial coefficient, a property which is determined by the behavior of precisely two water molecules.

The computation of the third virial coefficient poses far more serious difficulties. First, the integral in (6.23) is difficult to carry out numerically since it involves a twelvefold integration. Second, we cannot, at present, estimate the contribution from nonadditivity of the total potential to the integral.

It has recently been established that three-body forces have an important effect on the value of $B_3(T)$, even for nonpolar molecules [see Mason and Monchick (1967) for a review of this topic]. Finally, it should be noted that accurate values of the third virial coefficients are very difficult to obtain experimentally. This topic has been discussed in detail by Kell *et al.* (1968) and Eisenberg and Kauzmann (1969).

For all the above reasons, it is safe to conclude that computations of the third virial coefficient and comparison with experimental values are, at present, impractical for a test of the intermolecular potential chosen for water. In fact, all of the attempts made so far in this direction have failed to reproduce even qualitatively the temperature dependence of $B_3(T)$. For a review of some of these computations, see Ben-Naim and Stillinger (1972).

## 6.6. SURVEY OF THEORIES OF WATER

It is difficult to trace the historical development of the theory of water. One of the earliest documents attempting an explanation of some anomalous properties of water is Röntgen's (1892) article, which postulated that liquid water consists of two kinds of molecules, one of which is referred to as an "ice-molecule." Röntgen himself admitted that his explanation of the properties of water, using the so-called mixture-model approach, had been known in the literature for some time, but he could not point out its originator. An interesting review of the theories of water until 1927 was presented by Chadwell (1927). Most of the earlier theories were concerned with association complexes, or polymers of water molecules. There has been little discussion on the structural features of these polymers.

A major advance in the theory of water was initiated by Bernal and Fowler (1933). Conceptually, their approach stemmed from older theories, and postulated that liquid water may be viewed as a mixture of various components. However, the novelty of their approach was their detailed specification of the structure of the various components. In particular, they

## Liquid Water

stressed the importance of the local tetrahedral geometry of water molecules in the liquid state. In Bernal and Fowler's words, "We conclude, therefore, that the unique properties of water are due to a structure of the molecule which permits it to form in solid and liquid phases an extended electropolar complex characterized by tetrahedral (four) coordination."

This structural feature has affected almost all subsequent attempts to theorize about the properties of liquid water and aqueous solutions. Bernal and Fowler were also the first to attempt a theoretical calculation of the radial distribution function of water. Such an undertaking could not have been feasible without a detailed characterization of the structure of the various components presumed to exist in the liquid.

Until very recently, most of the theoretical approaches to liquid water have been based on an ad hoc model for the whole system. Extensive reviews of this topic are now available (Samoilov and Nosova, 1965; Némethy, 1965; Wicke, 1966; Drost-Hansen, 1967; Berendsen, 1967; Erlander, 1968; Eisenberg and Kauzmann, 1969; Frank, 1970; Krindel and Eliezer, 1971; Fletcher, 1971; Horne, 1972; and Franks, 1972). We shall present here only a very short survey of the main developments in the theoretical approaches to liquid water.

Samoilov (1946, 1957) advocated the idea that liquid water could be viewed essentially as ice $I_h$, with part of its cavities being filled by water molecules. This model may be referred to as an interstitial model for water. Numerous further developments of this model have taken place, due especially to Russian scientists (Yashkichev and Samoilov, 1962; Samoilov, 1963; Gurikov, 1963, 1965, 1966, 1968a, 1968b; Krestov, 1964; Vdovenko et al., 1966, 1967a, 1967b; Mikhailov, 1967, 1968; and Narten et al., 1967).

Pauling (1960) rejected the idea that liquid water contains a significant number of aggregates with quartzlike structures as proposed previously by Bernal and Fowler (1933). Instead, he proposed to view liquid water as a hydrate of itself. The idea is based on the well-known fact that molecules such as xenon, chlorine, and methane form clathrate compounds with water having a well-defined crystalline structure [for details, see Pauling (1960) and Frank and Quist (1961)]. Why not assume, then, that the same structure could host a water, instead of a nonelectrolyte, molecule? Frank and Quist (1961) undertook a quantitative development of Pauling's model [the essential features of their treatment, as well as a similar one by Mikhailov (1967), are discussed in detail in the next section].

A new development in the theory of water (as well as aqueous solutions) was then advanced by Némethy and Scheraga (1962). Their basic idea stems from Frank and Wen's picture of the "flickering clusters" of water molecules. Frank and Wen suggested (Frank and Wen, 1957; Frank, 1958) that the formation of hydrogen bonds should be a cooperative phenomenon. That is, a water molecule participating in a hydrogen bond is more likely to participate in a second hydrogen bond, and so forth. Therefore, they argued, if clusters of hydrogen-bonded molecules are formed, then these must have, with high probability, a compact form, i.e., the maximum number of hydrogen bonds per water molecule in the cluster. This idea was exploited by Némethy and Scheraga in constructing the partition function for water, and is outlined briefly below.

Suppose we adopt a definition of the concept of a hydrogen bond. Then, for each configuration $\mathbf{X}^N$, we can classify the water molecules according to the number of bonds in which they participate. Such a classification is, in principle, identical to the construction of a quasicomponent distribution function (Chapter 5) and hence, it is formally exact. (See also Section 6.10.) Another central concept of this theory is that of "energy levels" assigned to each of the species. In a qualitative manner, these energy levels may be viewed as the average binding energies of each species, again a well-defined quantity. The main approximation of the theory is the assumption of a factorizable partition function of the form

$$Q(T, V, N) = \sum_{\{N_k\}} N! \prod_{k=0}^{4} (N_k!)^{-1} [f_k \exp(-\beta E_k)]^{N_k} \qquad (6.25)$$

where $N_k$ ($k = 0, 1, 2, 3, 4$) is the number of molecules participating in $k$ hydrogen bonds. The factor $f_k$ bears the rotational, vibrational, and translational partition functions of a single molecule, $E_k$ is the corresponding energy level, and the summation is carried over all the compatible sets of numbers $\{N_k\} = \{N_0, N_1, N_2, N_3, N_4\}$. Once the partition function is written down and the appropriate parameters are assigned numerical values, the extraction of thermodynamic quantities becomes straightforward. It is worth noting that this theory has achieved remarkable agreement between predicted and experimental quantities.

As is the case for all other models of water, it is difficult to assess the validity of the assumptions underlying the choice of the particular partition function (6.25). Similar works using the mixture-model approach have accomplished a similar agreement with experimental results [for instance, models worked out by Samoilov (1946), Grjotheim and Krogh-Moe (1956), Wada (1961), Marchi and Eyring (1964), Jhon et al. (1966), Hagler et al. (1972)]. We elaborate in detail on the merits of the simplest mixture-model approach in Section 6.8.

Lennard-Jones and Pople (1951) and Pople (1951) maintained that only very few hydrogen bonds are broken upon the melting of ice. Pople (1951) developed this idea further and suggested that most hydrogen bonds should be viewed as distorted, or bent, rather than broken. Since there is as yet no precise, nor universally accepted, definition of the concept of hydrogen bonds, the distinction between broken or bent bonds is essentially a question of semantics. We return to this point in Sections 6.10 and 6.11.

In the next two sections, we undertake to describe the general theoretical reasoning underlying the mixture-model approach to water. In Section 6.7, we describe a special group of mixture models, which may be called interstitial models, and we outline the theory common to all of the models in this group. In Section 6.8, we discuss the general consequences of the two-structure models, without commitment to any specific model.

A comment regarding nomenclature is now in order. We distinguish between a one-component approach and a mixture-model approach to water. Within the latter, we further distinguish between continuous and discrete mixture-model approaches, according to the nature of the parameter used for the classification procedure (for instance, $v$ and $K$ in Chapter 5). All of these are equivalent and formally exact. The various ad hoc mixture models for water may be viewed as approximate versions of the general mixture-model approach. This will be discussed in Section 6.9.

Recently, a new approach to the study of liquid water was advocated by Ben-Naim and Stillinger (1972), which consists in the adoption of a model on a molecular level. This approach conforms to the traditional trend in the modern theories of simple fluids. Of course, because of the tremendous difficulties that such an approach is bound to encounter, the results are still far from being complete or satisfying. Nevertheless, interesting developments are presently taking place, some of which will be described in Sections 6.11 and 6.12.

Most of the discussions thus far have been focused on various "models for water," which we have classified into two categories: models for the water molecule, and models for the whole system.

The ultimate aim of the study of "models of water" is to achieve agreement with experimental quantities. However, because of the great complexity of the system, many models have been proposed for the study of just a few aspects of the behavior of liquid water or its solutions. Examples of these are one-dimensional models (Bell, 1969; Lovett and Ben-Naim, 1969) or two-dimensional models (Bell and Lavis, 1970; Ben-Naim, 1972c,d) having some features in common with water, and designed to simulate one or a few properties of the system. Such models may be referred to as "models of phenomena," rather than "models of liquid water." Their aim is certainly not to simulate the precise behavior of liquid water. They help, however, in gaining a deeper insight into some of the outstanding properties of both pure water and its solutions.

## 6.7. A PROTOTYPE OF AN INTERSTITIAL MODEL FOR WATER

The purpose of this section is threefold. First, it presents a prototype of an interstitial model having features in common with the models proposed for water, notably, the Samoilov (1957) and Pauling (1960) models. These have been worked out in considerable detail by Frank and Quist (1961) and Mikhailov (1967). Second, this model demonstrates some general aspects of the mixture-model approach to the theory of water, for which explicit expressions for thermodynamic quantities in terms of molecular properties may be obtained. Finally, the detailed study of this model has a didactic virtue, being an example of a simple and solvable model.

The interstitial model has a serious drawback, however, which is similar to the shortcomings of applying a lattice model for a fluid in general. It is therefore important to make a clear-cut distinction between results that strictly pertain to the model and results that have more general validity. We elaborate on this point throughout this section. The reader is urged to review Section 5.6, where the basic concepts relevant to the following considerations can be found.

### 6.7.1. Description of the Model

The system consists of $N_W$ water molecules at a specified temperature $T$ and pressure $P$. A total of $N_L$ molecules participate in the formation of a regular lattice with a well-defined structure which, for our purposes, need not be specified. (It may be an ice $I_h$ structure, as in the Samoilov model, or a clathrate type of structure, as in the Pauling model.) The lattice is presumed to contain empty spaces, or holes. We let $N_0$ be the number of holes per lattice molecule, i.e., $N_0 N_L$ is the total number of holes formed by $N_L$ lattice molecules. We assume that the system is macroscopically large, so that surface effects are negligible; hence, $N_0$ is considered to be independent of $N_L$. In the case of ice $I_h$, we have $N_0 = \frac{1}{2}$. (Each hole is surrounded by 12 molecules, but each molecule participates in six holes.)

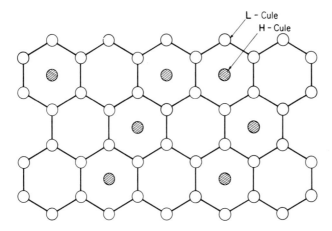

Fig. 6.15. Schematic illustration of an interstitial model for water in two dimensions. $L$-cules are molecules that build up the lattice, whereas the $H$-cules hold interstitial sites in the lattice.

Therefore, we can assign $N_0 = \frac{1}{2}$ holes per lattice molecule.) This value was used by Mikhailov (1967) for the Samoilov model. In the Pauling model, the assignment of $N_0$ depends on whether we count all the holes in the lattice as suitable for accommodating water molecules, or only the larger cavities. The choice of Frank and Quist (1961) was $N_0 = \frac{12}{46} \approx \frac{1}{4}$.

The remaining $N_H = N_W - N_L$ molecules are assumed to occupy the holes, and may be referred to as interstitial molecules. A schematic illustration of such a model is shown in Fig. 6.15.

To keep the complexity of the model at a minimum, we require the following simplifying assumptions: (1) All the holes have the same structure.[3] (2) A hole can accommodate at most one water molecule in such a way so as not to distort the lattice structure to a significant extent. (3) The interstitial molecules do not "see" each other, i.e., there is no direct interaction between interstitial molecules in adjacent holes. Hence, occupancy of a certain hole does not affect the chances of an adjacent hole being empty or filled. (4) The lattice molecules are assumed to hold the equilibrium lattice points, and vibrational excitation is negligible. (5) An interstitial molecule is assumed to be situated in a fixed position in the hole.

Clearly, the condition $N_H \leq N_0 N_L$ must be satisfied. Alternatively, the mole fraction of interstitial molecules is restricted to vary between the

---

[3] Frank and Quist (1961) considered two kinds of holes in the lattice, only one of which was occupied by interstitial molecules.

limiting values

$$0 \leq x_H = N_H/N_W \leq N_0/(1 + N_0) \quad (6.26)$$

For instance, in the model depicted in Fig. 6.15, $N_0 = \frac{1}{2}$ (i.e., each lattice particle belongs to three holes, but each hole is built up from six particles, hence each particle contributes half a hole), so that the mole fraction of interstitial molecules cannot exceed $\frac{1}{3}$.

The total energy of the system $E_T$ is composed of only the interaction energies among the molecules (the internal energies of the molecules can be collected to form the internal partition function of a single molecule; this, however, will not be added to our study of the present model):

$$E_T = N_L E_L + N_H E_H \quad (6.27)$$

where $E_L$ is the lattice energy per lattice molecule and $E_H$ is the interaction energy of an interstitial molecule with its surroundings.

Thus, the canonical partition function for our model is

$$Q(T, V, N_W) = \sum_{\text{configurations}} \exp[-\beta(N_L E_L + N_H E_H)]$$

$$= \sum_{E_T} \Omega(E_T, V, N_W) \exp[-\beta(N_L E_L + N_H E_H)] \quad (6.28)$$

where, in the second form on the rhs, we have collected all the terms in the partition function corresponding to the same value of the total energy $E_T$, which is $\Omega$-fold degenerate. Clearly, the total energy depends only on the number $N_L$ ($E_L$ and $E_H$ are assumed fixed by the model, and $N_H = N_W - N_L$). Hence, to compute the degeneracy of a given energy $E_T$, it is sufficient to compute the number of ways we can arrange $N_H$ interstitial molecules in $N_0 N_L$ holes. This number is[4]

$$\Omega(E_T, V, N_W) = (N_0 N_L)!/[N_H! (N_0 N_L - N_H)!] \quad (6.29)$$

Furthermore, in this particular model, the total volume $V$ can be written as

$$V = N_L V_L \quad (6.30)$$

---

[4] The first interstitial molecule can be put in any of the $N_0 N_L$ holes, the second in any of $N_0 N_L - 1$ holes, etc. The total number of such arrangements is $(N_0 N_L)(N_0 N_L - 1)$ $\cdots (N_0 N_L - N_H + 1)$. However, because of the equivalence of the molecules, we must divide this number by $N_H!$ to obtain the number of distinguishable arrangements, given in (6.29).

## Liquid Water

where $V_L$ is the volume of the system per lattice molecule. (Interstitial molecules do not contribute to the volume.) The condition (6.30) considerably simplifies the evaluation of the partition function in (6.28). If we determine the total volume $V$, then from (6.30), the number $N_L$ is determined as well. This, in turn, determines the total energy through (6.27). Therefore, the partition function (6.28) reduces to a single term, namely

$$Q(T, V, N_W) = \frac{(N_0 N_L)!}{N_H! (N_0 N_L - N_H)!} \exp[-\beta(N_L E_L + N_H E_H)] \quad (6.31)$$

Next, we write the $T$, $P$, $N_W$ partition function (this is more convenient for later applications in which various thermodynamic quantities are evaluated):

$$\Delta(T, P, N_W) = \sum_{V=V_{\min}}^{V_{\max}} Q(T, V, N_W) \exp(-\beta PV)$$

$$= \sum_{N_L = N_{L_{\min}}}^{N_W} \frac{(N_0 N_L)!}{N_H! (N_0 N_L - N_H)!}$$

$$\times \exp[-\beta(N_L E_L + N_H E_H + P N_L V_L)] \quad (6.32)$$

where, in the second form on the rhs, we have transformed the summation over the possible volumes of the system to a sum over $N_L$ [which determines the volume through (6.30)]. The summation is carried out from the minimum value of $N_{L_{\min}} = N_W/(1 + N_0)$ [see (6.26)] to the maximum value of $N_W$. This completes our description of the model.

### 6.7.2. General Features of the Model

Let us now consider some general features of the present model. At the very outset, we emphasize that the basic variables of our system are $T$, $P$, and $N_W$, i.e., we have a *one-component* system, and there is no trace of a mixture-model approach. The variables $N_L$ and $N_H$ in (6.32) play the role of convenient intermediary variables. Once we have carried out the summation in (6.32), the dependence of the partition function on $N_L$ disappears. Nevertheless, the nature of the model suggests a new way of looking at this one-component system; namely, we decide to refer to a lattice molecule as an *L*-cule and to an interstitial molecule as an *H*-cule (the letters $L$ and $H$ in the present context are chosen to remind us of water in "lattice" and water in "holes." The same notation will be used in Section 6.8 to indicate the type of environment, that of "low" local density and "high" local density, respectively.) Once we have made this classification, we may adopt the point of view that our system is a mixture of *two components*, *L*- and *H*-cules. As we emphasized in Section 5.6, such a point of view does not affect the physical properties of the system.[5]

Let us rewrite (6.32) in the form

$$\Delta(T, P, N_W) = \sum_{N_L} \Delta(T, P, N_W; N_L) \qquad (6.33)$$

where each summand $\Delta(T, P, N_W; N_L)$ is the partition function of a system in which the conversion reaction $L \rightleftarrows H$ has been "frozen in," so that $N_L$ attains a precise value (for more details, see Section 5.6).

Using a connection between this partition function and the Gibbs free energy (Section 1.4), we can rewrite (6.33) as

$$\exp[-\beta G(T, P, N_W)] = \sum_{N_L} \exp[-\beta G(T, P, N_W; N_L)] \qquad (6.34)$$

We now seek the maximum term in the sum on the rhs of (6.34), or equivalently, the term for which the Gibbs free energy $G(T, P, N_W; N_L)$ is minimum. From (6.32) and (6.34), we get

$$-\beta G(T, P, N_W; N_L) = -\beta(N_L E_L + N_H E_H + PN_L V_L) \\ + \ln[(N_0 N_L)!/N_H!(N_0 N_L - N_H)!] \qquad (6.35)$$

The condition for an extremum of $G$ with respect to the variable $N_L$ (keeping $T, P, N_W$ constant) is

$$0 = \partial(-\beta G)/\partial N_L = -\beta(E_L - E_H + PV_L) + N_0 \ln(N_0 N_L) + \ln N_H \\ - (N_0 + 1) \ln(N_0 N_L - N_H) \qquad (6.36)$$

We have used the Stirling approximation for the factorials in (6.35).

We denote by $N_L^*$ (and $N_H^* = N_W - N_L^*$) the value of $N_L$ for which $G$ is minimum. Condition (6.36) can be rewritten as an equilibrium condition (compare with Section 5.6)

$$\frac{(N_0 N_L^*)^{N_0} N_H^*}{(N_0 N_L^* - N_H^*)^{N_0+1}} = \exp[\beta(E_L - E_H + PV_L)] \equiv K(T, P) \qquad (6.37)$$

---

[5] Mikhailov (1967) worked out a detailed statistical mechanical theory of the Samoilov model. In particular, he made an estimate of the temperature dependence of the mole fraction of the lattice molecules. It is puzzling to find a comment in this article saying: "We do not consider Samoilov's ice-like model to be one of the so-called two-structure water models," since by the very fact that we choose to classify molecules into two or more species, we have already made the choice of the mixture-model approach (in the sense of Chapter 5).

# Liquid Water

The conventional mole fractions of $L$-cules and $H$-cules are

$$x_L = N_L/N_W, \qquad x_H = 1 - x_L \tag{6.38}$$

However, the equilibrium condition (6.37) can be written more conveniently in terms of the "mole fractions" of empty and occupied holes, which are defined, respectively, as

$$y_0 = (N_0 N_L - N_H)/N_0 N_L, \qquad y_1 = N_H/N_0 N_L \tag{6.39}$$

With these variables, the equilibrium condition (6.37) is

$$y_1^*/(y_0^*)^{N_0+1} = K(T, P) \tag{6.40}$$

which has a simple interpretation as an equilibrium constant for the "reaction"

$$(N_0 + 1)[\text{empty holes}] \rightleftarrows [\text{occupied holes}] \tag{6.41}$$

The stoichiometry of this reaction is understood as follows: In order to create one *mole* of occupied holes, we must cancel one mole of empty holes. In addition, in order to fill these holes, we need one mole of molecules, which come from the lattice; hence, $N_0$ empty holes must also be destroyed. Altogether, we need $N_0 + 1$ moles of empty holes to be converted to one mole of occupied holes.

Once we have determined $N_L^*$ by (6.36), we can replace the sum in (6.34) by a single term

$$\exp[-\beta G(T, P, N_W)] = \sum_{N_L} \exp[-\beta G(T, P, N_W; N_L)]$$
$$= \exp[-\beta G(T, P, N_W; N_L^*)] \tag{6.42}$$

where the second equality is understood to hold in the sense discussed in great detail in Section 5.6. We recall that $N_L^*$ in (6.42) is not a "new" independent variable, but is, in principle, dependent on $T, P, N_W$ through the equilibrium condition (6.37). In the general case, it is difficult to solve (6.37) explicitly. As a simple case, we assume that $N_0 = 1$ and rewrite (6.37) in terms of the mole fraction of $L$-cules, $x_L = N_L/N_W$:

$$x_L^*(1 - x_L^*)/(2x_L^* - 1)^2 = K \tag{6.43}$$

The solution of (6.43) is

$$x_L^* = \frac{1}{2} \pm \frac{1}{2}\left(\frac{1}{4K+1}\right)^{1/2} \tag{6.44}$$

There are two possible solutions, only one of which is physically acceptable. In order to determine the correct sign, we can check one limiting case. For instance, for given $T$ and $P$, if $E_L \ll E_H$, then $K \to 0$. In this case, $x_L^*$ must tend to unity (i.e., $L$-cules are more favorable). From (6.44), we get

$$x_L^* \xrightarrow{K \to 0} (1 \pm 1)/2 \qquad (6.45)$$

Hence, the plus sign is the acceptable one in (6.44).

It is also instructive to examine other limiting cases of (6.44). (1) For fixed molecular parameters ($E_L$, $E_H$, and $V_L$), if $T$ increases (at a fixed pressure $P$) then from (6.37), $K \to 1$ and $x_L^*$ tends to a limiting value of $x_L^* = \frac{1}{2}(1 + 1/\sqrt{5})$. (2) For fixed molecular parameters and temperature, the limiting pressure dependence is $K \to \infty$ as $P \to \infty$. Hence, $x_L^* \to \frac{1}{2}$, meaning that for $P \to \infty$, all the holes are occupied [remember that in this example, $N_0 = 1$, hence $x_L^* = \frac{1}{2}$ corresponds to $x_H^* = \frac{1}{2}$, which, by (6.26), means that all the holes are occupied].

### 6.7.3. First Derivatives of the Free Energy

For a given $N_L$ and $N_H$, the Gibbs free energy is given by (6.35), which can be simplified by using the Stirling approximation

$$G(T, P, N_W; N_L) = N_L E_L + N_H E_H + P N_L V_L - kT[(N_0 N_L) \ln(N_0 N_L)$$
$$- N_H \ln N_H - (N_0 N_L - N_H) \ln(N_0 N_L - N_H)] \qquad (6.46)$$

This function has a single minimum for $N_L = N_L^*$, as shown previously. The system described by the variables $T$, $P$, $N_W$, and $N_L$ must be "frozen in" with respect to the conversion $L \rightleftarrows H$, otherwise $N_L$ is not an independent variable.

For convenience of notation, we use the set of variables $T$, $P$, $N_L$, and $N_H$ (with the condition $N_L + N_H = N_W$) for the "frozen-in" system. The chemical potentials of the components $L$ and $H$ are

$$\mu_L = (\partial G/\partial N_L)_{T,P,N_H}$$
$$= E_L + PV_L - kT[N_0 \ln(N_0 N_L) - N_0 \ln(N_0 N_L - N_H)]$$
$$= E_L + PV_L - kTN_0 \ln[N_0 x_L/(N_0 x_L - x_H)] \qquad (6.47)$$

$$\mu_H = (\partial G/\partial N_H)_{T,P,N_L}$$
$$= E_H - kT[\ln(N_0 N_L - N_H) - \ln N_H]$$
$$= E_H + kT \ln[x_H/(N_0 x_L - x_H)] \qquad (6.48)$$

Some points are worth attention in (6.47) and (6.48).

1. The mixture of $L$- and $H$-cules is not ideal in any of the senses discussed in Chapter 4. In particular, it is not symmetric ideal. Note that $L$ and $H$ are both *water molecules*. However, by their very definition, they differ markedly in *local* environment; hence, it is unlikely that they obey the condition for symmetric ideal solutions. This point will be discussed further in Section 6.8, when we analyze a more general mixture-model approach to water. [See also the discussion following (6.52).]

2. The chemical potentials $\mu_L$ and $\mu_H$ are definable only in the "frozen-in" system, since we require $N_H$ and $N_L$ to be kept constant in (6.47) and (6.48), respectively. In the case for which $N_L = N_L^*$ and $N_H = N_H^*$, we get [note (6.37)]

$$\mu_L = \mu_H \tag{6.49}$$

Conversely, the condition (6.49) implies the equilibrium condition (6.37). Hence, (6.37) and (6.49) are equivalent.

3. As $x_H \to 0$ (or $x_L \to 1$), we get

$$\mu_L \to E_L + PV_L, \quad \mu_H \to E_H + kT\ln(x_H/N_0) = E_H + kT\ln y_1 \tag{6.50}$$

This is the case when we tend to a pure lattice $L$. The chemical potential $\mu_L$ tends to the enthalpy (see below) of pure $L$, whereas the chemical potential of $H$ tends to minus infinity. Such a system, of course, cannot be at equilibrium; it may be attained in the "frozen-in" system. (Note that in this case, $H$ forms a dilute ideal solution in the system.)

4. For $x_H \geq N_0 x_L$, we get a negative number in the logarithms in (6.47) and (6.48). This occurs when all the holes are filled and therefore $x_H$ cannot be increased further. The model "breaks down" for such a case, as indeed can be seen from the condition we imposed in (6.26).

The above points were discussed with respect to the specific model under consideration. However, similar comments apply to any mixture-model approach. For instance, the fact that the mixture is not symmetric ideal will be discussed in more general terms in Section 6.8. The condition of the equivalence of (6.49) and (6.37) holds in more general mixture models (a simpler example was discussed in Section 5.6), and the last two comments reflect the fact that in some mixture models, the components may not exist in a pure state (see further discussion in Section 5.7). Of course, the specific form of the above conditions applies to the specific model, and different models will lead to different *forms* of, say, the equilibrium constant in (6.37) or the limiting behavior in (6.50).

The total entropy of the system is obtained by differentiating (6.46) with respect to temperature; for convenience, we use the set of variables $T$, $P$, $N_L$, and $N_H$ with $N_L + N_H = N_W$:

$$-S = \left(\frac{\partial G}{\partial T}\right)_{P,N_W}$$

$$= \left(\frac{\partial G}{\partial T}\right)_{P,N_L,N_H} + \left[\left(\frac{\partial G}{\partial N_L}\right)_{T,P,N_H} - \left(\frac{\partial G}{\partial N_H}\right)_{T,P,N_L}\right]\left(\frac{\partial N_L}{\partial T}\right)_{P,N_W}$$

$$= \left(\frac{\partial G}{\partial T}\right)_{P,N_L,N_H} \tag{6.51}$$

The derivative of $G(T, P, N_L, N_H)$ is assumed to be evaluated at the point $N_L = N_L^*$, $N_H = N_H^*$. Two cases should be distinguished. If the system is "frozen in," then $N_L$ is fixed; hence, $\partial N_L/\partial T = 0$ and we get the second form on the rhs of (6.51). If, on the other hand, the system is at equilibrium, then $N_L$ is a function of temperature [through the condition (6.37)], so that $\partial N_L/\partial T \neq 0$. But in this case, $\mu_L = \mu_H$; hence, we again get the second form on the rhs of (6.51). This exemplifies a very general principle, that the value of the entropy is unaffected by "freezing in" the equilibrium. (This principle is sometimes stated as follows: Thermodynamic quantities have the same value in the presence or absence of a catalyst for the particular reaction under consideration. This is not true for all thermodynamic quantities, as will be shown below and in the next chapter.) Therefore, in order to compute the entropy of the system, we can differentiate the function $G(T, P, N_L, N_H)$ as if $N_L$ and $N_H$ were independent variables, provided that we put the values $N_L = N_L^*$ and $N_H = N_H^*$ in the final result.

From (6.46), we get by differentiation with respect to temperature

$$S = k[(N_0 N_L) \ln(N_0 N_L) - N_H \ln N_H - (N_0 N_L - N_H) \ln(N_0 N_L - N_H)]$$
$$= -k N_0 N_L (y_0 \ln y_0 + y_1 \ln y_1) \tag{6.52}$$

The second form on the rhs of (6.52) is interesting from the formal point of view. We recall that $y_0$ and $y_1$ are the "mole fractions" of empty and occupied holes [see (6.39)]. Therefore, the total entropy per mole of holes, $S/N_0 N_L$, has the same appearance as the ideal entropy of mixing empty and occupied holes. Although we will not make any use of this fact, it is interesting to trace the origin of this result to the assumption of independence of the holes. However, the empty and occupied holes cannot be

# Liquid Water

viewed as chemical entities and, as we noted earlier, the mixture of $L$ and $H$ in general is not an ideal one.

In a very similar fashion, by differentiating (6.46) with respect to pressure, we obtain the volume

$$V = \left(\frac{\partial G}{\partial P}\right)_{T,N_W} = \left(\frac{\partial G}{\partial P}\right)_{T,N_L,N_H} = N_L V_L \qquad (6.53)$$

which is an obvious result for this model.

The enthalpy and the internal energy are similarly obtained:

$$H = G + TS = N_L E_L + N_H E_H + P N_L V_L \qquad (6.54)$$

$$E = H - PV = N_L E_L + N_H E_H \qquad (6.55)$$

All the above quantities are independent of the "presence of a catalyst," i.e., the same value is obtained whether we measure, say, the volume at equilibrium, or if we "freeze in" the equilibrium (remove the catalyst). This result is very general for any mixture model, or any mixture of chemically reacting species. The result is not valid for all thermodynamic quantities, as is shown below.

## 6.7.4. Second Derivatives of the Free Energy

Among the second-order derivatives of the free energy, we distinguish between two subgroups: those that are definable only in the "frozen-in" system and those that may be defined in either the "frozen-in" or the equilibrated system. We start with quantities belonging to the first group. The partial molar entropies[6] of $L$ and $H$ are obtained by differentation, with respect to temperature, of the corresponding chemical potentials in (6.47) and (6.48):

$$\bar{S}_L = -(\partial \mu_L/\partial T)_{P,N_L,N_H} = kN_0 \ln[N_0 N_L/(N_0 N_L - N_H)] \qquad (6.56)$$

$$\bar{S}_H = -(\partial \mu_H/\partial T)_{P,N_L,N_H} = k \ln[(N_0 N_L - N_H)/N_H] \qquad (6.57)$$

Note that in our model, the quantities $E_L$, $E_H$, and $V_L$ are assumed to be temperature independent; hence, the partial molar entropies are

---

[6] One should refer strictly to partial *molecular*, rather than partial *molar* quantities. The distinction between the two must be made clear by the context in which they are used.

determined essentially by the mole fractions of the two species. An interesting quantity which will be needed later is the difference $\bar{S}_L - \bar{S}_H$, which, at the point of equilibrium, $N_L = N_L^*$, $N_H = N_H^*$, is

$$\bar{S}_L - \bar{S}_H = k \ln \frac{(N_0 N_L^*)^{N_0} N_H^*}{(N_0 N_L^* - N_H^*)^{N_0+1}} = k \ln K = \frac{E_L - E_H + PV_L}{T} \quad (6.58)$$

It is important to recognize that at any given $N_L$ and $N_H$, the values of $\bar{S}_L$ and $\bar{S}_H$ as well as of $\bar{S}_L - \bar{S}_H$ depend on the composition of the system, which is arbitrary (but within the limits of the model). In particular, the sign of $\bar{S}_L - \bar{S}_H$ may be either positive or negative. On the other hand, if we evaluate $\bar{S}_L - \bar{S}_H$ at the particular point $N_L = N_L^*$ and $N_H = N_H^*$, then we see that this difference depends only on the molecular parameters of the model.

Other partial molar quantities are obtainable from (6.47) and (6.48) from simple thermodynamic relations:

$$\bar{H}_L = \mu_L + T\bar{S}_L = E_L + PV_L, \qquad \bar{H}_H = E_H \quad (6.59)$$

$$\bar{V}_L = (\partial \mu_L/\partial P)_{T,N_L,N_H} = V_L, \qquad \bar{V}_H = 0 \quad (6.60)$$

$$\bar{E}_L = \bar{H}_L - P\bar{V}_L = E_L, \qquad \bar{E}_H = \bar{H}_H - P\bar{V}_H = E_H \quad (6.61)$$

Note that on the rhs of (6.59)–(6.61), we have molecular quantities, such as $E_L$, $E_H$, and $V_L$, whereas on the lhs, we have partial molecular quantities.

We now consider some other quantities which are definable in both the "frozen-in" and the equilibrated systems, but we shall always evaluate the quantities at $N_L = N_L^*$ and $N_H = N_H^*$.

The temperature dependence of the volume is

$$\left(\frac{\partial V}{\partial T}\right)_{P,N_W,\text{eq}} = \left(\frac{\partial V}{\partial T}\right)_{P,N_L,N_H}$$
$$+ \left[\left(\frac{\partial V}{\partial N_L}\right)_{T,P,N_H} - \left(\frac{\partial V}{\partial N_H}\right)_{T,P,N_L}\right]\left(\frac{\partial N_L}{\partial T}\right)_{P,N_W,\text{eq}}$$
$$(6.62)$$

The above splitting in the temperature dependence of the volume is very general and applies to any mixture model. If the system is "frozen in," then $N_L$ is constant and only the first term on the rhs contributes to (6.62). If, however, the system is at equilibrium with respect to the conversion $L \rightleftarrows H$, then $N_L$ (strictly $N_L^*$) is temperature dependent and we have two contributions to the thermal expansion.

## Liquid Water

The two contributions to (6.62) can be interpreted by use of the following "thought experiment." Suppose we change the temperature by $dT$, and follow the corresponding change in the volume, which can be written as

$$(dV)_{\text{total}} = (dV)_{N_L, N_H} + (\bar{V}_L - \bar{V}_H)\, dN_L \tag{6.63}$$

First, we measure the change of volume in the "frozen-in" system (i.e., in the absence of the catalyst), and get the value $(dV)_{N_L, N_H}$. In the next step, we introduce the catalyst to the system, which generally produces an additional change in the volume, given by the second term on the rhs of (6.63). In our particular model, (6.62) can be written explicitly. Note that in this model, the total volume $V$ is $N_L V_L$, and $V_L$ is assumed to be temperature independent. The quantity $\partial N_L / \partial T$ in (6.62) can be evaluated directly from the equilibrium condition (6.37). Alternatively, we can use the identity (Section 5.10)

$$\left(\frac{\partial N_L}{\partial T}\right)_{P, N_W, \text{eq}} = \frac{\bar{H}_L - \bar{H}_H}{T(\mu_{LL} - 2\mu_{LH} + \mu_{HH})} \tag{6.64}$$

In our model, we get from (6.47) and (6.48)

$$\mu_{LL} - 2\mu_{LH} + \mu_{HH} = \frac{kTN_0}{x_L x_H (N_0 N_L - N_H)} \tag{6.65}$$

Hence, using (6.59) and (6.60), we get, for the equilibrium point,

$$\left(\frac{\partial V}{\partial T}\right)_{P, N_W, \text{eq}} = \frac{V_L(E_L - E_H + PV_L)(N_0 N_L^* - N_H^*) x_L^* x_H^*}{kT^2 N_0} \tag{6.66}$$

Since $V_L > 0$, $N_0 N_L \geq N_H$, and $x_L x_H \geq 0$, the sign of (6.66) is determined by the quantity $E_L - E_H + PV_L$. Now, in most interstitial models, $E_L - E_H$ is of the order of about $-10^3$ cal/mole. At 1 atm and with $V_L$ of the order of 18 cm$^3$/mole, we have $PV_L \approx 18/41 = 0.44$ cal/mole. Thus, the $PV_L$ term is negligible compared to $E_L - E_H$ and, therefore, the sign of (6.66) is negative. This is the essence of the "explanation" of the temperature dependence of the volume of water. If we assume that $V_L$ is temperature dependent, so that instead of (6.66) we have

$$\left(\frac{\partial V}{\partial T}\right)_{P, N_W, \text{eq}} = N_L \left(\frac{\partial V_L}{\partial T}\right)_{P, N_L, N_H} + (\bar{V}_L - \bar{V}_H)\left(\frac{\partial N_L}{\partial T}\right)_{P, N_W, \text{eq}} \tag{6.67}$$

and if we assume that the first term on the rhs of (6.67) is positive (i.e., the lattice expands upon increasing the temperature), then we have competition between two terms of opposite signs, which happen to cancel out at about 4°C. Of course, by a judicious choice of the molecular parameters [as was actually done by Frank and Quist (1961) and Mikhailov (1967)], we can force the model to show a minimum of volume at about 4°C. This numerical aspect of the problem will not concern us here. We also note that in this model, as the pressure increases, the positive term $PV_L$ will be large enough to dominate the sign of the rhs of (6.66). In that case, both terms on the rhs of (6.67) will be positive, and hence we observe a "normal" temperature dependence of the volume.

The heat capacity at constant pressure is given by

$$C_P = \left(\frac{\partial H}{\partial T}\right)_{P,N_W,\text{eq}} = \left(\frac{\partial H}{\partial T}\right)_{P,N_L,N_H} + (\bar{H}_L - \bar{H}_H)\left(\frac{\partial N_L}{\partial T}\right)_{P,N_W,\text{eq}} \quad (6.68)$$

This is the general expression for the heat capacity in the mixture-model approach (two components). Note that the first term on the rhs of (6.68) is the heat capacity of the system that would have been measured in the "frozen-in" system. The second term is the contribution to the heat capacity due to "relaxation" to the final equilibrium state. The arguments leading to this interpretation are the same as those discussed following Eq. (6.63). In our particular model, the first term on the rhs of (6.68) is zero and, for the second term, we have

$$\begin{aligned} C_P &= (E_L - E_H + PV_L)\left(\frac{\partial N_L}{\partial T}\right)_{P,N_W,\text{eq}} \\ &= \frac{(E_L - E_H + PV_L)^2(N_0 N_L{}^* - N_H{}^*)x_L{}^* x_H{}^*}{kT^2 N_0} \end{aligned} \quad (6.69)$$

In this particular model, the heat capacity is due to thermal "excitation" from the $H$ to the $L$ state. Note that here the relaxation term is positive.[7] We will see in Section 6.8 that this result is of general validity.

As a final example of quantities in this group, we compute the isothermal compressibility for this model:

$$\begin{aligned} \varkappa_T &= -\frac{1}{V}\left(\frac{\partial V}{\partial P}\right)_{T,N_W,\text{eq}} \\ &= -\frac{1}{V}\left[\left(\frac{\partial V}{\partial P}\right)_{T,N_L,N_H} + (\bar{V}_L - \bar{V}_H)\left(\frac{\partial N_L}{\partial P}\right)_{T,N_W,\text{eq}}\right] \end{aligned} \quad (6.70)$$

---

[7] In this case, the total heat capacity must be positive because of the condition of thermal stability of the system. Here we refer specifically to the relaxation term which, as we will see in Section 6.8, must also be positive.

Using the identity (Section 5.10)

$$\left(\frac{\partial N_L}{\partial P}\right)_{T,N_W,\text{eq}} = \frac{-(\bar{V}_L - \bar{V}_H)}{\mu_{LL} - 2\mu_{LH} + \mu_{HH}}$$

$$= \frac{-(\bar{V}_L - \bar{V}_H)(N_0 N_L^* - N_H^*)x_L^* x_H^*}{kTN_0} \quad (6.71)$$

relation (6.70) reduces to

$$\varkappa_T = V_L^2(N_0 N_L^* - N_H^*)x_L^* x_H^*/kTVN_0 \quad (6.72)$$

which is always positive.[8]

We conclude this section with a general comment on interstitial models. The study of such models is useful and quite rewarding in that it gives an insight into the possible mechanism by which water reveals its anomalous behavior. One should not overstress, however, the significance of the numerical results obtained from the model as an indication of the extent of "reality" of the model. It is possible, by a judicious choice of the molecular parameters, to obtain thermodynamic results which are in agreement with experimental values measured for real water. Such an agreement can be achieved by quite different models. The important point is the qualitative explanation that the model is capable of offering of the various properties of water. In the next section, we discuss a general two-structure model approach and use arguments which are independent of any specific model. It is useful, however, to have in the background a simple and solvable model where every point can be checked and clarified in an explicit manner —a task which, in general, is impossible for more complex models.

## 6.8. APPLICATION OF AN EXACT TWO-STRUCTURE MODEL (TSM)

The simplest version of the mixture-model approach and, in fact, the most useful one is the so-called "two-structure model" (TSM). We studied one specific example of a TSM in the previous section. Many ad hoc models of liquid water fall into the realm of the TSM [notable examples are those of Samoilov (1946), Hall (1948), Grjotheim and Krogh-Moe (1954), Pauling (1960), Wada (1961), Danford and Levy (1962), Krestov (1964), Marchi

---

[8] Note again that the total compressibility must be positive, from the general condition of mechanical stability of the system. In this particular model, the total compressibility is identified with the relaxation term for the conversion between the two species.

and Eyring (1964), Davis and Litovitz (1965), Vdovenko *et al.* (1966, 1967a, 1967b), Walrafen (1968)].

In this section, we lay the general foundation of the idea of the TSM for liquids, and particularly for liquid water. We explore both the usefulness and the limitations of this approach. Of course, because of its generality, we cannot expect to pursue the study to the point where comparison with experimental results is possible. This must be done by invoking a specific ad hoc model for the system. The latter may be viewed as an approximate version of the exact TSM, a topic which will concern us in the next section.

We start with some examples of exact TSM's that are derivable from any one of the quasicomponent distribution functions discussed in Chapter 5. Consider, for instance, the distribution based on the concept of coordination number (CN) (Section 5.2). We have denoted by $x_C(K)$ the mole fraction of molecules having a CN equal to $K$. The vector $\mathbf{x}_C = (x_C(0), x_C(1), \ldots)$ describes the *composition* of the system when viewed as a mixture of quasicomponents, distinguishable according to the CN of the molecules. Instead, we may distinguish between two groups of molecules: those whose CN is smaller than or equal to some number, say $K^*$, and those whose CN is larger than $K^*$. The two corresponding mole fractions are

$$x_L = \sum_{K=0}^{K^*} x_C(K), \qquad x_H = \sum_{K=K^*+1}^{\infty} x_C(K) \qquad (6.73)$$

$x_L$ may be referred to as the mole fraction of molecules with low ($L$) local density, and $x_H$ as that of molecules with relatively high ($H$) local density. The new vector composed of two components $(x_L, x_H)$ is also a quasicomponent distribution function, and gives the composition of the system when viewed as a mixture of two components, which we may designate as $L$- and $H$-cules. Starting with the same vector $\mathbf{x}_C = (x_C(0), x_C(1), \ldots)$, we may, of course, derive many other TSM's differing from the one in (6.73). A possibility which may be useful for liquid water is

$$x_A = x_C(4), \qquad x_B = 1 - x_A \qquad (6.74)$$

where we distinguish between molecules with CN equal to four in one group ($A$), and place all other molecules in the second group ($B$).

As a second example, consider the quasicomponent distribution function, based on the concept of binding energy (BE) (Section 5.2). We recall that the vector (or the function) $\mathbf{x}_B$ gives the composition of the system when viewed as a mixture of molecules differing in their BE. Thus, $x_B(\nu) \, d\nu$

**Liquid Water**

is the mole fraction of molecules with BE between $v$ and $v + dv$. A possible TSM constructed from this function is [we drop the subscript $B$ in $x_B(v)$]

$$x_A = \int_{-\infty}^{v^*} x(v)\, dv, \qquad x_B = 1 - x_A \tag{6.75}$$

where $x_A$ is the mole fraction of molecules whose BE is below a certain value $v^*$ and $x_B$ is the mole fraction whose BE is larger than $v^*$. Again, we have a new quasicomponent distribution function composed of two components $(x_A, x_B)$.

The above examples illustrate the general procedure by which we construct a TSM from any quasicomponent distribution function. From now on, we assume that we have made a classification into two components, $L$ and $H$, without referring to a specific example. The arguments we use will be independent of any specific classification procedure. We will see that in order for such a TSM to be useful in interpreting the properties of water, we must assume that each component in itself behaves "normally" (in the sense discussed below). The anomalous properties of water are then interpreted in terms of "structural changes" that take place in the liquid.

Let $N_L$ and $N_H$ be the equilibrium numbers of $L$- and $H$-cules, respectively, and $N_W$ be the total number of water molecules in the system, $N_W = N_L + N_H$. Viewing the system as a mixture of two components, we write the volume of the system, using the Euler theorem, as

$$V = N_L \bar{V}_L + N_H \bar{V}_H \tag{6.76}$$

where $\bar{V}_L$ and $\bar{V}_H$ are the partial molar (or molecular) volumes of $L$ and $H$, respectively. The total differential of $V(T, P, N_L, N_H)$ is

$$dV = \left(\frac{\partial V}{\partial T}\right)_{P,N_L,N_H} dT + \left(\frac{\partial V}{\partial P}\right)_{T,N_L,N_H} dP + \left(\frac{\partial V}{\partial N_L}\right)_{T,P,N_H} dN_L$$

$$+ \left(\frac{\partial V}{\partial N_H}\right)_{T,P,N_L} dN_H \tag{6.77}$$

The temperature dependence of the volume, along the equilibrium line with respect to the reaction $L \rightleftarrows H$ (keeping $P$ and $N_W$ constant), is

$$\left(\frac{\partial V}{\partial T}\right)_{P,N_W,\text{eq}} = \left(\frac{\partial V}{\partial T}\right)_{P,N_L,N_H} + (\bar{V}_L - \bar{V}_H)\left(\frac{\partial N_L}{\partial T}\right)_{P,N_W,\text{eq}} \tag{6.78}$$

The first term on the rhs of (6.78) gives the temperature dependence of the volume of the "frozen-in" system and, in view of (6.76), can be written as

$$\left(\frac{\partial V}{\partial T}\right)_{P,N_L,N_H} = N_L\left(\frac{\partial \bar{V}_L}{\partial T}\right)_{P,N_L,N_H} + N_H\left(\frac{\partial \bar{V}_H}{\partial T}\right)_{P,N_L,N_H} \quad (6.79)$$

This quantity may be referred to as the "static" part of the total temperature dependence of the volume. It must be stressed that the splitting of $(\partial V/\partial T)$ in (6.78) into two terms depends strongly on the classification procedure employed to define the species $L$ and $H$. Therefore, any reference to a "static" part has meaning only for the particular classification that has been employed.

The second term on the rhs of (6.78) may be referred to as the relaxation part of the total temperature dependence of the volume. It is quite clear that this term contains the contribution to the temperature dependence of the volume that is associated with the "structural changes" in the system. Note that in the context of this section, the term "structural changes" refers only to the interconversion between the two species $L$ and $H$ arising from the change in temperature.

We denote by

$$\alpha^* = \frac{1}{V}\left(\frac{\partial V}{\partial T}\right)_{P,N_L,N_H} \quad (6.80)$$

the static part of the coefficient of thermal expansion. Using the identities (5.127) and (5.133), we can rewrite the coefficient of thermal expansion as

$$\alpha = \frac{1}{V}\left(\frac{\partial V}{\partial T}\right)_{P,N_W,\text{eq}} = \alpha^* + \frac{(\bar{V}_L - \bar{V}_H)(\bar{H}_L - \bar{H}_H)x_L x_H \eta}{kT^2} \quad (6.81)$$

where

$$\eta = \varrho_L + \varrho_H + \varrho_L\varrho_H(G_{LL} + G_{HH} - 2G_{LH}) \quad (6.82)$$

In order to use (6.81) for interpreting the anomalous negative temperature dependence of the volume of water between 0 and 4°C, we must choose a classification procedure in such a way so that (1) the static term $\alpha^*$ behaves normally, i.e., each of the derivatives in (6.79) is positive; such an assumption is invoked either explicitly or implicitly in the various ad hoc TSM's for water; (2) the relaxation term in (6.81) must be negative and large (in absolute magnitude) enough to overcompensate for the positive value of the static part.

Regarding the first condition, there is no general rule that guarantees that $\alpha^*$ is normal. In the simplified ad hoc TSM's, this assumption seems

# Liquid Water

to be quite natural. For instance, if one species is icelike and the second one is, say, "monomeric," then it is natural to assume that each of the partial molar volumes has a "normal" temperature dependence.

The analysis of the second condition leads to somewhat more interesting insights as to the origin of negative temperature dependence of the volume. Consider first the sign of the relaxation term in (6.81). Clearly, in order that this term be negative, we must have (note that $\eta > 0$; see Section 4.5)

$$(\bar{V}_L - \bar{V}_H)(\bar{H}_L - \bar{H}_H) < 0 \tag{6.83}$$

This means that $\bar{V}_L - \bar{V}_H$ and $\bar{H}_L - \bar{H}_H$ must have opposite signs. Next, in order to get a large negative value for the relaxation term in (6.81), two conditions must be fulfilled: (1) The two components $L$ and $H$ must not be very similar, otherwise the differences $\bar{V}_L - \bar{V}_H$ and $\bar{H}_L - \bar{H}_H$ will be very small. (2) Neither of the mole fractions $x_L$ or $x_H$ may be too small, otherwise the product $x_L x_H$ will be small, and we will get a small relaxation term in (6.81).

Let us further examine the above two conditions. Suppose we define $L$ and $H$ in such a way that $L$ has a low local density and $H$ has a relatively high local density [see (6.73) as an example]. Then, we expect that $\bar{V}_L - \bar{V}_H > 0$. Using the Kirkwood–Buff expressions for $\bar{V}_L$ and $\bar{V}_H$ (Section 4.5) and the identities (5.134) and (5.135), we can express the difference in the partial molar volumes in terms of molecular distribution functions, namely

$$\bar{V}_L - \bar{V}_H = \frac{\varrho_H G_{HH} - \varrho_H G_{LH} - \varrho_L G_{LL} + \varrho_L G_{HL}}{\eta} = \frac{\varrho_W}{\eta}(G_{WH} - G_{WL}) \tag{6.84}$$

Thus, the requirement that $\bar{V}_L - \bar{V}_H$ be positive is equivalent to the statement that the overall excess of *water* molecules around $H$ is larger than the overall excess of *water* molecules around $L$. This is precisely the meaning of the requirement that $H$ is a molecule with a relatively higher local density than $L$.

It is clear that fulfillment of one condition, say $\bar{V}_L - \bar{V}_H > 0$, can be achieved by proper *definition* of the two components. The question that arises is: Under what circumstances is the second condition, $\bar{H}_L - \bar{H}_H < 0$, also fulfilled?

In order to understand the characteristic nature of the packing of water molecules in the liquid state, consider first a simple fluid such as argon. Suppose that we define two components $L$ and $H$ having a low and high local density, respectively. For instance, we use (6.73) with $K^* = 6$. In this case, we have $\bar{V}_L - \bar{V}_H > 0$, by virtue of the definition of the two components. However, since an argon molecule, having more than six neighbors, is likely to have a lower partial molar enthalpy, we also expect that $\bar{H}_L - \bar{H}_H > 0$ (i.e., the conversion of a molecule from $H$ to $L$ is accompanied by an *increase* in enthalpy[9]). Such a splitting would not work for our purposes, i.e., (6.83) is not fulfilled. A different choice that may give the correct signs is obtained with, say, $K^* = 13$. Clearly, in this case, we also have $\bar{V}_L - \bar{V}_H > 0$. (The $H$ component has a very high local density.) In addition, because of this particular choice of $H$, it is clear that the transformation $H \to L$ is likely to involve a *decrease* in enthalpy (since most of the neighbors must exert strong repulsive forces on the $H$-cule; therefore its transfer to an $L$ releases energy). Hence, $\bar{H}_L - \bar{H}_H$ is likely to be negative. This choice seems to satisfy our conditions (6.83) for a negative relaxation term. However, with the particular choice of $K^* = 13$, it is very likely that $x_H \approx 0$ (a CN higher than 12 must be a very rare event!). Hence, the product $x_L x_H$ is very small and the whole relaxation term in (6.81) will be negligibly small, though having a correct sign.

The situation in water is different since we know that the strong directional forces (hydrogen bonds) are responsible for maintaining low local density. Hence, it seems possible to define two components $L$ and $H$ in such a way so that

$$\bar{V}_L - \bar{V}_H > 0, \qquad \bar{H}_L - \bar{H}_H < 0 \tag{6.85}$$

and, at the same time, none of the mole fractions $x_L$ and $x_H$ is too small. If this is indeed possible, we will have a large, negative relaxation term in (6.81) which will dominate the temperature dependence of the volume. Of course, in order to make a more quantitative statement about the magnitudes of $\bar{V}_L - \bar{V}_H$, $\bar{H}_L - \bar{H}_H$, and $x_L$, we must specialize to a particular ad hoc model. In Section 6.11, we present further discussion of the relation between local density and binding energy, which is relevant to the question of the sign and magnitude of the relaxation term in (6.81).

---

[9] We do not have a simple relation between the difference in the enthalpies $\bar{H}_L - \bar{H}_H$ and the generalized molecular distribution functions. Therefore, the conclusion drawn here is basically intuitive. Furthermore, from the behavior of the function $x_{B,C}(\nu, K)$ discussed in Chapter 5, we concluded that in simple fluids, high local density is coupled with strong binding energies, hence it is logical to conclude that a conversion $H \to L$ is associated with an increase in energy. One may conclude also that at 1 atm, this reaction also involves an increase in enthalpy.

**Liquid Water**

Consider next the heat capacity (at constant pressure) which, in the TSM formalism, can be obtained as follows. The total enthalpy of the system is

$$H = N_L \bar{H}_L + N_H \bar{H}_H \qquad (6.86)$$

and the heat capacity is given by

$$C_P = \left(\frac{\partial H}{\partial T}\right)_{P,N_W,\text{eq}}$$

$$= \left(N_L \frac{\partial \bar{H}_L}{\partial T} + N_H \frac{\partial \bar{H}_H}{\partial T}\right) + (\bar{H}_L - \bar{H}_H)\left(\frac{\partial N_L}{\partial T}\right)_{P,N_W,\text{eq}} \qquad (6.87)$$

where, again, we have split the heat capacity into two terms. The first is the heat capacity of the "frozen-in" system, and the second is the corresponding relaxation term. We stress again that the split into these two terms depends on the choice of the particular classification procedure employed to define $L$ and $H$.

Using the identity (5.127), we can rewrite the relaxation term in (6.87) as

$$(\bar{H}_L - \bar{H}_H)\left(\frac{\partial N_L}{\partial T}\right)_{P,N_W,\text{eq}} = \frac{(\bar{H}_L - \bar{H}_H)^2}{T(\mu_{LL} - 2\mu_{LH} + \mu_{HH})}$$

$$= \frac{(\bar{H}_L - \bar{H}_H)^2 x_L x_H \eta V}{kT^2} \qquad (6.88)$$

Clearly, the relaxation term in (6.88) is always positive, independent of the definition of $L$ and $H$. Of course, the magnitude of the relaxation term depends on the particular choice of the classification procedure. In order to "explain" the high value of the heat capacity of liquid water, one assumes that the "static" term, i.e., the first term on the rhs of (6.87), has a "normal" value. The excess heat capacity is then attributed to the relaxation term. In order for the latter to be large, we must have two components which differ appreciably in their partial molar enthalpy [otherwise, $(\bar{H}_L - \bar{H}_H)^2$ cannot be large] and none of the mole fractions $x_L$ and $x_H$ can be too small.

It should be noted that the requirements for a large value of the relaxation term of the heat capacity are somewhat weaker than those needed for a large relaxation term of the thermal expansion coefficient. It is likely that the peculiar relation between $\bar{V}_L - \bar{V}_H$ and $\bar{H}_L - \bar{H}_H$, supplemented by the requirement that $x_L$ and $x_H$ not be too small, is one of the most strikingly unique features of liquid water. On the other hand, the requirement that $(\bar{H}_L - \bar{H}_H)^2$ be large is certainly a result of the existence of strong hydrogen bonds in water. Such a property is not unique to water, and indeed, liquid ammonia, for example, also has a relatively high heat capacity, which probably has the same origin as that for liquid water.

As a final example, we present the form of the isothermal compressibility in the TSM:

$$\varkappa_T = -\frac{1}{V}\left(\frac{\partial V}{\partial P}\right)_{T,N_W,\mathrm{eq}}$$

$$= -\frac{1}{V}\left[N_L\frac{\partial \bar{V}_L}{\partial P} + N_H\frac{\partial \bar{V}_H}{\partial P}\right] - \frac{1}{V}(\bar{V}_L - \bar{V}_H)\left(\frac{\partial N_L}{\partial P}\right)_{T,N_W,\mathrm{eq}} \quad (6.89)$$

The relaxation term can be rewritten, using identity (5.128), as

$$-\frac{1}{V}(\bar{V}_L - \bar{V}_H)\left(\frac{\partial N_L}{\partial P}\right)_{T,N_W,\mathrm{eq}} = -\frac{(\bar{V}_L - \bar{V}_H)^2}{V(\mu_{LL} - 2\mu_{LH} + \mu_{HH})}$$

$$= -\frac{(\bar{V}_L - \bar{V}_H)^2 x_L x_H \eta}{kT} \quad (6.90)$$

We note that this term is always negative, independent of the particular definitions of $L$ and $H$.

One general comment which is applicable to all the above discussions is concerned with the nature of the mixture of $L$ and $H$. The question is whether the mixture may be assumed to be ideal and, if so, in what sense. First, suppose that we have defined $L$ and $H$ in such a way that one component is very diluted in the other. This can be easily achieved. For instance, if we take $K^* = 12$ in (6.73), then it is likely that $x_H \approx 0$. Hence, such a solution will be dilute ideal and we get

$$\mu_{LL} - 2\mu_{LH} + \mu_{HH} = kT/N_W x_L x_H \quad (6.91)$$

However, the fact that one component is *very* dilute in the other implies that the product $x_L x_H$ is very small. Hence, all the relaxation terms in (6.81), (6.88), and (6.90) are very small, and the whole treatment is rendered useless.

# Liquid Water

A second possibility is to define $L$ and $H$ as similar,[10] so that the solution is symmetric ideal, i.e.,

$$G_{LL} + G_{HH} - 2G_{LH} = 0 \qquad (6.92)$$

Hence, $\eta = \varrho_L + \varrho_H = \varrho_W$ [in (6.81), (6.88), and (6.90)]. The latter form of the relaxation term has been used by many authors (e.g., Wada, 1961; Davis and Litovitz, 1966; Davis and Jarzynski, 1967; Szkatula and Fulinski, 1967; Walrafen, 1968; and Angell, 1971).

Some care must be exercised in applying the symmetric ideal assumption to the TSM of water. We saw that in order to get large relaxation terms in (6.81), (6.88), and (6.90), we must assume that the two components $L$ and $H$ are markedly different. This, in general, will contradict the assumption of similarity in the sense of (6.92).

There is a subtle and interesting reason why the ideality assumption has been used so often without its validity being questioned. We recall that condition (6.92) is essentially a condition on the similarity of the *local environments* of $L$ and $H$. In a real mixture of two components $A$ and $B$, similarity of the *molecules* $A$ and $B$ implies similarity in their local environments as well. In the MM approach for water, we start with two components which are *identical* in their chemical composition (both $L$ and $H$ are water molecules!). Therefore, it is very tempting to assume that these two components also form a symmetric ideal solution. However, since by their definitions we require the two components to be different in their local environments, the validity of (6.92) cannot be guaranteed. Of course, one cannot make a precise statement as to the extent of dissimilarity that may be permitted and yet have condition (6.92) fulfilled. Such an assessment must be carried out using a specific model for water.

We now consider a specific example of a TSM for a system for which we know the form of the function $x_B(\nu)$. The example is in a sense a generalization of the case given in (5.119). We discuss the heat capacity. A similar treatment applies to any of the other quantities discussed in this section.

---

[10] It is very easy to construct a TSM in which the two components are similar. For instance, consider the distribution function $x_B(\nu)$, and choose a sufficiently dense division of the $\nu$ axis by the points $\nu_1, \nu_2, \nu_3, \ldots$ . We define

$$x_\alpha = \sum_{i=\text{odd}} \int_{\nu_i}^{\nu_{i+1}} x_B(\nu)\, d\nu, \qquad x_\beta = 1 - x_\alpha$$

Clearly, if the intervals $\nu_{i+1} - \nu_i$ become very small, the two components, $\alpha$ and $\beta$, become very similar.

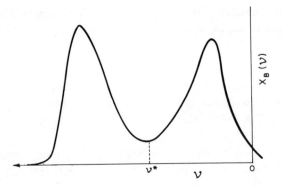

Fig. 6.16. A possible form of the function $x_B(v)$ suggesting a "natural" cutoff point $v^*$ for the construction of a two-structure model.

For illustrative purposes suppose that $x_B(v)$ has a form similar to the function depicted in Figure 6.16. Here, there is a "natural" choice of a cutoff point $v^*$ for constructing a TSM, by defining, as in (6.75),

$$x_\alpha = \int_{-\infty}^{v^*} x_B(v)\,dv, \qquad x_\beta = 1 - x_\alpha \qquad (6.93)$$

The heat capacity, say at constant volume, is [see (5.116) and (5.119)]

$$\begin{aligned} C_V &= \left(\frac{\partial E}{\partial T}\right)_{V,N} = NC^K + \frac{1}{2}\int_{-\infty}^{\infty} v\,\frac{\partial N_B^{(1)}(v)}{\partial T}\,dv \\ &= NC^K + \frac{1}{2}\frac{\partial}{\partial T}\left[\int_{-\infty}^{v^*} vN_B^{(1)}(v)\,dv + \int_{v^*}^{\infty} vN_B^{(1)}(v)\,dv\right] \\ &= NC^K + \frac{1}{2}\frac{\partial}{\partial T}[N_\alpha \bar{v}_\alpha + N_\beta \bar{v}_\beta] \\ &= NC^K + \left(\frac{1}{2}N_\alpha \frac{\partial \bar{v}_\alpha}{\partial T} + \frac{1}{2}N_\beta \frac{\partial \bar{v}_\beta}{\partial T}\right) + \frac{1}{2}(\bar{v}_\alpha - \bar{v}_\beta)\left(\frac{\partial N_\alpha}{\partial T}\right)_{V,N} \quad (6.94) \end{aligned}$$

The first form on the rhs of (6.94) is the one discussed in Chapter 5. That is, all the contributions to the heat capacity originating from intermolecular interactions are viewed as "structural changes" which, in the present case, involve a total redistribution of molecules among the various $v$ components, caused by the change of temperature. In the second form on the rhs of (6.94), we split the integral into two parts, induced by the choice of the cutoff point $v^*$. The number of molecules in each group is

## Liquid Water

$N_\alpha = Nx_\alpha$ and $N_\beta = Nx_\beta$, respectively. Thus, we have a TSM consisting of the two species $\alpha$ and $\beta$. Next, we define the average binding energy of an $\alpha$- and a $\beta$-cule by

$$\bar{v}_\alpha = \frac{1}{N_\alpha} \int_{-\infty}^{v^*} v N_B^{(1)}(v)\, dv, \qquad \bar{v}_\beta = \frac{1}{N_\beta} \int_{v^*}^{\infty} v N_B^{(1)}(v)\, dv \qquad (6.95)$$

Using these definitions, we get the third form on the rhs of (6.94). We now carry out the differentiation of each quantity in the brackets to get the last form on the rhs of (6.94). The second term on the rhs of (6.94) involves a shift in location of the average BE of the $\alpha$ and $\beta$ species, respectively. The third term is the contribution to the heat capacity due to thermal excitation between the two groups $\alpha$ and $\beta$. We can now present a general argument. When might a TSM be useful to explain the high value of the heat capacity of water? This occurs whenever the average values of $\bar{v}_\alpha$ and $\bar{v}_\beta$ do not change appreciably with temperature, in which case the major contribution to the heat capacity comes from rearrangements between the two groups $\alpha$ and $\beta$. Furthermore, in order to get a large contribution to this term, we must have a large difference $\bar{v}_\alpha - \bar{v}_\beta$, and the areas under the two peaks, i.e., $x_\alpha$ and $x_\beta$, must be of comparable size. In fact, in most ad hoc TSM's, such assumptions are introduced in an implicit manner by postulating that the two components behave normally; the anomalously large value of the heat capacity is attributed, to a large extent, to the structural changes between the two components.

If these conditions are fulfilled, we can rewrite (6.94) as

$$C_V = C_V^* + \tfrac{1}{2}(\bar{v}_\alpha - \bar{v}_\beta)(\partial N_\alpha/\partial T)_{V,N} \qquad (6.96)$$

In $C_V^*$, we have collected the contributions to the heat capacity due to the internal degrees of freedom of the molecules, and the contributions due to the shifts of the average values $\bar{v}_\alpha$ and $\bar{v}_\beta$. This example also serves to demonstrate that the split into a "static" and "relaxation" term, as, say, in (6.87), has no absolute significance. It is merely a convenient way of regrouping various contributions to the heat capacity. A similar analysis can be carried out for each of the thermodynamic quantities treated in this section.

Again we note that the heat capacity is determined by both the singlet and the pair distribution functions, $N_B^{(1)}(v)$ and $N_{B;B}^{(2)}(v, v')$ (see Chapter 5). Therefore, one cannot make a precise analysis of the heat capacity by studying only the singlet distribution function $N_B^{(1)}(v)$.

A splitting similar to that carried out in (6.94) can be made into three or more terms whenever we know that $x_B(v)$ has two or more natural cutoff points (see example given in Section 6.11). In such a case, the whole treatment can be extended to include a third or a fourth component for our consideration. In fact, some authors have indicated the need for such an extension (e.g., Frank and Quist, 1961; Wicke, 1966), but there has been no detailed examination of its consequences.

In concluding this section, mention must be made of a vigorous debate on the question of whether water may or may not be viewed as a mixture of various species. [For detailed elaboration on this question, the reader is referred to Buijs and Choppin (1963), Wall and Hornig (1965), Wicke (1966), Falk and Ford (1966), Walrafen (1966, 1968), Luck and Ditter (1967), Kell (1972), and Davis and Jarzynski (1972).] The major argument raised against the mixture-model approach is that no direct experimental evidence exists to support it. Therefore, one should view the whole approach as basically speculative. In the process of constructing the mixture-model approach in this section (based on Chapter 5), we have made no reference to experimental arguments. Therefore, experimental information cannot be used as evidence either to support or refute this approach. There is, however, a viewpoint from which to criticize specific ad hoc models for water. These can be viewed as approximate versions of the general and exact mixture-model formalism, and will be discussed in the next section.

## 6.9. EMBEDDING AD HOC MODELS FOR WATER IN THE GENERAL FRAMEWORK OF THE MIXTURE-MODEL APPROACH

In Chapter 5, we elaborated on the general aspects of the mixture-model (MM) approach to the theory of liquids. In this section, we present various ad hoc models for water as approximate versions of the general MM approach. We illustrate the point by a few examples.

The simplest ad hoc models are the two-structure models (TSM), such as those of Samoilov (1946), Pauling (1960), Wada (1961), and Frank and Quist (1961). The central assumption of the model is the distinction between two species, one having a relatively low local density ($L$) and the second having a relatively high local density ($H$).

Such models can be easily embedded within the MM, constructed from either $x_C(K)$ or from $x_\psi(\phi)$ defined in Section 5.2. For concreteness,

Liquid Water

suppose that we choose an integral number $K^*$; then, we can define the two components as

$$x_L = \sum_{K=0}^{K^*} x_C(K), \qquad x_H = \sum_{K=K^*+1}^{\infty} x_C(K) \qquad (6.97)$$

With a choice of, say, $K^* = 4$, we closely approach the idea underlying the simplest TSM. Similarly, by choosing a real, positive number $\phi^*$, we can define two new components

$$x_L = \int_0^{\phi^*} x_\psi(\phi) \, d\phi, \qquad x_H = \int_{\phi^*}^{\infty} x_\psi(\phi) \, d\phi \qquad (6.98)$$

In each case, $x_L$ is the mole fraction of molecules with a relatively low local density. Such models are exact whenever we exhaust all the possible components. This is equivalent to saying that we fulfill the normalization condition

$$x_L + x_H = 1 \qquad (6.99)$$

An approximate MM is obtained whenever we decide to ignore some of the species. For instance, suppose that instead of (6.97), we decide that the two most important species are

$$x_L = x_C(4), \qquad x_H = x_C(12) \qquad (6.100)$$

Here, the component $L$ may be referred to as the "icelike" component (of course it is not necessary to commit oneself to the ice structure or to any other particular structure) and $H$ is the close-packed species. Such a model would certainly fail to fulfill the normalization condition (6.99), since it is inconceivable that only coordination numbers of four and 12 exist. Nevertheless, such a simplified model may be useful if these two species are indeed the dominant ones.

As a second example, we consider the Némethy and Scheraga (1962) model. As we discussed in Section 6.6 (and again in Section 6.10), the classification of water molecules into five species, according to the number of hydrogen bonds in which they participate, can be made exact (contingent upon a precise definition of the concept of hydrogen bond, and acknowledging that a water molecule can participate in at most four bonds). However, the important parameters that enter into the theory are the "energy levels" which are assigned to the different species. In a qualitative manner, these energy levels can be embedded in the distribution function $x_B(\nu)$, defined in Section 5.2. The Némethy and Scheraga model may be

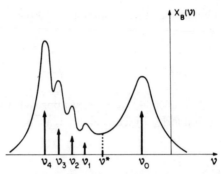

Fig. 6.17. A possible form of the function $x_B(\nu)$ for water. The various peaks on the left side of the curve correspond to molecules with 1–4 hydrogen bonds. The peak on the right side may correspond to nonbonded molecules. The details of the curve are purely imaginary and have been introduced for illustration only. The point $\nu^*$ may serve as a cutoff point. The arrows at the points $\nu_0$, $\nu_1$, $\nu_2$, $\nu_3$, and $\nu_4$ correspond to an idealized mixture model for water.

viewed as a replacement of the exact distribution function $x_B(\nu)$ by an approximate one, namely

$$x_B(\nu) = \sum_{i=0}^{4} x_i \, \delta(\nu - \nu_i) \tag{6.101}$$

where $\nu_i$ corresponds to the "energy level" of the $i$th species (i.e., a molecule connected by $i$ hydrogen bonds), and $x_i$ is the mole fraction of such molecules. A replacement of the form (6.101) may be motivated once we know the exact form of the function $x_B(\nu)$. As an illustration, suppose we are given the function $x_B(\nu)$ for water,[11] such as the one depicted in Figure 6.17. The five delta functions in (6.101) are represented by the five arrows at the various points $\nu_i$. Clearly, if the actual distribution function $x_B(\nu)$ has five very sharp peaks at $\nu_0, \nu_1, \ldots, \nu_4$, then the replacement made in (6.101) may be a good approximation. Such a replacement may also be a plausible starting point for a theory of water, if one recognizes that these five species are the most important ones in determining the properties of liquid water.

---

[11] This function is not yet known for liquid water. Therefore, the features of the function depicted in Fig. 6.17 are purely hypothetical. A similar function for a two-dimensional system of waterlike particles is discussed in Section 6.11.

# Liquid Water

A better replacement for (6.101) could be

$$x_B(\nu) = \sum_{i=0}^{4} x_i G(\nu - \nu_i) \quad (6.102)$$

where $G(\nu - \nu_i)$ is a Gaussian function centered at $\nu_i$. Indeed, an extension of the Némethy and Scheraga model, which replaces the discrete distribution of energy levels by a continuous distribution (or band) of energy levels, has been presented by Vand and Senior (1965).

We now proceed to demonstrate the general procedure by which one can construct either an exact or approximate TSM based on the concept of hydrogen bonds (HB). Suppose we are given a rule by which we can tell whether or not a pair of water molecules at configuration $(\mathbf{X}_i, \mathbf{X}_j)$ is hydrogen bonded. We now define the characteristic function (see next section for more details)

$$B(\mathbf{X}_i, \mathbf{X}_j) = \begin{cases} 1 & \text{if } i \text{ and } j \text{ are hydrogen bonded} \\ 0 & \text{otherwise} \end{cases} \quad (6.103)$$

The property (in the sense of Chapter 5) that is now used to construct a quasicomponent distribution function is the number of hydrogen bonds connected to the $i$th molecule, defined by

$$H_i(\mathbf{X}^N) = \sum_{\substack{j=1 \\ j \neq i}}^{N} B(\mathbf{X}_i, \mathbf{X}_j) \quad (6.104)$$

The corresponding "counting function" [in the following equations the subscript H stands for hydrogen bonds]

$$N_{\text{H}}^{(1)}(\mathbf{X}^N, K) = \sum_{i=1}^{N} \delta[H_i(\mathbf{X}^N) - K] \quad (6.105)$$

is the number of water molecules participating in precisely $K$ hydrogen bonds at a given configuration $\mathbf{X}^N$ (for more details, see Section 5.2). The average number of molecules participating in $K$ hydrogen bonds is

$$N_{\text{H}}^{(1)}(K) = \langle N_{\text{H}}^{(1)}(\mathbf{X}^N, K) \rangle \quad (6.106)$$

which is a generalized singlet molecular distribution function. The corresponding mole fractions are defined by

$$x_{\text{H}}(K) = N_{\text{H}}^{(1)}(K)/N = \int \cdots \int d\mathbf{X}^N P(\mathbf{X}^N) \, \delta[H_1(\mathbf{X}^N) - K] \quad (6.107)$$

with the normalization condition

$$\sum_{K=0}^{4} x_H(K) = 1 \qquad (6.108)$$

From (6.107), it is clear that the definition of $x_H(K)$ is equivalent to a division of the entire configurational space into subspaces. Each subspace corresponds to all the configurations for which a selected molecule, say 1, has a precise number of hydrogen bonds $K$. Summing over $K$ in (6.108) is equivalent to integrating over the entire configurational space. This is the essence of the requirement that the classification procedure be exhaustive. The entire configurational space is split into five subspaces, the union of which exhausts the entire configurational space. Now, suppose we wish to regroup the particles into two classes, those that are nonbonded ($K = 0$) and all the others. A different possibility is grouping those with $K > 3$ and all the others. These two possibilities provide *exact* TSM's derived from the distribution function $x_H(K)$. The third possibility is to construct an *approximate* TSM. For instance, we may feel that the most important components are the nonbonded ($K = 0$) and the fully hydrogen-bonded ($K = 4$) molecules, with the appropriate mole fractions being $x_H(0)$ and $x_H(4)$, respectively. Clearly, these two mole fractions do not fulfill the normalization condition $x_H(0) + x_H(4) = 1$ [see (6.108)]. In this respect, such a model may not be a *good* approximation. Nevertheless, if it turns out that these two species are really the most important ones in determining the unique behavior of water, it may still be a *useful* model. This kind of reasoning is the underlying justification for using simplified TSM's for water.

## 6.10. A POSSIBLE DEFINITION OF THE "STRUCTURE OF WATER"

The "structure of water" is a most ubiquitous and controversial concept in the literature concerning liquid water. In the solid, there exists a clear-cut structure due to the periodicity of the geometry of the unit cell. There is no such structure in the liquid, though in some cases, one refers to the "local structure" of a liquid as revealed by the function $g(R)$. For water, there is widespread usage of the concept of "structure" as a measure of the degree of crystallinity, or the extent to which the structure of ice persists after melting. Having this qualitative idea in mind, we seek to construct a precise definition of the structure which conforms with the

# Liquid Water

common usage of this concept. We base our definition on the concept of hydrogen bond (HB). The latter, itself, calls for a precise definition. A proper discussion of a hydrogen bond between a pair of water molecules must be made in terms of the total interaction energy between the pair of molecules, which, in essence, is a quantum mechanical problem. We shall adopt a simplified definition of the hydrogen bond, and, hence, of the structure of water, based on the *configuration* of a pair of particles. Motivation for our definition stems from the recognition of the fact that a hydrogen bond is formed only along four particular directions, so that the geometry of the pair of molecules participating in a hydrogen bond is determined within a relatively narrow range of distances and angles. We refer to the notation of Section 6.4 for the following definition. Consider the *set* of all configurations of the pair of molecules $i$ and $j$ which conforms to the following conditions

$$C_{\alpha\beta}^{HB}(i \to j) = \{|\ R_{ij} - R_H\ | < \varepsilon_1,\ |\ \mathbf{h}_{i\alpha} \cdot \mathbf{u}_{ij} - 1\ | < \varepsilon_2,\ |\ \mathbf{l}_{j\beta} \cdot \mathbf{u}_{ij} + 1\ | < \varepsilon_2\} \tag{6.109}$$

where $\varepsilon_1$ and $\varepsilon_2$ are small numbers chosen at will [for instance, they may be identical with $\sigma'$ and $\sigma$ of Eq. (6.16)]. In the brackets on the rhs of (6.109), we stipulate that the distance between the centers of the molecules is close to $R_H$, where $R_H$ may be taken to equal the O–O distance in ice, 2.76 Å. The vector $\mathbf{h}_{i\alpha}$ ($\alpha = 1, 2$) points almost along the $\mathbf{u}_{ij}$ direction, whereas the vector $\mathbf{l}_{j\beta}$ ($\beta = 1, 2$) points almost along the direction $-\mathbf{u}_{ij}$. If the conditions in the brackets on the rhs of (6.109) are fulfilled, we say that molecule $i$ is bonded (as a donor) to the molecule $j$ (the acceptor).

The set of all configurations for which $i$ is bonded to $j$ is obtained by taking the union of four sets, allowing $\alpha$ and $\beta$ to be either 1 or 2:

$$C^{HB}(i \to j) = \bigcup_{\alpha,\beta=1}^{2} C_{\alpha\beta}^{HB}(i \to j) \tag{6.110}$$

Two molecules $i$ and $j$ are said to be bonded if either $i$ is bonded to $j$ or $j$ is bonded to $i$. The set of all such configurations is

$$C^{HB}(i, j) = C^{HB}(i \to j) \cup C^{HB}(j \to i) \tag{6.111}$$

The measure of the set $C^{HB}(i, j)$ is determined by the choice of the parameters $\varepsilon_1$ and $\varepsilon_2$. For instance, if we take a very small value of $\varepsilon_2$, then the hydrogen bond as defined in (6.109) will be almost linear (in the sense that the angle O–H$\cdots$O is almost 180°). On the other hand, if we increase the value of $\varepsilon_2$, we also include distorted, or bent, bonds in our definition.

We can now proceed to define more precisely the characteristic function in (6.103), namely

$$B(\mathbf{X}_i, \mathbf{X}_j) = \begin{cases} 1 & \text{if } (\mathbf{X}_i, \mathbf{X}_j) \in C^{\text{HB}}(i, j) \\ 0 & \text{otherwise} \end{cases} \quad (6.112)$$

The vector $\mathbf{x}_\text{H}$ defined in (6.107) gives the distribution of molecules with different numbers of hydrogen bonds.

The concept of the structure of water is usually identified as the concentration of icelike components in a TSM. For example, in the TSM defined in (6.100), $x_L$ may serve as a measure of the structure of the system. A more general definition could be the average number of hydrogen bonds in the system. If $N_\text{H}^{(1)}(K)$ is the average number of molecules participating in $K$ hydrogen bonds, then the average number of hydrogen bonds in the system is

$$N_{\text{HB}} = \tfrac{1}{2} \sum_{K=0}^{4} K N_\text{H}^{(1)}(K) \quad (6.113)$$

The idea behind this definition is that the more hydrogen bonds in the system, the more structured it is. The factor $\tfrac{1}{2}$ has been included in (6.113) to avoid counting each HB twice. Relation (6.113) can be transformed into an equivalent form, using the pair distribution function,

$$\begin{aligned}
N_{\text{HB}} &= \tfrac{1}{2} \sum_{K=0}^{4} K \int \cdots \int d\mathbf{X}^N\, P(\mathbf{X}^N) \sum_{i=1}^{N} \delta[H_i(\mathbf{X}^N) - K] \\
&= \tfrac{1}{2} \int \cdots \int d\mathbf{X}^N\, P(\mathbf{X}^N) \sum_{i=1}^{N} \sum_{K=0}^{4} K\, \delta[H_i(\mathbf{X}^N) - K] \\
&= \tfrac{1}{2} \int \cdots \int d\mathbf{X}^N\, P(\mathbf{X}^N) \sum_{i=1}^{N} H_i(\mathbf{X}^N) \\
&= \tfrac{1}{2} \int \cdots \int d\mathbf{X}^N\, P(\mathbf{X}^N) \sum_{\substack{i=1 \\ i \neq j}}^{N} \sum_{j=1}^{N} B(\mathbf{X}_i, \mathbf{X}_j) \\
&= \tfrac{1}{2} \int d\mathbf{X}_1 \int d\mathbf{X}_2\, \varrho^{(2)}(\mathbf{X}_1, \mathbf{X}_2) B(\mathbf{X}_1, \mathbf{X}_2) \quad (6.114)
\end{aligned}$$

In the second form on the rhs of (6.114), we have interchanged the order of integration and summation. In the third form, we have used the basic property of the Kronecker delta function. In the fourth form, we have used the definition of $H_i(\mathbf{X}^N)$ in (6.104). Now, we recognize that the integral is an average of a pairwise function, for which the general theorem of Section 3.2 is applicable, and we get the last form for $N_{\text{HB}}$ expressed in terms of the pair distribution function. The last form was suggested recently

## Liquid Water

by Ben-Naim (1972e). A somewhat different definition employs the strength of interaction between pairs of particles [suggested by Ben-Naim and Stillinger (1972)]. Instead of the characteristic function defined in terms of the geometry of the configuration $\mathbf{X}_i$, $\mathbf{X}_j$, we define a new characteristic function in terms of the interaction energy between the pair. Thus

$$D(\mathbf{X}_i, \mathbf{X}_j) = \begin{cases} 1 & \text{if } U(\mathbf{X}_i, \mathbf{X}_j) < E_{\text{HB}} \\ 0 & \text{if } U(\mathbf{X}_i, \mathbf{X}_j) \geq E_{\text{HB}} \end{cases} \quad (6.115)$$

Hence, the quantity

$$N^*_{\text{HB}} = \tfrac{1}{2} \int d\mathbf{X}_1 \int d\mathbf{X}_2 \, \varrho^{(2)}(\mathbf{X}_1, \mathbf{X}_2) D(\mathbf{X}_1, \mathbf{X}_2) \quad (6.116)$$

is the average number of pairs whose interaction energy is less than $E_{\text{HB}}$. As in the definition of $B(\mathbf{X}_i, \mathbf{X}_j)$ in (6.112), where we used the arbitrary parameters $\varepsilon_1$ and $\varepsilon_2$, we have here a single parameter $E_{\text{HB}}$ that is chosen at will.

It is worth noting that the two definitions of the "structure of water" given in (6.113) and in (6.116) involve averages in the configurational space only. Time does not feature explicitly in these definitions, nor is the average duration of a hydrogen bond of relevance in this context. We refer the reader to Eisenberg and Kauzmann (1969) for a different discussion of the concept of structure.

As a final comment, we stress that the above definitions of the structure may be evaluated by any of the numerical methods of computing average quantities. These, however, are not directly measurable quantities. Of course, some measurements may indicate changes of structure. For instance, one measures the viscosity of water as a function of temperature, and infers that the decrease in viscosity is a manifestation of the decrease in structure of the system. Such an interpretation, though probably correct in a qualitative sense, cannot be made precise unless we know the functional dependence of viscosity on structure. Similar conclusions apply to any other relation between a measurable quantity and the structure.

### 6.11. WATERLIKE PARTICLES IN TWO DIMENSIONS

The study of waterlike particles in two dimensions has two important merits. First, it can be viewed as a prelude to the study of the more difficult three-dimensional cases. Second, the study of nonsimple particles by the available statistical mechanical methods is interesting in its own right. For instance, information on the full pair correlation function for nonspherical particles cannot be obtained from X-ray diffraction data. Therefore, it is of value to investigate the properties of this function by theoretical means.

A system of nonspherical particles in two dimensions is, from the computational point of view, an intermediate case between spherical particles (in either two or three dimensions) and nonspherical particles in three dimensions. The pair potential in this case depends on three coordinates (see below), compared with six in the three-dimensional case. Some very useful information on the numerical procedure, on the problem of convergence, and so forth, can thus be gained in a system which is relatively simpler than the three-dimensional case. We shall also present some results on the generalized molecular distribution function, which thus far are available only in two dimensions, yet are of relevance to the case of real liquid water.

### 6.11.1. The Physical Model

The guiding idea underlying the choice of a pair potential for waterlike particles stems from our conclusion in Section 6.8 (which will again be cited in Chapters 7 and 8) that one of the most important features of the mode of packing of water molecules in the liquid state is the unique coupling between binding energy and local density. More specifically, molecules that are bound most strongly to their environment (presumably by hydrogen bonds) are the ones that, on the average, have a relatively low local density.

The particles comprising our system are basically spherical disks. The intermolecular potential function operating between two particles is a superposition of two functions: a Lennard-Jones type, depending on the distance only, and a "hydrogen-bond-like" potential, which depends on the distance as well as on the relative orientation of the pair of particles. The construction of the latter potential is, with some simplifications, similar to the construction of the potential function given in (6.16).

To each particle we affix three unit vectors emanating from the center of the particle and pointing in three selected directions, along which the particles can form "hydrogen bonds." This is similar to the situation we described in Section 6.4 for real water molecules, where we recognized four preferable directions in each molecule. In this preliminary stage, one can also drop the distinction between donor and acceptor molecules for the formation of a hydrogen bond. Hence, all three directions are assumed to be equivalent. [For more details, see Ben-Naim (1971c, 1972c,d).] Figure

# Liquid Water

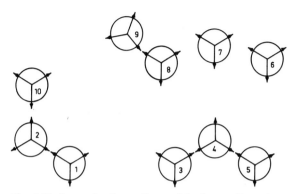

Fig. 6.18. A sample of waterlike particles in two dimensions. The circles indicate the Lennard-Jones diameter of the particles. The arrows attached to each particle are unit vectors along which a "hydrogen bond" may be formed. Particles 1 and 2 are considered to be bonded. Particles 2 and 10 are correctly oriented but too far to form a bond. Particles 8 and 9 are almost at the correct configuration for bond formation. Particles 3, 4, and 5 are connected successively by hydrogen bonds.

6.18 shows a sample of such particles in two dimensions. For each particle, we observe the three "selected directions" indicated by arrows. The qualitative nature of the hydrogen bond potential can be described with the help of a few examples, shown in the figure. For instance, particles 1 and 2 are assumed to be in the correct configuration to form a hydrogen bond (i.e., the distance and relative orientation are the "correct" ones for such bonding; a more precise statement is given below). Particles 2 and 10 are correctly oriented to form a hydrogen bond but their separation is too large. On the other hand, particles 8 and 9 are almost at the correct distance and orientation. Such a case may be viewed as a bent hydrogen bond which will be assigned a weaker interaction energy than in the case of a "straight" hydrogen bond (see below). Particles 3, 4, and 5 are connected successively by hydrogen bonds.

The above qualitative description of the bonding between the particles is now translated into more precise language. Let $x_i$, $y_i$ be the Cartesian coordinates of the center of the $i$th particle. Let $\mathbf{i}_k$ ($k = 1, 2, 3$) be the $k$th unit vector of the $i$th particle. The angle between the unit vectors is $2\pi/3$. The orientation of the $i$th particle is given by the angle $\alpha_i$ between the vector $\mathbf{i}_1$ and the positive direction of the $x$ axis. This coordinate system is shown in Fig. 6.19. We denote by the vector $\mathbf{X}_i = (x_i, y_i, \alpha_i)$ the full configuration

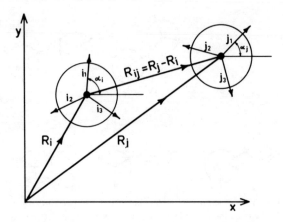

Fig. 6.19. Coordinate system for a pair of particles $i$ and $j$. $\mathbf{i}_k$ and $\mathbf{j}_k$ ($k = 1, 2, 3$) are unit vectors along which a bond may be formed. The vectors $\mathbf{R}_i$ and $\mathbf{R}_j$ are the locational vectors for the centers of particles $i$ and $j$, respectively. $\alpha_i$ is the angle between the vector $\mathbf{i}_1$ and the positive direction of the $x$ axis.

of the $i$th particle. The intermolecular potential function for a pair of particles $i$ and $j$ in the configuration $(\mathbf{X}_i, \mathbf{X}_j)$ is

$$U(\mathbf{X}_i, \mathbf{X}_j) = U_{\mathrm{LJ}}(R_{ij}) + U_{\mathrm{HB}}(\mathbf{X}_i, \mathbf{X}_j) \qquad (6.117)$$

where

$$U_{\mathrm{LJ}}(R) = 4\varepsilon_W[(\sigma_W/R)^{12} - (\sigma_W/R)^6] \qquad (6.118)$$

and

$$U_{\mathrm{HB}}(\mathbf{X}_i, \mathbf{X}_j) = \varepsilon_{\mathrm{H}} G_{\sigma'}(R_{ij} - R_{\mathrm{H}}) \sum_{k,l=1}^{3} G_\sigma(\mathbf{i}_k \cdot \mathbf{u}_{ij} - 1) G_\sigma(\mathbf{j}_l \cdot \mathbf{u}_{ij} + 1) \qquad (6.119)$$

Here, $\varepsilon_W$ and $\sigma_W$ are the Lennard-Jones parameters of the waterlike particles, and $\varepsilon_{\mathrm{H}}$ is the maximum attainable HB energy (for more details, see Section 6.4). Clearly, if $\sigma'$ and $\sigma$ are small enough, then the hydrogen bond is defined for a very restricted set of configurations. If, on the other hand, they are large, then the hydrogen bond is defined in a very broad sense. Thus, the choice of $\sigma'$ and $\sigma$ determines the extent of stretching and bending of the hydrogen bond. [More details on the choice of the various parameters for this model are given by Ben-Naim (1971c, 1972c,d).]

From the form of the potential function (6.117), we expect that if $\varepsilon_{\mathrm{H}}/kT$ becomes very large, the system will tend to form an extended net-

**Liquid Water**

work of successive hydrogen-bonded molecules, as depicted in Fig. 6.20. This structure is open, akin to the open structure of ice, and is expected at very low temperatures (for a fixed value of $\varepsilon_H$). At high temperatures, due to thermal agitation, we may find clusters of bonded molecules with random sizes and shapes. Some pairs of particles, such as 8 and 9 in Fig. 6.18, may be considered to be either in a broken or a bent hydrogen bond. (The distinction between broken and bent hydrogen bonds is a matter of definition; see also Section 6.10. In fact, even in real water, no sharp distinction exists between the two concepts.)

As a measure of the "structure of the system," we may use a quantity defined in a similar manner to the quantity $N_{HB}$ in Section 6.10. First, we define the set of configurations

$$C^{HB}(i,j) = \bigcup_{k,l=1}^{3} \{|\, R_{ij} - R_H\, | < \delta_1,\ |\, \mathbf{i}_k \cdot \mathbf{u}_{ij} - 1\, | < \delta_2,\ |\, \mathbf{j}_l \cdot \mathbf{u}_{ij} + 1\, | < \delta_2\}$$

(6.120)

The curly brackets stand for the set of configurations in which $R_{ij}$ is close to $R_H$, $\mathbf{i}_k$ is close to the direction of $\mathbf{u}_{ij}$, and $\mathbf{j}_l$ is close to the direction $-\mathbf{u}_{ij}$. The set $C^{HB}(i,j)$ is the union of all possible sets of configurations in which

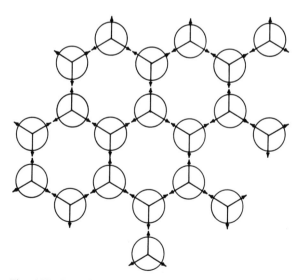

Fig. 6.20. Extended network of hydrogen-bonded particles in two dimensions. In the liquid state, one may encounter aggregates of connected molecules, with random size and shape, of similar packing geometry.

a bond is formed between the $k$th direction of particle $i$ and the $l$th direction of particle $j$.

The characteristic function for the set (6.120) is defined as in Eq. (6.112); hence, the average number of hydrogen bonds per particle is

$$n_{\text{HB}} = N_{\text{HB}}/N = (1/2N) \int d\mathbf{X}_1 \int d\mathbf{X}_2\, \varrho^{(2)}(\mathbf{X}_1, \mathbf{X}_2) B(\mathbf{X}_1, \mathbf{X}_2) \quad (6.121)$$

where $N$ is the total number of particles in the system. Note that the measure of the set $C^{\text{HB}}(i, j)$ is determined by the numbers $\delta_1$ and $\delta_2$. For simplicity, in the following illustrations, we choose $\delta_1 = \sigma'$, $\delta_2 = \sigma$, where $\sigma'$ and $\sigma$ are the same parameters chosen for the Gaussian functions in (6.119). This completes our description of the model. We now give a brief survey of some numerical results pertaining to this system.

### 6.11.2. Application of the Percus–Yevick Integral Equation

The Percus–Yevick equation, adapted to the two-dimensional system of nonspherical particles, is

$$y(\mathbf{X}_1, \mathbf{X}_2) = 1 + \frac{\varrho}{2\pi} \int y(\mathbf{X}_1, \mathbf{X}_3) f(\mathbf{X}_1, \mathbf{X}_3)$$
$$\times [\, y(\mathbf{X}_3, \mathbf{X}_2) f(\mathbf{X}_3, \mathbf{X}_2) + y(\mathbf{X}_3, \mathbf{X}_2) - 1]\, d\mathbf{X}_3 \quad (6.122)$$

where

$$f(\mathbf{X}_i, \mathbf{X}_j) = \exp[-\beta U(\mathbf{X}_i, \mathbf{X}_j)] - 1 \quad (6.123)$$

$$y(\mathbf{X}_i, \mathbf{X}_j) = g(\mathbf{X}_i, \mathbf{X}_j) \exp[\beta U(\mathbf{X}_i, \mathbf{X}_j)] \quad (6.124)$$

The details of the derivation of the Percus–Yevick equation and the numerical procedure for its solution are highly technical and will not be presented here. [Details may be found in Appendices 9-D and 9-E and in Ben-Naim (1971c, 1972c,d).] We note, however, that each pairwise function in (6.122) depends only on three coordinates, which we can choose as follows: $R$ is the distance between the centers of the two particles, $R = |\mathbf{R}_j - \mathbf{R}_i|$, and $\alpha_i$ is the angle between the vector $\mathbf{i}_1$ and the direction of $\mathbf{R}_{ij} = \mathbf{R}_j - \mathbf{R}_i$, measured counterclockwise. The full pair correlation function is thus a function of three variables, $g(R, \alpha_1, \alpha_2)$. Because of the special symmetry of the pair potential, it is clear that all of the pairwise functions, such as $U$, $y$, or $g$, will be invariant to a rotation of the particle

# Liquid Water

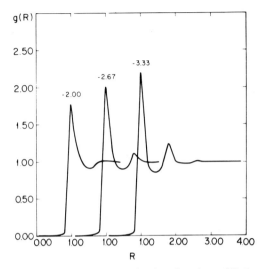

Fig. 6.21. The radial distribution function $g(R)$ for the "waterlike" particles with the parameters listed in (6.126). The value of the HB energy parameter $\varepsilon_H/kT$ is indicated next to each curve. The density for all the curves is $\varrho = 0.6$.

by an angle of $2\pi/3$. Hence, it is sufficient to describe the behavior of the pair correlation function in the range

$$0 \leq R \lesssim 4\sigma_W, \qquad 0 \leq \alpha_1, \alpha_2 \leq 2\pi/3 \qquad (6.125)$$

Figure 6.21 shows the angle-average pair correlation function for a system of waterlike particles with the following parameters[12]:

$$R_H = \sigma_W = 1.0, \quad \varepsilon/kT = 0.1, \quad \varepsilon_H/kT = -2.0, -2.67, -3.33 \qquad (6.126)$$

The total density is $\varrho = 0.6$. The angle-average pair correlation function is defined by

$$g(R) = \left(\frac{3}{2\pi}\right)^2 \int_0^{2\pi/3} d\alpha_1 \int_0^{2\pi/3} d\alpha_2\, g(R, \alpha_1, \alpha_2) \qquad (6.127)$$

Note that because of the symmetry of the function $g$ for rotations by $2\pi/3$, it is sufficient to integrate for each angle in the range $0 \leq \alpha \leq 2\pi/3$.

---

[12] As in the previous illustrations, we use here a dimensionless unit of length for $R_H$ and $\sigma_W$; hence, the density $\varrho$ is the number of particles in the corresponding unit of area.

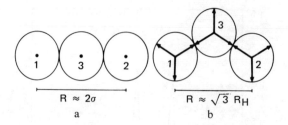

Fig. 6.22. The configuration of particles 1 and 2 at a separation corresponding to the second peak of the radial distribution function (a) for spherical particles and (b) for nonspherical particles.

The most important feature of the curves in Fig. 6.21 is the second peak at $R = 1.8$, which is roughly the distance between the second hydrogen-bonded neighbors (the exact value being $R = \sqrt{3}\, R_\mathrm{H}$). This should be compared with the second peaks of the pair correlation function for spherical particles occurring at about $R \approx 2$ (see Chapter 2 for more details). This finding has direct bearing on the form of the pair correlation function of real water. We stressed in Section 6.3 that the second peak of $g(R)$ of water at $R \approx 4.6$ Å is a result of the strong directional forces (hydrogen bonds), which dictate a particular mode of packing around a given molecule. Similarly, in the two-dimensional, waterlike particles, we see that as $\varepsilon_\mathrm{H}/kT$ increases, the predominant factor that determines the propagation of the correlation between particles becomes the hydrogen bond potential, and not the mere "filling of space" mechanism discussed in Chapters 2 and 4. The configuration pertaining to the second nearest neighbors is depicted in Fig. 6.22.

The information provided by $g(R)$ is evidently very restricted, and a richer content is conveyed by the full pair correlation function. We present here for illustrative purposes only the most important aspect of this topic. We confine ourselves to five angles[13] between 0 and $2\pi/3$:

$$\alpha = 0°,\ 30°,\ 60°,\ 90°,\ 120° \qquad (6.128)$$

The pair correlation function at $R = R_\mathrm{H} = 1$ is given by the following

---

[13] This particular set of angles was chosen by Ben-Naim (1971c, 1972c,d). A larger set of angles involves a very long computing time, which, at present, makes the computation unfeasible. Nevertheless, this restricted set of angles is sufficient for demonstrating the most important aspect of the angle dependence of $g(R, \alpha_1, \alpha_2)$.

## Liquid Water

matrix [the illustration here corresponds to the parameters listed in (6.126) with $\varepsilon_H/kT = -3.33$ and $\varrho = 0.6$]

$$[g(R = 1.0, \alpha_1, \alpha_2)] = \begin{vmatrix} 1.4 & 0.7 & 19.2 & 0.7 & 1.4 \\ 1.4 & 0.9 & 1.1 & 0.9 & 1.4 \\ 1.5 & 1.0 & 1.0 & 1.0 & 1.5 \\ 1.4 & 0.9 & 1.1 & 0.9 & 1.4 \\ 1.4 & 0.7 & 19.2 & 0.7 & 1.4 \end{vmatrix} \quad (6.129)$$

where the rows and columns correspond to the five values of $\alpha_1$ and $\alpha_2$ given in (6.128). The most prominent feature of this matrix is the high value of the correlation function for $\alpha_1 = 0°$ and $\alpha_2 = 60°$ (as well as the configuration $\alpha_1 = 120°$ and $\alpha_2 = 60°$), which corresponds to the most probable configuration of the pair of molecules at $R = R_H = 1$. This is shown in Fig. 6.23(a). The average value of the entries of this matrix gives the strong peak of $g(R)$ at $R = 1$. The next interesting distance is $R = \sqrt{3} R_H$, which is the distance between two particles bonded through an intermediate molecule, as shown in Fig. 6.23(b). The computed values of the matrix elements at $R = \sqrt{3} R_H \approx 1.8 R_H$ are

$$[g(R = 1.8, \alpha_1, \alpha_2)] = \begin{vmatrix} 1.1 & 1.1 & 0.7 & 1.1 & 1.1 \\ 1.2 & 3.0 & 1.0 & 1.0 & 1.2 \\ 1.1 & 0.8 & 0.5 & 0.8 & 1.1 \\ 1.2 & 1.0 & 1.0 & 3.0 & 1.2 \\ 1.1 & 1.1 & 0.7 & 1.1 & 1.1 \end{vmatrix} \quad (6.130)$$

The largest entry in this matrix is for $\alpha_1 = 30°$ and $\alpha_2 = 30°$ (and similarly, for $\alpha_1 = 90°$ and $\alpha_2 = 90°$), which corresponds to the configuration shown in Fig. 6.23(b). Again, we note that the average value of the entries of this matrix produces the strong peak of $g(R)$ at $R \sim 1.8 R_H$.

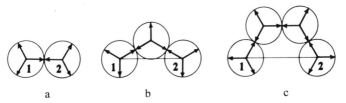

Fig. 6.23. Various ideal geometries of nearest neighbors for "waterlike" particles. The separation between particles 1 and 2 is (a) $R = R_H$; (b) $R = \sqrt{3} R_H$; (c) $R = 2 R_H$.

The third distance of interest is at $R = 2R_H$, for which we have

$$[g(R = 2.0, \alpha_1, \alpha_2)] = \begin{vmatrix} 1.3 & 1.1 & 0.9 & 1.1 & 1.3 \\ 1.2 & 0.9 & 0.8 & 1.0 & 1.2 \\ 1.4 & 0.8 & 0.7 & 0.8 & 1.4 \\ 1.2 & 1.0 & 0.8 & 0.9 & 1.2 \\ 1.3 & 1.1 & 0.9 & 1.1 & 1.3 \end{vmatrix} \quad (6.131)$$

Here, we have a relatively slight preference for the configuration $\alpha_1 = 60°$ and $\alpha_2 = 0°$ (or $\alpha_1 = 60°$ and $\alpha_2 = 120°$), corresponding to the configuration depicted in Fig. 6.23(c). The average value of the entries of this matrix is not large enough to produce a maximum in $g(R)$ in this case.

There is one important conclusion that can be drawn from the study of the pair correlation function for two-dimensional waterlike particles which is relevant to the study of liquid water. If strong directional forces, or bonds, are operative at some selected directions, then the correlation between the positions of two particles is propagated mainly through a chain of bonds, and less by the "filling of space"—a characteristic feature of the mode of packing of simple fluids.

### 6.11.3. Application of Monte Carlo Method

We present here an example of complementary information on the system of waterlike particles in two dimensions, obtained by the standard Monte Carlo method. The model is the same as above, but we focus our attention mainly on the singlet generalized molecular distribution functions (Chapter 5). Figure 6.24 shows a sample of 36 waterlike particles. The molecular parameters chosen for this particular illustration are

$$\varepsilon_W/kT = 0.5, \quad \sigma_W = 0.7, \quad R_H = 1.0, \quad \varepsilon_H/kT = -3.0, -5.0, -8.0 \quad (6.132)$$

The total number density is $\varrho = 0.9$ and the radius of the first coordination sphere was chosen as $R_C = 1.3\sigma_W$. The parameters $\sigma'$ and $\sigma$ in (6.119) were chosen after some experimentation with the results obtained from the computations. The arguments underlying the choice of these parameters are discussed in the section following (6.119) and we will not be concerned with them here [for details, see Ben-Naim (1973b)].

Figure 6.25 shows the radial distribution function for our waterlike particles with different HB energies as in (6.132). The peak at about $R = 0.8$ is the "normal" peak one would expect for a system of *spherical* particles without the HB potential. As we increase the HB energy, the peak at

# Liquid Water

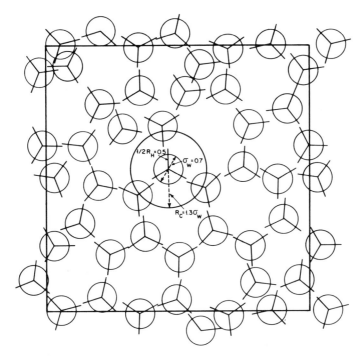

Fig. 6.24. A sample of 36 waterlike particles with parameters given in (6.132). The diameter of the particles is $\sigma_W = 0.7$, the HB length is $R_H = 1.0$, and the first coordination radius is $R_C = 1.3\sigma_W$. Note the open mode of packing of fully bonded particles.

$R \approx R_H = 1.0$ grows and becomes the dominant feature of the curve. The peak at $R \approx 0.8$ persists, however, even when the HB energy is quite large. The reason is interesting, and also has some relevance to the study of liquid water. The choice of the particular density $\varrho = 0.9$ was made on the following basis. One has to make a distinction between two types of "close-packing" densities. The first is the maximum density of spheres (disks) of diameter $\sigma_W = 0.7$, which is $\varrho_{cp} = 2/\sqrt{3\sigma^2} \approx 2.36$. The second is the density of the two-dimensional lattice (Fig. 6.20) of bonded particles, which in our case is $\varrho_{lattice} = 0.77/R_H^2 = 0.77$. Thus, the density $\varrho = 0.9$ is such that we are far below the close-packing density, and at the same time slightly above the density of the regular lattice (as with liquid water at $t \approx 0°C$). As a result, some of the particles must find "interstitial" positions in the holes formed by the network of bonded particles. That is, the particles are forced, to a certain extent, to approach each other to within a distance of the order of $R \approx 0.7$.

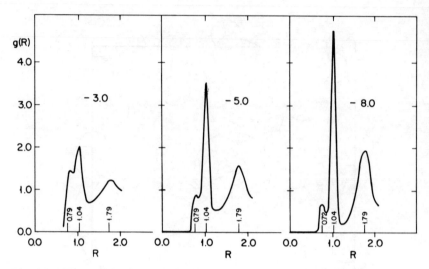

Fig. 6.25. The spatial pair correlation function for a system of waterlike particles, with parameters given in (6.132). The HB energy $\varepsilon_H/kT$ is indicated next to each curve. The locations of the various maxima are indicated on the abscissa.

The second feature of the curves in Fig. 6.25 is the peak at about $R = \sqrt{3}R_H \approx 1.8$, which indicates a large probability of finding pairs of particles bonded through an intermediary particle, as depicted in Fig. 6.22.

In Fig. 6.26, the singlet generalized molecular distribution functions $x_C(K)$ are plotted for the three cases listed in (6.132). The most prominent feature of these curves is the shift to the left of the most probable coordina-

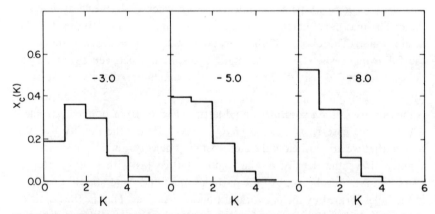

Fig. 6.26. The singlet distribution function $x_C(K)$ for the system of nonspherical particles, with parameters given in (6.132). The HB energy $\varepsilon_H/kT$ is indicated next to each curve.

# Liquid Water

tion number as the HB energy increases. Because of the particular choice of $R_H = 1.0 > \sigma_W = 0.7$, as we increase the strength of the hydrogen bond, more pairs are forced to attain a distance of $R = R_H$ from each other and, hence, the local density around each particle will, on the average, decrease. Hence, at $\varepsilon_H/kT = -8.0$, we find that the most probable coordination number is zero. [The actual numerical value of the coordination number depends, of course, on the choice of $R_C$, defined in the paragraph following (6.132). The important point is the general shift in location of the maximum in these curves rather than the absolute numerical magnitudes.]

Figure 6.27 shows the distribution function for the binding energy $x_B(\nu)$ (Chapter 5) for the parameters in (6.132). The curves in this figure should be compared with the corresponding curves of $x_B(\nu)$ for simple particles (Section 5.3). The most striking difference between the two sets of curves is the appearance of several well-resolved peaks in the present case, compared with a single peak in the simple spherical case.

For a relatively low HB energy ($\varepsilon_H/kT = -3.0$), we find essentially two peaks in $x_B(\nu)$, one corresponding to singly bonded molecules, and the second to nonbonded molecules. As we increase the HB energy to

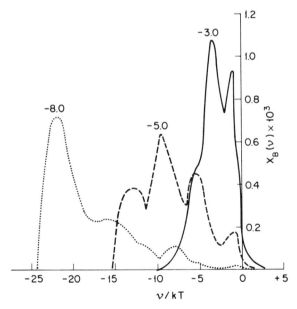

Fig. 6.27. The singlet distribution functions $x_B(\nu)$ for the system of nonspherical particles, with parameters given in (6.132). The HB energy $\varepsilon_H/kT$ is indicated next to each curve.

$\varepsilon_H/kT = -5.0$, we get a clear resolution into four peaks, roughly corresponding to molecules with zero, one, two, and three hydrogen bonds. The form of this curve may suggest a few cutoff points [at the locations of the minima $v_1^*$, $v_2^*$, $v_3^*$ of $x_B(v)$] by the use of which we can define four species, and the corresponding mole fractions

$$x_3 = \int_{-\infty}^{v_1^*} x_B(v)\, dv, \qquad x_2 = \int_{v_1^*}^{v_2^*} x_B(v)\, dv,$$

$$x_1 = \int_{v_2^*}^{v_3^*} x_B(v)\, dv, \qquad x_0 = \int_{v_3^*}^{\infty} x_B(v)\, dv \tag{6.133}$$

Such a splitting into four quasicomponents can serve as a rigorous basis for a mixture-model approach for this fluid. This has a direct relevance to the theory of real liquid water. Suppose we know the function $x_B(v)$ for water; then the validity of various ad hoc models can be assessed according to the form of this function.

As the HB energy becomes very large ($\varepsilon_H/kT = -8.0$), most of the waterlike particles tend to engage in three hydrogen bonds; hence we get a strong peak at about $v/kT \approx -24$, with small peaks corresponding to particles with two, one, and zero bonds.

Another feature of the mode of packing of waterlike particles akin to the behavior of liquid water is demonstrated by the joint singlet generalized molecular distribution function, constructed by combining the binding energy and coordination number (Fig. 6.28). The values of $x_{B;C}(v, K)\, \Delta v$ are the mole fractions of particles having coordination numbers equal to $K$ *and* binding energies between $v$ and $v + \Delta v$. This function serves to illustrate the coupling between local density[14] and binding energy. We recall that in Chapter 5, we studied the same function for simple spherical particles in two dimensions. Particularly, we noted that, in general, high local density is coupled with strong binding energy, which is referred to as the normal behavior of a liquid. In water, as was discussed in Section 6.8, we expect to have anomalous coupling of low local density with strong binding energy. This feature is demonstrated by the waterlike particles in Fig. 6.28. (Com-

---

[14] As discussed in Chapter 5, a better function conveying such information is $x_{\psi;B}(\phi, v)$. However, the computation of the coordination number is much easier than is that of the volume of the Voronoi polygon of each particle. Both of these properties provide a measure of the local density around particles in the liquid. [For more details see Ben-Naim (1973b).]

# Liquid Water

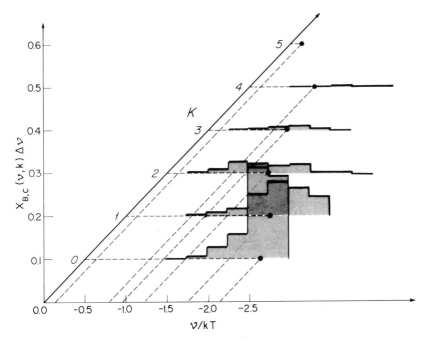

Fig. 6.28. Schematic illustration of the function $x_{B,C}(v, K)\,\Delta v$ for waterlike particles, with parameters given in (6.132), but with HB energy $\varepsilon_H/kT = -8.0$. Note the difference between this figure and the corresponding figure for spherical particles (Fig. 5.8).

pare with the corresponding functions for spherical particles given in Section 5.3.) Figure 6.29 shows $\langle v(K)\rangle/kT$, defined in Section 5.3, which should be compared with the corresponding figure, Fig. 5.9, for spherical particles.

Finally, a general comment regarding the numerical procedure is in order. The Monte Carlo method was described in Section 2.7 as a process that gives preference to steps in which the total energy of the system decreases. If the intermolecular attractive forces are very large (compared with $kT$, as is the case for some of the model particles discussed in this section, and is certainly the case in real liquid water), then once a "bond" is formed, it will persist for many configurations. Breaking the bond requires an appreciable increase of the potential energy, and therefore such a step is acceptable only with an extremely low probability. Hence, convergence of the Monte Carlo method for such systems is expected to be very slow. Indeed, for a two-dimensional system of Lennard-Jones particles with, say, $\varepsilon/kT \approx 1$, one gets a fairly good convergence after about $5 \times 10^5$

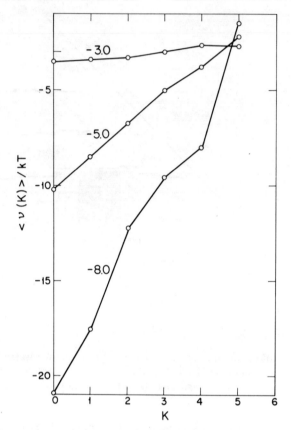

Fig. 6.29. The conditional average binding energy of particles as a function of their coordination number for waterlike particles with parameters given in (6.132). The HB energy $\varepsilon_H/kT$ is indicated next to each curve. Note that the slope increases when the hydrogen bond becomes stronger. For a weak hydrogen bond, the binding energy is almost independent of the coordination number. Compare with Fig. 5.9 for the spherical case.

configurations. For waterlike particles with, say, $\varepsilon_H/kT \approx -5.0$, however, one needs about three times as many configurations for the same degree of convergence.[15]

---

[15] This comment applies to the standard Monte Carlo procedure, in which a *single* particle is moved at each step. A possible improved procedure for strongly interacting particles would be to move a few particles simultaneously, so that the energy change at each step would result from both the forming and breaking of "bonds."

## 6.12. WATERLIKE PARTICLES IN THREE DIMENSIONS

Simulation of the behavior of water by waterlike particles in three dimensions has all the merits discussed in the previous section. In addition, this type of computation, which may be referred to as the ab initio approach to liquid water, is of importance in establishing the most appropriate effective pair potential for water molecules. On the other hand, simulations in the three-dimensional case vastly increase the computer time required to execute a typical computation. In particular, because of the strong attractive forces operating among water molecules, the convergence of the numerical methods is usually slower than in the case of particles with relatively weak attractive forces. This aspect was discussed in the previous section, but it pertains to the three-dimensional case equally well.

Three well-established techniques in the theory of simple fluids have been adapted recently to the study of waterlike particles in three dimensions. The effective pair potential which has been tried in all of the methods is essentially the one based on the Bjerrum model described in Section 6.4.

### 6.12.1. Application of the Monte Carlo Technique

Barker and Watts (1969) published a preliminary report on the computations of energy, heat capacity, and the radial distribution function for waterlike particles. The potential function used for these calculations is similar[16] to the one discussed in Section 6.4; however, instead of a smooth "switching function," they used a hard-sphere cutoff at 2 Å so that the point charges could not approach each other to zero separation.

The standard Monte Carlo technique was applied to 64 waterlike particles at 25°C. About $2\times10^5$ configurations were generated, and the average energy and heat capacity computed for this system are $-8.36$ kcal/mole and 20.5 cal/mole deg, compared with the experimental values of $-8.12$ kcal/mole and 18.0 cal/mole deg, respectively. The agreement was judged to be satisfactory. Figure 6.30 shows the computed radial distribution function and the experimental curve obtained by Narten *et al.* (1967). Here, agreement with the experimental results is quite poor. First, the average coordination number (computed for $0 \leq R \leq 3.5$ Å) was found to be 6.4, compared with the experimental value of about 4.4. More important, however, is the location of the second peak of $g(R)$, which was

---

[16] This potential was devised by Rowlinson (1951) for the computation of the virial coefficients of water.

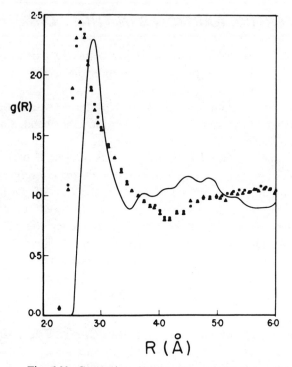

Fig. 6.30. Comparison between the experimental radial distribution function for water at 25°C (full curve) and the corresponding results from Monte Carlo computations (triangles and squares). [Redrawn with changes from Barker and Watts (1969).]

found to fall at about 5.8 Å, more than twice the distance of the first peak found at about 2.6 Å. This is far from the typical peak at 4.6 Å in the experimental curve, which was discussed at great length in Section 6.3.

It is difficult to trace the origin of these discrepancies between the computed and experimental radial distribution function. It is possible that the pair potential does not produce enough preference for the tetrahedral geometry (hence, the coordination number and the location of the second peak do not show the characteristic values as in water), or that the numerical procedure was not run with a sufficient number of particles and configurations. [Sixty four particles in three dimensions corresponds to four particles in one dimension, which is quite a small number. Furthermore, the number of configurations, of the order of $10^5$, is quite small to ensure convergence. See also the discussion in Section 6.11.] Some preliminary recent Monte Carlo results were reported by Popkie *et al.* (1973).

## 6.12.2. Application of the Percus–Yevick Equation

An approximate version of the Percus–Yevick (PY) equation has been applied for the pair potential based on the Bjerrum model (Ben-Naim, 1970a). The essence of the approximation is the following. We write the pair correlation function in the form

$$g(\mathbf{X}_1, \mathbf{X}_2) = y(\mathbf{X}_1, \mathbf{X}_2) \exp[-\beta U(\mathbf{X}_1, \mathbf{X}_2)] \qquad (6.134)$$

and assume that all of the angle dependence of $g$ is borne by the pair potential $U$, whereas the function $y$ is assumed to depend on the distance $R = |\mathbf{R}_2 - \mathbf{R}_1|$ only. This assumption has been partially justified in a two-dimensional system (Ben-Naim, 1972c,d).

We will not go through the details of the computations because they are quite lengthy and highly technical, and because the results obtained have been only partially successful. We present here one result that is of significance.

Consider the pair potential given in Section 6.4 [Eq. (6.20)], which we now write as

$$U(\mathbf{X}_1, \mathbf{X}_2; \lambda) = U_{\text{LJ}}(R_{12}) + \lambda S(R_{12}) U_{\text{HB}}(\mathbf{X}_1, \mathbf{X}_2) \qquad (6.135)$$

where $\lambda$, $0 \leq \lambda \leq 1$, is a coupling parameter. We begin the iteration process of solving the integral equation (see Appendix 9-E) by putting $\lambda = 0$ and solving for the Lennard-Jones potential $U_{\text{LJ}}(R_{12})$. Next, we increase the value of $\lambda$, as a result of which the contribution due to the HB potential increases. It is found that no convergent solution of the integral equation can be obtained with $\lambda = 1$. In fact, as one increases $\lambda$ to about $\lambda \approx 0.3$, convergence of the solution becomes extremely slow. Therefore, extending the computation beyond this value becomes unfeasible. Figure 6.31 shows one interesting feature of the coupling of the HB part of the pair potential. The dotted line is the solution for the Lennard-Jones potential (i.e., $\lambda = 0$). The other three curves show the response of the function $g(R)$ to increasing the value of $\lambda$ ($0 \leq \lambda \leq 0.3$). Note a shift of the second peak leftward from the initial value of 5.6 Å to the final value of 4.8 Å (at $\lambda \approx 0.3$). [The latter should be compared with the value of 4.6 Å in the experimental curve of $g(R)$.] This result indicates that as one couples the HB part of the potential, the tendency to form a tetrahedral geometry increases, and therefore the characteristic peak at about 4.6 Å is developed. Other features of the radial distribution function obtained from these computations are not in agreement with the experimental results. (For instance, the average coordination number obtained for the case $\lambda \approx 0.3$ was about 7.4, compared with the experimental value of about 4.4.)

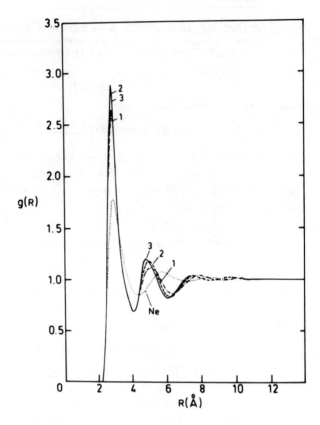

Fig. 6.31. The response of the radial distribution function $g(R)$ to increasing $\lambda$ in Eq. (6.135). The dotted curve corresponds to $\lambda = 0$. Curves 1, 2, and 3 correspond to increasing values of $\lambda$ ($0 \leq \lambda \leq 0.3$).

It is possible that the PY integral equation, though very successful for simple fluids, is inadequate for a system of particles interacting with very strong attractive forces. In addition, the solution of the full PY equation is still unfeasible for nonspherical particles, and only an approximate version of this equation has been applied in this work. [For more details, see Ben-Naim (1970a).]

### 6.12.3. Application of the Molecular Dynamics Method

The most successful and complete study of the behavior of a sample of waterlike particles in three dimensions has been carried out by Rahman and Stillinger (Rahman and Stillinger, 1971, 1973; Stillinger and Rahman, 1972). They applied the conventional molecular dynamics method for a system of 216 particles interacting via the effective pair potential based on the Bjerrum model (Section 6.4). The molecular dynamics method has an important advantage over the Monte Carlo method: Whereas the latter is devised for computing average quantities for a system in equilibrium only, in the former, it is possible to compute transport properties as well (see Section 2.7 for more details).

The sample of 216 particles (six particles per one dimension) was studied at a number density of $\varrho = 3.344 \times 10^{22}$ cm$^{-3}$ (or mass density of 1 g/cm$^3$). The typical time increment was chosen to be $\Delta t \approx 4.6 \times 10^{-16}$ sec, and altogether about $10^4$ steps were taken (half of that time was considered as the time required for the system to "age," and the averages were computed from the subsequent period of $5 \times 10^3 \, \Delta t$). The temperature was computed from the average kinetic energy of the molecules. (Note that the molecular dynamics method is concerned with a microcanonical system; the temperature is inferred from the average kinetic energy after the equilibrium distribution of the kinetic energy has been established.) The average potential energy corresponding to the system at a temperature of 34.3°C was found to be $-9.184$ kcal/mole, which compares nicely with the experimental value of $-9.84$. A less satisfactory value was obtained for the heat capacity. The average contribution to the heat capacity due to intermolecular interactions (which may be referred to as the structural part of the heat capacity; see also Chapter 5 and Section 6.8) was estimated between the temperatures $-8.2$ and 34.3°C to be 21.4 cal/mole deg. This should be compared with the experimental value of about 12 cal/mole deg for water. (The latter is obtained from a total heat capacity $C_V$ of about 18 cal/mole deg in this range of temperature, minus the contribution due to translation and rotation, which is $3R \approx 6$ cal/mole deg.) As in the other simulation techniques, it is difficult to trace the exact source of this discrepancy, whether from shortcomings of the model or from inaccuracy of the computations.

Rahman and Stillinger have carried out a very thorough investigation of the pair correlation function and its temperature dependence. Figure 6.32 shows the function $g_{OO}(R)$ for three temperatures (corresponding to $-8.2$, 34.3, and 314.8°C. These values correspond to the potential function as given in Section 6.4. (In the course of this study, it was found that a rescaling of the potential by a factor of about 1.06 improves the results. Such a rescaling of the potential function also affects the computed temperature of the system under consideration.)

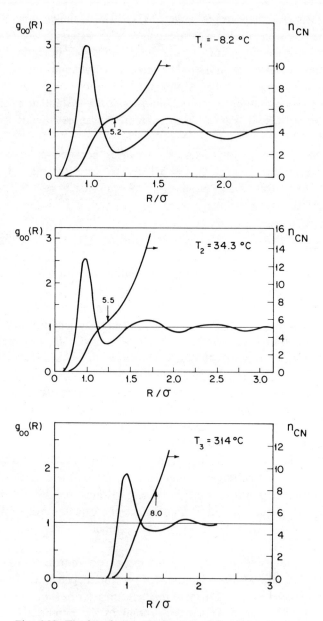

Fig. 6.32. The function $g_{00}(R)$ for three different temperatures, as indicated next to each curve, computed by the method of molecular dynamics. The monotonic-increasing curves give the "running coordination number" $n_{CN}$ [see (6.136)]. [Redrawn with changes from Stillinger and Rahman (1972).]

# Liquid Water

All of the features of the radial distribution function were simulated by the model particles. There is a sharp first peak at $R^* = R/\sigma = 0.975$, corresponding to $R = 2.75$ Å (with $\sigma = 2.82$ Å). The second peak occurs at $R^* = 1.69$, corresponding to $R = 4.76$ Å, which is a little above the experimental value of about $R = 4.6$ Å. Also, the average coordination number, computed up to the first minimum (following the first maximum), is 5.5, which is somewhat higher than the experimental value of about 4.4. Figure 6.32 also shows the "running coordination number," i.e., the function

$$n_{\text{CN}}(R) = \varrho \int_0^R g_{\text{OO}}(r) 4\pi r^2 \, dr \qquad (6.136)$$

The temperature dependence of the radial distribution function $g_{\text{OO}}(R)$ obtained from these computations was also in qualitative agreement with the experimental results (see Section 6.3). We note especially that at the highest temperature ($t = 314.8°C$), the second peak of $g_{\text{OO}}(R)$ has decreased appreciably and a clear-cut shift to the right is observed, indicating a destruction of the characteristic tetrahedral geometry of the packing in liquid water. One feature that further supports this conclusion is the average coordination number, computed up to the first minima, which is 5.2, 5.5, and 8.0 at the three temperatures, respectively. For comparison, the computed curve for argon (Rahman, 1964) is shown in Fig. 6.33. Both the location of the second peak and the average coordination number

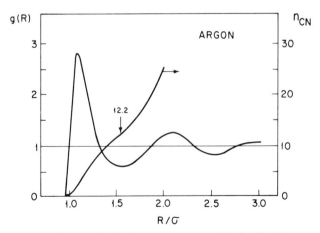

Fig. 6.33. The radial distribution function $g(R)$ for liquid argon computed by Rahman (1964) by the method of molecular dynamics. The Lennard-Jones parameters $\varrho\sigma^3 = 0.81$ and $\varepsilon/kT = 1.35$. [Redrawn with changes from Rahman and Stillinger (1971).]

reflect the "normal" packing of simple molecules, contrasting the behavior of water.

There is a wealth of other information, on both equilibrium and kinetic properties, of this sample of waterlike particles obtained by the molecular dynamics computations which are not reproduced here. The interested reader is referred to the original articles for further details.

## 6.13. CONCLUSION

We have described two main lines of development in the theory of liquid water. The first, and the older, was founded on the mixture-model approach (Chapter 5) to liquids, which offers certain approximate or ad hoc models for the fluids as a whole. The second approach may be referred to as the ab initio method, based on first principles of statistical mechanics. In the past, these two lines of development were thought to be conflicting and a vigorous debate has taken place on this issue. As we have stressed throughout this and the previous chapter, both approaches can be developed from first principles, and, in fact, provide complementary information on this liquid. Once we attempt to pursue this theory along either route, we must introduce serious approximations. Therefore, it is very difficult to establish a clear-cut preference for one approach or the other.

The term ab initio as applied to various computations on a sample of waterlike particles is somewhat misleading if it is understood in the same sense that it is used in quantum chemistry. In the latter, one undertakes a computation based on first principles (of quantum mechanics) on a well-defined system of particles (the electrons and the nuclei of the molecule), the identity of which is rarely a subject of controversy. A similar situation occurs for simple fluids, where again one uses a model of particles which is believed to be quite realistic. The situation in water is that we are still in the early stages of characterizing the elementary particles constituting our system. It is for this reason that we have consistently referred in this chapter to "waterlike" particles rather than to water molecules. The question still remains: To what extent are these particles good representatives of water molecules? The main source of difficulty is, no doubt, our incomplete knowledge of the analytical form of the pair potential and the uncertainty as to the role of higher-order potentials in determining the behavior of liquid water.

Another general comment regarding the two approaches is their ability to "explain" the anomalous behavior of liquid water and its solutions. We have seen that the mixture-model approach provides a qualitative explanation of some of the outstanding properties of water; this will be demonstrated in the case of solutions as well. The results of the computer simulations are, after all, a sort of experiment on waterlike particles. Therefore, it must be borne in mind that performing such an experiment either in the laboratory or by a computer tells us *what* water is, not *why* it behaves in a particular manner. These reflections lead us to the conclusion that even when all the details of the potential function are solved and all the computations performed on the computer provide results of the utmost accuracy, there will still be a need for a simple, qualitative, and visualizable interpretation of the unusual properties of aqueous fluids. No doubt, this will be furnished by the mixture-model approach.

*Chapter 7*
# Water with One Simple Solute Particle

## 7.1. INTRODUCTION

The study of very dilute solutions of simple nonpolar solutes is of interest for various reasons. First, these solutions reveal some anomalous properties in comparison with corresponding nonaqueous solutions, and have therefore presented an attractive challenge to scientists. Second, the study of dilute aqueous solutions, viewed as systems deviating slightly from pure water, provides some further information which is helpful in the study of water itself. Finally, understanding the thermodynamics of these systems is a first, indispensable step in the study of the problem of hydrophobic interaction, a topic of prominence especially in biochemical systems and to which we devote the next chapter.

Regarding the theoretical development of the study of dilute aqueous solutions, one cannot expect the situation to be better than that for pure water. In fact, progress in this field is contingent upon progress in the theory of water. Therefore, it is no wonder that most of the theories developed for these systems are rather speculative. It is difficult to make a critical assessment of the various assumptions and approximations that have been introduced, either explicitly or implicitly, in these theories.

In this chapter, we survey some of the theories that have been suggested for dilute aqueous solutions. As in the previous chapter, we elaborate on the molecular foundations of each of the approaches, and discuss some of the approximations that are introduced, in order to pursue the theory to the point where thermodynamic results can be computed.

Consider a dilute solution of $N_S$ molecules (or moles) of solute $S$ and $N_W$ molecules (or moles) of water $W$ at a given temperature $T$ and pressure $P$. In most of this chapter, we use these variables, since they are the most common ones in practice. We assume that the system is macroscopic, but that $N_S \ll N_W$. Suppose we are interested in the response of the volume of the system to an addition of solute. We write the expansion ($T, P, N_W$ constant)

$$V(T, P, N_W, N_S) = V(T, P, N_W) + \left(\frac{\partial V}{\partial N_S}\right)N_S + \frac{1}{2}\left(\frac{\partial^2 V}{\partial N_S^2}\right)N_S^2 + \cdots \quad (7.1)$$

where $V(T, P, N_W)$ is the volume of pure water, and the successive terms bear the corrections to the volume produced by addition of the solute.

We define the partial molar[1] volume at infinite dilution (i.e., at the limit of zero solute concentration) by

$$\bar{V}_S^\circ = \lim_{N_S \to 0} \left(\frac{\partial V}{\partial N_S}\right)_{T,P,N_W} \quad (7.2)$$

Thus, the expansion (7.1) to first order in $N_S$ is written as

$$V(T, P, N_W, N_S) = V(T, P, N_W) + \bar{V}_S^\circ N_S + \cdots \quad (7.3)$$

We see that the study of the first-order deviation of the volume of pure water is equivalent to the study of the partial molar volume of the solute at infinite dilution. Similarly, we may study the first-order expansion for, say, the energy, heat capacity, compressibility, etc.:

$$E(T, P, N_W, N_S) = E(T, P, N_W) + \bar{E}_S^\circ N_S + \cdots \quad (7.4)$$

$$C_P(T, P, N_W, N_S) = C_P(T, P, N_W) + \bar{C}_{P,S}^\circ N_S + \cdots \quad (7.5)$$

$$\varkappa_T(T, P, N_W, N_S) = \varkappa_T(T, P, N_W) + \bar{\varkappa}_{T,S}^\circ N_S + \cdots \quad (7.6)$$

where the coefficients of the linear terms are defined as the infinite dilution limit of the corresponding derivative, as in (7.2).

It turns out that some of these coefficients do not exist. For instance, if we were to expand the Gibbs free energy as in (7.4)–(7.6), we should write

$$G(T, P, N_W, N_S) = G(T, P, N_W) + \left(\frac{\partial G}{\partial N_S}\right)_{T,P,N_W} N_S + \cdots \quad (7.7)$$

---

[1] Note again that we refer to partial *molar* quantities, although in some cases we are discussing the corresponding partial molecular quantities.

Water with One Simple Solute Particle    311

But the coefficient of the "linear" term diverges and, in fact, we know precisely the way it diverges (see Chapter 4 for more details)

$$\mu_S = \left(\frac{\partial G}{\partial N_S}\right)_{T,P,N_W} \xrightarrow{N_S \to 0} \mu_S^\circ(T,P) + kT \ln \frac{N_S}{V} \quad (7.8)$$

where $V$ is the (average) volume of the system and $\mu_S^\circ(T,P)$ is the standard chemical potential, introduced in Chapter 4.

As $N_S \to 0$, the chemical potential diverges to minus infinity and the expansion (7.7) is invalid.[2] Nevertheless, the situation is not too bad since we know that the divergent term in (7.8) is the same independent of the type of solute or solvent. This enables us to circumvent the difficulty by various means.

1. Instead of studying the limiting behavior of $\mu_S$ as $N_S \to 0$, we can remove the singular part and study the limiting behavior of the quantity $\mu_S - kT \ln(N_S/V)$, which is simply the standard chemical potential

$$\mu_S^\circ = \lim_{N_S \to 0} [\mu_S - kT \ln(N_S/V)]$$

In fact, the above relation is the only proper definition of the standard chemical potential within the realm of thermodynamics.

2. A different and usually more practical way of overcoming the difficulty is to look at differences in chemical potentials of the same solute in different phases. Since the singular part of the chemical potential is the same in the two phases, we can perform an expansion similar to (7.7) but for the difference of the Gibbs free energy of the two phases, in which case the coefficient of the linear term is finite. For instance, if we have two solvents $W_1$ and $W_2$, we can write

$$G(T,P,N_{W_1},N_S) - G(T,P,N_{W_2},N_S)$$
$$= G(T,P,N_{W_1}) - G(T,P,N_{W_2}) + (\mu_{S,1}^\circ - \mu_{S,2}^\circ)N_S + \cdots \quad (7.9)$$

where $\mu_{S,1}^\circ$ and $\mu_{S,2}^\circ$ are the standard chemical potentials of $S$ in the two solvents.

3. A third possibility, though uncommon, has some theoretical advantages. Instead of adding solute molecules to the system and examining the change in the thermodynamic quantities, we add a single solute molecule

---

[2] Note, however, that the product $\mu_S N_S$ tends to zero as $N_S \to 0$. Indeed,

$$\lim_{N_S \to 0} (N_S \ln N_S) = \lim_{N_S \to 0} \frac{\partial (\ln N_S)/\partial N_S}{\partial (N_S^{-1})/\partial N_S} = \lim_{N_S \to 0} \frac{N_S^{-1}}{-N_S^{-2}} = 0$$

to a *fixed* position and seek the corresponding change in the thermodynamic quantities. The chemical potential in, say (7.7), is now replaced by the pseudo-chemical potential, and therefore no singularities occur in the new expansion

$$G(T, P, N_W, N_S = 1; \mathbf{R}_S) = G(T, P, N_W) + \bar{\mu}_S \qquad (7.10)$$

Note that (7.10) has been used in Section 3.6 to *define* the pseudo-chemical potential. The only difference is that we are now concerned with the $T, P, N$ ensemble. (See also Appendix 9-F.)

In most practical cases, one encounters differences of chemical potentials; thus the difficulty of divergence does not arise. We also recall that the difference in the standard chemical potential of $S$ in two solvents equals the difference in the pseudo-chemical potential of $S$ in these two solvents.

There are essentially two approaches to the theory of aqueous solutions of simple solutes; both split the process of dissolution into two parts, which is believed to simplify the interpretation of the corresponding thermodynamic quantities. The first method consists in creating a suitable cavity for accommodating the solute and then introducing the solute into the cavity. (This process is meaningful only when the two steps are carried out at a fixed position. See Section 3.11.) The second splitting is induced by the mixture-model approach to the theory of liquid water. It consists first in introducing the particle to a "frozen-in" system, and then letting the system relax to its new equilibrium state. This splitting can be carried out for either a free solute particle or a solute at a fixed position in the solute. We shall develop both theories in the subsequent sections.

## 7.2. SURVEY OF EXPERIMENTAL OBSERVATIONS

In this section, as in most of the remaining sections of this chapter, we confine ourselves to the discussion of aqueous solutions of the simplest solutes, e.g., inert gases or simple hydrocarbons.

Eley (1939) was the first to investigate systematically the anomalous behavior of aqueous solutions of inert gases. We shall describe in Section 7.3 his contribution to the theory of these systems. Here, we survey the experimental facts. Only representative examples are given to illustrate the outstanding behavior of aqueous solutions. [A review article concerned mainly with the experimental facts has been published by Battino and Clever (1966). Other reviews are by Ben-Naim (1972e) and Franks (1973a).]

1. The solubility of a gas will always be given in terms of the Ostwald absorption coefficient. This is defined as follows: Let $V^g$ be the volume of the gas (at a given $T$ and $P$) dissolved in a given volume $V^l$ of liquid (at the same $T$ and $P$). The Ostwald absorption coefficient is defined by

$$\gamma = V^g/V^l \qquad (7.11)$$

This is a practical definition in terms of directly measurable quantities. A more useful form, equivalent to (7.11), is obtained as follows. Let $\varrho_S{}^l$ and $\varrho_S{}^g$ be the molar concentration of the solute $S$ in the liquid and gaseous phases at equilibrium, respectively. Then, since the number of moles of solute in the liquid $\varrho_S{}^l V^l$ originated from the gaseous phase, we have the equality

$$\varrho_S{}^g V^g = \varrho_S{}^l V^l \qquad (7.12)$$

Hence, from (7.11) and (7.12), we get

$$\gamma = (\varrho_S{}^l/\varrho_S{}^g)_{\text{eq}} \qquad (7.13)$$

where we stress the fact that the ratio of the solute concentrations is taken at *equilibrium*. In the last form, $\gamma$ is viewed as a distribution coefficient of $S$ between the two phases. This form is directly related to the standard free energy of solution (see Chapter 4 for more details)

$$\Delta\mu_S{}^\circ(g \to l) = -kT \ln(\varrho_S{}^l/\varrho_S{}^g)_{\text{eq}} \qquad (7.14)$$

There exists a variety of other "units" by which solubility data may be reported; these and the interconversion formulas are discussed in the literature (Hildebrand and Scott, 1950; Friend and Adler, 1957; Himmelblau, 1959; and Battino and Clever, 1966). In this chapter, we adhere to the application of the Ostwald absorption coefficient as a measure of the solubility.

Figure 7.1 shows the solubility of argon (at 25°C and 1 atm) in water and in a series of alcohols. The important point to be noted is that the solubility of argon in water cannot be obtained by extrapolation from a series of alcohols using the molecular weight as an extrapolation parameter. It is clear that the solubility of argon in water is markedly smaller than in the alcohols. It should be noted, though, that the low solubility of gases in water (in terms of $\gamma$) is not a property of water exclusively; there are other liquids in which the solubility of argon is of the same order of mag-

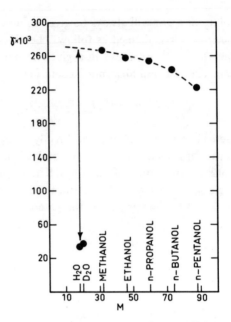

Fig. 7.1. Ostwald absorption coefficient ($\gamma \times 10^3$) for argon at 25°C in $H_2O$, $D_2O$, and some alcohols, as a function of the molecular weight of the solvent. The anomalous low solubility in water is indicated by the vertical line connecting the experimental value with the extrapolated value. [Redrawn from Ben-Naim (1972e).]

nitude [e.g., the value of $\gamma$ for argon in ethylene glycol at 25°C is about 0.035, almost the same as the value for argon in water at this temperature (Ben-Naim, 1968)]. Further data on the solubility of argon in water and organic liquids are provided in Table 7.1.

2. The second striking difference between the solubility of gases in water and in other solvents is the temperature dependence of the solubility. Figure 7.2 illustrates that difference for argon in water and in methanol. The steep decrease in the solubility (in terms of $\gamma$) as a function of temperature is very characteristic of water. In other solvents, the solubility may either increase or decrease with temperature; in both cases, it occurs with a relatively small slope. Figure 7.3 includes some further information on the temperature dependence of the solubility of methane (in terms of $\Delta \mu_S^\circ$) in water and in a few other solvents. The difference in the temperature dependence of the solubility is also discernible from Table 7.1.

## Table 7.1
## Ostwald Absorption Coefficient $\gamma$ for Argon in Water and in Some Organic Liquids at Two Temperatures[a]

| Solvent | 15°C | 25°C |
|---|---|---|
| Benzene | 0.232 | 0.240 |
| Cyclohexane | 0.330 | 0.334 |
| n-Hexane | 0.474 | 0.472 |
| n-Heptane | 0.411 | 0.415 |
| n-Octane | 0.355 | 0.367 |
| 3-Methylheptane | 0.360 | 0.377 |
| 2,3-Dimethylhexane | 0.381 | 0.377 |
| 2,4-Dimethylhexane | 0.400 | 0.400 |
| 2,2,4-Trimethylpentane | 0.436 | 0.431 |
| n-Nonane | 0.340 | 0.338 |
| n-Decane | 0.315 | 0.311 |
| n-Dodecane | 0.271 | 0.275 |
| n-Tetradecane | 0.247 | 0.250 |
| Fluorobenzene | 0.291 | 0.298 |
| Chlorobenzene | 0.202 | 0.204 |
| Bromobenzene | 0.153 | 0.157 |
| Iodobenzene | 0.104 | 0.109 |
| Toluene | 0.240 | 0.249 |
| Nitrobenzene | 0.100 | 0.105 |
| Water | 0.0396 | 0.0341 |

[a] Data taken from Clever et al. (1957), Saylor and Battino (1958), and Ben-Naim (1965b).

It is found (Himmelblau, 1959) that as the temperature increases beyond, say 100°C, the solubility of the gas in water passes through a minimum and then takes a positive slope, as in "normal" solvents.

The temperature dependence of the solubility is better expressed in terms of the standard entropy of solution. In Chapter 4, we introduced two definitions of the standard entropy of solutions corresponding to the processes I and II. The pertinent relations are

$$\Delta S_S^\circ(\text{I}) = -\frac{\partial}{\partial T}[\Delta \mu_S^\circ(\text{I})] - kT\frac{\partial}{\partial T}\left(\ln \frac{\varrho^g}{\varrho^l}\right) \quad (7.15)$$

$$\Delta S_S^\circ(\text{II}) = -\frac{\partial}{\partial T}[\Delta \mu_S^\circ(\text{I})] \quad (7.16)$$

where $\varrho^g$ and $\varrho^l$ are the densities of the two "solvents" in the gaseous and

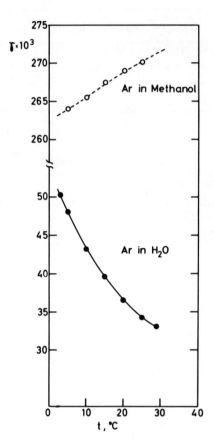

Fig. 7.2. Temperature dependence of $\gamma$ for argon in water and in methanol. [Redrawn from Ben-Naim (1972e).]

liquid phases, respectively. Note that since the differentiations in (7.15) and (7.16) are at constant pressure and number of particles, the second term on the rhs of (7.15) is essentially the difference in the coefficient of thermal expansion of the two phases (for more details, see Section 4.11).

Table 7.2 presents some thermodynamic data for the solution of methane in water and in nonaqueous solvents. The most outstanding feature is the relatively large negative entropy and enthalpy of solution in water. In all of the following theoretical developments, we refer to the values of $\Delta S_S^{\circ}(II)$ and $\Delta H_S^{\circ}(II)$ as the standard entropy and enthalpy of solution, respectively. The connecting relations to the more common standard states are given in Section 4.11.

# Water with One Simple Solute Particle 317

The continuous passage from water to a "normal" solvent is exemplified in Figs. 7.4–7.6, where the free energy, the entropy, and the enthalpy of a solution of methane are plotted as a function of the mole fraction of ethanol. Note that the anomalously low entropy and enthalpy of solution are confined to the water-rich region of the water–ethanol mixtures.

3. The partial molar volume of gases (at infinite dilution) is usually smaller in water than in other liquids. Table 7.3 presents a few representative examples of these quantities.

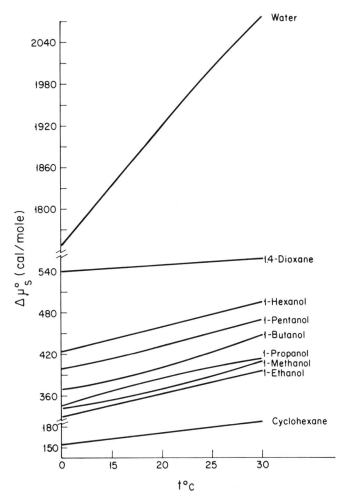

Fig. 7.3. Temperature dependence of $\Delta \mu_s^\circ$ for methane in water and some nonaqueous solvents.

## Table 7.2

**Values of the Standard Free Energy, Entropy, Enthalpy, and Heat Capacity of Methane in Water and in Some Nonaqueous Solvents at Two Temperatures**[a]

| Solvent | t, °C | $\Delta \mu_S°$, cal/mole | $\Delta S_S°$(II), e.u. | $\Delta H_S°$(II), cal/mole | $\Delta C_S°$(II), cal/mole deg |
|---|---|---|---|---|---|
| Water | 10 | 1747 | −18.3 | −3430 | 53 |
|  | 25 | 2000 | −15.5 | −2610 |  |
| Heavy water | 10 | 1703 | −19.2 | −3740 | 52 |
| ($D_2O$) | 25 | 1971 | −16.5 | −2940 |  |
| Methanol | 10 | 343 | −2.6 | −390 | −21 |
|  | 25 | 390 | −3.7 | −710 |  |
| Ethanol | 10 | 330 | −3.2 | −570 | −5 |
|  | 25 | 380 | −3.5 | −650 |  |
| 1-Propanol | 10 | 345 | −4.3 | −880 | 25 |
|  | 25 | 400 | −3.0 | −500 |  |
| 1-Butanol | 10 | 369 | −2.8 | −420 | −33 |
|  | 25 | 430 | −4.5 | −910 |  |
| 1-Pentanol | 10 | 399 | −3.3 | −530 | −7 |
|  | 25 | 450 | −3.6 | −630 |  |
| 1,4-Dioxane | 10 | 538 | −0.8 | +310 | −6 |
|  | 25 | 553 | −1.1 | +220 |  |
| Cyclohexane | 10 | 154 | −1.9 | −390 | 11 |
|  | 25 | 179 | −1.4 | −230 |  |

[a] All values refer to process II (Section 4.11), i.e., the process of transferring one solute from a fixed position in the gas to a fixed position in the liquid at constant $T$ and $P$. [Data from Yaacobi and Ben-Naim (1974).]

4. The partial molar heat capacity of gases is usually larger in water than in other solvents. This quantity is obtained from the second derivative of experimental curves and therefore is generally not very accurate. Nevertheless, the difference between water and other solvents is considered to be quite clear-cut. As an example, the partial molar heat capacity of argon in water at room temperature is about 50 cal/mole deg, whereas in ethanol, methanol, or $p$-dioxane, it is almost zero. Table 7.2 includes some information on the partial molar heat capacity of methane in water and in nonaqueous solvents.

# Water with One Simple Solute Particle

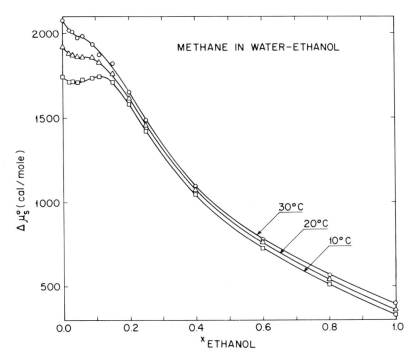

Fig. 7.4. Standard free energy of solution $\Delta\mu_s^\circ$ (in cal/mole) of methane as a function of mole fraction of ethanol at three temperatures.

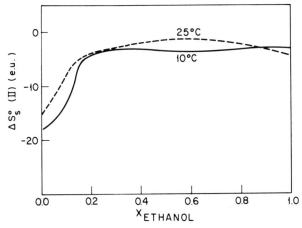

Fig. 7.5. Standard entropy of solution $\Delta S_s^\circ(\mathrm{II})$ (in cal/mole deg) of methane as a function of mole fraction of ethanol at two temperatures.

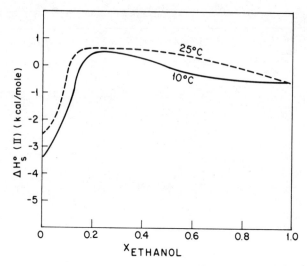

Fig. 7.6. Standard enthalpy of solution $\Delta H_S°(II)$ (in kcal/mole) of methane as a function of mole fraction of ethanol at two temperatures.

We have presented in this section only a very limited number of examples from a vast literature on this subject. The purpose has been to familiarize the reader with some of the basic facts on these systems, not to survey the literature. A more complete source of information, with references to original contributions in this field, can be found in Franks' books (1973a,b).

Table 7.3

**Partial Molar Volume (cm$^3$/mole) of Argon, Methane, and Ethane in Water, Benzene, and Carbon Tetrachloride at 25°C[a]**

| Solvent | Ar | CH$_4$ | C$_2$H$_6$ |
|---|---|---|---|
| Water | 32 | 37 | 41 |
| Benzene | 43 | 52 | 73 |
| Carbon tetrachloride | 44 | 52 | 61 |

[a] Data from Horiuti (1931), Hildebrand and Scott (1950), Namiot (1961), and Smith and Walkley (1962).

## 7.3. "HARD" AND "SOFT" PARTS OF THE DISSOLUTION PROCESS

Perhaps the earliest attempt to interpret the anomalous properties of aqueous solutions of inert gases was undertaken by Eley (1939, 1944). [Earlier work of Körösy (1937) and Uhlig (1937) should also be noted.] Eley suggested splitting each thermodynamic quantity associated with the dissolution process into two parts: forming a cavity, and then introducing the solute into this cavity. At the very outset, it is important to stress that the formation of a cavity, in the present context, is understood to refer to a *fixed* position (more details are given in Section 3.11). Otherwise, most of the quantities associated with the creation of a cavity are rendered meaningless.

The interpretation of the large, negative enthalpy (or energy[3]) of solution of gases in water is interpreted by Eley as follows. In normal fluids, one assumes that the energy of formation of a cavity is positive, i.e., one has to invest energy in order to push away the solvent molecules from the cavity. The energy of introducing the solute into the cavity is negative, since now the interaction between the solute and its environment is attractive. Furthermore, if we assume that the two processes require energies of a similar order of magnitude, the resulting enthalpy of solution would have a small positive or negative value.

In water, Eley postulated that the energy required to build up a cavity is almost zero, since "natural" cavities are already present in the open structure of water. (This is true within the lattice-type model used by Eley, or if one refers to a particular hole in the solid ice. The argument is not valid for a liquid, where the "natural" holes cannot be taken for the cavities at *fixed* positions; see also Section 3.11.) Hence, the only contribution to the energy of solution is expected to come from the interaction between the solute and the cavity, which is assumed to be negative. This is Eley's explanation of the negative value of the standard enthalpy of solution of inert gases in water. Eley also attempted to explain the large, negative entropy of solution by assuming that the gas molecules are distributed in a restricted number of holes in water. A similar computation based on an interstitial model for water will be discussed in Section 7.5.

[3] The difference between the standard energy and enthalpy of solution for process II (see Section 7.2) is quite small. As an example, for argon in water at 25°C, the standard enthalpy of solution is $\Delta H_S{}^\circ(\text{II}) \approx -2000$ cal/mole. The value of $P \Delta V_S{}^\circ(\text{II})$ at 1 atm is of the order of about 0.8 cal/mole. Therefore, the difference between $\Delta H_S{}^\circ(\text{II})$ and $\Delta E_S{}^\circ(\text{II})$ is, in general, negligible.

We now turn to a more detailed elaboration of the two parts of the dissolution process from the molecular point of view. The importance of this particular splitting is twofold. First, substantial evidence exists showing that the anomalous properties of water are already revealed in the first step, the creation of a cavity. Second, powerful theoretical tools have been developed recently to compute the work required to create such a cavity. This will be described later in this section.

For convenience, we consider a system in the $T$, $V$, $N$ ensemble. (A similar treatment can be carried out in the $T$, $P$, $N$ ensemble; see, for example, Appendix 9-F.) The chemical potential of the solute $S$ at infinite dilution in water is given by

$$\mu_S = A(T, V, N, N_S = 1) - A(T, V, N) \qquad (7.17)$$

for which the statistical mechanical expression is

$$\exp(-\beta\mu_S) = Q(T, V, N, N_S = 1)/Q(T, V, N) \qquad (7.18)$$

As in Section 3.5, we introduce a coupling parameter $\xi$ defined by

$$U(\xi) = U_N(\mathbf{X}_1, \ldots, \mathbf{X}_N) + \xi \sum_{j=1}^{N} U(\mathbf{R}_S, \mathbf{X}_j) \qquad (7.19)$$

where $\mathbf{X}_i$ denotes the configuration of the $i$th water molecule ($i = 1, 2, \ldots, N$) and $\mathbf{R}_S$ is the location of the solute particle $S$. We assume that the solute particle is spherical and that its internal degrees of freedom do not contribute to the partition function of the system (i.e., we put $q_S = 1$). $U(\mathbf{R}_S, \mathbf{X}_i)$ is the solute–solvent pair potential.

Using this coupling parameter, we can rewrite the chemical potential of the solute as

$$\mu_S = kT \ln(\varrho_S \Lambda_S^3) + \int_0^1 d\xi \int d\mathbf{X}\, U(\mathbf{R}_S, \mathbf{X}) \varrho(\mathbf{X}/\mathbf{R}_S, \xi) \qquad (7.20)$$

In this particular case, we have $\varrho_S = V^{-1}$, since only one solute particle is present in the system.

It is clear that insofar as we are interested in the solvation properties of the system, the first term on the rhs of (7.20) carries no relevant information. In fact, it would be the same for different solvents at the same solute density $\varrho_S$. Therefore, we focus our attention on the pseudo-chemical potential, defined by

$$\bar{\mu}_S = A(T, V, N, N_S = 1, \mathbf{R}_S) - A(T, V, N) \qquad (7.21)$$

## Water with One Simple Solute Particle

The corresponding expression for $\bar{\mu}_S$ in terms of molecular distribution functions is

$$\bar{\mu}_S = \int_0^1 d\xi \int d\mathbf{X}\, U(\mathbf{R}_S, \mathbf{X})\varrho(\mathbf{X}/\mathbf{R}_S, \xi) \tag{7.22}$$

We recall that the integral in (7.22) is obtained by coupling the solute particles at the fixed position $\mathbf{R}_S$ to the rest of the system. The coupling process is accomplished by increasing the parameter $\xi$ from zero to unity. We now repeat the same coupling process, as in Sections 3.5 and 3.6, but in two steps, corresponding to the formation of the cavity and introduction of the solute into it.

Let us assume that the solute–solvent pair potential $U(\mathbf{R}_S, \mathbf{X}_W)$ can be split into two parts,

$$U(\mathbf{R}_S, \mathbf{X}_W) = U^H(R) + U^S(\mathbf{R}_S, \mathbf{X}_W) \tag{7.23}$$

where $R = |\mathbf{R}_W - \mathbf{R}_S|$ is the distance between the centers of the solute and the solvent molecules. $U^H(R)$ is the "hard" part of the pair potential, defined by

$$U^H(R) = \begin{cases} \infty & \text{for } R \leq \sigma_{SW} \\ 0 & \text{for } R > \sigma_{SW} \end{cases} \tag{7.24}$$

In order for the definition (7.24) to be valid, we must assume the existence of a hard-core diameter $\sigma_{SW}$, which is effectively the distance of minimum approach. We have also assumed that this "hard" part of the potential is dependent only on the distance $R$ and not on the relative orientation of the pair of particles. Once $U^H(R)$ has been defined, the "soft" part of the pair potential is defined through (7.23) [assuming that the total pair potential $U(\mathbf{R}_S, \mathbf{X}_W)$ is given]. The latter may, in general, be dependent on the orientation of the solvent molecule. With a judicious choice of the hard-core diameter [say, by taking $\sigma_{SW} = (1/2)(\sigma_S + \sigma_W)$, with $\sigma_S$ and $\sigma_W$ the effective hard-core diameter of the solute and solvent molecules, respectively], we can get a soft potential which is attractive for most of the configurations of the pair of particles.

We now use a double coupling process as follows. We define an auxiliary potential function [analogous to (7.19)] by

$$U(\xi_1, \xi_2) = U_N(\mathbf{X}_1, \ldots, \mathbf{X}_N) + \xi_1 \sum_{i=1}^{N} U^H(\mathbf{R}_S, \mathbf{R}_i) + \xi_2 \sum_{i=1}^{N} U^S(\mathbf{R}_S, \mathbf{X}_i) \tag{7.25}$$

The state $\xi_1 = \xi_2 = 0$ corresponds to the uncoupled particle [i.e., $\xi = 0$ in

(7.19)], and the state $\xi_1 = \xi_2 = 1$ corresponds to the fully coupled particle [i.e., $\xi = 1$ in (7.19)]. With the potential function of (7.25), we define the auxiliary configurational partition function

$$Z(\xi_1, \xi_2) = \int \cdots \int d\mathbf{X}^N \exp[-\beta U(\xi_1, \xi_2)] \qquad (7.26)$$

Note that integration is carried out only over all the configurations of the solvent particles; the position of the solute is assumed to be fixed at $\mathbf{R}_S$.

The ordinary coupling process employs one parameter which "builds up" the solute particle into the system. Here, instead, we build up the particle in two steps; first, we couple only the hard part of the pair potential, and *then* complete the process by coupling the soft part of the pair potential.

The corresponding mathematical steps are similar to the ones carried out in Sections 3.5 and 3.6; therefore we omit excessive details.

We start with the expression for the pseudo-chemical potential of the solute

$$\exp(-\beta \bar{\mu}_S) = \frac{\int \cdots \int d\mathbf{X}^N \exp[-\beta U(\xi_1 = 1, \xi_2 = 1)]}{\int \cdots \int d\mathbf{X}^N \exp[-\beta U(\xi_1 = 0, \xi_2 = 0)]} \qquad (7.27)$$

After rearranging, we get

$$\begin{aligned}
-\beta \bar{\mu}_S &= \ln Z(\xi_1 = 1, \xi_2 = 1) - \ln Z(\xi_1 = 0, \xi_2 = 0) \\
&= [\ln Z(\xi_1 = 1, \xi_2 = 0) - \ln Z(\xi_1 = 0, \xi_2 = 0)] \\
&\quad + [\ln Z(\xi_1 = 1, \xi_2 = 1) - \ln Z(\xi_1 = 1, \xi_2 = 0)] \\
&= \int_0^1 d\xi_1 \frac{\partial \ln Z(\xi_1, \xi_2 = 0)}{\partial \xi_1} + \int_0^1 d\xi_2 \frac{\partial \ln Z(\xi_1 = 1, \xi_2)}{\partial \xi_2}
\end{aligned} \qquad (7.28)$$

In the second form on the rhs of (7.28), we have added and subtracted the quantity $\ln Z(\xi_1 = 1, \xi_2 = 0)$, and in the third form, we have used the definition of the integral of a derivative. We now carry out the differentiation of $Z(\xi_1, \xi_2)$ in (7.26) with respect to $\xi_1$ and $\xi_2$. The steps are identical to those of Section 3.5, the result being

$$\frac{\partial \ln Z(\xi_1, \xi_2 = 0)}{\partial \xi_1} = -\beta \int d\mathbf{X}_W \, U^H(\mathbf{R}_S, \mathbf{R}_W) \varrho(\mathbf{X}_W/\mathbf{R}_S, \xi_1, \xi_2 = 0)$$

$$\frac{\partial \ln Z(\xi_1 = 1, \xi_2)}{\partial \xi_2} = -\beta \int d\mathbf{X}_W \, U^S(\mathbf{R}_S, \mathbf{X}_W) \varrho(\mathbf{X}_W/\mathbf{R}_S, \xi_1 = 1, \xi_2)$$

(7.29)

Combining with (7.28), we get the final expression

$$\bar{\mu}_S = \int_0^1 d\xi_1 \int d\mathbf{X}_W \, U^H(\mathbf{R}_S, \mathbf{R}_W) \varrho(\mathbf{X}_W/\mathbf{R}_S, \xi_1, \xi_2 = 0)$$

$$+ \int_0^1 d\xi_2 \int d\mathbf{X}_W \, U^S(\mathbf{R}_S, \mathbf{X}_W) \varrho(\mathbf{X}_W/\mathbf{R}_S, \xi_1 = 1, \xi_2) \qquad (7.30)$$

It is important to bear in mind the order of the coupling processes. First, we increase $\xi_1$ from zero to unity, keeping $\xi_2 = 0$ (i.e., the soft part is "switched off") and *then* we increase $\xi_2$ from zero to unity, after $\xi_1$ has attained its maximal value $\xi_1 = 1$.

The singlet conditional distribution functions in (7.29) and (7.30) have an obvious meaning. For instance, $\varrho(\mathbf{X}_W/\mathbf{R}_S, \xi_1, \xi_2 = 0)$ denotes the singlet conditional distribution function for solvent molecules, given a solute particle at $\mathbf{R}_S$ coupled to the extent $\xi_1$ and $\xi_2 = 0$.

Now we can split the total work of introducing a solute particle into the system into three parts:

$$\bar{\mu}_S = \bar{\mu}_S{}^H + \bar{\mu}_S{}^S + kT \ln(\varrho_S \Lambda_S{}^3) \qquad (7.31)$$

The first term on the rhs of (7.31) corresponds to the work of creating a cavity at a fixed position; the *size* of the cavity is determined by the hard-core diameter that we have chosen in (7.24). The second term corresponds to the work of "switching on" the soft part of the potential function. (Note the double meaning of the letter *S*, as a subscript for "solute" and as a superscript for "soft.") The last term is the work associated with the release of the constraint imposed on the solute to a fixed position (this has been discussed in great detail in Section 3.11).

It is very important to recognize that the splitting of the dissolution process has been carried out for the pseudo-chemical potential, not the full chemical potential. The quantity $\bar{\mu}_S{}^H$ is the work required to create a cavity of radius $\sigma_{SW}$ at a fixed position. This is equivalent to introducing a hard-sphere particle of diameter $\sigma_{HS}$ (so that $2\sigma_{SW} = \sigma_{HS} + \sigma_W$) at a fixed position. The equivalence of these two processes is meaningless upon removal of the requirement of a fixed position.

Once we have the splitting of the chemical potential as in (7.31), we can proceed to evaluate other thermodynamic quantities by differentiation with respect to temperature (and pressure, when working in the $T, P, N$ ensemble). In this way, we get a splitting of any thermodynamic quantity (associated with process II) into "hard" and "soft" parts.

It is difficult to make a firm statement as to the usefulness of this splitting in understanding the properties of water, there being as yet no reliable theoretical estimates of the relative contributions of the two terms to the various thermodynamic quantities. Nevertheless, there remains one relevant question which is of interest: Are the anomalous properties of water already revealed for hard-sphere solutes? This question is of considerable theoretical significance, since a hard-sphere solute may be viewed as the simplest solute we can put into the solvent.

Progress in this field has been made possible due to the development of the scaled particle theory (SPT), which provides a prescription for calculating the work of creating a cavity in liquids. We will not describe the SPT in detail; only the essential result relevant to our problem will be quoted [see Reiss *et al.* (1959, 1960), Helfand *et al.* (1960), Reiss (1966), and, for more recent developments, Tully-Smith and Reiss (1970), Reiss and Tully-Smith (1971) and Stillinger (1973).]

Let $\sigma_W$ and $\sigma_S$ be the effective diameters of the solvent and the solute molecules, respectively. A suitable cavity for accommodating such a solute must have a radius of $\sigma_{WS} = \frac{1}{2}(\sigma_W + \sigma_S)$. The SPT provides an approximate expression for the work required to create a cavity of radius $\sigma_{WS}$ at a fixed position in the liquid, which reads

$$\bar{\mu}_S{}^H = K_0 + K_1 \sigma_{WS} + K_2 \sigma_{WS}^2 + K_3 \sigma_{WS}^3 \qquad (7.32)$$

where the coefficients $K_i$ are given functions of the temperature, pressure, and the solvent density $\varrho_W$:

$$K_0 = kT[-\ln(1-y) + 4.5z^2] - \tfrac{1}{6}\pi P \sigma_W{}^3 \qquad (7.33)$$

$$K_1 = -(kT/\sigma_W)(6z + 18z^2) + \pi P \sigma_W{}^2 \qquad (7.34)$$

$$K_2 = (kT/\sigma_W{}^2)(12z + 18z^2) - 2\pi P \sigma_W \qquad (7.35)$$

$$K_3 = 4\pi P/3 \qquad (7.36)$$

where we have used

$$y = \pi \varrho_W \sigma_W{}^3/6, \qquad z = y/(1-y) \qquad (7.37)$$

We shall not discuss the SPT in any detail. However we note that the SPT is not a pure molecular theory in the following sense. A molecular theory is supposed to provide, say, the Gibbs free energy as a function of $T, P, N$ as well as of the molecular parameters of the system. Once this function is available, the density of the system can be computed from the relation $\varrho^{-1} = (\partial \mu/\partial P)_T$ (with $\mu = G/N$). The SPT utilizes the effective diameter of the solvent molecules as the only molecular parameter (which, indeed, is the case for a hard-sphere fluid) and, in addition to the specification of $T$ and $P$, the solvent density $\varrho_W$ is also employed. The latter, being a measurable quantity, carries with it implicitly any other molecular properties of the system.

Table 7.4 shows some values for the free energy $\bar{\mu}_S{}^H$, enthalpy $\bar{H}_S{}^H$, and entropy $\bar{S}_S{}^H$ of cavity formation, computed by Pierotti (1963, 1965) for water and benzene at 25°C and at 1 atm. The effective diameters of argon and water for these calculations are $\sigma_S = 3.4$ Å and $\sigma_W = 2.75$ Å, respectively. (The subscript $S$ stands for "solute." Here, for argon, the superscript $H$ stands for the "hard" part of the corresponding thermodynamic quantity, and the bar is added to remind us that these quantities are computed from the pseudo-chemical potential, i.e., they all refer to a fixed position in the solvent.)

Perhaps the most striking figure in Table 7.4 is the relatively low entropy of cavity formation compared to that for benzene. This means that the anomalous properties of water are already revealed for the simplest possible solute, a hard sphere. The hard part of the enthalpy of solution is positive here, though much smaller than the corresponding value for benzene. This indicates that the "switching on" of the soft part of the interaction energy contributes significantly to the enthalpy of solution.

Some general comments regarding the application of the SPT to water are now in order. First, the SPT was originally devised for treating a fluid of hard spheres, or simple nonpolar fluids. The extension of the theory to complex fluids such as water is questionable. Second, the SPT employs an effective diameter of the solvent as the only molecular parameter. It is very

Table 7.4

Computed Values of $\bar{\mu}_S{}^H$, $\bar{H}_S{}^H$, and $\bar{S}_S{}^H$ for Argon in Water and Benzene at 25°C and 1 atm[a]

| Solvent | $\bar{\mu}_S{}^H$, cal/mole | $\bar{H}_S{}^H$, cal/mole | $\bar{S}_S{}^H$, e.u. |
|---|---|---|---|
| Water | 4430 | 690 | $-12.5$ |
| Benzene | 3610 | 3520 | $-0.3$ |

[a] From Pierotti (1965).

likely that this diameter is temperature dependent. It is not clear, however, which kind of temperature dependence should be assumed for $\sigma_W$. [This topic has been discussed by Ben-Naim and Friedman (1967) and Pierotti (1967).] Finally, we stress again that the SPT is not a pure molecular theory. Even if it can predict the properties of aqueous solutions of gases, it is still incapable of providing an "explanation" of these properties. This is an inherent drawback of the theory, since it cannot tell us *why* aqueous solutions behave in such a peculiar way as compared with other fluids.

## 7.4. APPLICATION OF A TWO-STRUCTURE MODEL

The idea that a solute changes the "structure" of the solvent is very old. As an example of the application of this idea, we refer to Chadwell's explanation of the following puzzling observation (Chadwell, 1927): The addition of solutes such as ether or methyl acetate to water was found to *decrease* the compressibility of the system, in spite of the fact that the compressibilities of these pure solutes are about three times larger than the compressibility of pure water. It has been postulated that water contains two species, say monomeric water molecules and polymers of water molecules. Addition of a solute causes a shift toward the component that has a lower compressibility; hence a qualitative "explanation" of the experimental observation is provided.[4] Similar attempts to explain the effect of solute on viscosity, dielectric relaxation, self-diffusion, and many other properties have been suggested in the literature.

There are, essentially, two fundamental questions that have been the subject of extensive research. The first is concerned with the type of structure that water molecules are assumed to form around the solute molecules. Progress in this field was mainly due to comparison of the thermodynamics of dissolution of gases in water with the thermodynamics of gas-hydrate formation [see, for example, Glew (1962), Namiot (1961), and a review by Ben-Naim (1972e)]. The second problem is concerned with the mechanism by which a simple solute such as argon enhances the "structure of the solvent."

Perhaps one of the most striking pieces of evidence that a simple solute has a significant structural effect on water comes from a comparison of the

---

[4] This explanation is not satisfactory, however. Suppose we know that the solute shifts the equilibrium toward the component with lower compressibility. There is still an unknown effect on the relaxation term which cannot be predicted even in a qualitative manner. We shall discuss this point at the end of this section [following Eq. (7.60)].

entropy of solution of argon and KCl in water. Consider the standard entropy of solution of KCl and of argon at 25°C (Frank and Evans, 1945; Friedman and Krishnan, 1973)

$$\Delta S^\circ_{KCl} = -51.9 \text{ e.u.}, \quad 2\Delta S^\circ_{Ar} = -60.4 \text{ e.u.}$$

(The standard states are, in the gas, hypothetical 1 atm, in solution, hypothetical 1 $M$ concentration. In the present context, the precise meaning of these will be of no concern to us.)

The fact that the entropy of solution of two moles of argon is more negative than that for KCl is quite surprising. We know that the electric field on the surface of an ion is large enough so that the ordering of the dipole of the water molecules around the ion is an understandable phenomenon. The puzzling observation is that the "ordering" effect seems to be even larger when we introduce argon into water. Most of the remainder of this chapter will be devoted to elucidating the effect of simple nonpolar molecules on the structure of water.

The effect of the solute on the structure of water has been investigated by numerous authors. In most studies, one assumes an ad hoc mixture model for water and then examines the shift in chemical equilibrium between the various species involved. [Examples of such studies are: Frank and Quist (1961), Namiot (1961), Grigorovich and Samoilov (1962), Yashkichev and Samoilov (1962), Wada and Umeda (1962 ), Némethy and Scheraga (1962), Ben-Naim (1965a,b), Mikhailov (1968), Mikhailov and Ponomareva (1968), Frank and Franks (1968).]

In this section, we formulate the general aspect of the application of the simplest mixture-model (MM) approach to water. We shall use an exact two-structure model (TSM), as introduced in Section 6.8. In the next section, we illustrate the application of a prototype of an interstitial model for water to solutions and, in Section 6.7, we discuss the application of a more general MM approach to this problem.

Consider a system of $N_W$ water molecules and $N_S$ solute molecules at a given temperature $T$ and pressure $P$. (We later confine ourselves to dilute solutions, $N_S \ll N_W$; in this section, the treatment is more general.) We henceforth assume that $T$ and $P$ are constant and therefore omit them from our notation.

Let $N_L$ and $N_H$ be the average number of $L$-cules and $H$-cules obtained by any classification procedure (Chapter 5). For the purpose of this section, we need not specify the particular choice of the two species; therefore our treatment will be very general. For concreteness, one may think of a TSM constructed from the vector $\mathbf{x}_C$ (see Section 6.8), i.e.,

$$N_L = N \sum_{K=0}^{K^*} x_C(K), \qquad N_H = N \sum_{K=K^*+1}^{\infty} x_C(K), \qquad N_W = N_L + N_H \quad (7.38)$$

Any extensive thermodynamic quantity can be viewed either as a function of the variables $(T, P, N_W, N_S)$ or the variables $(T, P, N_L, N_H, N_S)$.

Because of the equilibrium condition

$$\mu_L(T, P, N_L, N_H, N_S) = \mu_H(T, P, N_L, N_H, N_S) \quad (7.39)$$

we can, in principle, solve (7.39) in terms of the original set of variables of the system, i.e.,

$$N_L = f(T, P, N_W, N_S), \qquad N_H = N_W - N_L \quad (7.40)$$

An example of such a solution has been discussed in Section 6.7. In this section, we adopt the MM approach, and view the variables $N_L$ and $N_H$ as virtually independent. Of course, at the end of any computation, one must substitute the values of $N_L$ and $N_H$ that fulfill the equilibrium condition (7.39).

Consider the volume as an example of an extensive variable. The total differential of the volume ($T$, $P$ constant) is

$$dV = \left(\frac{\partial V}{\partial N_S}\right)_{N_L, N_H} dN_S + \left(\frac{\partial V}{\partial N_L}\right)_{N_S, N_H} dN_L + \left(\frac{\partial V}{\partial N_H}\right)_{N_S, N_L} dN_H \quad (7.41)$$

We write

$$V_S^* = \left(\frac{\partial V}{\partial N_S}\right)_{N_L, N_H}, \qquad \bar{V}_L = \left(\frac{\partial V}{\partial N_L}\right)_{N_S, N_H}, \qquad \bar{V}_H = \left(\frac{\partial V}{\partial N_H}\right)_{N_S, N_L} \quad (7.42)$$

and rewrite relation (7.41), using the condition $dN_L + dN_H = 0$, as

$$dV = V_S^* \, dN_S + (\bar{V}_L - \bar{V}_H) \, dN_L \quad (7.43)$$

As we have noted before, $N_L$ is not a real independent variable, and since we have kept $T$, $P$, and $N_W$ constant in (7.43), there remains only one independent variable, $N_S$. In fact, recognizing that $N_L$ is a function of $T$, $P$, $N_W$, and $N_S$, through (7.39) or (7.40), we can rewrite (7.43) as

$$dV = V_S^* \, dN_S + (\bar{V}_L - \bar{V}_H)\left(\frac{\partial N_L}{\partial N_S}\right)_{N_W, \text{eq}} dN_S \quad (7.44)$$

## Water with One Simple Solute Particle

where now the variation is only in the number of solute molecules (we have appended the subscript eq to stress that this derivative is taken along the equilibrium line for the "reaction" $L \rightleftarrows H$).

The splitting of $dV$ in (7.44) is characteristic of the application of the MM approach to the theory of solutions. It corresponds to splitting the dissolution process into two steps. First, we add $dN_S$ moles (or molecules) to the system in the "frozen-in" state (i.e., keeping $N_L$ and $N_H$ fixed). The measurable change in volume is the first term on the rhs of (7.44). Next, we release the constraint imposed by fixing $N_L$ and $N_H$, i.e., adding a hypothetical catalyst to our system. The system, in general, then relaxes to a new equilibrium position, and the corresponding change in volume is the second term on the rhs of (7.44).

The ordinary partial molar volume of the solute $S$ is thus

$$\bar{V}_S \equiv \left(\frac{\partial V}{\partial N_S}\right)_{N_W,\text{eq}} = V_S^* + (\bar{V}_L - \bar{V}_H)\left(\frac{\partial N_L}{\partial N_S}\right)_{N_W,\text{eq}} \qquad (7.45)$$

In (7.45), we have also added eq in the notation for the partial molar volume to stress the difference between the derivative along an equilibrium line and the derivative when the equilibrium is "frozen in" (such a notation is, of course, not necessary if we use the variables $T$, $P$, $N_W$). In the following, for simplicity of notation, we drop the subscript eq from the derivative along the equilibrium line. The notation "$N_W$ constant" will henceforth stand for both "$N_W$ constant" and "equilibrium condition." We shall also refer to the two terms on the rhs of (7.45) as the "static" and "relaxation" parts of the partial molar volume of $S$. It must be borne in mind that such a splitting depends on the particular choice of the two components $L$ and $H$. On the other hand, the partial molar volume $\bar{V}_S$ is, of course, a quantity which is definable in terms of the variables $T$, $P$, $N_W$, and $N_S$ and is independent of the choice of the two components.

As a more vivid example of the splitting of $\bar{V}_S$ in (7.45), consider the following system. Let $\alpha$ and $\beta$ be two solvents in two compartments at the same $T$ and $P$. In each compartment, we have two solutes, water $W$ and solute $S$. We now choose the following classification. Each water molecule in the left compartment is called an $L$-cule and each in the right compartment an $H$-cule. Furthermore, we assume that there exists a membrane permeable to both $W$ and $S$, but not to $\alpha$ and $\beta$. At equilibrium, we find an average number $N_L$ of water molecules on one side and $N_H$ on the other side.

In such a system we may ask: What is the total change of volume accompanying the addition of one mole of solute $S$? This is the analog of $\bar{V}_S$. The total change of volume $\bar{V}_S$ can be built up in two steps. First, one "closes" the membrane with respect to $W$ only and adds the solute $S$; then, one opens the membrane so that the system attains a new equilibrium position. The two contributions to the change of volume are the two terms in (7.45).

The analogy between aqueous solutions and the system described above is perfect if we agree to adopt a more complicated classification procedure (instead of "being in the left or right compartment") and envisage a hypothetical inhibitor to the reaction $L \rightleftarrows H$ instead of the membrane.

The two contributions to the change of volume in (7.44) are described graphically in Fig. 7.7. We start with pure water, having the volume $V_W{}^p$, and add solute molecules. The volume of the system changes along the curve denoted by eq. Now, suppose that at some point we "freeze in" the equilibrium $L \rightleftarrows H$ and continue to add solute. The volume will change along a new path, which we denote by fr. The two components of the total change of volume $dV$ are the two segments $a$ and $b$ shown in the figure. This splitting will be shown in a more explicit manner in Section 7.5, where we shall work out an interstitial model for aqueous solutions. (Note that the slopes of the curves need not be positive, nor have the same sign.)

A similar splitting can be carried out for other partial molar quantities. For instance, for the enthalpy, entropy, and the free energy, we have

$$\bar{H}_S = \left(\frac{\partial H}{\partial N_S}\right)_{N_W} = \left(\frac{\partial H}{\partial N_S}\right)_{N_L, N_H} + (\bar{H}_L - \bar{H}_H)\left(\frac{\partial N_L}{\partial N_S}\right)_{N_W} = H_S{}^* + \Delta H_S \quad (7.46)$$

$$\bar{S}_S = \left(\frac{\partial S}{\partial N_S}\right)_{N_W} = \left(\frac{\partial S}{\partial N_S}\right)_{N_L, N_H} + (\bar{S}_L - \bar{S}_H)\left(\frac{\partial N_L}{\partial N_S}\right)_{N_W} = S_S{}^* + \Delta S_S{}^r \quad (7.47)$$

$$\mu_S = \left(\frac{\partial G}{\partial N_S}\right)_{N_W} = \left(\frac{\partial G}{\partial N_S}\right)_{N_L, N_H} + (\mu_L - \mu_H)\left(\frac{\partial N_L}{\partial N_S}\right)_{N_W} = \mu_S{}^* + \Delta \mu_S{}^r \quad (7.48)$$

In each case, the first term is the static term and the second the relaxation term corresponding to the particular partial molar quantity.

An important general result follows from the condition of chemical equilibrium $\mu_L = \mu_H$, as a result of which we have the equality

$$\mu_S = \mu_S{}^* \quad (7.49)$$

This is a unique feature of the partial molar free energy of the solute $S$. It says that whatever classification into two components we have chosen,

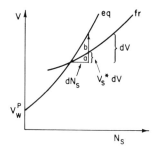

Fig. 7.7. Schematic description of the dependence of the volume of a system on $N_S$. The two curves are for the equilibrated (eq) and the "frozen-in" (fr) systems. The intersection point of the two curves is the point at which we "freeze in" the equilibrium $L \rightleftarrows H$.

the chemical potential does not receive any contribution from a relaxation term (this result holds also for a multicomponent mixture model, discrete or continuous). In other words, the presence or absence of a catalyst for the "reaction" $L \rightleftarrows H$ does not affect the chemical potential of the solute, although it may appreciably affect other partial molar quantities (a specific example is worked out in Section 7.5).

A graphic description of this is shown schematically in Fig. 7.8, where we see that if at any point we "freeze in" the equilibrium, the new curve for the free energy will have the same slope as the equilibrium curve at the point at which we have "frozen in" the system. This should be compared with the situation in Fig. 7.7, where the slopes of the two curves at the point at which we "freeze in" the system are different. (It is easily shown,

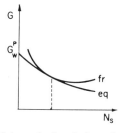

Fig. 7.8. Schematic description of the dependence of the Gibbs free energy of the system on $N_S$. The two curves eq and fr have the same slopes at the point at which we "freeze in" the equilibrium $L \rightleftarrows H$.

using arguments similar to those given in Section 5.10, that the curve fr is always above the curve eq.)

One direct consequence of relation (7.49) is that the solubility of $S$ does not change if we "freeze in" the equilibrium. This is clear since the solubility of $S$ is governed by its chemical potential. Another consequence of considerable importance is the so-called "entropy–enthalpy compensation" law (Lumry and Rajender, 1970). This follows directly from the equilibrium condition

$$\bar{H}_L - T\bar{S}_L = \mu_L = \mu_H = \bar{H}_H - T\bar{S}_H \qquad (7.50)$$

or, equivalently,

$$\Delta H_S^r = T \Delta S_S^r \qquad (7.51)$$

The relaxation part of the partial molar enthalpy is equal to the relaxation part of $T\bar{S}_S$. Thus, the fact that the entropy of solution of $S$ in water is large and negative may be partially due to the large contribution from the relaxation term in (7.47). The latter *cannot*, however, be used to explain the large positive value of $\Delta\mu_S^\circ$ (i.e., the low solubility). Thus, if we write $\Delta\mu_S^\circ = \Delta H_S^\circ - T \Delta S_S^\circ$, only the static parts of $\Delta H_S^\circ$ and $\Delta S_S^\circ$ contribute to the determination of the value of $\Delta\mu_S^\circ$; the relaxation parts exactly compensate each other. (This point has been overlooked in the past, leading to an erroneous interpretation of the positive value of $\Delta\mu_S^\circ$ in water.)

Thus far, we have not specified our components, and all the relations we have written may be applied to any TSM. The traditional interpretation of the large and negative enthalpy and entropy of solution of nonelectrolytes in water is the following: One identifies the $L$ form with the hydrogen-bonded molecules and the $H$ form with the nonbonded molecules. Hence, it is expected that $\bar{H}_L - \bar{H}_H$ will be negative. If, in addition, one postulates that the solute $S$ stabilizes the $L$ form, then we have a negative contribution from the relaxation term to the enthalpy [as well as to the entropy, by virtue of (7.51)]. In the next section, we present a simple example showing such a stabilization effect; a more general argument, independent of any ad hoc model for water, is presented in Section 7.6.

The assertion that a simple solute such as argon stabilizes the component that is fully hydrogen bonded is somewhat contradictory to our intuition. Suppose we add large quantities of say, alcohol, to water. It is clear that the very fact that water becomes diluted in the water–alcohol mixture brings about a dissociation of the hydrogen bonds between water molecules. At a very high dilution of water in any solvent, we expect very little concentration of fully hydrogen-bonded molecules. Therefore, we

# Water with One Simple Solute Particle

conclude that if such a stabilization effect exists, it must occur only in the range of very dilute solutions of a solute in water.

There exists an immense body of experimental results which supports the assertion that nonelectrolytes in very dilute solutions have a stabilizing effect on the "structure of water." We shall be concerned with the molecular reasons for the existence of such an effect in the next two sections. We note here that experimental evidence has come from a wide variety of measurements, such as NMR (Glew, 1968; Wen and Hertz, 1972; Zeidler, 1973), dielectric properties (Haggis *et al.*, 1952; Hasted, 1973), effect of solute on the temperature of maximum density (Wada and Umeda, 1962a,b), and many others. [For reviews, see Franks and Ives (1966), Horne (1972), and Franks (1973a).] In fact, the large negative entropy and enthalpy of solution of gases in water was originally used as indicative evidence that these solutes stabilize the structure of water.

There are other thermodynamic quantities which, though dependent on the structural changes in the solvent, are not easily explainable in these terms. An important example is the partial molar heat capacity (PMHC) of the solute $S$, which can be expressed in either one of the following ways:

$$\bar{C}_S = \left(\frac{\partial \bar{H}_S}{\partial T}\right)_{N_W, N_S} = \left(\frac{\partial C}{\partial N_S}\right)_{T, N_W} \tag{7.52}$$

(where $C$ stands for the heat capacity of the system at constant pressure). Using a TSM for water, one can formally write, as in (7.46)–(7.48),

$$\bar{C}_S = C_S{}^* + (\bar{C}_L - \bar{C}_H)(\partial N_L/\partial N_S)_{N_W} \tag{7.53}$$

where

$$C_S{}^* = (\partial C/\partial N_S)_{N_L, N_H} \tag{7.54}$$

The difficulty of interpreting the large value of $\bar{C}_S$ within the TSM is that the quantity $C$ itself can be split into two terms, the static and relaxation terms (Chapter 6). Therefore, the quantity $C_S{}^*$ already includes contributions due to structural changes in the solvent. The general expression for the PMHC of $S$ is quite involved, and is omitted here (see Ben-Naim, 1970b). Instead, we present a simple example to demonstrate an important result. Suppose that the mixture of $L$ and $H$ is ideal (in any of the senses discussed in Chapter 4). Also, for simplicity, we assume that $S$ is very dilute in water. The total heat capacity of the system can be written as (Section 6.8)

$$C = C^* + (\bar{H}_L - \bar{H}_H)^2 x_L x_H N_W / kT^2 \tag{7.55}$$

where $C^*$ is given by

$$C^* = N_S\left(\frac{\partial H_S^*}{\partial T}\right)_{N_L,N_H,N_S} + N_L \bar{C}_L + N_H \bar{C}_H \qquad (7.56)$$

The PMHC of $S$ in the limit $\varrho_S \to 0$ is

$$\bar{C}_S{}^\circ = \left(\frac{\partial H_S^*}{\partial T}\right)_{N_L,N_H,N_S} + (\bar{C}_L - \bar{C}_H)\left(\frac{\partial N_L}{\partial N_S}\right)_{N_W}$$

$$+ \frac{(\bar{H}_L - \bar{H}_H)^2}{kT^2}(x_H - x_L)\left(\frac{\partial N_L}{\partial N_S}\right)_{N_W} \qquad (7.57)$$

The first term on the rhs of (7.57) is essentially the contribution of the kinetic degrees of freedom of $S$ to the PMHC. The second term is expected to be small since $\bar{C}_L$ and $\bar{C}_H$ are expected to be of similar magnitude. The important contribution is probably due to the last term on the rhs of (7.57). The sign of this term depends on the relative amounts of $L$ and $H$. Thus, even when we know for sure that $S$ stabilizes the $L$ form, we can still say nothing about the sign of the PMHC of $S$ in water.[5] A detailed and explicit example is worked out in the next section.

A very similar treatment applies to the effect of $S$ on the compressibility of the system. Again, the general expression is quite cumbersome, and we present here only the results for the ideal solution.

The total volume is

$$V = N_S V_S^* + N_L \bar{V}_L + N_H \bar{V}_H \qquad (7.58)$$

Instead of dealing directly with the compressibility, it will be easier to discuss the pressure dependence of the volume, defined by

$$D = -V\varkappa_T = \left(\frac{\partial V}{\partial P}\right)_{T,N_W}$$

$$= N_S \frac{\partial V_S^*}{\partial P} + N_L \frac{\partial \bar{V}_L}{\partial P} + N_H \frac{\partial \bar{V}_H}{\partial P} + (\bar{V}_L - \bar{V}_H)\left(\frac{\partial N_L}{\partial P}\right)_{N_W} \qquad (7.59)$$

Taking the limit of dilute solution of $S$ in water and using the expression

---

[5] A traditional interpretation of the high value of the partial molar heat capacity of nonelectrolytes in water has been based on the fact that $S$ stabilizes the structure of water, or increases the amount of the "icelike" form. Since there is more icelike material available for melting, then one concludes that the total heat capacity of the system should increase. The above example shows that such a conclusion cannot be drawn even for the *sign* of the partial molar heat capacity.

(6.90) for ideal solutions, we can write the corresponding "partial molar quantity" of $S$ as

$$\bar{D}_S{}^\circ = \frac{\partial V_S{}^*}{\partial P} + (\bar{D}_L - \bar{D}_H)\left(\frac{\partial N_L}{\partial N_S}\right)_{N_W}$$

$$+ \frac{(\bar{V}_L - \bar{V}_H)^2 (x_H - x_L)}{kT} \left(\frac{\partial N_L}{\partial N_S}\right)_{N_W} \quad (7.60)$$

We can now go back to Chadwell's explanation of the puzzling observation on the effect of solutes on the compressibility of water. The first term on the rhs of (7.60) can be viewed as the internal contribution of the solute to the compressibility. The second and third terms give the contribution due to structural changes in the solvent. One clearly sees that even if we know for sure that $S$ stabilizes one of the forms $L$ or $H$, this information is insufficient to draw conclusions as to the effect of $S$ on the compressibility of the system. Of course, the general expression is more complicated than that given in (7.60). However, this example is sufficient to demonstrate the complexity of the expressions for quantities which are second derivatives of the free energy.

## 7.5. APPLICATION OF AN INTERSTITIAL MODEL

We extend here the application of the interstitial model for water to aqueous solutions of simple solutes. The merits of this model are essentially the same as those discussed in Section 6.7; this is the simplest model that contains elements in common with similar models worked out by various authors (Frank and Quist, 1961; Yashkichev and Samoilov, 1962; Bulsaeva and Samoilov, 1963; Ben-Naim, 1965a; Malenkov, 1966; Mikhailov, 1968; Mikhailov and Ponomareva, 1968; Frank and Franks, 1968). This is also the only model proposed for water that can be solved exactly, and, therefore, various general results of the mixture-model formalism can be obtained in an explicit manner.

The basic assumptions of the model have been introduced in Section 6.7. To adapt the model for aqueous solutions, we further assume that $N_S$ solute molecules occupy the interstitial positions in the framework built up by the $L$-cules. Only one new molecular parameter is introduced in the new model, i.e., the interaction energy between the solute $S$ and its surroundings, which we denote by $E_S$.

The total energy of the system is [compare with (6.27)]

$$E_T = N_L E_L + N_H E_H + N_S E_S \tag{7.61}$$

and the corresponding isothermal–isobaric partition function is

$$\Delta(T, P, N_W, N_S)$$
$$= \sum_{N_L = N_{L_{\min}}}^{N_W} \frac{(N_0 N_L)! \exp[-\beta(N_L E_L + N_H E_H + N_S E_S + P N_L V_L)]}{N_S! \, N_H! \, (N_0 N_L - N_H - N_S)!} \tag{7.62}$$

The combinatorial factor in (7.62) is the number of ways in which one can place $N_H$ $H$-cules and $N_S$ solute molecules into $N_0 N_L$ holes. The summation is carried over all possible volumes of the system, which, by virtue of the assumptions of the model, is the same as a summation over all $N_L$. The condition $N_H \leq N_0 N_L$ for pure water is replaced by the condition

$$N_H + N_S \leq N_0 N_L \tag{7.63}$$

Hence, the minimum value of $N_L$ is $N_{L_{\min}} = (N_W + N_S)/(N_0 + 1)$.

As in the case of the one-component system, we take the maximal term in the sum (7.62), from which we obtain the equilibrium condition

$$\frac{N_H^*(N_0 N_L^*)^{N_0}}{(N_0 N_L^* - N_H^* - N_S)^{N_0+1}} = \exp[\beta(E_L - E_H + P V_L)] = K(T, P) \tag{7.64}$$

The "mole fractions" of empty holes, holes occupied by $H$-cules, and holes occupied by $S$ are

$$y_0 = \frac{N_0 N_L - N_H - N_S}{N_0 N_L}, \qquad y_1 = \frac{N_H}{N_0 N_L}, \qquad y_S = \frac{N_S}{N_0 N_L} \tag{7.65}$$

With these mole fractions, the equilibrium condition is

$$y_1^*/(y_0^*)^{N_0+1} = K(T, P) \tag{7.66}$$

which has the same form as in the one-component case [Eq. (6.40)].

We now evaluate some thermodynamic quantities of this system. The total Gibbs free energy of the system with given values of $N_L$, $N_H$ and $N_S$ is

$$G(T, P, N_L, N_H, N_S) = N_L E_L + N_H E_H + N_S E_S + P N_L V_L$$
$$- kT[N_0 N_L \ln(N_0 N_L) - N_H \ln N_H - N_S \ln N_S$$
$$- (N_0 N_L - N_H - N_S) \ln(N_0 N_L - N_H - N_S)] \tag{7.67}$$

## Water with One Simple Solute Particle

As in the one-component system, this function has a minimum at a point $N_L = N_L^*$ (and $N_H = N_H^*$) which satisfies the condition (7.64). The mole fractions of $L$, $H$, and $S$ are defined by

$$x_L = N_L/N_W, \qquad x_H = N_H/N_W, \qquad x_S = N_S/N_W \qquad (7.68)$$

Note that for any $x_S$, the addition of solute will always lead to an increase in the mole fraction of $L$-cules. Such an effect can be ascribed to a stabilization effect of the $L$-cules by the solute molecules. In this particular model, the molecular reason for such a stabilization effect is quite obvious. Since we permit $S$ to hold only interstitial sites, their presence compels the $H$-cules to vacate some of the holes and transform into $L$-cules. This statement will be given a formal proof below.

The total entropy of the system is obtained by differentiating (7.67) with respect to temperature

$$S = -kN_0 N_L (y_0 \ln y_0 + y_1 \ln y_1 + y_S \ln y_S) \qquad (7.69)$$

which, as in the one-component system, has a form similar to the entropy of "mixing" empty holes, holes with $S$, and holes with $H$-cules.

The total volume, enthalpy, and energy of the system are readily obtainable from (7.67) as

$$V = N_L V_L, \ H = N_L E_L + N_H E_H + N_S E_S + P N_L V_L, \ E = H - PV \quad (7.70)$$

Note that in order to obtain the values of these quantities at equilibrium, one must substitute $N_L = N_L^*$ and $N_H = N_H^*$.

The chemical potentials of the three components are

$$\mu_L = (\partial G/\partial N_L)_{N_H, N_S}$$
$$= E_L + PV_L - kT[N_0 \ln(N_0 N_L) - N_0 \ln(N_0 N_L - N_H - N_S)] \quad (7.71)$$

$$\mu_H = (\partial G/\partial N_H)_{N_L, N_S}$$
$$= E_H - kT[\ln(N_0 N_L - N_H - N_S) - \ln N_H] \qquad (7.72)$$

$$\mu_S = (\partial G/\partial N_S)_{N_W} = (\partial G/\partial N_S)_{N_L, N_H}$$
$$= E_S + kT[\ln N_S - \ln(N_0 N_L - N_H - N_S)] \qquad (7.73)$$

Note that $\mu_L$ and $\mu_H$ are definable only for a "frozen-in" system (i.e., we must keep $N_H$ and $N_L$ constant in the respective definitions). On the other hand, the chemical potential of $S$ is definable in both the equilibrated and the "frozen-in" systems; the equality between the two holds for the case in which $N_L = N_L^*$ and $N_H = N_H^*$ [i.e., when $\mu_L = \mu_H$ or, equivalently, when the equilibrium condition (7.64) is fulfilled].

From (7.71)–(7.73), we can compute the partial molar quantities of the various components. We first evaluate the corresponding quantities for $L$ and $H$

$$\bar{S}_L = -(\partial \mu_L/\partial T)_{N_L,N_H} = kN_0[\ln(N_0N_L) - \ln(N_0N_L - N_H - N_S)] \quad (7.74)$$

$$\bar{S}_H = -(\partial \mu_H/\partial T)_{N_L,N_H} = k[\ln(N_0N_L - N_H - N_S) - \ln N_H] \quad (7.75)$$

$$\bar{H}_L = E_L + PV_L, \qquad \bar{H}_H = E_H \quad (7.76)$$

$$\bar{V}_L = V_L, \qquad \bar{V}_H = 0 \quad (7.77)$$

$$\bar{E}_L = E_L, \qquad \bar{E}_H = E_H \quad (7.78)$$

Note the difference between partial molar (or molecular) quantities on the lhs of (7.76)–(7.78) and the molecular parameters of the model on the rhs. Furthermore, we note that all the partial molar quantities of $L$ and $H$ are definable only in the "frozen-in" system.

We now turn to the corresponding quantities of the solute, definable in either the "frozen-in" or the equilibrated system (see Section 7.4 for more details). Before doing this, we first evaluate the following derivative:

$$\begin{aligned}\left(\frac{\partial N_L}{\partial N_S}\right)_{N_W} &= -(\mu_{LL} - 2\mu_{LH} + \mu_{HH})^{-1}\left[\frac{\partial(\mu_L - \mu_H)}{\partial N_S}\right]_{N_L,N_H} \\ &= -\frac{x_L x_H(N_0N_L - N_H - N_S)}{kT(N_0 + x_S(N_0x_H - x_L)]} \frac{-kT(N_0 + 1)}{N_0N_L - N_H - N_S} \\ &= \frac{x_L x_H(N_0 + 1)}{N_0 + x_S(N_0 x_H - x_L)} \end{aligned} \quad (7.79)$$

where, in the first step, we used the thermodynamic identity (5.129). All the derivatives on the lhs can now be evaluated in the "frozen-in" system by direct differentiation of (7.71) and (7.72). Although we have evaluated this quantity for a "frozen-in" system, at the end of our computation, we must substitute $x_L = x_L^*$ and $x_H = x_H^*$, which satisfies the equilibrium condition (7.64) and, in terms of the mole fractions, is written as

$$\frac{x_H^*(N_0 x_L^*)^{N_0}}{(N_0 x_L^* - x_H^* - x_S)^{N_0+1}} = K(T, P) \quad (7.80)$$

Once we have computed $\partial N_L/\partial N_S$, we can write down all the partial molar quantities of the solute (each quantity must be evaluated at the

## Water with One Simple Solute Particle

point $N_L = N_L^*$ and $N_H = N_H^*$)

$$\bar{S}_S = S_S^* + (\bar{S}_L - \bar{S}_H)\left(\frac{\partial N_L}{\partial N_S}\right)_{N_W} = -k[\ln N_S - \ln(N_0 N_L - N_H - N_S)]$$
$$+ \left(\frac{\partial N_L}{\partial N_S}\right)_{N_W} k \ln K \qquad (7.81)$$

$$\bar{H}_S = H_S^* + (\bar{H}_L - \bar{H}_H)\left(\frac{\partial N_L}{\partial N_S}\right)_{N_W} = E_S + (E_L - E_H + PV_L)\left(\frac{\partial N_L}{\partial N_S}\right)_{N_W} \qquad (7.82)$$

$$\bar{V}_S = V_S^* + (\bar{V}_L - \bar{V}_H)\left(\frac{\partial N_L}{\partial N_S}\right)_{N_W} = V_L\left(\frac{\partial N_L}{\partial N_S}\right)_{N_W} \qquad (7.83)$$

In this particular model, the static part of the partial molar volume is zero (i.e., solutes in holes do not contribute to the total volume of the system); hence, all of $\bar{V}_S$ is made up of structural changes in the solvent. Note also that the "entropy–enthalpy compensation" law is fulfilled in this model [see Eq. (7.51)].

Figure 7.9 shows some of the thermodynamic quantities of this model as a function of the solute concentration $x_S$. The slopes of $G$, $H$, and $V$ give the corresponding partial molar quantities. We have also indicated the "frozen-in" curve obtained by "freezing in" the equilibrium at $x_S = 0.2$.

As an illustrative example, consider the case in which $N_0 = 1$ and we take the limit $x_S \to 0$. The corresponding value of $\bar{H}_S$ is

$$\bar{H}_S^\circ = E_S + (E_L - E_H + PV_L) 2 x_L^* x_H^* \qquad (7.84)$$

For this particular case, we can solve the equilibrium condition (7.64) and express $x_L$ and $x_H$ (at equilibrium) in terms of the equilibrium constant $K$. The result is (see Section 6.7)

$$x_L^* = \frac{1}{2} + \frac{1}{2}\left(\frac{1}{4K+1}\right)^{1/2} \qquad (7.85)$$

Hence,

$$x_L^* x_H^* = \frac{K}{(4K+1)} \qquad (7.86)$$

and (7.84) is rewritten as

$$\bar{H}_S^\circ = E_S + (kT \ln K)[2K/(4K+1)] \qquad (7.87)$$

The important point to be noted in (7.87) is the fact that the relaxation term is independent of any property of the *solute*. Therefore, in principle, the two terms in (7.87) may be of a different order of magnitude.

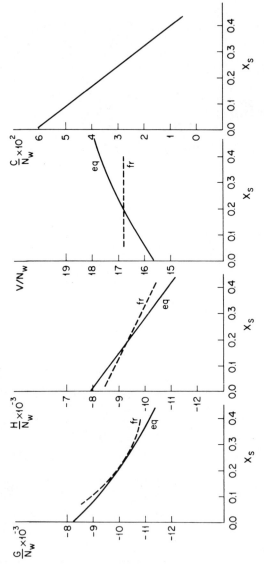

Fig. 7.9. Dependence of the free energy, enthalpy, volume, and heat capacity on the mole fraction of the solute $x_S$. The computations were carried out for a particular choice of molecular parameters for the interstitial model. The numerical values obtained for the thermodynamic quantities are of no importance for the discussions in this chapter.

Perhaps it is appropriate at this junction to pause and reflect on the usefulness of using a two-structure model for understanding the properties of aqueous solutions. Consider for concreteness the enthalpy of a solution of a solute $S$ in the specific model considered above (i.e., $N_0 = 1$ and $x_S \to 0$). The intuitive expectation is that the enthalpy of solution should be dominated by the interaction energy of $S$ with the cavity, $E_S$. A measurement may show that $\bar{H}_S$ is much more negative [note that the second term in (7.87) is negative, since we chose $E_L - E_H + PV_L < 0$]. We would be puzzled by this finding, were we not aware of any other processes which may have taken place in the solvent. Of course, if we know the details of the molecular model, and if it is simple enough, we can solve for $\bar{H}_S$ and arrive at (7.87) without ever mentioning the mixture-model approach to this system. On the other hand, if we do adopt the mixture-model approach, then we arrive at the same conclusion (7.87), which may be more *convenient* for comprehending the two contributions to $\bar{H}_S$, one from direct interaction of $S$ with the solvent and one from structural rearrangement in the solvent induced by the addition of the solute. This is the main virtue of using the mixture-model approach to liquid water.

As a final application of the interstitial model, consider the partial molar heat capacity of $S$ at infinite dilution. From (7.84), we get

$$\bar{C}_S^\circ = \left(\frac{\partial \bar{H}_S^\circ}{\partial T}\right)_{P,N_W,N_S} = \frac{-2(E_L - E_H + PV_L)^2 x_L^* x_H^* (x_L^* - x_H^*)^2}{kT^2} \tag{7.88}$$

which is always negative. Here we have an example where $S$ *stabilizes* the $L$ form (i.e., the one with the lower partial molar enthalpy), yet there is no guarantee of a positive value for the partial molar heat capacity of $S$. Therefore, the sign and magnitude of the partial molar heat capacity of nonelectrolytes are not explainable by this model. This point has already been discussed in more general terms in Section 7.4.

## 7.6. THE PROBLEM OF STABILIZATION OF THE STRUCTURE OF WATER

One of the most widely applied concepts in the study of aqueous solutions is the "structural change" of the solvent brought on by the addition of a solute. In most cases, one measures a specific quantity for pure water and for aqueous solutions, and then a qualitative inference is made on the change in the structure of water caused by that particular solute.

For instance, if a solute increases the viscosity of water, it seems logical to attribute this effect to the increase in the structure of the solvent.

Bernal and Fowler (1933) introduced the idea of "structural temperature," which is defined as follows. Suppose one measures some physical property $\eta$ for pure water and for an aqueous solution, both at the same temperature $T$. We denote the change in this property by

$$\Delta\eta = \eta(T, N_S) - \eta(T, N_S = 0) \qquad (7.89)$$

Next, consider the change of the same property $\eta$ for pure water caused by a change in temperature,

$$\Delta\eta' = \eta(T', N_S = 0) - \eta(T, N_S = 0) \qquad (7.90)$$

The "structural temperature" of the solution, with $N_S$ moles of $S$ at temperature $T$, is defined as the temperature $T'$ for which we have the equality

$$\Delta\eta = \Delta\eta' \qquad (7.91)$$

Note that the definition of the "structural temperature" as given here does not make use of the concept of the structure of water. The idea underlying this definition is qualitatively clear. It is believed that, however we choose to define the "structure of water," this quantity must be a monotonically decreasing function of the temperature. Therefore, changes in the structure can be detected on a corresponding temperature scale.

The concept of "structural temperature," though a useful operational definition of the "structure of water," is not entirely satisfactory, since it is generally dependent on the property $\eta$. Using different properties for the same solution, we can get different structural temperatures.

In Chapter 5, we elaborated on one possible definition of the structure of water, which may also be applied to aqueous solutions of simple solutes.

In this section, we discuss a more general problem. Suppose we classify molecules into quasicomponents by any one of the classification procedures. We then select one of these species and inquire about the change in its concentration upon the addition of a solute. As a particular example, we may choose one of the species to be the fully hydrogen-bonded molecules; hence, its concentration can serve as a measure of the structure of the solvent.

Let $L$ and $H$ be two quasicomponents obtained by any classification procedure. The corresponding average numbers of $L$- and $H$-cules are $N_L$ and $N_H$, respectively. We assume that the temperature and the pressure are always constant. The quantity of interest is, then, the derivative

# Water with One Simple Solute Particle

$(\partial N_L/\partial N_S)_{T,P,N_W}$. We say that the component $L$ is "stabilized" by $S$ if this derivative is positive, i.e., $N_L$ increases upon the addition of the solute.

We now explore the general and exact conditions under which a stabilization of $L$ by $S$ occurs. We can then specialize to a particular choice of the component $L$ and speculate on the possibility of stabilization of the structure of water by the solute $S$.

A convenient starting relation is the thermodynamic identity [Eq. (5.129)]

$$(\partial N_L/\partial N_S)_{N_W} = -(\mu_{LL} - 2\mu_{LH} + \mu_{HH})^{-1}(\mu_{LS} - \mu_{HS}) \qquad (7.92)$$

where $\mu_{ij} = \partial^2 G/\partial N_i \, \partial N_j$. The advantage of using relation (7.92) is that it transforms a derivative in the equilibrated system [the lhs of (7.92)] into derivatives in the "frozen-in" system; the latter are expressible in terms of molecular distribution functions through the Kirkwood–Buff theory of solution (Section 4.5).

Before we plunge into the tangles of the mathematical derivation, it is instructive to elaborate on the qualitative physical ideas we will be using. Suppose we start with an equilibrated system for which we have the condition

$$\mu_L(N_L, N_H, N_S) = \mu_H(N_L, N_H, N_S) \qquad (7.93)$$

As an auxiliary device, we envisage a catalyst which, when absent from the system, causes the reaction $L \rightleftarrows H$ to become "frozen in." Now, at some given equilibrium state, we take our catalyst out, and add $dN_S$ moles of the solute $S$. Generally, at this stage, we have the inequality

$$\mu_L(N_L, N_H, N_S + dN_S) \neq \mu_H(N_L, N_H, N_S + dN_S) \qquad (7.94)$$

For concreteness, suppose that at this stage we have in (7.94) the inequality $\mu_L < \mu_H$. Then, if we reintroduce our catalyst, water molecules will "flow" from the state of the high to the state of the low chemical potential. This means that the overall effect of adding $dN_S$ is an increase in number of $L$-cules. The hypothetical catalyst was invented as an intermediary device; the important quantity is the final change in $N_L$ brought on by the change in $N_S$. The splitting of the process into two parts shows that in order to examine the quantity $(\partial N_L/\partial N_S)_{N_W}$ at equilibrium, one can equivalently examine the change in $\mu_L - \mu_H$ in the "frozen-in" system. This is the content of identity (7.92).

Let us now apply the Kirkwood–Buff theory to reexpress the rhs of (7.92) in terms of molecular distribution functions.

The basic relation that we need is (4.60),

$$\mu_{\alpha\beta} = \frac{kT}{V} \frac{B^{\alpha\beta}}{|\mathbf{B}|} - \frac{\bar{V}_\alpha \bar{V}_\beta}{V \varkappa_T} \qquad (7.95)$$

where the determinant $|\mathbf{B}|$ for the three-component system is

$$|\mathbf{B}| = \begin{vmatrix} \varrho_S + \varrho_S^2 G_{SS} & \varrho_S \varrho_L G_{SL} & \varrho_S \varrho_H G_{SH} \\ \varrho_L \varrho_S G_{LS} & \varrho_L + \varrho_L^2 G_{LL} & \varrho_L \varrho_H G_{LH} \\ \varrho_H \varrho_S G_{HS} & \varrho_H \varrho_L G_{HL} & \varrho_H + \varrho_H^2 G_{HH} \end{vmatrix} \qquad (7.96)$$

from which we derive the various cofactors $B^{\alpha\beta}$.

From here on, we specialize to the limiting case of $\varrho_S \to 0$. In a formal way, we can write the quantity (7.92) for any $\varrho_S$ in terms of the $G_{\alpha\beta}$. However, the general case is very complicated. As a matter of fact, in this chapter, we are interested only in very dilute solutions, which, indeed, aqueous solutions of inert gases are.

In the limit $\varrho_S \to 0$, the partial molar volumes $\bar{V}_L$ and $\bar{V}_H$ and the compressibility of the system tend to their values in pure water, with composition $N_L$ and $N_H$, i.e.,

$$\bar{V}_L = [1 + \varrho_H(G_{HH} - G_{LH})]/\eta \qquad (7.97)$$

$$\bar{V}_H = [1 + \varrho_L(G_{LL} - G_{LH})]/\eta \qquad (7.98)$$

$$\varkappa_T^* = \zeta/kT\eta \qquad (7.99)$$

where $\zeta$ and $\eta$ are given by [see Eqs. (4.66) and (4.67)]

$$\eta = \varrho_L + \varrho_H + \varrho_L\varrho_H(G_{LL} + G_{HH} - 2G_{LH}) \qquad (7.100)$$

$$\zeta = 1 + \varrho_L G_{LL} + \varrho_H G_{HH} + \varrho_L\varrho_H(G_{LL}G_{HH} - G_{LH}^2) \qquad (7.101)$$

Note that in (7.99) we have the *static* part of the compressibility. This is because we are now working in the "frozen-in" system, where $N_L$ and $N_H$ are assumed to behave as independent variables. Note also that $\bar{V}_L$ and $\bar{V}_H$ are definable only in the "frozen-in" system, and therefore there is no need to add an asterisk to these symbols.

We will also need the limiting value of the *static* partial molar volume of the solute $S$. This is obtained for a system of any number of $C$ components as follows. From Eqs. (4.51), (4.54), and (4.58) and the application of the Gibbs–Duhem relation, we find

$$0 = \sum_{\beta=1}^{C} \varrho_\beta \mu_{\alpha\beta} = \frac{kT}{V|\mathbf{B}|} \sum_{\beta=1}^{C} \varrho_\beta B^{\alpha\beta} - \frac{\bar{V}_\alpha}{V \varkappa_T} \sum_{\beta=1}^{C} \varrho_\beta \bar{V}_\beta \qquad (7.102)$$

### Water with One Simple Solute Particle

Now we multiply (7.102) by $\varrho_\alpha$ and sum over all the species $\alpha$ to obtain (note that $\sum_{\beta=1}^{C} \varrho_\beta \bar{V}_\beta = 1$)

$$\frac{kT}{V |\mathbf{B}|} \sum_{\alpha=1}^{C} \sum_{\beta=1}^{C} \varrho_\alpha \varrho_\beta B^{\alpha\beta} = \frac{1}{V \varkappa_T} \qquad (7.103)$$

which provides a general expression for the isothermal compressibility of a multicomponent system. Substituting $\varkappa_T$ in (7.102), we get an expression for the partial molar volume

$$\bar{V}_\alpha = \left( \sum_{\beta=1}^{C} \varrho_\beta B^{\alpha\beta} \right) \Big/ \left( \sum_{\alpha=1}^{C} \sum_{\beta=1}^{C} \varrho_\alpha \varrho_\beta B^{\alpha\beta} \right) \qquad (7.104)$$

For our three-component system of $L$, $H$, and $S$, we get from (7.104)

$$V_S^* = \frac{\varrho_S B^{SS} + \varrho_L B^{LS} + \varrho_H B^{HS}}{\varrho_S^2 B^{SS} + \varrho_L^2 B^{LL} + \varrho_H^2 B^{HH} + 2\varrho_S\varrho_L B^{LS} + 2\varrho_S\varrho_H B^{HS} + 2\varrho_L\varrho_H B^{LH}} \qquad (7.105)$$

Note that here the asterisk in $V_S^*$ is essential. Taking the appropriate cofactors from (7.96), and retaining only linear terms in $\varrho_S$ in the numerator and the denominator of (7.104), we get

$$\lim_{\substack{\varrho_S \to 0}} V_S^* = \left\{ \varrho_S \varrho_L \varrho_H \begin{vmatrix} 1+\varrho_L G_{LL} & \varrho_H G_{LH} \\ \varrho_L G_{LH} & 1+\varrho_H G_{HH} \end{vmatrix} - \varrho_L^2 \varrho_H \varrho_S \begin{vmatrix} G_{LS} & \varrho_H G_{LH} \\ G_{HS} & 1+\varrho_H G_{HH} \end{vmatrix} \right.$$

$$+ \varrho_H^2 \varrho_L \varrho_S \begin{vmatrix} G_{SL} & G_{SH} \\ 1+\varrho_L G_{LL} & \varrho_L G_{LH} \end{vmatrix} \right\}$$

$$\times \left\{ \varrho_L^2 \varrho_H \varrho_S \begin{vmatrix} 1 & 0 \\ \varrho_H G_{HS} & 1+\varrho_H G_{HH} \end{vmatrix} + \varrho_H^2 \varrho_S \varrho_L \begin{vmatrix} 1 & 0 \\ G_{LS} & 1+\varrho_L G_{LL} \end{vmatrix} \right.$$

$$\left. - 2\varrho_L^2 \varrho_H \varrho_S \begin{vmatrix} 1 & 0 \\ G_{LS} & \varrho_H G_{LH} \end{vmatrix} \right\}^{-1}$$

$$= [\zeta - \varrho_L G_{LS}(1 + \varrho_H G_{HH} - \varrho_H G_{LH})$$

$$- \varrho_H G_{HS}(1 + \varrho_L G_{LL} - \varrho_L G_{HL})]/\eta$$

$$= kT\varkappa_T^* - \varrho_L G_{LS}\bar{V}_L - \varrho_H G_{HS}\bar{V}_H \qquad (7.106)$$

The last relation for the limiting value of the *static* partial molar volume is also the limiting value of the partial molar volume of a solute $S$ in any three-component system.

Next, we express the derivatives on the rhs of (7.92) in terms of the $G_{\alpha\beta}$. From (7.95), we have

$$\mu_{LS} - \mu_{HS} = \frac{kT}{V}\left[\frac{B^{LS} - B^{HS}}{|\mathbf{B}|} - \frac{V_S^*(\bar{V}_L - \bar{V}_H)}{kT\varkappa_T^*}\right] \quad (7.107)$$

Using the notation

$$\Delta_{LH}^S = G_{LS} - G_{HS} \quad (7.108)$$

and the identity

$$\bar{V}_L\bar{V}_H = (\zeta - \eta G_{LH})/\eta^2 \quad (7.109)$$

we get the limiting form of (7.107)

$$\lim_{\varrho_S \to 0}(\mu_{LS} - \mu_{HS}) = -(kT/V)(\bar{V}_L - \bar{V}_H + \varrho_W \Delta_{LH}^S/\eta) \quad (7.110)$$

where $\varrho_W = \varrho_L + \varrho_H$. Note that the limiting behavior of the determinant $|\mathbf{B}|$ is

$$|\mathbf{B}| \xrightarrow{\varrho_S \to 0} \varrho_S\varrho_L\varrho_H\zeta \quad (7.111)$$

Using relations (5.133) and (7.110), we get for (7.92)

$$\lim_{\varrho_S \to 0}(\partial N_L/\partial N_S)_{N_W} = x_L x_H[\eta(\bar{V}_L - \bar{V}_H) + \varrho_W \Delta_{LH}^S)] \quad (7.112)$$

Applying relation (6.84), we can rewrite (7.112) as

$$\lim_{\varrho_S \to 0}[\partial N_L/\partial N_S)_{N_W} = \varrho_W x_L x_H[(G_{WH} - G_{WL}) + (G_{LS} - G_{HS})] \quad (7.113)$$

Relation (7.113) is very general. First, it applies to any two-component system at chemical equilibrium,[6] and to any classification procedure we have chosen to identify the two quasicomponents. Second, because of the application of the Kirkwood–Buff theory of solutions, we do not have to restrict ourselves to any assumptions on the total potential energy of the system. Furthermore, the quantities $G_{\alpha\beta}$ appearing here depend on the *spatial* pair correlation functions $g_{\alpha\beta}(R)$, even though we may be dealing with nonspherical particles.

As we stressed in the beginning of this section, we still have a "degree of freedom," which is the choice of the components $L$ and $H$. This is post-

---

[6] Here we refer also to a system of two real components in chemical equilibrium, such as two isomeric forms of a molecule.

poned to a later stage when we interpret the relaxation terms of the various partial molar quantities introduced in Section 7.4.

Let us consider some general implications of (7.113) with regard to the conditions for the stabilization of $L$ by $S$.

1. If either $x_L$ or $x_H$ is very small, then the whole rhs of (7.113) is small and we cannot get a large stabilization effect. If we choose, for instance, $L$ to be strictly icelike molecules, then it is likely that $x_L$ will be small; hence, a small stabilization effect will be expected.

2. If we choose two components $L$ and $H$ which are similar in the sense that

$$G_{LS} \approx G_{HS} \quad \text{and} \quad G_{WH} \approx G_{WL} \quad (7.114)$$

then we again end up with a small stabilization effect. This is an important finding, since in some ad hoc models it is assumed that the mixture of $L$ and $H$ is symmetric ideal. Such an assumption, although it does not necessarily imply extreme similarity of the two components (see Section 4.6), should be avoided. [We will see that in order to get a large relaxation term in Eqs. (7.45)–(7.47), we must choose two components that differ greatly. The situation is similar to that in the discussion of pure water, Section 6.8.]

3. Suppose that $x_L x_H$ is not too small and that $L$ and $H$ differ appreciably. The rhs of (7.113) will tend to zero as $\varrho_W \to 0$. Note that all the $G_{\alpha\beta}$ tend to a constant value in the limit of zero total density.

4. Suppose that $\varrho_W$ is very large, approaching the close-packing density. It is likely, then, that either condition 1 or 2 will become effective. If we choose two components which differ appreciably (in their local properties), then it is likely that the concentration of one of them will be very small. On the other hand, if we choose two components having comparable mole fractions, it is likely that they will be quite similar in their local environments. Hence, in this case, we cannot get a large stabilization effect. The arguments given in this paragraph are basically intuitive and do not follow directly from (7.113). They reflect the idea that as the density becomes very high, the fluctuations in the local properties of the molecules must be very small. For instance, all the particles will have a coordination number of about 12.

All the considerations made thus far apply to any fluid. We have seen that a large stabilization effect is attainable only under very restricted conditions. Water, as one of its unique features, may conform to all the necessary conditions leading to a relatively large stabilization effect.

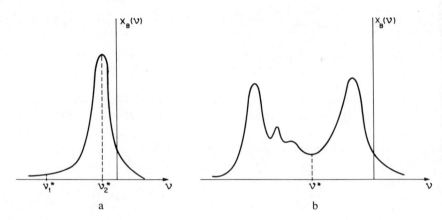

Fig. 7.10. A schematic comparison of the form of the function $x_B(\nu)$ (a) for a normal fluid; (b) for water. Various possible cutoff points, $\nu_1^*$, $\nu_2^*$, and $\nu^*$, are indicated, by the use of which a two-structure model can be defined.

Figure 7.10 depicts a possible form of the function $x_B(\nu)$ for water and a simple fluid. (While the details of the form of these functions are hypothetical the drawings have been influenced by the knowledge of the form of this function for spherical and waterlike particles in two dimensions, as presented in Chapter 6.) In a simple fluid, the distribution $x_B(\nu)$ is expected to be concentrated under one narrow peak. Hence, choice of a cutoff point at, say, $\nu_1^*$ will produce two dissimilar components. However, one of the components will have a very low concentration, thus fulfilling the first of the above conditions. On the other hand, choice of a cutoff point at, say, $\nu_2^*$ will produce two components with almost equal concentrations. These will be quite similar, hence fulfilling the second condition. In both cases, we must end up with a small stabilization effect.

Liquid water, because of the strong directional forces, is expected to have a more complicated function $x_B(\nu)$, as depicted in Fig. 7.10(b). Some examples of these functions for waterlike particles in two dimensions were presented in Section 6.11. In this particular example, we can find a cutoff point $\nu^*$ that induces a classification into two very dissimilar components in such a way that the two corresponding mole fractions are of comparable magnitude.

It is of some interest to note that the rhs of (7.113) contains two contributions to the stabilization effect. The first depends only on the properties of the solvent $G_{WH} - G_{WL}$, and not on the type of solute. The second term depends on the relative "overall affinity" of the solute toward the two components $L$ and $H$.

### Water with One Simple Solute Particle

Let us now make a specific choice of two components for water, which seems to be the most useful one for interpreting the thermodynamic behavior of aqueous solutions. We use the singlet distribution function $x_C(K)$, based on coordination number CN, and define the two mole fractions

$$x_L = \sum_{K=0}^{4} x_C(K), \qquad x_H = \sum_{K=5}^{\infty} x_C(K) \qquad (7.115)$$

Clearly, we may refer to $L$ and $H$ as the components of relatively low ($L$) and high ($H$) local densities. Another useful choice could be based on the singlet distribution function for molecules with a different number of hydrogen bonds (see Section 6.10). The important thing in this particular choice is that by the definition of the two components, we should have

$$\bar{V}_L - \bar{V}_H > 0 \quad \text{or equivalently} \quad G_{WH} - G_{WL} > 0 \qquad (7.116)$$

We recall that $\varrho_W(G_{WH} - G_{WL})$ measures the excess of *water* molecules around $H$ as compared to the excess of water molecules around $L$ (for more details, see Chapter 4). This means that if we can find two components which are very different and for which $x_L x_H$ is close to its maximum value, then we have already guaranteed a positive term for the stabilization effect (7.113) which is independent of the solute. In fact, one may think of an ideal solute which does not interact with the solvent at all. In such a case, we get $G_{LS} - G_{HS} = 0$, and the whole stabilization effect becomes a property of the solvent *only*. It will be positive for any classification for which (7.116) is fulfilled.

The next question concerns the sign of $G_{LS} - G_{HS}$ for simple solutes. From its definition, we have

$$\varrho_S \Delta_{LH}^S = \varrho_S \int_0^\infty [g_{LS}(R) - 1] 4\pi R^2 \, dR - \varrho_S \int_0^\infty [g_{HS}(R) - 1] 4\pi R^2 \, dR$$

$$= \varrho_S \int_0^\infty [g_{LS}(R) - g_{HS}(R)] 4\pi R^2 \, dR \qquad (7.117)$$

From (7.117), we see that $\varrho_S \Delta_{LH}^S$ measures the average overall excess of $S$ molecules in the neighborhood of $L$ relative to $H$. Again, from the definition of the two components in (7.115), we expect that an $L$ molecule, being in a low local density region, will let more solute molecules enter its environment. Therefore $\Delta_{LH}^S$ is likely to be positive.

A somewhat different argument may be given as follows. Since both $L$ and $H$ are water molecules, we can assign to each a hard-core diameter $\sigma_W$. If $\sigma_S$ is the hard-core diameter of the solute, then the integral in (7.117) can be replaced by

$$\Delta^S_{LH} \approx \int_{\sigma_{WS}}^{\infty} [g_{LS}(R) - g_{HS}(R)] 4\pi R^2 \, dR \qquad (7.118)$$

where $\sigma_{WS} = \tfrac{1}{2}(\sigma_W + \sigma_S)$. Next, we assume that the potential of average force between solute and solvent molecules is small compared to $kT$ in the region $R \geq \sigma_{WS}$, i.e.,

$$g_{LS}(R) = \exp[-\beta W_{LS}(R)] \approx 1 - \beta W_{LS}(R) \qquad (7.119)$$

$$g_{HS}(R) = \exp[-\beta W_{HS}(R)] \approx 1 - \beta W_{HS}(R) \qquad (7.120)$$

Hence, (7.118) is rewritten approximately as

$$\Delta^S_{LH} \approx \beta \int_{\sigma_{WS}}^{\infty} [W_{HS}(R) - W_{LS}(R)] 4\pi R^2 \, dR \qquad (7.121)$$

Since $H$, by definition, must have more water molecules in its neighborhood than $L$, it is clear that at short distances, near $\sigma_{WS}$, $W_{HS}(R)$ will exhibit a repulsive behavior relative to $W_{LS}(R)$. Therefore, we expect $W_{HS}(R) - W_{LS}(R)$ for $R \gtrsim \sigma_{WS}$ to be positive. Furthermore, since we know the general form of the function $W(R)$ (Section 2.6), we expect that the region $\sigma_{WS} \lesssim R \lesssim 2\sigma_{WS}$ will give the major contribution to the integral (7.121).

To amplify the above argument, suppose, for simplicity, that $L$ has a very low local density and $H$ a very high local density. The quantity $W_{HS}(R)$ is the work required to bring $S$ from infinity to a distance $R$ from an $H$-cule. Hence, $W_{HS}(R) - W_{LS}(R)$ is the work required to transfer an $S$ molecule from a distance $R$ from $L$ to a distance $R$ from $H$. If we consider the range $\sigma_{WS} \leq R \leq 2\sigma_{WS}$, then we expect this work to be positive (i.e., the solute is attracted by the $L$-cule, but because of the high local density around $H$, it will be repelled from the environment of an $H$-cule; this is shown schematically in Fig. 7.11). Furthermore, since this range of distance is expected to give the major contribution to the integral in (7.121), the whole term $\Delta^S_{LH}$ is expected to be positive as well.

Thus far, we have shown that if we choose two components in such a way so that one has a relatively low local density ($L$), then it is likely that this component will be stabilized by the addition of a solute $S$. This argument applies for any fluid. It is expected, though, to be particularly large for water, which may be one of the outstanding features of the mode of packing of water molecules. More important, however, is the peculiar, and probably unique, coupling between low local density and strong binding

Water with One Simple Solute Particle 353

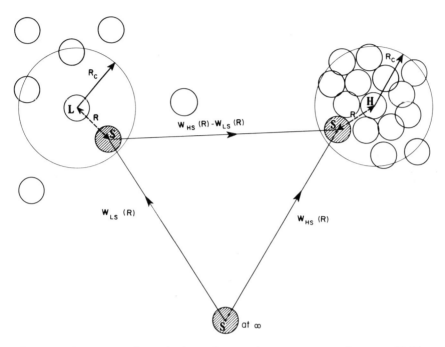

Fig. 7.11. The process of transferring $S$ from environment $L$ to environment $H$. The work corresponding to this process is $W_{HS}(R) - W_{LS}(R)$.

energy. The latter is essential to an interpretation of the large, negative enthalpy and entropy of solution discussed in Section 7.4. Consider, for example, the relaxation term of $\bar{H}_S$ [because of the compensation relation (7.51), the same argument applies to the relaxation part of $\bar{S}_S$]

$$\Delta H_S^r = (\bar{H}_L - \bar{H}_H)(\partial N_L/\partial N_S)_{N_W} \qquad (7.122)$$

Thus, if $L$ is chosen as in (7.115), so that it will be stabilized by $S$, and if the reaction $H \to L$ involves a decrease in the enthalpy of the system (presumably because of formation of hydrogen bonds), then we get a negative relaxation part for $\bar{H}_S$. We recall that the coupling of low local density with strong binding energy was found to be an essential argument in the explanation of the unique temperature dependence of the volume of water at $0 \lesssim t \lesssim 4°C$.

Although we have not explicitly introduced the concept of hydrogen bonds in the present discussion, it is clear that the $L$ component may be identified with, say, fully hydrogen-bonded molecules. In such a case, the addition of solute $S$ is likely to enhance the formation of hydrogen bonds in the system. Other solvents, such as ammonia or hydrogen fluoride, have strong hydrogen bonds. However, a fully hydrogen-bonded molecule is not expected to exist in a low local density; hence, a stabilization of hydrogen-bonded molecules by simple solutes is unlikely to occur in these fluids.

## 7.7. APPLICATION OF A CONTINUOUS MIXTURE-MODEL APPROACH

In this section, we examine the general aspects of the application of the mixture-model (MM) formalism. We have already seen that the splitting of partial molar quantities into static and relaxation terms is totally dependent on the choice of the classification procedure. Here, we elaborate on the conditions under which such a splitting may be useful to the theory of aqueous solutions.

Let $N_W(\alpha)\, d\alpha$ be the average number (in the $T$, $P$, $N_W$, $N_S$ ensemble) of *water* molecules which are distinguished by some local property having the numerical value between $\alpha$ and $\alpha + d\alpha$. Similarly, $N_S(\beta)\, d\beta$ is the average number of solute molecules classified according to some other property, having a numerical value between $\beta$ and $\beta + d\beta$. [Here, $\alpha$ and $\beta$ are continuous parameters; in this section, we will not use the notation $\beta = (kT)^{-1}$.] The two normalization conditions are

$$N_W = \int N_W(\alpha)\, d\alpha, \qquad N_S = \int N_S(\beta)\, d\beta \tag{7.123}$$

where the integration in each case extends over all the possible values of $\alpha$ and $\beta$, respectively. As in Chapter 5, we use the symbols $\mathbf{N}_W$ and $\mathbf{N}_S$ to denote the *whole* functions whose components are $N_W(\alpha)$ and $N_S(\beta)$, respectively. Viewing the system as a mixture of quasicomponents, we apply the generalized Euler theorem to express any extensive thermodynamic quantity, say the energy $E$, as

$$E(\mathbf{N}_W, \mathbf{N}_S) = \int \bar{E}_W(\mathbf{N}_W, \mathbf{N}_S, \alpha) N_W(\alpha)\, d\alpha + \int \bar{E}_S(\mathbf{N}_W, \mathbf{N}_S, \beta) N_S(\beta)\, d\beta \tag{7.124}$$

[See Section 5.8 for more details.]

The partial molar quantities in (7.124) are obtained by functional differentiation of $E(\mathbf{N}_W, \mathbf{N}_S)$:

$$\bar{E}_W(\mathbf{N}_W, \mathbf{N}_S, \alpha) = \frac{\delta E(\mathbf{N}_W, \mathbf{N}_S)}{\delta N_W(\alpha)}, \qquad \bar{E}_S(\mathbf{N}_W, \mathbf{N}_S, \beta) = \frac{\delta E(\mathbf{N}_W, \mathbf{N}_S)}{\delta N_S(\beta)} \tag{7.125}$$

### Water with One Simple Solute Particle

The experimental partial molar energy of the solute $S$ is given by

$$\bar{E}_S = \left(\frac{\partial E}{\partial N_S}\right)_{N_W} = \int \frac{\partial \bar{E}_W(\mathbf{N}_W, \mathbf{N}_S, \alpha)}{\partial N_S} N_W(\alpha)\, d\alpha$$

$$+ \int \frac{\partial \bar{E}_S(\mathbf{N}_W, \mathbf{N}_S, \beta)}{\partial N_S} N_S(\beta)\, d\beta$$

$$+ \int \bar{E}_W(\mathbf{N}_W, \mathbf{N}_S, \alpha) \frac{\partial N_W(\alpha)}{\partial N_S}\, d\alpha$$

$$+ \int \bar{E}_S(\mathbf{N}_W, \mathbf{N}_S, \beta) \frac{\partial N_S(\beta)}{\partial N_S}\, d\beta \quad (7.126)$$

This is the most general expression for $\bar{E}_S$ (or any other partial molar quantity) in the MM formalism. We have applied the MM approach here to both the solute and the solvent. The first two terms on the rhs of (7.126) may be referred to as the "static" terms, whereas the last two are the corresponding "relaxation" terms, for the solvent and solute, respectively.

It is instructive to demonstrate that with a specific choice of a quasicomponent distribution function (QCDF), we can express $\bar{E}_S$ as a pure relaxation term.[7] To do this, we specialize to the case of very dilute solutions of $S$ in $W$, and also assume pairwise additivity of the total potential. We define the following two QCDF's for $W$ and $S$ molecules:

$$N_W(\nu) = N_W \int dV \int d\mathbf{X}^N\, P(\mathbf{X}^N, V)\, \delta[B_1^{ww}(\mathbf{X}^N) - \nu] \quad (7.127)$$

$$N_S(\nu) = N_S \int dV \int d\mathbf{X}^N\, P(\mathbf{X}^N, V)\, \delta[B_1^{sw}(\mathbf{X}^N) - \nu] \quad (7.128)$$

where $\mathbf{X}^N$ denotes the configuration of the whole system of $N = N_S + N_W$ molecules. $P(\mathbf{X}^N, V)$ is the probability density in the $T, P, N_W, N_S$ ensemble. The binding energies of $S$ and $W$ are defined as follows:

$$B_1^{ww}(\mathbf{X}^N) = \sum_{j=2}^{N_W} U^{ww}(\mathbf{X}_1, \mathbf{X}_j) \quad (7.129)$$

$$B_1^{sw}(\mathbf{X}^N) = \sum_{j=1}^{N_W} U^{sw}(\mathbf{X}_1, \mathbf{X}_j) \quad (7.130)$$

---

[7] It is trivial to express $\bar{E}_S$ as a pure static term. For instance, taking a two-structure model as in Section 7.6, with one component extremely dilute in the other, e.g., $x_L \to 0$, the relaxation term will tend to zero.

where $U^{ww}$ is the solvent–solvent and $U^{sw}$ the solute–solvent pair potential. [The symbol $X_1$ in (7.129) stands for the configuration of a solvent molecule, whereas in (7.130), $X_1$ stands for the configuration of a solute molecule.]
The total energy of the system is written as

$$E = N_W \varepsilon_W^K + N_S \varepsilon_S^K + \tfrac{1}{2} \int \nu N_W(\nu) \, d\nu + \int \nu N_S(\nu) \, d\nu \quad (7.131)$$

Note that since $S$ is very dilute in $W$, we have neglected terms that include solute–solute interactions. $\varepsilon_i^K$ is the average internal energy per molecule of species $i$. Using the normalization conditions

$$N_W = \int N_W(\nu) \, d\nu, \qquad N_S = \int N_S(\nu) \, d\nu \quad (7.132)$$

we rewrite (7.131) as

$$E = \int [\varepsilon_W^K + \tfrac{1}{2}\nu] N_W(\nu) \, d\nu + \int [\varepsilon_S^K + \nu] N_S(\nu) \, d\nu \quad (7.133)$$

In this representation, we identify the partial molar energies of the various quasicomponents as

$$\bar{E}_W(N_W, N_S, \nu') = \delta E/\delta N_W(\nu') = \varepsilon_W^K + \tfrac{1}{2}\nu' \quad (7.134)$$

$$\bar{E}_S(N_W, N_S, \nu'') = \delta E/\delta N_S(\nu'') = \varepsilon_S^K + \nu'' \quad (7.135)$$

The most significant aspect of this particular classification is that the partial molar quantities given in (7.134) and (7.135) are independent of the composition ($N_W$ and $N_S$). Therefore, using the general expression (7.126) for this particular choice, we get

$$\bar{E}_S = \int [\varepsilon_W^K + \tfrac{1}{2}\nu][\partial N_W(\nu)/\partial N_S] \, d\nu + \int [\varepsilon_S^K + \nu][\partial N_S(\nu)/\partial N_S] \, d\nu \quad (7.136)$$

In this form, $\bar{E}_S$ is viewed as a pure relaxation quantity. It includes a redistribution among the $\nu$-cules of the solvent and solutes. In practical applications of the idea of relaxation, we focus our attention on the structural changes in the solvent. To do this, let us rewrite (7.136) in a somewhat different form. From the normalization conditions (7.132), we get [noting that the derivatives in (7.136) are at constant $N_W$]

$$\int \varepsilon_W^K [\partial N_W(\nu)/\partial N_S] \, d\nu = 0, \qquad \int \varepsilon_S^K [\partial N_S(\nu)/\partial N_S] \, d\nu = \varepsilon_S^K \quad (7.137)$$

## Water with One Simple Solute Particle

and write

$$\int v[\partial N_S(v)/\partial N_S]\, dv = (\partial/\partial N_S)\left[N_S \int v x_S(v)\, dv\right] \xrightarrow{\varrho_S \to 0} \langle B_S^\circ \rangle \quad (7.138)$$

where $\langle B_S^\circ \rangle$ is the *average* binding energy of the solute $S$ to the solvent at infinite dilution (it is only in the limit of $\varrho_S \to 0$ that this quantity becomes independent of $N_S$). Hence, using (7.137) and (7.138), we rewrite (7.136) as

$$\bar{E}_S^\circ = \varepsilon_S^K + \langle B_S^\circ \rangle + \tfrac{1}{2}\int v[\partial N_W(v)/\partial N_S]\, dv \quad (7.139)$$

Thus, the partial molar energy of $S$ at infinite dilution, $\bar{E}_S^\circ$, consists of the average kinetic energy, the average binding energy of the solute to the solvent, and a contribution due to "structural changes" in the solvent brought about by the addition of the solute.

For the standard energy of solution, $\Delta E_S^\circ = E_S^{\circ l} - E_S^{\circ g}$, we have

$$\Delta \bar{E}_S^\circ = \langle B_S^\circ \rangle + \tfrac{1}{2}\int v[\partial N_W(v)/\partial N_S]\, dv \quad (7.140)$$

(we have assumed that in the gaseous phase, the partial molar energy is equal to $\varepsilon_S^K$). A special case of (7.140) is for hard-sphere (HS) solutes, for which the average binding energy $\langle B_{HS}^\circ \rangle$ must be zero. Hence, the standard energy of solution is

$$\Delta \bar{E}_{HS}^\circ = \tfrac{1}{2}\int v[\partial N_W(v)/\partial N_S]\, dv \quad (7.141)$$

Thus, the standard energy of solution of a hard sphere is viewed here as arising only from structural changes in the solvent. This brings us back to Eley's speculation that in water, one may assume that $\Delta \bar{E}_{HS}^\circ$ is zero, because of the existence of "natural holes" in the water. Clearly, such a conjecture is erroneous if the dissolution of a hard sphere (or creation of a cavity) induces large structural changes in the solvent. The latter point was overlooked by Eley.

We now examine some special cases of (7.140) or (7.141).

1. The standard energy of solution of an HS solute in an HS solvent, $\Delta \bar{E}_{HS}^\circ$, must be zero, since (see Section 5.3)

$$(\partial/\partial N_S)\left[\int v N_W(v)\, dv\right] = 0 \quad (7.142)$$

2. The standard energy of solution of $S$ in $S$ is equal to half the average binding energy of $S$ to its environment, i.e.,

$$\bar{E}_S^\circ = \tfrac{1}{2}\langle B_S^\circ\rangle \qquad (7.143)$$

This can be shown by direct computation of the energy change for the transfer of an $S$ molecule from the gaseous phase to pure liquid $S$. Clearly, the relaxation term in (7.140) must be equal to $-\tfrac{1}{2}\langle B_S^\circ\rangle$ in order to produce (7.143). This is intuitively understandable since in defining the binding energy of the solvent in (7.129) we count only interactions among solvent molecules. Therefore, addition of a solute to the system does not contribute to the binding energy of a solvent molecule. Hence we expect that the average binding energy of the solvent molecules will be shifted to the right, i.e., $-\tfrac{1}{2}\langle B_S^\circ\rangle$ is, in general, a positive quantity.

3. Suppose that $N_W(\nu)$ is concentrated in two very sharp peaks,[8] a limiting case of which is

$$N_W(\nu) = N_1\,\delta(\nu - \nu_1) + N_2\,\delta(\nu - \nu_2) \qquad (7.144)$$

where $\nu_1$ and $\nu_2$ are constant values, and $N_1$ and $N_2$ are the average numbers of particles in each "state." From (7.140), we get

$$\Delta\bar{E}_S^\circ = \langle B_S^\circ\rangle + \tfrac{1}{2}(\partial/\partial N_S)\int \nu N_W(\nu)\,d\nu$$

$$= \langle B_S^\circ\rangle + \tfrac{1}{2}(\nu_1 - \nu_2)(\partial N_1/\partial N_S)_{N_W} \qquad (7.145)$$

Here, the relaxation term is due to "excitation" of solvent molecules from one state to the second, caused by the addition of $S$. We recall that the second term on the rhs of (7.145) is expected to be large if $N_1$ and $N_2$ are of comparable magnitude and if the separation between $\nu_1$ and $\nu_2$ is also large. Although this particular example is very artificial, it serves to demonstrate an important aspect of solution thermodynamics, namely that the two contributions to the standard energy of solution in (7.145) can, in principle, be of different orders of magnitude. For instance, $S$ may interact very weakly with the solvent, so that $\langle B_S^\circ\rangle \approx 0$, but it may induce structural changes involving large changes in energy. We stressed this possibility when applying the interstitial model in Section 7.5.

---

[8] An extension of this example may be identified with the Némethy and Scheraga (1962) model, where it is assumed that there exist five sharp states. See Chapter 6 for more details.

## Water with One Simple Solute Particle

4. We now consider a possible distribution function $x_W(\nu)$ for water. A schematic illustration was given in Fig. 7.10 (see also Section 6.11). Let $\nu^*$ be the cutoff point for constructing a two-structure model (TSM) as follows:

$$x_1 = \int_{-\infty}^{\nu^*} x_W(\nu)\, d\nu, \qquad x_2 = 1 - x_1 \qquad (7.146)$$

With this definition, we can rewrite the general expression for $\Delta \bar{E}_S^\circ$ in (7.140) as

$$\Delta \bar{E}_S^\circ = \langle B_S^\circ \rangle + \tfrac{1}{2}(\partial/\partial N_S)\left[\int_{-\infty}^{\nu^*} \nu N_W(\nu)\, d\nu + \int_{\nu^*}^{\infty} \nu N_W(\nu)\, d\nu\right]$$
$$= \langle B_S^\circ \rangle + \tfrac{1}{2}(\partial/\partial N_S)(N_1 \bar{\nu}_1 + N_2 \bar{\nu}_2)$$
$$= \{\langle B_S^\circ \rangle + \tfrac{1}{2}[(N_1\, \partial \bar{\nu}_1/\partial N_S) + (N_2\, \partial \bar{\nu}_2/\partial N_S)]\} + \tfrac{1}{2}(\bar{\nu}_1 - \bar{\nu}_2)(\partial N_1/\partial N_S) \qquad (7.147)$$

All the derivatives in (7.147) are at $N_W$ constant. In the second form on the rhs, we have used the average binding energies of the two components, defined as

$$\bar{\nu}_1 = N_1^{-1}\int_{-\infty}^{\nu^*} \nu N_W(\nu)\, d\nu, \qquad \bar{\nu}_2 = N_2^{-1}\int_{\nu^*}^{\infty} \nu N_W(\nu)\, d\nu \qquad (7.148)$$

In the third form on the rhs of (7.147), we have carried out the differentiation with respect to $N_S$. We can now reinterpret the total structural changes originally written in (7.140) as consisting of three contributions. The first two involve the effect of $N_S$ on the "locations" of the two peaks, i.e., the shifts of the average values of $\bar{\nu}_1$ and $\bar{\nu}_2$. The third term conveys the structural changes between the two components, i.e., the change in the relative area under the two peaks of the distribution function $x_W(\nu)$.

We can now make a general statement on the conditions required for a useful TSM. If the curve of $x_W(\nu)$ is concentrated at about two values, $\bar{\nu}_1$ and $\bar{\nu}_2$, which are well separated in such a way that $x_1$ and $x_2$ have comparable magnitudes, then it is expected that a major contribution to the energy of solution will come from structural changes between the two components; this contribution may involve energies of different orders of magnitude compared with the interaction of the solute with the solvent. (An explicit example was discussed in Section 7.5.) The shifts in the locations of $\bar{\nu}_1$ and $\bar{\nu}_2$ due to the addition of $S$ can be collected into a modified static term, so that (7.147) is finally written as

$$\Delta \bar{E}_S^\circ = \Delta E_S^* + \tfrac{1}{2}(\bar{\nu}_1 - \bar{\nu}_2)(\partial N_1/\partial N_S)_{N_W} \qquad (7.149)$$

We now briefly mention a similar treatment of the partial molar volume of the solute. Consider the quasicomponent distribution function based on the volume of the Voronoi polyhedra (VP) (Chapter 5). Let $N_W(\phi)$ and $N_S(\phi)$ be the corresponding singlet distribution functions. The total volume of the system is written as[9]

$$V(T, P, \mathbf{N}_W, \mathbf{N}_S) = \int \phi N_W(\phi) \, d\phi + \int \phi N_S(\phi) \, d\phi \qquad (7.150)$$

Note that in constructing the VP of either $S$ or $W$, we use the centers of *all* the molecules in the system. The partial molar volume of $S$ is

$$\bar{V}_S = (\partial V/\partial N_S)_{N_W} = \int \phi [\partial N_W(\phi)/\partial N_S] \, d\phi + \int \phi [\partial N_S(\phi)/\partial N_S] \, d\phi \qquad (7.151)$$

which is made up of relaxation terms only. At infinite dilution of $S$ in $W$, we have

$$\lim_{\varrho_S \to 0} (\partial/\partial N_S) \left[ N_S \int \phi x_S(\phi) \, d\phi \right] \equiv \langle \phi_S^\circ \rangle \qquad (7.152)$$

Hence

$$\bar{V}_S^\circ = \langle \phi_S^\circ \rangle + \int \phi [\partial N_W(\phi)/\partial N_S] \, d\phi \qquad (7.153)$$

i.e. the partial molar (strictly, molecular) volume of $S$ at infinite dilution is the average volume of the VP of the solute, plus the structural change of the solvent induced by the solute.

## 7.8. CONCLUSION

The nature of the "structural changes" in water induced by the addition of solutes seems to hold the clue to understanding the properties of aqueous solutions. Concepts such as "structure making" and "structure breaking" have been used for many years to interpret various experimental results. With the development of high-speed computers, we are now in a position to examine such conjectures by the available simulation techniques.

We recall that both the "structure of water" and the corresponding "structural changes" were defined as ensemble averages. Therefore, these quantities are amenable to direct computation by both the Monte Carlo

---

[9] In order to minimize the complexity of the notation, we use here the same symbols $\mathbf{N}_W$ and $\mathbf{N}_S$ as in (7.124), but with different meanings.

and the molecular dynamics methods. No work has yet been published along these lines, but surely such studies will be carried out in the near future, ultimately leading to a better understanding of the importance of structural changes in the solvent.

We have seen that structural changes induced by simple solutes are likely to involve a net expansion of the system. Therefore, computations carried out in constant-volume systems do not seem to be appropriate. Instead, the newly developed simulation techniques in the $T, P, N$ ensemble seem to be more promising (Wood, 1968).

The "stabilization effects" in systems at constant volume and at constant pressure are related to each other by

$$\left(\frac{\partial N_L}{\partial N_S}\right)_{T,P,N_W} = \left(\frac{\partial N_L}{\partial N_S}\right)_{T,V,N_W} + \left(\frac{\partial N_L}{\partial V}\right)_{T,N_W,N_S} \left(\frac{\partial V}{\partial N_S}\right)_{T,P,N_W} \quad (7.154)$$

If $L$ represents the component with a relatively low local density, then the second term on the rhs of (7.154) is likely to be positive. Hence, the quantity $(\partial N_L/\partial N_S)_{T,V,N_W}$ may turn out to be either very small or even negative. As we have seen throughout this chapter, it is the stabilization effect at constant pressure that enters into the various thermodynamic quantities of solution. Therefore, the natural arena for studying the structural changes in the solvent should be the $T, P, N$ ensemble.

*Chapter 8*

# Water with Two or More Simple Solutes. Hydrophobic Interaction (HI)

## 8.1. INTRODUCTION

The limiting dilute ideal solution is characterized by the complete absence of "solute–solute interaction."[1] From the molecular point of view, the properties of such systems are determined by the behavior of a system of one solute in a solvent, a topic dealt with in Chapter 7. The next natural system to be studied is that of two solutes in a solvent. In a formal way, the situation is analogous to the case of a low-density gas, for which we can write the virial expansion

$$P/kT = \varrho + B_2\varrho^2 + B_3\varrho^3 + \cdots \quad (8.1)$$

where $\varrho$ is the total number density and $B_K$ is the $K$th virial coefficient.

Similarly, for dilute solutions of a solute $S$, we write the virial expansion for the osmotic pressure as

$$\pi/kT = \varrho_S + B_2{}^*\varrho_S{}^2 + B_3{}^*\varrho_S{}^3 + \cdots \quad (8.2)$$

---

[1] In this book, we reserve the term "solute–solute interaction" to describe the direct interaction between the solutes. Here, we use this term in its more common sense. See Section 4.9 for more details.

where $\varrho_S$ is the number density of the solute and $B_K^*$ is the $K$th virial coefficient of the osmotic pressure.

The complete analogy between various expansions in the total density in the gaseous phase, and expansions in the solute density in solution, has been developed by McMillan and Mayer (1945). In this chapter, we will be interested in the expansion (8.2) up to only the second-order term. For this purpose, we use the statistical mechanical expression for $B_2^*$ as obtained from the Kirkwood–Buff theory in Section 4.9:

$$B_2^* = \lim_{\varrho_S \to 0} \left\{ -\tfrac{1}{2} \int_0^\infty [g_{SS}(R) - 1] 4\pi R^2 \, dR \right\} \tag{8.3}$$

where $g_{SS}(R)$ is the pair correlation function for the solutes. [In the case of nonspherical solutes, (8.3) still holds with the understanding that $g_{SS}(R)$ is the spatial correlation function, obtained after proper averaging over all the orientations.]

The corresponding expression for the second virial coefficient in the gaseous phase is

$$B_2 = \lim_{\varrho \to 0} \left\{ -\tfrac{1}{2} \int_0^\infty [g(R) - 1] 4\pi R^2 \, dR \right\} \tag{8.4}$$

where we know that the low-density limit of the pair correlation function for spherical particles is

$$g(R) \xrightarrow{\varrho \to 0} \exp[-\beta U(R)] \tag{8.5}$$

In solutions, we have no simple, explicit expression for the limiting form of $g_{SS}(R)$. In fact, we saw in Chapter 4 that this function depends strongly on the molecular properties of the solvent.

In this chapter, we are mainly concerned with comparing small deviations from dilute ideal behavior in water and in nonaqueous solutions. A systematic study of this topic has never been undertaken, either because of lack of stimulating motivation or because of experimental difficulties. New interest in this topic has been aroused only recently with the realization that a central problem in biochemistry, the so-called hydrophobic interaction, can be intimately related to the problem of small deviations from dilute ideal solutions. This brought a new impetus to the study of this entire area.

The hydrophobic interaction (HI) problem can be viewed as consisting of one aspect of a more general topic, namely the role of water as a unique medium for supporting most biochemical processes [for reviews, see Edsall and Wyman (1958), Kayushin (1966), Drost-Hansen (1971, 1972)]. In this chapter, we stress a somewhat different point of view. We consider the phenomenon of HI as one aspect of the anomalous properties of liquid water, and do not elaborate on its biological significance.

There are actually several phenomena currently referred to by the term HI. The common element in all of these stems from the picturesque view that simple nonpolar solutes "hate" or "have a phobia" for water, hence they tend to avoid as much as possible being exposed to the aqueous environment. However, there exist more than slight differences in nuance between some of these phenomena, a fact which will become clearer in the following sections. Figure 8.1 is a schematic illustration of various processes believed to be partially governed by HI [for reviews, see Kauzmann (1959), Némethy (1967), and Tanford (1968, 1970)].

It is well known that a particular conformation of a biopolymer maintains its stability only in aqueous solutions. Addition of, say, 20% alcohol causes a conformational change which eventually leads to the process of denaturation. The latter is a very complex process and involves the combination of many factors such as hydrogen bonding, ionic interaction, and van der Waals interaction between the various residues of the polymer. It has been conjectured that the tendency of the nonpolar groups (such as methyl or ethyl groups attached to the amino acids) to avoid the aqueous environment is one of the major reasons for the stabilization of the native conformation of the biopolymer. This is shown schematically in the first process depicted in Fig. 8.1. Here, we stress an extreme example where the polymer is folded in such a way that the side-chain nonpolar groups are *completely* removed from the aqueous medium and transferred to the interior of the polymer, where they are exposed to an environment similar to that of a typical nonpolar solvent.

Kauzmann (1959) has advocated the comparison between the process described above, where a nonpolar group is transferred from the aqueous environment to a relatively nonpolar one, and the process of transferring a solute from water to a nonaqueous solvent. This topic will be discussed further in Section 8.2.

The other extreme case of HI is depicted at the bottom of Fig. 8.1. This is the process of dimerization, i.e., the tendency of simple solutes to adhere to each other in aqueous media. The precise meaning of this tendency will be discussed in the following sections. Here, we stress the fact that as the two solute particles approach each other, they must squeeze out the water molecules between them, as a result of which the "dimer" is surrounded by water molecules to a smaller extent compared with the two separate monomers. As was pointed out by Némethy and Scheraga

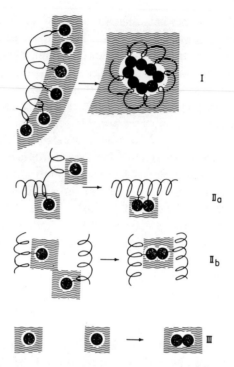

Fig. 8.1. Schematic description of various processes referred to in connection with hydrophobic interaction. (I) Side-chain, nonpolar groups transferred from an aqueous environment into the interior of the polymer. ($II_a$, $II_b$) Intermediate cases where the nonpolar groups are *partially* removed from their aqueous environment. (III) What can be thought of as the most elementary process involving HI, i.e., two simple nonpolar solutes forming a dimer. All the processes take place in a solvent. Heavily shaded circles are the nonpolar groups.

(1962), the process of dimerization can be viewed as a partial reversal of the solubility of the solute in water.

Other intermediary processes involving HI may be the binding of a substrate to an enzyme, association between different polymers, micelle formation, and so forth.

In most of this chapter, we refer to the process of "dimerization" (in a modified sense—see Section 8.3) as the fundamental one in the study of HI. In Section 8.8, we find, in a rather subtle way, the connection between this process and the transfer of a solute from water to a nonaqueous solvent.

Most of the experimental evidence regarding HI is rather indirect and somewhat speculative. In the next section, we present a short survey of the relevant experimental facts. More experimental data will be presented later, after we develop a connection between the strength of the HI and experimental quantities.

Further discussions on this problem are given by Kirkwood (1954), Klotz (1958), Klotz and Luborsky (1959), Kauzmann (1959), Ramachandran and Sasisekharan (1970), and Tanford (1968, 1970).

## 8.2. SURVEY OF EXPERIMENTAL EVIDENCE ON HYDROPHOBIC INTERACTION

1. Perhaps the simplest indication that nonpolar solutes such as argon or methane "dislike" the aqueous environment is found in the values of the standard free energy of transfer of these solutes from a nonaqueous solvent to water. Consider, for example, the standard free energy of transfer of methane from ethanol to water at 10°C ($T, P$ constant)

$$\Delta \mu_t^\circ = \Delta \mu_S^\circ(\text{in water}) - \Delta \mu_S^\circ(\text{in ethanol}) = 1747 - 325 = 1422 \text{ cal/mole} \quad (8.6)$$

What does this quantity mean? In Chapter 4, we gave three possible interpretations of this quantity: (1) the free energy of transferring one mole of methane from ethanol to water, at $T$ and $P$ constant, and at $\varrho_S(\text{water}) = \varrho_S(\text{alcohol})$, i.e., the molar concentrations of methane being the same in the two phases; (2) the free energy of transferring one $S$ molecule from a fixed position in ethanol to a fixed position in water (multipled by Avogadro's number); (3) for $S$ in equilibrium between the two phases,

$$\Delta \mu_t^\circ = -RT \ln[\gamma_S(\text{water})/\gamma_S(\text{alcohol})]_{\text{eq}} \quad (8.7)$$

where $\gamma_S$ is the Ostwald absorption coefficient (Section 7.2).

Relation (8.7), together with the information in (8.6), can be interpreted in terms of the extent to which the solute molecules favor one environment over the other. Such an interpretation has been used widely (e.g., Kauzmann, 1959; Wishnia, 1962, 1963; Wishnia and Pinder, 1964, 1966; Nozaki and Tanford, 1963, 1965; Wetlaufer et al., 1964; Barone et al., 1966).

We have already stressed the fact that the above process does not provide any direct indication of the extent of adherence between two solutes in an aqueous environment. In Section 8.6, we present an example in which such an inference can indeed be misleading.

2. The second experimental quantity that, in principle, is suitable for the study of HI, in the sense of attraction between a pair of particles, is the dimerization constant, or the corresponding standard free energy of dimerization of nonelectrolytes in various solvents.

The most extensive study along this line has been carried out by Scheraga and his collaborators (Schrier *et al.*, 1964; Schneider *et al.*, 1965; Moon *et al.*, 1965; Kunimitsu *et al.*, 1968). As a concrete example, consider the dimerization constants for association of carboxylic acids

$$2R\text{–}COOH \to (R\text{–}COOH)_2 \qquad (8.8)$$

for which we can write the equilibrium constant

$$K_D = [(R\text{–}COOH)_2]/[R\text{–}COOH]^2 \qquad (8.9)$$

where the square brackets stand for the molar concentration (it is assumed that the system is dilute ideal with respect to both the monomer and the dimer). There are two possible dimers of these carboxylic acids, one involving a single hydrogen bond between the two monomers and the other—the cyclic dimer—involving two hydrogen bonds:

$$(8.10)$$

The experimental findings are that the dimerization constants for a homologous series of carboxylic acids in water are much larger than the corresponding values in nonpolar solvents. For example, $K_D$ for acetic acid in water is about 0.16 liter/mole, whereas in $CCL_4$ the value is

$4.2 \times 10^{-4}$ liter/mole (Katchalsky *et al.*, 1951; Wenograd and Spurr, 1957). Furthermore, it is found that $K_D$ in water increases with the chain length of the hydrocarbon residue (Mukerjee, 1965), whereas in nonpolar solvents, it is far less sensitive to variations of the hydrocarbon residue. These facts were utilized to suggest that the dimers in water are of the form depicted on the lhs of (8.10), and therefore the interaction between the nonpolar groups, R, contributes to the dimerization constant, whereas for nonpolar solvents, the rhs dimer was proposed, where these groups hardly "see" each other. [See also Murty (1971).]

Schrier *et al.* (1964) have suggested the following method of extracting information on HI from these data. Suppose that the standard free energy of dimerization ($\Delta\mu_D{}^\circ = -RT \ln K_D$) could be split into two contributions,

$$\Delta\mu_D{}^\circ = \Delta\mu_{\text{HI}}^\circ + \Delta\mu_{\text{HB}}^\circ \qquad (8.11)$$

where the first term on the rhs of (8.11) is the contribution due to the hydrophobic interaction, and the second term is due to the hydrogen bond formation. Furthermore, one assumes that the standard free energy of dimerization of formic acid is devoid of any contribution other than that from the hydrogen bond formation, i.e.,

$$\Delta\mu_D{}^\circ(\text{H–COOH}) = \Delta\mu_{\text{HB}}^\circ(\text{H–COOH}) \qquad (8.12)$$

The third assumption is that variations in the values of $\Delta\mu_D{}^\circ$ are attributed only to variations in $\Delta\mu_{\text{HI}}^\circ$, i.e., the contribution of the quantity $\Delta\mu_{\text{HB}}^\circ$ is independent of the chain length of the hydrocarbon.

Hence, the following equality is assumed to hold for any R:

$$\Delta\mu_{\text{HB}}^\circ(\text{R–COOH}) = \Delta\mu_{\text{HB}}^\circ(\text{H–COOH}) \qquad (8.13)$$

From (8.11)–(8.13), we conclude that

$$\Delta\mu_{\text{HI}}^\circ(\text{R–COOH}) = \Delta\mu_D{}^\circ(\text{R–COOH}) - \Delta\mu_D{}^\circ(\text{H–COOH}) \qquad (8.14)$$

On the basis of (8.14), Schrier *et al.* found the values summarized in Table 8.1 for the contribution to $\Delta\mu_D{}^\circ$ from HI. From the values in this table, we can conclude that the HI increases with the chain length of the hydrocarbon residue. (It is regrettable, however, that no systematic comparison with other solvents has been carried out to establish the unique behavior of liquid water.) The assumptions leading to (8.14), though very appealing on intuitive grounds, have never been supported by sound ar-

**Table 8.1**

Average Values of $\Delta\mu_{\mathrm{HI}}^{\circ}$(R–COOH) in kcal/mole, Computed from (8.14) at 25°C[a]

| | |
|---|---|
| $CH_3$–COOH | −0.8 |
| $CH_3CH_2$–COOH | −1.1 |
| $CH_3CH_2CH_2$–COOH | −1.4 |
| $C_6H_5CH_2$–COOH | −1.6 |

[a] Schrier et al. (1964).

guments, and, therefore, the values in the table should be viewed as only a tentative indication of the existence of HI.

Two other comments, independent of the assumptions made above, are concerned with the suitability of this particular system for the study of HI. First, our intent in this chapter is to study the HI between two simple nonpolar solutes in water. The fact that we looked at molecules possessing polar functional groups, such as the carboxyl group, may lead to erroneous conclusions. The reason for this is that the two nonpolar groups "see" each other through a medium which is already perturbed by the presence of the polar groups. Therefore, the extent of attraction between the alkyl groups must be different from that in pure water. One may argue that the environment around side-chain nonpolar groups attached to the biopolymer is also not immersed in pure water. [This point has been stressed by Wetlaufer et al. (1964), Tanford (1962), and Nozaki and Tanford (1963), and has been reviewed by Némethy (1967).] This is certainly true. However, once we have isolated the problem of HI, we are left with the problem of two simple solutes in a solvent. Such a study must precede the next step, in which various perturbation effects on HI are examined.

The second comment refers to the geometry of the dimers, which, in this case, is dictated to a large extent by the formation of the hydrogen bond between the carboxylic groups. This geometry may not conform to the one which actually occurs in biopolymers, and, therefore, conclusions from this study may not be directly relevant to the HI in actual polymers.

3. The third type of experimental evidence relevant to the problem of HI comes from the analysis of the second virial coefficient $B_2^*$ in the density expansion of the osmotic pressure, carried out by Kozak et al.

(1968). Consider the statistical mechanical expression for $B_2^*$ given in (8.3), which, for simplicity of notation, is written as

$$B_2^* = -\tfrac{1}{2} \int_0^\infty [g_{SS}^\circ(R) - 1] 4\pi R^2 \, dR \qquad (8.15)$$

where $g_{SS}^\circ$ denotes the limit of the solute–solute pair correlation function at infinite dilution. We first outline the process of extracting information on HI from (8.15) for simple spherical solutes. The more general case is quite involved and requires further assumptions. Suppose that $\sigma$ is the effective diameter of the solute particles; then, for $R \leq \sigma$, the pair correlation function is practically zero and we can rewrite (8.15) as

$$B_2^* = -\tfrac{1}{2} \int_0^\sigma (-4\pi R^2) \, dR - \tfrac{1}{2} \int_\sigma^\infty [g_{SS}^\circ(R) - 1] 4\pi R^2 \, dR$$
$$= \tfrac{1}{2}(4\pi\sigma^3/3) - A_2^* \qquad (8.16)$$

In this way, we have split $B_2^*$ into two contributions. The first results from the strongly repulsive region $R \leq \sigma$, and is proportional to the volume of the solute. The second can be identified qualitatively with the attractive region, $R > \sigma$. This last statement is not exact, since $g_{SS}^\circ(R)$ oscillates in this region. The precise regions of attraction and repulsion can be identified through the form of the potential of average force between the two solute particles. A schematic description of this function is given in Fig. 8.2. A positive slope of $W_{SS}(R)$ corresponds to attraction, a negative slope corresponds to repulsion (more details are given in Section 2.6). Thus, although it is clear that the first term in (8.16) results from a repulsive

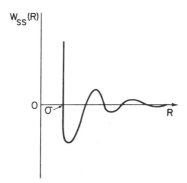

Fig. 8.2. Schematic description of the function $W_{SS}(R)$. Positive slope corresponds to attraction and negative slope to repulsion.

region, the second term includes alternating regions of attraction and repulsion. The important point in this splitting is the distinction between a term that is purely a property of the solute (and therefore has the same value in any solvent) and a term that depends on the solute as well as on the solvent properties.

Since the major contribution to $A_2{}^*$ comes from the area under the first peak of $g_{SS}^0(R)$, it is expected that this quantity is positive. In a qualitative manner, it can serve as a measure of the overall correlation, or "affinity," of one solute with respect to another in the region $\sigma < R < \infty$. This is the basis of the work of Kozak *et al.* In practice, one cannot get values of $B_2{}^*$ for simple solutes such as argon or methane, since their solubilities are very low. Thus, one has to adopt the above method for more complex solutes such as alcohols, amines, amino acids, etc. The generalization of the splitting (8.16) to the last-named molecules is not simple, however. First, there is no clear-cut distinction between a purely repulsive region, which produces a term proportional to the volume of the solute molecules, and an attractive region. Kozak *et al.* assumed that instead of the first term on the rhs of (8.16), one may take a quantity which is proportional to the partial molar volume of the solute at infinite dilution. The idea is that such a quantity will account for the term that is proportional to the volume of the solute, an assumption which, in general, cannot be justified if large structural changes occur in the solvent. In such a case, there may be no simple relation between $\bar{V}_S{}^\circ$ and the volume of the solute [see Eq. (7.153)].

The second difficulty involves the interpretation of the second term in (8.16) as a measure of the extent of HI. This term will now include many regions of attraction and repulsion, some of which are due to the interaction between the polar groups such as OH, COOH, or $NH_2$. The identification of the contribution from the HI to $A_2{}^*$ is very difficult. In spite of these difficulties, computed values of $A_2{}^*$ for a homologous series of, say, alcohols, show that it is positive and the value increases with the chain length of the solute. This indicates a trend similar to that for the results obtained from the study of dimerization constants discussed above. Unfortunately, there has been no comparative study of the same quantities in nonaqueous solvents—a study which is indispensable for establishing the unique character of aqueous solutions.

We have thus far surveyed some of the experimental facts having the most direct bearing on the problem of HI. There is much indirect experimental evidence from which it is quite difficult to extract quantitative information, though it provides a useful qualitative indication of the existence of HI. Examples are the comparison of the behavior of various synthetic polymers, such as polymethacrylic acid with polyacrylic acid (Eliassaf and Silberberg, 1959; Priel and Silberberg, 1970), or the study of alternating copolymers with various side-chain nonpolar groups (Dubin and Strauss, 1970). In the rest of this chapter, we focus our attention on the simplest "reaction" having direct bearing on the HI problem. Our emphasis will be on the study of the properties of water as a medium for HI. Therefore, the particular solutes chosen for such a study will be of secondary importance.

Finally, we note that there exists some controversy regarding the term HI, or hydrophobic bond (Hildebrand, 1968; Nemethy et al., 1968; Ben-Naim, 1971a). In the next section, we shall define precisely the concept to which we assign the term HI. Although we will be mainly concerned with aqueous solutions, we will use the same notation when discussing nonaqueous solutions.

## 8.3. FORMULATION OF THE PROBLEM OF HYDROPHOBIC INTERACTION (HI)

In this section, we formulate the problem of HI in statistical mechanical terms, and develop the various concepts in the $T$, $V$, $N$ ensemble. The corresponding treatment in the $T$, $P$, $N$ ensemble is slightly more complicated, and is deferred to Section 8.10.

Consider a system of $N$ solvent molecules (we sometimes use the subscript $W$ for water or any other solvent molecule) and two simple, spherical solute molecules at fixed positions $\mathbf{R}_1$ and $\mathbf{R}_2$, the system being at a given volume $V$ and temperature $T$. For simplicity of notation, we reserve the first two indices, $i = 1, 2$, for the solutes, and the remaining indices, $i = 3, 4, \ldots, N + 2$, for the solvent molecules. The Helmholtz free energy of such a system is given by

$$\exp[-\beta A^l_{N+2}(\mathbf{R}_1, \mathbf{R}_2)] = (q_W{}^N q_S{}^2/N!) \int \cdots \int d\mathbf{X}^N \exp[-\beta U_N(\mathbf{X}^N)$$
$$- \beta U(\mathbf{R}_1, \mathbf{R}_2) - \beta U(\mathbf{X}^N/\mathbf{R}_1, \mathbf{R}_2)] \qquad (8.17)$$

where $q_W$ and $q_S$ are the internal partition functions for a single solvent and solute molecule, respectively. We have also included in $q_W$ the momentum partition function and a normalization constant $8\pi^2$. Note that since the two solute particles are assumed to have a fixed position, $q_S$ does not include a momentum partition function. The total potential function of the whole system is

$$U_{N+2}(\mathbf{X}^N, \mathbf{R}_1, \mathbf{R}_2) = U_N(\mathbf{X}^N) + U(\mathbf{R}_1, \mathbf{R}_2) + U(\mathbf{X}^N/\mathbf{R}_1, \mathbf{R}_2) \quad (8.18)$$

where $U_N(\mathbf{X}^N)$ denotes the potential energy of interaction among the $N$ solvent molecules for the specific configuration $\mathbf{X}^N$. The direct pair potential between the two solute particles is denoted by $U(\mathbf{R}_1, \mathbf{R}_2)$ and the total interaction between the solvent molecules at $\mathbf{X}^N$ and the solute particles at $\mathbf{R}_1, \mathbf{R}_2$ is denoted by $U(\mathbf{X}^N/\mathbf{R}_1, \mathbf{R}_2)$. The integration in (8.17) extends over all the configurations of the solvent molecules $\mathbf{X}^N = \mathbf{X}_3, \mathbf{X}_4, \ldots, \mathbf{X}_{N+2}$. Note also that the assumption made in (8.18) does not involve pairwise additivity for the solvent molecules, a fact of considerable importance in the study of aqueous solutions. Later, we will require, however, that the last term in (8.18) be represented by a sum of the form

$$U(\mathbf{X}^N/\mathbf{R}_1, \mathbf{R}_2) = \sum_{k=1}^{2} \sum_{i=3}^{N+2} U(\mathbf{X}_i, \mathbf{R}_k) \quad (8.19)$$

where $U(\mathbf{X}_i, \mathbf{R}_k)$ is the solvent–solute pair potential.[2] The assumption (8.19) is far weaker than the one involving pairwise additivity for the solvent molecules. In fact, we later use (8.19) for such solutes as hard-sphere particles, for which this relation becomes exact.

The Helmholtz free energy of the two solute particles at fixed positions $\mathbf{R}_1$ and $\mathbf{R}_2$ in a vacuum is

$$\exp[-\beta A_2{}^g(\mathbf{R}_1, \mathbf{R}_2)] = q_S{}^2 \exp[-\beta U(\mathbf{R}_1, \mathbf{R}_2)] \quad (8.20)$$

and the free energy of the pure solvent is

$$\exp(-\beta A_N{}^l) = (q_W{}^N/N!) \int \cdots \int d\mathbf{X}^N \exp[-\beta U_N(\mathbf{X}^N)] \quad (8.21)$$

Consider the work required to transfer the two solute particles from fixed positions $\mathbf{R}_1, \mathbf{R}_2$ in a vacuum to the same fixed positions in the liquid[3] (the subscript tr stands for transfer):

$$\Delta A_{\text{tr}}(\mathbf{R}_1, \mathbf{R}_2) = A^l_{N+2}(\mathbf{R}_1, \mathbf{R}_2) - A_N{}^l - A_2{}^g(\mathbf{R}_1, \mathbf{R}_2) \quad (8.22)$$

---

[2] Note that we use the same letter, $U$, for different pair potentials. We refrain from using additional subscripts to indicate the species involved, whenever this is clear from the argument. Here, $\mathbf{X}_i$ is always the configuration of a *solvent* molecule, whereas $\mathbf{R}_k$ refers to a solute molecule.

[3] We use here $\mathbf{R}_1$ and $\mathbf{R}_2$ for the positions of the solute particles in the two phases. The important requirement is, of course, that the distance $R = |\mathbf{R}_2 - \mathbf{R}_1|$ be the same in the two phases.

## Water with Two or More Simple Solutes

The corresponding statistical mechanical expression is

$$\exp[-\beta \Delta A_{\mathrm{tr}}(\mathbf{R}_1, \mathbf{R}_2)] = \frac{\int \cdots \int d\mathbf{X}^N \exp[-\beta U_N(\mathbf{X}^N) - \beta U(\mathbf{X}^N/\mathbf{R}_1, \mathbf{R}_2)]}{\int \cdots \int d\mathbf{X}^N \exp[-\beta U_N(\mathbf{X}^N)]}$$

$$= \langle \exp[-\beta U(\mathbf{X}^N/\mathbf{R}_1, \mathbf{R}_2)] \rangle \qquad (8.23)$$

where on the rhs we have an average over all the configurations of the solvent molecules. It is important to realize that the internal properties and the pair potential of the solutes do not appear in (8.23). The effect of the solutes enters only through the "potential field" that they produce on the system, i.e., $U(\mathbf{X}^N/\mathbf{R}_1, \mathbf{R}_2)$. This property will later be exploited to get a relation between HI and experimental quantities. Relation (8.23) may also be viewed as a simple generalization of the work required to transfer a single solute particle from a fixed position in the gas to a fixed position in the liquid.

Next, consider the work required to bring the two solute particles from fixed positions at infinite separation from each other, to the final configuration $\mathbf{R}_1, \mathbf{R}_2$, the process being carried out within the liquid with $T$ and $V$ constant. From (8.17), we have

$$\exp\{-\beta[A^l_{N+2}(\mathbf{R}_1, \mathbf{R}_2) - A^l_{N+2}(R = \infty)]\}$$
$$= \frac{\int \cdots \int d\mathbf{X}^N \exp[-\beta U_N(\mathbf{X}^N) - \beta U(\mathbf{R}_1, \mathbf{R}_2) - \beta U(\mathbf{X}^N/\mathbf{R}_1, \mathbf{R}_2)]}{\int \cdots \int d\mathbf{X}^N \exp[-\beta U_N(\mathbf{X}^N) - \beta U(\mathbf{X}^N/R = \infty)]} \qquad (8.24)$$

where $R = \infty$ indicates a configuration $\mathbf{R}_1, \mathbf{R}_2$ for which

$$R = |\mathbf{R}_2 - \mathbf{R}_1| = \infty.$$

For any configuration of the two solutes $\mathbf{R}_1$ and $\mathbf{R}_2$, we define the quantity $A^{\mathrm{HI}}(\mathbf{R}_1, \mathbf{R}_2)$ by

$$\exp[-\beta A^{\mathrm{HI}}(\mathbf{R}_1, \mathbf{R}_2)] = \int \cdots \int d\mathbf{X}^N \exp[-\beta U_N(\mathbf{X}^N) - \beta U(\mathbf{X}^N/\mathbf{R}_1, \mathbf{R}_2)] \qquad (8.25)$$

and rewrite (8.24) as

$$\Delta A^l(R) = A^l_{N+2}(\mathbf{R}_1, \mathbf{R}_2) - A^l_{N+2}(R = \infty)$$
$$= U(\mathbf{R}_1, \mathbf{R}_2) + [A^{\mathrm{HI}}(\mathbf{R}_1, \mathbf{R}_2) - A^{\mathrm{HI}}(R = \infty)]$$
$$= U(\mathbf{R}_1, \mathbf{R}_2) + \delta A^{\mathrm{HI}}(\mathbf{R}_1, \mathbf{R}_2) \qquad (8.26)$$

Thus, the total work $\Delta A^l(R)$ of bringing the two solutes from fixed positions at infinite separation to a distance $R$ is split in (8.26) into two parts. The first is the direct work against the solute–solute pair potential, and the second is the indirect work resulting from the presence of the solvent.

We can now formulate the problem of HI in terms of the function $\Delta A^l(R)$. Suppose we compare this function for a given pair of solutes in two different solvents, water and nonaqueous solvent. The question is whether this function has a notably lower minimum, at a separation $R \approx \sigma$ ($\sigma$ being the diameter of the solute), in water as compared to other solvents.

A glance at (8.26) shows that any information on the direct pair potential $U(\mathbf{R}_1, \mathbf{R}_2)$ is totally irrelevant to our problem. Thus, although it is true that direct van der Waals forces contribute to the total work $\Delta A^l(R)$, this part is common in all the solvents, and, therefore, if there exists any anomaly in water, it must show up in the indirect part of the work $\delta A^{\mathrm{HI}}(\mathbf{R}_1, \mathbf{R}_2)$. It was Kirkwood (1954) who first stressed the point that more attention should be paid to the role of the solvent, i.e., water, than to the direct van der Waals attraction, in establishing any excessive tendency for the two solutes to adhere to each other in aqueous media. It is for this reason that we have used the notation $\delta A^{\mathrm{HI}}(\mathbf{R}_1, \mathbf{R}_2)$ in (8.26) to stress that it is this quantity that should be studied in connection with the problem of HI.[4]

We now turn to a different, though equivalent, formulation of the problem of HI. To do this, we use the following two identities derived in Section 2.6:

$$\exp\{-\beta[A^l_{N+2}(R) - A^l_{N+2}(R = \infty)]\} = g_{SS}(R)/g_{SS}(R = \infty) \quad (8.27)$$

$$g_{SS}(R) = y_{SS}(R)\exp[-\beta U(R)] \quad (8.28)$$

where, for simplicity, we put $R = |\mathbf{R}_2 - \mathbf{R}_1|$. In the first relation, we may take $g_{SS}(R = \infty) = 1$, assuming that we have already taken the infinite-system-size limit (for details, see Section 3.9). Relation (8.27) tells us that in order to study the work required to bring the two solute particles from infinite separation to a distance $R$, it is sufficient to study the pair correlation function $g_{SS}(R)$. The question can now be stated as follows: Having a system with two *free* solute particles, what is the probability density of finding these particles at a distance $R \approx \sigma$, and is this probability density notably larger in water than in other solvents?

---

[4] In this chapter, we use the superscript HI mainly for "hydrophobic interaction." Clearly, the quantities defined in (8.26) can also apply to any other solvent for which the term "solvophobic interaction" may be more appropriate. Nevertheless, for simplicity of notation, we continue to use the same symbol for these other solvents as well.

### Water with Two or More Simple Solutes

It is important to realize that the two formulations of the problem of HI have been referred to two different systems. In the first, we have two solute particles at *fixed* positions in the system and we search for the free energy of the whole system as a function of $R$. In the second, the two solute particles are assumed to be *free* so that they can wander about in the system, and the relevant question involves the probability of finding the two particles at some distance $R \approx \sigma$.

Relation (8.28) indicates a way to eliminate the direct pair potential $U(R)$. Suppose that we compare two solvents $l$ and $f$, and seek the pair correlation functions $g_{SS}^l(R)$ and $g_{SS}^f(R)$ for the same solute in the two solvents. From (8.28), we get

$$g_{SS}^l(R)/g_{SS}^f(R) = y_{SS}^l(R)/y_{SS}^f(R) \qquad (8.29)$$

Therefore, instead of looking at the ratio of the $g$'s, we can look at the ratio of the $y$'s. The latter is more convenient, since this function is well-behaved in the region $0 < R < \infty$, even for hard spheres (see Section 2.6).

Finally, from (8.26)–(8.28), we get the relation

$$y_{SS}(R) = \exp[-\beta\,\delta A^{\mathrm{HI}}(R)] \qquad (8.30)$$

In the following sections, we will find it more convenient to study the function $\delta A^{\mathrm{HI}}(R)$, which, by virtue of (8.28) and (8.30), can be transformed into $g_{SS}(R)$.

## 8.4. A POSSIBLE CONNECTION BETWEEN HI AND EXPERIMENTAL QUANTITIES

Consider the quantity $\delta A^{\mathrm{HI}}(\mathbf{R}_1, \mathbf{R}_2)$ defined in (8.26), to which we shall refer as a measure of the strength of the HI at the configuration $\mathbf{R}_1, \mathbf{R}_2$.

$$\begin{aligned}
\delta A^{\mathrm{HI}}(\mathbf{R}_1, \mathbf{R}_2) &= A_{N+2}^l(\mathbf{R}_1, \mathbf{R}_2) - A_{N+2}^l(R = \infty) - U(\mathbf{R}_1, \mathbf{R}_2) \\
&= [A_{N+2}^l(\mathbf{R}_1, \mathbf{R}_2) - A_N^l - A_2^g(\mathbf{R}_1, \mathbf{R}_2)] \\
&\quad - [A_{N+2}^l(R = \infty) - A_N^l - A_2^g(R = \infty)] \\
&= \Delta A_{\mathrm{tr}}(\mathbf{R}_1, \mathbf{R}_2) - \Delta A_{\mathrm{tr}}(R = \infty) \qquad (8.31)
\end{aligned}$$

In the second form on the rhs, we have added and subtracted $A_N^l$, and we have also used relation (8.20). The last form on the rhs of (8.31) employs the definition of $\Delta A_{\mathrm{tr}}$ in (8.22).

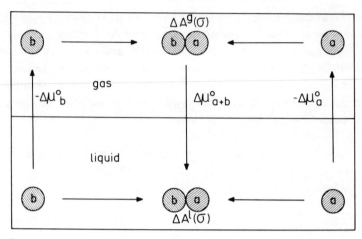

Fig. 8.3. An "ideal cycle" relevant to Eqs. (8.31) and (8.32). Two solute particles $a$ and $b$ are brought from infinite separation to a close distance $\sigma$. The corresponding free energy changes in the liquid and gaseous phases are denoted by $\Delta A^l(\sigma)$ and $\Delta A^g(\sigma)$, respectively. The cycle is closed by transferring all solutes from one phase to another. Note that $\Delta\mu_a^\circ$ and $\Delta\mu_b^\circ$ are measurable quantities but $\Delta\mu_{a+b}^\circ$ is not.

Relation (8.31) suggests an interesting possibility for a connection between HI and experimental quantities. Suppose, for concreteness, that $R = |\mathbf{R}_2 - \mathbf{R}_1| = \sigma$. Thus, $\Delta A_{\text{tr}}(R = \infty)$ is the free energy change for transferring two solute particles from fixed positions in the gas to fixed positions in the liquid. In both cases, the separation between the two particles is infinite. Similarly, $\Delta A_{\text{tr}}(R = \sigma)$ is the free energy change for transferring the pair of particles at distance $\sigma$, as a single entity, into the liquid. Figure 8.3 shows a cyclic process relevant to the above discussion. We consider two solutes $a$ and $b$ at fixed positions in the liquid but far apart, brought to a contact distance within the liquid. We denote the pair of solute particles at this separation by $a + b$. The corresponding total work associated with this process is given in (8.26). Alternatively, we may choose the "round trip" as indicated in the figure. First, we take the two solutes from the liquid into the gaseous phase, then we bring the two solutes to the distance $\sigma$, and finally we introduce the pair $a + b$ as a whole into the liquid. The total balance of work in the two routes is

$$\Delta A^l(\sigma) = \Delta A^g(\sigma) + \Delta\mu_{a+b}^\circ - \Delta\mu_a^\circ - \Delta\mu_b^\circ \tag{8.32}$$

or

$$\delta A^{\text{HI}}(\sigma) = \Delta\mu_{a+b}^\circ - \Delta\mu_a^\circ - \Delta\mu_b^\circ \tag{8.33}$$

# Water with Two or More Simple Solutes

The above relation has been "read off" by looking at the cyclic process in Fig. 8.3. We shall prove it in a more formal way later. Before doing this, it is appropriate to note that although $\Delta\mu_a^\circ$ and $\Delta\mu_b^\circ$ are measurable quantities, $\Delta\mu_{a+b}$ is not. The reason for this is that we do not have a "molecule" denoted by $a + b$. We therefore refer to the cycle in Fig. 8.3 as an "ideal cycle."

We now develop a formal and convenient expression for $\delta A^{\mathrm{HI}}(R)$. From (8.23) and (8.31), we get (in the following, we assume that $a$ and $b$ are the same kind of solute, denoted by $S$)

$$\exp[-\beta\, \delta A^{\mathrm{HI}}(\mathbf{R}_1, \mathbf{R}_2)]$$
$$= \langle\exp[-\beta U(\mathbf{X}^N/\mathbf{R}_1, \mathbf{R}_2)]\rangle/\langle\exp[-\beta U(\mathbf{X}^N/R = \infty)]\rangle \quad (8.34)$$

The most important aspect of this relation is the absence of the internal properties as well as of the direct pair potential for the two solute particles. One may therefore envisage a hypothetical system of pure solvent subjected to an "external" field of force originating from two fixed points and given by $U(\mathbf{X}^N/\mathbf{R}_1, \mathbf{R}_2)$. In such a system, $\delta A^{\mathrm{HI}}(\mathbf{R}_1, \mathbf{R}_2)$ is the work required to move the two sources of the field of force from infinite separation to the final positions $\mathbf{R}_1$ and $\mathbf{R}_2$. This point of view will be exploited further in the next section.

We now use a procedure similar to the one employed in Sections 3.5 and 3.6 to rewrite (8.34) in a more convenient form.

We introduce a coupling parameter $\xi$ which takes the values in the range $0 \leq \xi \leq 1$, and define the auxiliary potential function

$$U(\mathbf{X}^N, \mathbf{R}_1, \mathbf{R}_2, \xi) = U_N(\mathbf{X}^N) + \xi \sum_{k=1}^{2} \sum_{i=3}^{N+2} U(\mathbf{X}_i, \mathbf{R}_k) \quad (8.35)$$

Clearly, we have [see (8.19)]

$$U(\mathbf{X}^N, \mathbf{R}_1, \mathbf{R}_2, \xi = 0) = U_N(\mathbf{X}^N)$$
$$U(\mathbf{X}^N, \mathbf{R}_1, \mathbf{R}_2, \xi = 1) = U_N(\mathbf{X}^N) + U(\mathbf{X}^N/\mathbf{R}_1, \mathbf{R}_2)$$

Note that $\xi$ serves to "switch on" only the solute–solvent interaction, i.e., the field of force produced by the solutes. The direct, solute–solute pair potential does not have any role in this context.

We also introduce the auxiliary configurational partition function

$$Z(\mathbf{R}_1, \mathbf{R}_2, \xi) = \int \cdots \int d\mathbf{X}^N \exp[-\beta U(\mathbf{X}^N, \mathbf{R}_1, \mathbf{R}_2, \xi)] \quad (8.36)$$

so that (8.34) can be rewritten, noting (8.23), as

$$-\beta \, \delta A^{\text{HI}}(\mathbf{R}_1, \mathbf{R}_2) = \ln[Z(\mathbf{R}_1, \mathbf{R}_2, \xi = 1)/Z(\mathbf{R}_1, \mathbf{R}_2, \xi = 0)]$$
$$- \ln[Z(R = \infty, \xi = 1)/Z(R = \infty, \xi = 0)] \quad (8.37)$$

where again $R = \infty$ indicates two fixed positions but at infinite separation.
Following standard steps as in Section 3.5, we write

$$\ln[Z(\mathbf{R}_1, \mathbf{R}_2, \xi = 1)/Z(\mathbf{R}_1, \mathbf{R}_2, \xi = 0)]$$
$$= \int_0^1 d\xi \, \{\partial \ln Z(\mathbf{R}_1, \mathbf{R}_2, \xi)]/\partial \xi\}$$
$$= \int_0^1 d\xi \, Z(\mathbf{R}_1, \mathbf{R}_2, \xi)^{-1} \int \cdots \int d\mathbf{X}^N \exp[-\beta U(\mathbf{X}^N, \mathbf{R}_1, \mathbf{R}_2, \xi)]$$
$$\times \left[-\beta \sum_{k=1}^{2} \sum_{i=3}^{N+2} U(\mathbf{X}_i, \mathbf{R}_k)\right]$$
$$= -\beta \int_0^1 d\xi \sum_{k=1}^{2} \sum_{i=3}^{N+2} \int \cdots \int d\mathbf{X}^N P(\mathbf{X}^N/\mathbf{R}_1, \mathbf{R}_2, \xi) U(\mathbf{X}_i, \mathbf{R}_k)$$
$$= -\beta \int_0^1 d\xi \int d\mathbf{X}_W \left[\sum_{k=1}^{2} U(\mathbf{X}_W, \mathbf{R}_k)\right] \varrho(\mathbf{X}_W/\mathbf{R}_1, \mathbf{R}_2, \xi) \quad (8.38)$$

In the second form on the rhs, we have differentiated (8.36) with respect to $\xi$. In the third form, we have introduced the conditional probability density of observing a configuration $\mathbf{X}^N$, given two solute particles at $\mathbf{R}_1$ and $\mathbf{R}_2$ coupled to the extent $\xi$. The last form employs the definition of the conditional singlet distribution function for finding any solvent molecule at $\mathbf{X}_W$ given two solute particles at $\mathbf{R}_1$ and $\mathbf{R}_2$ coupled to the extent $\xi$. [Recall that the coupling parameter $\xi$ in (8.35) serves to "switch on" only the solute–solvent interaction.]

The last form in (8.38) is very similar to the coupling integral which appears in the chemical potential, e.g., (3.86). The only difference is that we have two, instead of one, field-of-force sources acting on our system.

In order to rewrite (8.37) in terms of integrals of the form (8.38), we also need the limit of (8.38) for $R = \infty$. This is obtained as follows. For any conditional distribution function, we can write [see, e.g., (2.114)]

$$\varrho(\mathbf{X}_W/\mathbf{R}_1, \mathbf{R}_2) = \frac{\varrho^{(3)}(\mathbf{X}_W, \mathbf{R}_1, \mathbf{R}_2)}{\varrho^{(2)}(\mathbf{R}_1, \mathbf{R}_2)} = \frac{\varrho(\mathbf{R}_1, \mathbf{R}_2/\mathbf{X}_W)\varrho^{(1)}(\mathbf{X}_W)}{\varrho^{(2)}(\mathbf{R}_1, \mathbf{R}_2)} \quad (8.39)$$

## Water with Two or More Simple Solutes

For $R \to \infty$, we have

$$\varrho^{(2)}(\mathbf{R}_1, \mathbf{R}_2) \to \varrho^{(1)}(\mathbf{R}_1)\varrho^{(1)}(\mathbf{R}_2) \qquad (R \to \infty)$$

$$\varrho(\mathbf{R}_1, \mathbf{R}_2/\mathbf{X}_W) \to \varrho(\mathbf{R}_1/\mathbf{X}_W)\varrho(\mathbf{R}_2/\mathbf{X}_W) \qquad (R \to \infty)$$

Hence,

$$\varrho(\mathbf{X}_W/\mathbf{R}_1, \mathbf{R}_2) \to \frac{\varrho(\mathbf{R}_1/\mathbf{X}_W)\varrho(\mathbf{R}_2/\mathbf{X}_W)\varrho^{(1)}(\mathbf{X}_W)}{\varrho^{(1)}(\mathbf{R}_1)\varrho^{(1)}(\mathbf{R}_2)} = \frac{\varrho(\mathbf{X}_W/\mathbf{R}_1)\varrho(\mathbf{X}_W/\mathbf{R}_2)}{\varrho^{(1)}(\mathbf{X}_W)} \qquad (8.40)$$

Relation (8.40) is now employed in (8.38) to obtain the limiting form of the integral

$$\lim_{R \to \infty} \left\{ \int_0^1 d\xi \int d\mathbf{X}_W \, [U(\mathbf{X}_W, \mathbf{R}_1) + U(\mathbf{X}_W, \mathbf{R}_2)]\varrho(\mathbf{X}_W/\mathbf{R}_1, \mathbf{R}_2, \xi) \right\}$$

$$= \int_0^1 d\xi \int d\mathbf{X}_W \, U(\mathbf{X}_W, \mathbf{R}_1)\varrho(\mathbf{X}_W/\mathbf{R}_1, \xi)\varrho(\mathbf{X}_W/\mathbf{R}_2, \xi)/\varrho^{(1)}(\mathbf{X}_W)$$

$$+ \int_0^1 d\xi \int d\mathbf{X}_W \, U(\mathbf{X}_W, \mathbf{R}_2)\varrho(\mathbf{X}_W/\mathbf{R}_1, \xi)\varrho(\mathbf{X}_W/\mathbf{R}_2, \xi)/\varrho^{(1)}(\mathbf{X}_W)$$

$$= \int_0^1 d\xi \int d\mathbf{X}_W \, U(\mathbf{X}_W, \mathbf{R}_1)\varrho(\mathbf{X}_W/\mathbf{R}_1, \xi)$$

$$+ \int_0^1 d\xi \int d\mathbf{X}_W \, U(\mathbf{X}_W, \mathbf{R}_2)\varrho(\mathbf{X}_W/\mathbf{R}_2, \xi) = 2\Delta\mu_S^\circ \qquad (8.41)$$

In the second form on the rhs, we have separated the integral into two integrals. We now recall that $\mathbf{R}_1$ and $\mathbf{R}_2$ are very far apart. Furthermore, the first integral receives a contribution only when $|\mathbf{R}_W - \mathbf{R}_1|$ is small enough so that $U(\mathbf{X}_W, \mathbf{R}_1)$ is nonzero. Therefore, the conditional distribution function $\varrho(\mathbf{X}_W/\mathbf{R}_2, \xi)$ is equal to $\varrho^{(1)}(\mathbf{X}_W)$. In other words, since the integration over $d\mathbf{X}_W$ extends effectively over a small region around $\mathbf{R}_1$, the fact that we have another particle fixed at $\mathbf{R}_2$ does not influence the function $\varrho(\mathbf{X}_W/\mathbf{R}_2, \xi)$, which attains its limiting value $\varrho^{(1)}(\mathbf{X}_W)$ in the first integral. Similarly, in the second integral, the integration extends over a small region around $\mathbf{R}_2$; hence, here we can replace $\varrho(\mathbf{X}_W/\mathbf{R}_1, \xi)$ by its limiting value $\varrho^{(1)}(\mathbf{X}_W)$. This leads to the third form on the rhs of (8.41). Next, we recognize that each of the two integrals is equal to the standard free energy of solution of the solute $S$.

Let us now collect the last results and substitute in (8.37):

$$\delta A^{\mathrm{HI}}(\mathbf{R}_1, \mathbf{R}_2) = \Delta A_{\mathrm{tr}}(\mathbf{R}_1, \mathbf{R}_2) - \Delta A_{\mathrm{tr}}(R = \infty)$$

$$= \int_0^1 d\xi \int d\mathbf{X}_W \, [U(\mathbf{X}_W, \mathbf{R}_1)$$

$$+ U(X_W, \mathbf{R}_2)]\varrho(\mathbf{X}_W/\mathbf{R}_1, \mathbf{R}_2, \xi) - 2\Delta\mu_S^\circ \quad (8.42)$$

It is worthwhile noting that the equality

$$\Delta A_{\mathrm{tr}}(R = \infty) = 2\Delta\mu_S^\circ \quad (8.43)$$

is valid only if we refer to the process of transferring a pair of particles from fixed positions in the gas to fixed positions in the liquid. This relation simply states that the work of transferring such a pair of particles is twice the work required to transfer one particle—provided that the separation between the two particles is very large. Such an equality does not exist if we remove the constraint imposed on the positions of the two particles. [More details on the latter are given by Ben-Naim (1971a, 1972a).]

As we have noted, relation (8.42) is *almost* a connection between HI and experimental quantities. The missing link is the integral, which, in general, is not a measurable quantity. But suppose that we could have formed a "dimer" $a + b$ by sticking the two solutes to each other as in Fig. 8.3. In such a case, the configuration of the "dimer" $(\mathbf{R}_1, \mathbf{R}_2)$ (with $R = \sigma$) can be denoted by $\mathbf{X}_D$, and the dimer–solvent interaction is

$$U(\mathbf{X}_W, \mathbf{X}_D) = U(\mathbf{X}_W, \mathbf{R}_1) + U(\mathbf{X}_W, \mathbf{R}_2) \quad (8.44)$$

Hence, the integral in (8.42) takes the form

$$\int_0^1 d\xi \int d\mathbf{X}_W \, U(\mathbf{X}_W, \mathbf{X}_D)\varrho(\mathbf{X}_W/\mathbf{X}_D, \xi) = \Delta\mu_D^\circ \quad (8.45)$$

where it is identified as the standard free energy of solution of the "dimer." The final result is

$$\delta A^{\mathrm{HI}}(R = \sigma) = \Delta\mu_D^\circ - 2\Delta\mu_S^\circ \quad (8.46)$$

a relation at which we previously arrived using arguments based on the cyclic process, as depicted in Fig. 8.3. It is worth noting that had we a means for measuring the concentration of the "dimers" $a + b$ in a dilute solution of a solute $S$, then (8.46) could have been evaluated through the

# Water with Two or More Simple Solutes

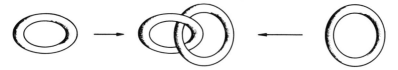

Fig. 8.4. Two large cyclic hydrocarbons ($C_nH_{2n}$, $n > 22$) are brought from infinity to a final configuration where they are interlocked to form a catenan molecule.

ratio of the equilibrium constant for dimerization[5] in the gas and in the liquid, namely

$$\Delta\mu_D^\circ - 2\Delta\mu_S^\circ = -kT\ln(\varrho_D/\varrho_S^2)_{eq}^l - kT\ln(\varrho_D/\varrho_S^2)_{eq}^g \quad (8.47)$$

where $\varrho_S$ and $\varrho_D$ are the concentrations of $S$ and $D$ at equilibrium.

Although we shall be mainly interested in the HI between simple solutes such as methane, argon, and the like, for which $\delta A^{HI}(\sigma)$ cannot be measured directly through (8.47), it is instructive at this stage to present an example of real molecules for which the "ideal cycle" depicted in Fig. 8.3 is applicable, and for which (8.47) can, in principle, be evaluated.

Let $a$ and $b$ be two large cycloparaffin molecules, $C_nH_{2n}$ (with say $n \geq 22$). A *catenan*[6] is formed by interlocking the two rings as shown in Fig. 8.4. The interesting feature of the molecule $a + b$ formed in this way is that it does not involve a direct chemical bond between $a$ and $b$. Therefore, if we perform the whole cycle, as in Fig. 8.3, with these two solutes, we can actually dissolve $a + b$ and, in principle, $\Delta\mu_{a+b}^\circ$ can be measured. Hence, the HI between the two cyclic molecules is given by

$$\delta A^{HI} = \Delta\mu_{a+b}^\circ - \Delta\mu_a^\circ - \Delta\mu_b^\circ \quad (8.48)$$

As we will see in Section 8.7, the quantity $\delta A^{HI}$ in (8.48) must be understood as an average over all possible configurations of the catenan molecule. Here we wished only to give an example of the applicability of relation (8.46).

---

[5] Note that application of (8.47) is different from the method of extracting information on HI discussed in Section 8.2. Relation (8.47) is based on the assumption (8.18). In particular, we assume that the direct solute–solute interaction is the same in all solvents. Such an assumption may be a good approximation for simple solutes, but not for solutes that bear highly polar groups, such as carboxylic acids.

[6] There are some fascinating problems regarding the preparation and properties of these molecules (Frisch and Wasserman, 1961; Wasserman, 1962).

We now briefly discuss a similar cycle in the $T, P, N$ ensemble. Consider the process of bringing two solute particles from fixed positions at infinite separation to some final separation $R$, the process being carried out at constant $T, P, N$. The total work is the change in the Gibbs free energy. As in (8.26), we write

$$\Delta G^l(R) = G^l_{N+2}(R) - G^l_{N+2}(R = \infty) = U(R) + \delta G^{\mathrm{HI}}(R) \quad (8.49)$$

where $\delta G^{\mathrm{HI}}(R)$ is referred to as the HI part of the total work.

Repeating almost the same process carried out in this section, we arrive at a final expression for $\delta G^{\mathrm{HI}}(R)$,

$$\delta G^{\mathrm{HI}}(R) = \Delta\mu^\circ_{a+b} - \Delta\mu^\circ_a - \Delta\mu^\circ_b \quad (8.50)$$

Thus, the strength of the HI can be measured either by $\delta A^{\mathrm{HI}}(R)$, for a process at constant volume, or by $\delta G^{\mathrm{HI}}(R)$, for the same process performed at constant pressure. We will see in Section 8.10 that the latter is advantageous for evaluating the entropy and enthalpy of the HI from experimental quantities.

## 8.5. AN APPROXIMATE CONNECTION BETWEEN HI AND EXPERIMENTAL QUANTITIES

In the previous section, we derived the following expression for the strength of the HI between two simple solutes $S$ at a distance $R = |\mathbf{R}_2 - \mathbf{R}_1|$ [see (8.34) and (8.42)]:

$$\delta A^{\mathrm{HI}}(R) = -kT \ln\langle\exp[-\beta U(\mathbf{X}^N/\mathbf{R}_1, \mathbf{R}_2)]\rangle - 2\Delta\mu_S^\circ$$

$$= \int_0^1 d\xi \int d\mathbf{X}_W \, [U(\mathbf{X}_W, \mathbf{R}_1)$$

$$+ U(\mathbf{X}_W, \mathbf{R}_2)]\varrho(\mathbf{X}_W/\mathbf{R}_1, \mathbf{R}_2, \xi) - 2\Delta\mu_S^\circ \quad (8.51)$$

We also observed that neither the internal properties nor the direct solute–solute pair potential appears in these expressions. This fact is utilized in this and the following sections to obtain some interesting relations between $\delta A^{\mathrm{HI}}(R)$ and experimentally measurable quantities. In order to appreciate the forthcoming arguments, one has to overcome a sort of psychological barrier which has to do with applying (8.51) for distances shorter than $\sigma$. To facilitate grasping the central idea of this section, it is useful to view (8.51) in a somewhat different light. Figure 8.5 shows two solute particles

# Water with Two or More Simple Solutes

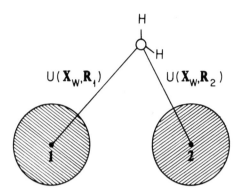

Fig. 8.5. Two solute particles at $R_1$ and $R_2$ and a water molecule at $X_W$. The only interactions appearing in Eq. (8.51) are $U(X_W, R_1)$ and $U(X_W, R_2)$.

at $R_1$ and $R_2$ and a water molecule at $X_W$. The only way the solvent "notices" the presence of the two solutes is through the "field of force" they produce—i.e., through $U(X^N/R_1, R_2)$ in the first form on the rhs of (8.51), or, equivalently, through $U(X_W, R_1) + U(X_W, R_2)$ in the second. (Clearly, the singlet conditional distribution function depends on this field of force.)

Once we recognize that the presence of the solutes is revealed only through the field of force they produce, we can forget about the real solutes and consider the following hypothetical system of $N$ solvent molecules subjected to the "external" field of force equal to $U(X^N/R_1, R_2)$. This point of view will be found to be quite rewarding. The main reason is that with two *real* solutes, we are limited to considering distances $R = |R_2 - R_1|$ which are larger than $\sigma$. With the two field-of-force sources, on the other hand, there are no restrictions on the distance $R$.

We recall that our main interest is focused on the properties of the *solvent* as a medium for HI. Therefore, from the theoretical point of view, since the parameter $\sigma$ is a property of the *solutes*, there is no reason to restrict ourselves to $R \gtrsim \sigma$, just because the two solute particles repel each other at $R \lesssim \sigma$. Relation (8.51) allows us to formulate the problem of HI to any distance $R \leq \sigma$.

Before elaborating on an example involving real solutes, it is instructive to work out a simple case. Consider a solvent of $N$ hard spheres of diameter $\sigma_W$. The two solute particles are also hard spheres, of diameter $\sigma_S$. Clearly, the distance between these two solute particles cannot be less than $\sigma_S$. However, if we focus our attention on the function $\delta A^{HI}(R)$, and since we know that this function is smooth and finite in all the region $0 \leq R < \infty$ [see relation (8.30), and further discussion on the properties of the function $y(R)$ in Section 2.6], we can inquire into the value of $\delta A^{HI}(R)$ for $R \leq \sigma$. Such a point of view is made possible by replacing the two solute particles by two cavities of appropriate size.

We recall that the work required to introduce a hard-sphere solute to a fixed position in a solvent is equal to the work of creating a cavity of a suitable radius at a fixed position. For a pair of hard spheres, a similar equivalence holds provided that we limit ourselves to distances larger than $\sigma_S$. The work required to move two hard spheres from $R = \infty$ to some finite distance $R > \sigma$ is the same as the work required to move the corresponding two cavities.

Figure 8.6(a) shows the two hard-sphere solutes at some distance $R > \sigma_S$. The solvent molecules are being influenced by the field of force produced by the two solutes, and are excluded from the region formed by the union of the two spheres of radius $(\sigma_S + \sigma_W)/2$ centered at $\mathbf{R}_1$ and $\mathbf{R}_2$. However, as far as the solvent molecules are concerned, the same situation occurs if we replace the two solute particles by two cavities of radius $(\sigma_S + \sigma_W)/2$. The work required to bring either the two solutes or the two cavities to the final position ($R > \sigma_S$) is the same, since at this distance the direct pair potential is zero.

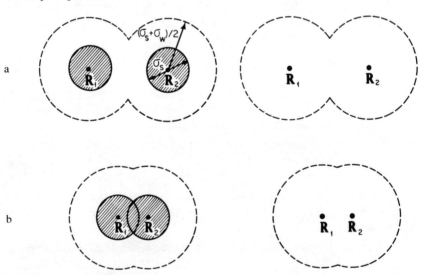

Fig. 8.6. (a) Two hard spheres at a distance $R > \sigma_S$ and the corresponding cavities. (b) Two hard spheres at an impossible distance $R < \sigma_S$, and the corresponding possible cavity.

The situation is markedly different for $R < \sigma_S$ (Fig. 8.6b). The total work required to bring the two solute particles to such a final state is evidently infinite. The reason for this is the infinitely strong repulsion the two solutes exert on each other at $R < \sigma_S$. However, if we eliminate the direct pair potential, then we can replace the two hard-sphere solutes by their appropriate cavities. The latter can be approached to within any distance $R < \sigma_S$ up to $R = 0$.

In the above example, the field of force produced by the two solutes can be described in terms of simple cavities. With a little imagination, we can generalize this idea to real solute particles. We recall that in (8.51) the solvent molecules "feel" the presence of the solute through the function

$$U(\mathbf{X}^N/\mathbf{R}_1, \mathbf{R}_2) = \sum_{k=1}^{2} \sum_{i=3}^{N+2} U(\mathbf{X}_i, \mathbf{R}_k) \quad (8.52)$$

We can imagine a system of pure solvent subjected to the "external" field of force, defined by (8.52), and originating from the two sources $\mathbf{R}_1$ and $\mathbf{R}_2$. The HI, as formulated in (8.51), is equal to the (total) work required to move the two sources from infinite separation to the final separation $R$. Since now we are not dealing with real solutes, we can move the two points $\mathbf{R}_1$ and $\mathbf{R}_2$ to any distance up to $R = 0$.

We now consider a specific example of real solutes. Consider two methane molecules. The standard free energy of solution of methane, $\Delta\mu_M^\circ$ is well known in water and in nonaqueous solvents (Chapter 7). Suppose that we start with two methane molecules at infinite separation and replace them by the corresponding field of force. Next, we move the centers of the field of force from infinity to the final distance $R = \sigma_1 = 1.533$ Å, $\sigma_1$ being the carbon–carbon distance in the ethane molecule.

The final situation is depicted in Fig. 8.7. The essential approximation introduced into the theory is that the field of force produced by two methane molecules at $R = \sigma_1$ is approximately the same as that produced by a single ethane molecule placed at the same positions (see Fig. 8.7b). Mathematically, this approximation is written as

$$U(\mathbf{X}_i, \mathbf{R}_1) + U(\mathbf{X}_i, \mathbf{R}_2) \approx U(\mathbf{X}_i/\mathbf{X}_E), \quad i = 3, 4, \ldots, N+2 \quad (8.53)$$

where $\mathbf{X}_E$ is the configuration of ethane for which one carbon is placed at $\mathbf{R}_1$ and the second carbon at $\mathbf{R}_2$, such that $|\mathbf{R}_2 - \mathbf{R}_1| = \sigma_1$. If this approximation is valid for all configurations $\mathbf{X}^N$ having nonzero probability, then we can replace the average quantity in (8.51) by

$$-kT \ln\langle\exp[-\beta U(\mathbf{X}^N/\mathbf{R}_1, \mathbf{R}_2)]\rangle = -kT \ln\langle\exp[-\beta U(\mathbf{X}^N/\mathbf{X}_E)]\rangle = \Delta\mu_E^\circ \quad (8.54)$$

Once we have made the replacement of the two sources of the field of force by a single molecule, we identify the average quantity on the rhs of (8.54) as the standard free energy of solution of the ethane molecule.

In the above arguments, we have used the first form on the rhs of Eq. (8.51). The same arguments can be applied to the second form on the rhs of (8.51). Here, we have the total interaction between a water molecule at configuration $\mathbf{X}_W$ and two sources of fields of force originating from $\mathbf{R}_1$ and $\mathbf{R}_2$. Clearly, if we find a single molecule having exactly the same field of force $U(\mathbf{X}_W, \mathbf{R}_1) + U(\mathbf{X}_W, \mathbf{R}_2)$, then the conditional singlet distribution function for the solvent molecules must be the same before and after the replacement has been made, i.e., the solvent molecules cannot tell which agent is used to produce the field of force. Perhaps the simplest and the most convincing argument for justifying the approximation (8.53) is to recall that this approximation becomes exact for hard-sphere solutes. Therefore an exact statement for hard spheres is likely to be a reasonable approximation for simple nonpolar solutes.

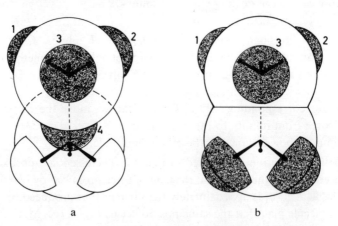

Fig. 8.7. (a) Two methane molecules at intermolecular distance $\sigma_1 = 1.533$ Å. All hydrogen atoms belonging to the upper molecule are shown. Hydrogen number 4 is directed toward the center of the second molecule. (The direct intermolecular forces between the two molecules are assumed to be "switched off," so that they can penetrate into each other to any chosen distance.) (b) One ethane molecule situated at the same space occupied previously by the two methane molecules.

# Water with Two or More Simple Solutes

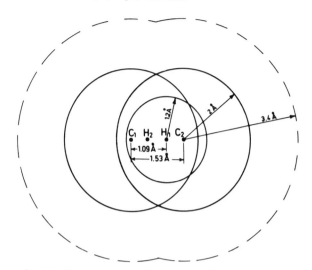

Fig. 8.8. Schematic, two-dimensional illustration of two methane molecules at distance 1.53 Å. The dashed line is the boundary of the excluded volume, calculated for the case of water as a solvent (with radius of 1.4 Å). The position of one of the hydrogens situated on the line connecting the centers of the two solutes is shown. The effective radius assigned to this hydrogen (1.2 Å) is indicated. It is clearly seen that in this configuration, a solvent molecule approaching the pair of methane particles will hardly notice the existence of the interior hydrogens. Therefore, replacing this pair of methane molecules by a single ethane molecule at the same configuration will produce little effect on the distribution of solvent molecules.

Let us further elaborate on the nature of approximation (8.53). Figure 8.8 shows a schematic description of two methane molecules at a distance of $R = \sigma_1$. The particular configuration of the two methane molecules is such that one hydrogen atom of each molecule is directed toward the second molecule. We note also that the methane molecule, the methyl group ($\cdot CH_3$), and the methylene group ($\cdot CH_2 \cdot$) are assigned almost the same van der Waals radius of about 2.0 Å. The C–H distance in the saturated alkanes is 1.093 Å and the van der Waals radius assigned to the hydrogen atom is about 1.2 A.

Figure 8.8 also shows the "excluded" volume for this pair of molecules.[7]

---

[7] The excluded volume has been drawn here for a solvent molecule of diameter $\sigma_W = 2.8$ Å. Clearly, if the solvent molecules are not spherical, then the form of the excluded volume depends on the orientation of the solvent molecule.

The dashed line is roughly the closest distance that the center of a water molecule can approach. Thus, if we replace the pair of methane molecules at this specific configuration by an ethane molecule, we expect that the geometry of the excluded region will not be changed.

The above consideration on the form of the excluded region accounts for the repulsive part of the solute–solvent interaction. We now consider the attractive part of the field of force produced by the pair of methane molecules. It is convenient to separate the total field of force produced by each methane molecule into two contributions: that produced by the methyl group and that by the inner hydrogen atom. We assume that the field of force produced by the methyl group of methane is the same as that by the methyl group of ethane. Furthermore, the two hydrogen atoms which we have placed along the C–C line are assumed to make a very small contribution to the total field of force in the region beyond the excluded volume and therefore this contribution is neglected.

It is important to stress that our replacement of the two methane molecules by one ethane has nothing to do with the *chemical reaction* $2CH_4 \rightarrow C_2H_6 + H_2$. In fact, we can imagine the process of bringing two argon molecules to a distance $R = \sigma_1$ and making the replacement by a single ethane molecule. Such a replacement can be valid if (8.53) is a good approximation, but we certainly have no "chemical reaction" $2Ar \rightarrow C_2H_6$.

Although it is possible, in principle, to find a different pair of solute particles $S$ so that the replacement $2S \rightarrow C_2H_6$ will provide a good approximation in (8.53), it is believed that the pair of methane molecules replaced by an ethane molecule is a particularly fortunate example, since it is known that the properties of the methyl radical are almost independent of the remaining part of the molecule, whether it is a hydrogen, a methyl group, or a larger alkyl group.

From (8.51), (8.53), and (8.54), we write the HI between two methane molecules at a distance $\sigma_1 = 1.533$ Å as

$$\delta A^{\text{HI}}(\sigma_1) = \Delta \mu_E^\circ - 2\Delta \mu_M^\circ \tag{8.55}$$

where we have measurable quantities on the rhs. We use (8.55) in the next section as a simple index for estimating the strength of HI in various solvents. In all of the examples presented there, we stress the difference between the value of $\delta A^{\text{HI}}(\sigma_1)$ in water and in other solvents. It should be noted that for the purpose of comparison of various solvents, the requirement is that the replacement approximation in (8.53) be equally valid in the two solvents. This is a far weaker requirement than to assume the validity of (8.53) for

# Water with Two or More Simple Solutes

each solvent separately. In Section 8.7, we show how, by the choice of bulkier solute molecules, one can improve approximation (8.53) to any required degree of accuracy.

All of the arguments presented thus far in this section have been concerned with relation (8.51), in which we have considered the properties of the fields of force produced by various solute molecules. It is now instructive to show a different, though equivalent, argument which involves only real solute molecules.

Consider a modification of the ideal cycle depicted in Fig. 8.3, which we shall refer to as a "real cycle" and which is shown in Fig. 8.9. Instead of bringing the two solute particles $a$ and $b$ to the final distance $R = \sigma$, we push the two solutes against the repulsive forces to a shorter distance of $R = \sigma_1$ (here, $\sigma_1$ is any distance shorter than $\sigma$). As in the ideal cycle, we compare the total work required for this process $\Delta A^l(\sigma_1)$ with the total work in the "round trip" process. First, we bring the two solutes into the gaseous phase, and then to a separation $\sigma_1$. The work required for the latter step is $\Delta A^g(\sigma_1)$. Next, we replace the pair of solute particles at $\sigma_1$ by a single molecule denoted by $a \cdot b$, which being a real molecule, can be dissolved into the liquid and then replaced by the two solute particles at $\sigma_1$. The replacements involve work $\varepsilon^g$ and $\varepsilon^l$, respectively.

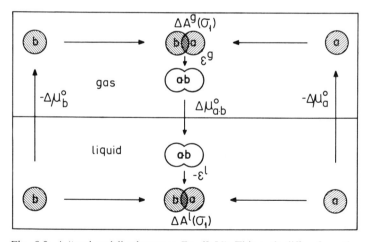

Fig. 8.9. A "real cycle" relevant to Eq. (8.56). This cycle differs from the ideal one (Fig. 8.3) in two respects. (1) The two solute particles $a$ and $b$ are brought to a new configuration which, in general, involves partial interpenetration. (2) The pair $a$ and $b$ is replaced by a single molecule denoted $a \cdot b$. Therefore, two new steps are added, involving the work $\varepsilon^l$ and $\varepsilon^g$.

The total balance of work is

$$\Delta A^l(\sigma_1) = -\Delta\mu_a{}^\circ - \Delta\mu_b{}^\circ + \Delta A^g(\sigma_1) + \varepsilon^g + \Delta\mu_{a\cdot b}^\circ - \varepsilon^l \quad (8.56)$$

After rearranging (8.56), we get

$$\delta A^{\mathrm{HI}}(\sigma_1) = \Delta A^l(\sigma_1) - \Delta A^g(\sigma) = \Delta\mu_{a\cdot b}^\circ - \Delta\mu_a{}^\circ - \Delta\mu_b{}^\circ + \varepsilon^g - \varepsilon^l \quad (8.57)$$

The approximation invoked now is that the work required to replace the pair of solutes at $\sigma_1$ in the gas by a single molecule $a \cdot b$ is approximately the same as the work required to achieve the same replacement in the liquid, i.e., the assumption is made that

$$\varepsilon^g - \varepsilon^l \approx 0 \quad (8.58)$$

Using (8.58), we get in (8.57) the required relation between HI and experimentally measurable quantities. Note that we have not used the concept of field of force in the above arguments. The idea is that the process of bringing the two solute particles to $\sigma_1$ is done twice, in the gas and in the liquid. Therefore, however large the direct work may be, this part is subtracted by forming the difference $\Delta A^l(\sigma_1) - \Delta A^g(\sigma_1)$ in (8.57) and we are left with the HI part of the work. The approximation (8.58) can be easily shown to be equivalent to the replacement approximation (8.53).

We consider the first argument involving "fields of force" to be more elegant since we are not required to subtract very large work terms to obtain the HI. In fact, we will see later that the replacement approximation can also be employed for zero separation $R = 0$. Using the real cycle in Fig. 8.9 for such a case is quite awkward, since it involves subtracting two almost infinite quantities to obtain the finite value of $\delta A^{\mathrm{HI}}(0)$.

## 8.6. FURTHER EXPERIMENTAL DATA ON HI

In Section 8.2 we surveyed some of the experimental evidence relevant to the problem of HI. With the construction of relation (8.55), we can use the quantity $\delta A^{\mathrm{HI}}(\sigma_1)$ as a simple and convenient index for comparison of the strength of HI in various solvents.[8] We note here that our attention will be focused not on the absolute magnitude of $\delta A^{\mathrm{HI}}(\sigma_1)$, but rather on the difference between these values in different solvents. In establishing the

---

[8] Note again the use of the term "hydrophobic interaction" for water as well as for nonaqueous solvents.

connection between HI and experimental quantities, it is sometimes more convenient to refer to the process of bringing the two solute particles from infinite separation to any distance $R$ at constant pressure rather than constant volume. The total work is given by

$$\Delta G^l(R) = U(R) + \delta G^{HI}(R) \qquad (8.59)$$

and we have the equality

$$\delta G^{HI}(R; T, P, N) = \delta A^{HI}(R; T, V, N) \qquad (8.60)$$

provided that $V$ in the constant-volume system is equal to the average volume $\langle V \rangle$ in the constant-pressure system (see also Appendix 9-F). Thus, insofar as we are concerned with estimates of the *strength* of the HI, we can use either $\delta A^{HI}(R)$ or $\delta G^{HI}(R)$. The latter is more convenient, however, for reporting the entropy and enthalpy of HI (Section 8.10). All the experimental quantities referred to in this section are computed either by relation (8.55) or, equivalently,[9] by

$$\delta G^{HI}(\sigma_1) = \Delta \mu_E{}^\circ - 2\Delta \mu_M{}^\circ \qquad (8.61)$$

Figure 8.10 gives the values of $\delta G^{HI}(\sigma_1)$ for various solvents. There are essentially two prominent differences between water and nonaqueous solvents which are worth noting.

1. The absolute magnitude of $\delta G^{HI}(\sigma_1)$ is larger in water than in the other solvents. The difference, say at 15°C, is quite significant and amounts to about 0.6 kcal/mole. This value is of the same order of magnitude as $kT$ at this temperature. One can conclude, therefore, that there exists an excess attraction between two methane molecules in water as compared with other solvents. The last statement can be reformulated in terms of ratio of probabilities. We use relation (8.30) for a pair of methane molecules in water and in, say, ethanol at 25°C, to obtain

$$\frac{y(\sigma_1, \text{ in water})}{y(\sigma_1, \text{ in ethanol})} = \exp(1.36) = 3.9 \qquad (8.62)$$

The interpretation of the above result is the following. For two methane molecules with an effective diameter of $\sigma \approx 3.82$ Å, the probability of

---

[9] No distinction is made between $\delta A^{HI}$ of Section 8.5 and $\delta G^{HI}$ of this section. Such a distinction is important for computation of other thermodynamic quantities, as will be shown in Section 8.10.

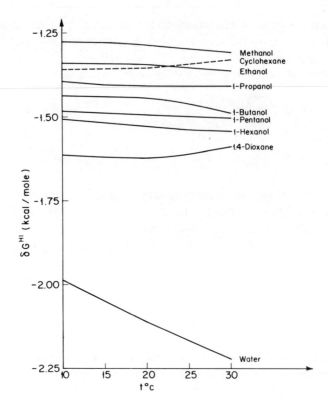

Fig. 8.10. Values of $\delta G^{\mathrm{HI}}(\sigma_1)$, defined in (8.61), in water and some nonaqueous solvents [data from Yaacobi and Ben-Naim (1974)].

finding them at about $R \approx 1.533$ Å is almost zero. However, since the ratio of such probabilities in two solvents is equal to the ratio of the quantities $y(\sigma_1)$, and since the latter does not include the direct pair potential operating between the two solutes, we can assign to this ratio the meaning of the relative probability of finding the two solutes at $R = \sigma_1$ in a system for which the direct solute–solute pair potential has been "switched off." We see from (8.62) that the probability of finding the two particles at a close distance in water is about four times larger than the corresponding value in ethanol. We stress, however, that from (8.62), one cannot reach any conclusion regarding the relative probabilities of finding the two solutes at contact distance $R = \sigma$. We return to this point in Section 8.9.

2. The second important difference between water and other solvents is in the temperature dependence of $\delta G^{\mathrm{HI}}(\sigma_1)$. It is seen that the HI in

water becomes stronger as the temperature increases; in other solvents, the temperature dependence is very small and is sometimes opposite in direction to that in water. The unique temperature dependence of the HI in water is quite remarkable, and in a sense it seems contradictory to our intuition. We return to a more detailed discussion of this subject in Section 8.10.

Table 8.2 presents some values of $\delta G^{HI}(\sigma_1)$ for water, heavy water, and other solvents. Note that the HI is slightly stronger in $H_2O$ than in $D_2O$, and that the corresponding values of $|\delta G^{HI}(\sigma_1)|$ are appreciably smaller in all of the solvents for which the relevant experimental quantities are available.

Adopting the quantity $\delta G^{HI}(\sigma_1)$ as a probe for the strength of HI in various solvents, we can follow its variation in a mixture of water and a nonaqueous solvent. Figure 8.11 shows such a variation for the system water–ethanol at two temperatures. Note that in the water-rich region, the values of $\delta G^{HI}(\sigma_1)$ are large (in absolute magnitude) and strongly temperature dependent. As the mole fraction of alcohol becomes larger than, say, $x \sim 0.2$, the variation of $\delta G^{HI}(\sigma_1)$ with the mole fraction becomes smooth and $|\delta G^{HI}(\sigma_1)|$ steadily decreases to the corresponding value in pure ethanol. Note the inversion of the temperature dependence at about $x \approx 0.5$.

The most remarkable feature of this figure is the maximum followed by a minimum in the curve at 10°C. As yet, there is no explanation for this phenomenon; a partial interpretation, in terms of a structural change in water, is provided in Section 8.11.

As an example of another application of the quantity $\delta G^{HI}(\sigma_1)$, we can look at the change in strength of the HI brought on by the addition of various solutes, a topic of considerable importance in the study of solute

**Table 8.2**

**Hydrophobic Interaction; $\delta G^{HI}(\sigma_1) = \Delta\mu_E^\circ - 2\Delta\mu_M^\circ$ Computed for a Pair of Methane Molecules at a Final Distance $\sigma_1 = 1.533$ Å in Various Solvents**[a]

| t, °C | $\delta G^{HI}(\sigma_1)$, kcal/mole | | | | | |
|---|---|---|---|---|---|---|
| | $H_2O$ | $D_2O$ | Methanol | Ethanol | 1,4-Dioxane | Cyclohexane |
| 10 | −1.99 | −1.94 | −1.28 | −1.34 | −1.61 | −1.36 |
| 25 | −2.17 | −2.14 | −1.29 | −1.36 | −1.61 | −1.34 |

[a] Data from Ben-Naim et al. (1973) and Ben-Naim and Yaacobi (1974).

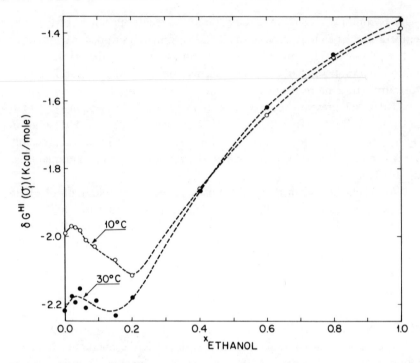

Fig. 8.11. Variation of $\delta G^{HI}(\sigma_1)$ with the mole fraction of ethanol in a water–ethanol system [data from Yaacobi and Ben-Naim (1974)].

effects on the conformational stability of various biopolymers. Table 8.3 presents a sample of data on the variation of $\delta G^{HI}(\sigma_1)$ in various aqueous solutions of electrolytes and nonelectrolytes. As can be seen from the table, almost all solutes added to water at low concentration cause an increase in the HI. An exceptional case in this table is the aqueous solution of ethanol ($x = 0.03$ and $t = 10°C$), for which the HI is slightly weaker than in pure water. (This can also be seen from Fig. 8.11, where in the range $0 \leq x \lesssim 0.03$, there is a decrease in HI as a function of the mole fraction of ethanol.)

We now conclude this section with a few examples showing that values of $\Delta \mu_S°$ and $\delta G^{HI}(\sigma_1)$ are not always compatible, since they may provide contradictory information on HI.

Let us consider first water and ethanol. The standard free energy of transferring a methane molecule from water to ethanol at $t = 10°C$ is

$$\Delta \mu_t°(\text{water} \to \text{ethanol}) = \Delta \mu_M°(\text{in ethanol}) - \Delta \mu_M°(\text{in water})$$
$$= -1.42 \text{ kcal/mole}$$

## Table 8.3
### Solute Effect on HI; Value of $\delta G^{HI}(\sigma_1)$ (in kcal/mole) for Various Aqueous Solutions at Two Temperatures[a]

| $t$, °C | $H_2O$ | Electrolytes | | | | | | |
|---|---|---|---|---|---|---|---|---|
| | | LiCl (1 $M$) | NaCl (1 $M$) | KCl (1 $M$) | CsCl (1 $M$) | $NH_4Cl$ (1 $M$) | NaBr (1 $M$) | NaI (1 $M$) |
| 10 | −1.99 | −2.11 | −2.17 | −2.16 | −2.14 | −2.09 | −2.18 | −2.19 |
| 25 | −2.17 | −2.29 | −2.34 | −2.33 | −2.32 | −2.26 | −2.34 | −2.33 |

| $t$, °C | Nonelectrolytes | | | | | | |
|---|---|---|---|---|---|---|---|
| | Ethanol ($x = 0.03$) | 1,4-Dioxane ($x = 0.03$) | Dimethyl sulfoxide ($x = 0.03$) | Sucrose (0.5 $M$) | Urea | | |
| | | | | | 1 $M$ | 2 $M$ | 7 $M$ |
| 10 | −1.97 | −2.05 | −2.04 | −2.12 | −2.04 | −2.11 | −2.38 |
| 25 | −2.19 | −2.20 | −2.22 | −2.30 | −2.21 | −2.26 | −2.49 |

[a] $x$ = mole fraction. From Ben-Naim and Yaacobi (1974).

This value can be interpreted in terms of the tendency of the methane molecule to favor the ethanol over the aqueous environment. Alternatively, we can say that methane tends to escape from water into ethanol. Consistent information on this tendency is obtained from the following two values (all values are for $\sigma_1$ at 10°C):

$$\delta G^{HI}(\text{in water}) = -1.99 \text{ kcal/mole}$$

$$\delta G^{HI}(\text{in ethanol}) = -1.39 \text{ kcal/mole}$$

meaning that the tendency to adhere to each other is stronger in water as compared to ethanol, i.e., the two methane molecules tend to avoid the aqueous environment more than they do the ethanol environment.

We now present a different example showing that the two "measures" $\Delta\mu_t^\circ$ and $\delta G^{HI}(\sigma_1)$ lead to different conclusions. Consider a mixture of water and ethanol with a mole fraction of ethanol of $x = 0.2$ (referred to below as "mixture"). For these solvents, we have (for methane at 10°C)

$$\Delta\mu_t°(\text{water} \to \text{mixture}) = \Delta\mu_M°(\text{in mixture}) - \Delta\mu_M°(\text{in water})$$
$$= 1.58 - 1.75 = -0.17 \text{ kcal/mole}$$

showing that the mixture is a slightly more favorable solvent to methane than is pure water. On the other hand,

$$\delta G^{\text{HI}}(\text{in water}) = -1.99 \text{ kcal/mole}$$
$$\delta G^{\text{HI}}(\text{in mixture}) = -2.11 \text{ kcal/mole}$$

showing that the tendency to adhere to each other is *stronger* in the mixture than in pure water.

The last example shows that $\Delta\mu_t°$, though useful as a measure for the extreme process of transferring a nonpolar group from an aqueous to a nonpolar environment, is, in general, not suitable for measuring the tendency of two solute particles to adhere to each other in various solvents. In Section 8.8, we encounter one special case for which $\Delta\mu_t°$ and $\delta G^{\text{HI}}$ provide identical information.

## 8.7. GENERALIZATIONS

In previous sections, we dealt with the HI between two simple solutes in various solvents. The method can be generalized along two directions; one studies the HI between more complex molecules and the second examines the HI among a set of, say, $M$ solute molecules, which may be simple or complex. In this section, we outline a few possible applications of the approximation discussed in Section 8.5.

### 8.7.1. HI among a Set of M Identical Spherical, Nonpolar Solute Particles

Consider the work required to bring $M$ solute particles which, for simplicity, are presumed to be spherical, from infinite separation to the final configuration $\mathbf{R}^M$:

$$\Delta A^l(\mathbf{R}^M) = A^l_{N+M}(\mathbf{R}^M) - A^l_{N+M}(\mathbf{R}^M = \infty) = U_M(\mathbf{R}^M) + \delta A^{\text{HI}}(\mathbf{R}^M) \quad (8.63)$$

where $\mathbf{R}^M = \infty$ indicates the configuration of the $M$ solute particles in the solvent, the separation between each pair being infinity, i.e.,

$$|\mathbf{R}_j - \mathbf{R}_i| = \infty, \quad \text{for } i = 1, \ldots, M, \quad j = 1, \ldots, M, \quad i \neq j \quad (8.64)$$

### Water with Two or More Simple Solutes

The splitting in (8.63) into direct and indirect work follows directly from the statistical mechanical expression for $\Delta A^l(\mathbf{R}^M)$, namely

$$\exp[-\beta\,\Delta A^l(\mathbf{R}^M)]$$
$$= \frac{\int\cdots\int d\mathbf{X}^N\,\exp[-\beta U_N(\mathbf{X}^N) - \beta U_M(\mathbf{R}^M) - \beta U(\mathbf{X}^N/\mathbf{R}^M)]}{\int\cdots\int d\mathbf{X}^N\,\exp[-\beta U_N(\mathbf{X}^N) - \beta U_M(\mathbf{R}^M = \infty) - \beta U(\mathbf{X}^N/\mathbf{R}^M = \infty)]} \quad (8.65)$$

where $U(\mathbf{X}^N/\mathbf{R}^M)$ is the total interaction energy between solvent molecules for the configuration $\mathbf{X}^N$ and the solute molecules for the configuration $\mathbf{R}^M$. [Clearly, relation (8.65) can be easily generalized to nonspherical solutes. Here, the notation is particularly simplified by using $\mathbf{R}$ and $\mathbf{X}$ for the solute and solvent configurations, respectively.]

The integrations in (8.65) extend over all the configurations of the *solvent* particles; hence, we can rewrite (8.65) as

$$\exp[-\beta\,\Delta A^l(\mathbf{R}^M)] = \exp[-\beta U_M(\mathbf{R}^M)]\,\frac{\langle\exp[-\beta U(\mathbf{X}^N/\mathbf{R}^M)]\rangle}{\langle\exp[-\beta U(\mathbf{X}^N/\mathbf{R}^M = \infty)]\rangle} \quad (8.66)$$

where we put $U_M(\mathbf{R}^M = \infty) = 0$. From (8.63) and (8.66), we identify the expression for the HI among the $M$ solute particles as

$$\exp[-\beta\,\delta A^{\mathrm{HI}}(\mathbf{R}^M)] = \langle\exp[-\beta U(\mathbf{X}^N/\mathbf{R}^M)]\rangle/\langle\exp[-\beta U(\mathbf{X}^N/\mathbf{R}^M = \infty)]\rangle \quad (8.67)$$

which is the generalization of relation (8.34).

We note again that (8.67) does not include any internal properties of the solutes or the direct interaction among the solute particles, $U_M(\mathbf{R}^M)$. Hence, we can interpret $\delta A^{\mathrm{HI}}(\mathbf{R}^M)$ as the *total* work required to move $M$ sources of fields of force from the configuration $\mathbf{R}^M = \infty$ to the final configuration $\mathbf{R}^M$.

By direct generalization of the arguments in Section 8.4 (either through the formal statistical mechanical expressions or by using a generalization of the cyclic process shown in Fig. 8.3), we can express $\delta A^{\mathrm{HI}}(\mathbf{R}^M)$ in terms of the free energy of transferring the solute particles from a fixed configuration in the gas to a fixed configuration in the liquid,

$$\delta A^{\mathrm{HI}}(\mathbf{R}^M) = \Delta A_{tr}(\mathbf{R}^M) - \Delta A_{tr}(\mathbf{R}^M = \infty) \quad (8.68)$$

which is a generalization of (8.31). Using arguments similar to those in Section 8.4, we can easily show that the work required to transfer $M$ identical solute particles from fixed positions in the gas to fixed positions in the liquid at infinite separation from each other is equal to $M$ times the work of transferring a single particle to a fixed position, namely

$$\Delta A_{\text{tr}}(\mathbf{R}^M = \infty) = M \Delta A_{\text{tr}}(\mathbf{R}_1) = M \Delta \mu_S^\circ \qquad (8.69)$$

where $\Delta \mu_S^\circ$ is the standard free energy of solution of the solute $S$.

As in Section 8.4, we note again that a relation such as (8.69) holds only for transferring to *fixed* positions and at infinite separations. The next step is to find a connection between $\Delta A_{\text{tr}}(\mathbf{R}^M)$ and experimental quantities. In order to do this, we must find a particular configuration, say $\mathbf{R}^{*M}$, of the $M$ solutes and one single molecule such that the field of force produced by the $M$ solute particles at $\mathbf{R}^{*M}$ is approximately equal to that produced by the single molecule for a given configuration $\mathbf{Y}$ (see Section 1.2 for the notation; here, $\mathbf{Y}$ denotes the configuration of a single molecule, including location, orientation, and conformation). Mathematically, the requirement is

$$U(\mathbf{X}^N/\mathbf{R}^{*M}) = U(\mathbf{X}^N/\mathbf{Y}) \qquad (8.70)$$

As an example, suppose we start with four methane molecules and bring them to a final configuration $\mathbf{R}^*$ in such a way that a replacement by one butane molecule in a configuration $\mathbf{Y}$ produces approximately the same field of force. If (8.70) is valid for all the configurations of the solvent molecules that have nonzero probability, then (8.67) can be rewritten as

$$\exp[-\beta \, \delta A^{\text{HI}}(\mathbf{R}^{*M})] = \langle \exp[-\beta U(\mathbf{X}^N/\mathbf{Y})] \rangle \exp(M\beta \, \Delta \mu_S^\circ) \qquad (8.71)$$

where $\Delta \mu_S^\circ$ is the standard free energy of solution of a single solute $S$. In order to achieve a connection between HI and relevant experimental quantities, we have to distinguish between two cases: when the replacing molecule does not possess internal rotations and when it does. In the first case, the average quantity on the rhs of (8.71) can be related to the standard free energy of solution of the single molecule, say propane, and we obtain the relation

$$\langle \exp[-\beta U(\mathbf{X}^N/\mathbf{Y})] \rangle = \exp(-\beta \, \Delta \mu_{M \cdot S}^\circ) \qquad (8.72)$$

The subscript $M \cdot S$ denotes a single molecule that replaces $M$ solute particles of type $S$ at the configuration $\mathbf{R}^{*M}$.

From (8.71) and (8.72), we get

$$\delta A^{\text{HI}}(\mathbf{R}^{*M}) = \Delta \mu_{M \cdot S}^\circ - M \Delta \mu_S^\circ \qquad (8.73)$$

where we have measurable quantities on the rhs. An example of a molecule in this group is propane.

## Water with Two or More Simple Solutes

The second case includes molecules that possess internal rotations, in which case (8.73) should be understood as an average over all the conformations of the molecule. It is convenient, in this case, to introduce an auxiliary quantity, the standard free energy of solution of the molecule in a specific configuration $\mathbf{Y}$,

$$\exp[-\beta \, \Delta\mu^\circ_{M \cdot S}(\mathbf{Y})] = \frac{\int \cdots \int d\mathbf{X}^N \exp[-\beta U_N(\mathbf{X}^N) - \beta U(\mathbf{X}^N/\mathbf{Y})]}{\int \cdots \int d\mathbf{X}^N \exp[-\beta U(\mathbf{X}^N)]} \quad (8.74)$$

This is the work required to transfer a single molecule, say butane, from a fixed configuration $\mathbf{Y}$ in the gas to a fixed configuration $\mathbf{Y}$ in the liquid. [Note that $\Delta\mu^\circ_{M \cdot S}(\mathbf{Y})$ is actually a function of the conformation of the molecule. If no internal rotations exist, then $\Delta\mu^\circ_{M \cdot S}$ is, in fact, independent of both the location and orientation of the molecule.]

The standard free energy of solution of the molecule $M \cdot S$ is obtained from (8.74) as an average over all configurations of the molecule, weighted by the internal rotation potential, which we denote by $U_M{}^*(\mathbf{Y})$,

$$\exp(-\beta \, \Delta\mu^\circ_{M \cdot S}) = \frac{\int d\mathbf{Y} \exp[-\beta U_M{}^*(\mathbf{Y}) - \beta \, \Delta\mu^\circ_{M \cdot S}(\mathbf{Y})]}{\int d\mathbf{Y} \exp[-\beta U_M{}^*(\mathbf{Y})]} \quad (8.75)$$

If $\Delta\mu^\circ_{M \cdot S}$ and $\delta A^{\text{HI}}(\mathbf{R}^{*M})$ are understood as average quantities in the sense of (8.75), then we can apply (8.73) to molecules possessing internal degrees of freedom as well. [For more details, see Ben-Naim (1971b, 1972a).]

Figures 8.12 and 8.13 show the HI among a group of $M$ methane molecules brought to the final configuration of the molecules indicated next to each curve. (Note that in these examples, only butane requires averaging over all configurations.)

Clearly, the values for the HI in water are considerably larger (in absolute magnitude) than the corresponding values in methanol. Also, the temperature dependence is distinctly different. In water, we find a clear-cut negative temperature dependence, whereas in methanol, there is a slight positive slope which is estimated to be within the range of experimental error (Ben-Naim, 1971b).

Tables 8.4 and 8.5 present further information on the HI among $M$ methane molecules brought to a final configuration of a suitable paraffin molecule. The first column shows the value of $M$, i.e., the number of carbon atoms in each molecule. The second column presents the total HI for the $M$ methane molecules. The third column indicates that the HI per molecule increases with $M$. Note that for a given $M$, the value of $\delta A^{\text{HI}}_M/M$ is insensitive

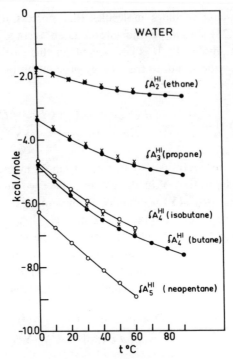

Fig. 8.12. Values of $\delta A_M^{HI}$ in water for various numbers of methane molecules brought to the final configuration as indicated next to each curve. Full circles are values computed from data by Morrison and Billet (1952). All other values were computed from data by Wetlaufer et al. (1964).

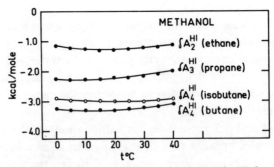

Fig. 8.13. The HI (or "solvophobic interaction") for methane molecules as a function of temperature in methanol. The configurations pertinent to the various curves are indicated in the figure.

### Table 8.4

Values of $\delta A_M^{HI}$ (in kcal/mole at 25°C) for the Interaction among $M$ Methane Molecules Brought to a Final Configuration Similar to an Existing Molecule Containing $M$ Carbons[a]

| Hydrocarbon | $M$ | $-\delta A_M^{HI}$ | $-\delta A_M^{HI}/M$ | $-\delta A_M^{HI}/M_B$ |
|---|---|---|---|---|
| Ethane | 2 | 2.16 | 1.08 | 2.16 |
| Propane | 3 | 4.01 | 1.33 | 2.00 |
| n-Butane | 4 | 5.87 | 1.47 | 1.96 |
| Isobutane | 4 | 5.70 | 1.42 | 1.90 |
| n-Pentane | 5 | 7.53 | 1.50 | 1.88 |
| Isopentane | 5 | 7.59 | 1.52 | 1.89 |
| 2,2-Dimethylpropane | 5 | 7.34 | 1.47 | 1.83 |
| n-Hexane | 6 | 9.43 | 1.57 | 1.88 |
| 2-Methylpentane | 6 | 9.48 | 1.58 | 1.89 |
| 3-Methylpentane | 6 | 9.41 | 1.57 | 1.88 |
| 2,2-Dimethylbutane | 6 | 9.39 | 1.58 | 1.88 |
| n-Heptane | 7 | 11.34 | 1.62 | 1.89 |
| 2,4-Dimethylpentane | 7 | 11.07 | 1.58 | 1.85 |
| n-Octane | 8 | 13.08 | 1.63 | 1.87 |
| 2,2,4-Trimethylpentane | 8 | 13.10 | 1.64 | 1.87 |

[a] In the last column, $M_B$ is the number of chemical bonds (or nearest neighbors) in the hydrocarbon. [Computations based on data from Wen and Hung (1970) and McAuliffe (1966).]

to the final configuration[10] (i.e., the isomers of $M \cdot S$). The last column in Table 8.4 contains the value of $\delta A_M^{HI}/M_B$, where $M_B$ is the number of C–C bonds in the molecule. There is a clear decrease in this value for the first four members. Beyond that, the values seem to converge to an almost constant value of about $-1.88$ kcal/mole. This remarkable fact indicates that the HI is almost additive insofar as we count only the number of nearest-neighbor groups in the paraffin molecules. Table 8.5 shows similar results for cycloparaffin molecules, where $M$ is the same as $M_B$.

[10] This insensitivity may be due to the insufficient accuracy of the data from which the two tables have been prepared. It seems that with more accurate measurements, one can find definite differences between various isomers of the paraffin molecules.

**Table 8.5**

**Values of $\delta A_M^{HI}$** [a]

| Cycloparaffin | $M$ | $-\delta A_M^{HI}$ | $-\delta A_M^{HI}/M$ |
|---|---|---|---|
| Cyclopropane | 3 | 3.49 | 1.16 |
| Cyclopentane | 5 | 8.80 | 1.76 |
| Cyclohexane | 6 | 10.72 | 1.78 |
| Cycloheptane | 7 | 13.17 | 1.88 |
| Cyclooctane | 8 | 15.13 | 1.89 |
| Methylcyclopentane | 6 | 10.37 | 1.73 |
| Methylcyclohexane | 7 | 12.25 | 1.75 |
| 1-cis-2-Dimethylcyclohexane | 8 | 14.40 | 1.80 |

[a] This table records values similar to those of Table 8.4, but for cyclic molecules. Here $M = M_B$. [From Ben-Naim (1972a).]

## 8.7.2. Attaching a Methyl Group to Various Molecules

We consider here "reactions" of the form

$$CH_4 + RH \to R\text{-}CH_3 \qquad (8.76)$$

The term "reaction" is used to describe the process of bringing a methane molecule from infinity to a final position next to a hydrocarbon molecule, where the pair is replaced by a single hydrocarbon denoted by R–CH$_3$. (Note that this is not a stoichiometric reaction in the usual sense.)

For the "reaction" (8.76), $\delta A^{HI}$ is written as

$$\delta A^{HI}(R\text{-}CH_3) = \Delta\mu^\circ(R\text{-}CH_3) - \Delta\mu^\circ(RH) - \Delta\mu^\circ(CH_4) \qquad (8.77)$$

where we have measurable quantities on the rhs. Note that in every case in which internal rotations are involved, one must interpret the appropriate quantity in (8.77) as an average over all the conformations of the molecule [in the sense of (8.75)].

Table 8.6 shows $\delta A^{HI}$ for some normal hydrogen molecules. Although the accuracy of the data on which the values of $\delta A^{HI}$ are based is not sufficient to draw definite conclusions on the effect of the various residues R in the hydrocarbon molecules, it is appropriate at this stage to point out two kinds of information that can, in principle, be obtained from "reactions"

## Table 8.6

**Values of $\delta A^{HI}$ for the Interaction between Methane and Various Normal Hydrocarbons[a]**

| $CH_4 + RH$ | $\rightarrow R\text{-}CH_3$ | $-\delta A^{HI}$, kcal/mole at 25°C |
|---|---|---|
| C + C | → C–C | 2.16 |
| C + C–C | → C–C–C | 1.85 |
| C + C–C–C | → C–C–C–C | 1.86 |
| C + C–C–C–C | → C–C–C–C–C | 1.66 |
| C + C–C–C–C–C | → C–C–C–C–C–C | 1.91 |
| C + C–C–C–C–C–C | → C–C–C–C–C–C–C | 1.91 |
| C + C–C–C–C–C–C–C | → C–C–C–C–C–C–C–C | 1.74 |
| C + C=C | → C–C=C | 2.01 |
| C + C–C=C | → C–C–C=C | 1.89 |
| C + C–C–C=C | → C–C–C–C=C | 1.72 |
| C + C–C–C–C=C | → C–C–C–C–C=C | 2.02 |
| C + C–C≡C | → C–C–C≡C | 1.68 |
| C + C–C–C≡C | → C–C–C–C≡C | 1.82 |
| C + C–C–C–C≡C | → C–C–C–C–C≡C | 1.69 |
| C + C–C–C–C–C≡C | → C–C–C–C–C–C≡C | 1.68 |
| C + C–C–C–C–C–C≡C | → C–C–C–C–C–C–C≡C | 1.93 |
| C + C–C–C–C–C–C–C≡C | → C–C–C–C–C–C–C–C≡C | 1.65 |

[a] From Ben-Naim (1972a).

of type (8.76). Let P be any group attached to a hydrocarbon molecule, and consider the following "reaction":

$$CH_4 + H\text{-}R_n\text{-}P \rightarrow CH_3\text{-}R_n\text{-}P \qquad (8.78)$$

i.e., we bring a methane molecule from infinity and attach it to the side of a paraffin molecule which has $n$ carbon atoms and is connected to a group P on its other side.

For a given group P (e.g., halogen or carboxylic or amino acid residue), one can study the HI as a function of the chain length $n$ of the paraffin molecule. Such a study can reveal the *range* of influence of the group P on the medium, i.e., for a large $n$, one expects the HI to be independent of P, but for small $n$, due to the perturbation effect of P on the solvent, the HI may vary considerably with $n$.

A different study would be to fix $n$ and examine the relative effects of different groups $P$ on the solvent as they are revealed in the values of the HI.

### 8.7.3. Attaching an Ethyl Group to Various Molecules

The "reaction" given in (8.76) can be generalized by bringing an ethane, rather than a methane, molecule from infinity and attaching it to various molecules. The relevant "reaction" is

$$CH_3CH_3 + RH \rightarrow R\text{-}CH_2CH_3 \qquad (8.79)$$

Table 8.7 shows $\delta A^{HI}$ for "reactions" of the type (8.79), for which

$$\delta A^{HI}(RCH_2CH_3) = \Delta\mu^\circ(RCH_2CH_3) - \Delta\mu^\circ(RH) - \Delta\mu^\circ(CH_3CH_3) \quad (8.80)$$

We note here that the study of "reactions" of the type (8.79) provides new and independent information on HI which is not contained in the study of "reactions" of the type (8.76). Table 8.7 shows that the values of $\delta A^{HI}$ for "reactions" in this class are systematically smaller than the corresponding values for "reactions" of the type (8.76). These findings are in accord with theoretical expectations that will be discussed in Section 8.9.

We conclude this section by pointing out that quite a variety of information on HI can be obtained by processing the same experimental data in various ways. For example, having the standard free energies of solution of methane, ethane, and butane, we can obtain the HI for three different "reactions," which we write schematically as

$$\begin{aligned} &C + C\text{-}C\text{-}C \rightarrow C\text{-}C\text{-}C\text{-}C \\ &C\text{-}C + C\text{-}C \rightarrow C\text{-}C\text{-}C\text{-}C \qquad (8.81) \\ &C + C + C + C \rightarrow C\text{-}C\text{-}C\text{-}C \end{aligned}$$

For larger and more complex molecules, many new possibilities exist. In this section, we have presented a few examples, mainly for normal hydrocarbons. Further examples for branched hydrocarbons are given by Ben-Naim (1972a).

## Table 8.7
### Values of $\delta A^{\mathrm{HI}}$ for Interaction between Ethane and Various Hydrocarbons[a]

| $CH_3-CH_3 + RH$ | $\rightarrow R-CH_2-CH_3$ | $-\delta A^{\mathrm{HI}}$, kcal/mole at 25°C |
|---|---|---|
| C–C + C | → C–C–C | 1.85 |
| C–C + C–C | → C–C–C–C | 1.55 |
| C–C + C–C–C | → C–C–C–C–C | 1.36 |
| C–C + C–C–C–C | → C–C–C–C–C–C | 1.40 |
| C–C + C–C–C–C–C | → C–C–C–C–C–C–C | 1.66 |
| C–C + C–C–C–C–C–C | → C–C–C–C–C–C–C–C | 1.49 |
| C–C + C=C | → C–C–C=C | 1.73 |
| C–C + C–C=C | → C–C–C–C=C | 1.45 |
| C–C + C–C–C=C | → C–C–C–C–C=C | 1.58 |
| C–C + C–C=C | → C–C–C=C–C | 1.77 |
| C–C + C–C–C=C–C | → C–C–C–C–C=C–C | 1.39 |
| C–C + C–C–C–C–C=C | → C–C–C–C–C–C–C=C | 1.28 |
| C–C + C≡C | → C–C–C≡C | 1.85 |
| C–C + C–C≡C | → C–C–C–C≡C | 1.34 |
| C–C + C–C–C≡C | → C–C–C–C–C≡C | 1.36 |
| C–C + C–C–C–C≡C | → C–C–C–C–C–C≡C | 1.21 |
| C–C + C–C–C–C–C≡C | → C–C–C–C–C–C–C≡C | 1.45 |
| C–C + C–C–C–C–C–C≡C | → C–C–C–C–C–C–C–C≡C | 1.42 |
| C–C + benzene | → ethylbenzene | 1.56 |

[a] From Ben-Naim (1972a).

## 8.8. HYDROPHOBIC INTERACTION AT ZERO SEPARATION

This section presents some information of theoretical value pertinent to the problem of HI. In the next section, we also indicate how such information can be employed to extract information on HI at "real" distances, but for the moment, we focus our attention on the formal aspect of the problem.

Consider first the case of two simple solute particles such as argon, methane, or even hard-sphere particles. We recall the general expression for the indirect part of the work required to bring the two solute particles from infinite separation to the final configuration $\mathbf{R}_1, \mathbf{R}_2$ (the process being carried out at constant $T, V, N$)

$$\delta A^{\mathrm{HI}}(\mathbf{R}_1, \mathbf{R}_2) = -kT \ln \langle \exp[-\beta U(\mathbf{X}^N/\mathbf{R}_1, \mathbf{R}_2)] \rangle - 2\Delta\mu_S^\circ \quad (8.82)$$

We have already noted that from the practical point of view, the interesting question concerning the HI is for $R = |\mathbf{R}_2 - \mathbf{R}_1| \approx \sigma$, with $\sigma$ the diameter of the solute. However, since we are now interested in the solvent properties, and since in (8.82) we have no trace of the direct pair potential for solute–solute interaction, we can search for the value corresponding to the HI at any distance $R < \sigma$. This section is devoted to the one particular distance $R = 0$.

We can apply a cyclic process similar to the one depicted in Fig. 8.9 and obtain $\delta A^{\mathrm{HI}}(0)$ as the difference between the work required to bring the two solutes to $R = 0$ in the liquid and in the gas. We will avoid such a cyclic process, however, since it compels us to subtract two quantities of the order of infinity to obtain a finite quantity. Instead, we adhere to the point of view suggested by (8.82), i.e., we regard $\delta A^{\mathrm{HI}}(0)$ as the *total* work required to bring the two fields of force from infinity to zero separation. In such a process, we do not encounter any strong repulsive forces that originate from the direct solute–solute interaction.

We start with the simplest solutes, hard-sphere particles. Clearly, two hard spheres of diameter $\sigma_{\mathrm{HS}}$ cannot be brought to a distance $R < \sigma$. However, in (8.82), we need only be concerned with the field of force produced by the two hard spheres; the latter may be referred to as two cavities in the solvent.[11]

Applying (8.82) to two cavities at zero separation, we get

$$\delta A^{\mathrm{HI}}(R = 0) = -kT \ln \langle \exp[-\beta U(\mathbf{X}^N/R = 0)] \rangle - 2\Delta\mu_{\mathrm{HS}}^\circ$$
$$= \Delta\mu_{\mathrm{HS}}^\circ - 2\Delta\mu_{\mathrm{HS}}^\circ = -\Delta\mu_{\mathrm{HS}}^\circ \quad (8.83)$$

The second form on the rhs follows from the fact that the field of force produced by one cavity at some point in the solvent is exactly the

---

[11] If $\sigma_W$ is the effective diameter of the solvent molecule, then the radius of the cavity is $(1/2)(\sigma_{\mathrm{HS}} + \sigma_W)$. However, if the solvent molecule is not spherical, the geometric form of the cavity depends on the orientation of the solvent molecule. We refer to a "cavity" in both cases to mean the actual field of force produced by the hard sphere.

## Water with Two or More Simple Solutes

same as the field of force produced by a "double cavity" at the same point. In other words, a cavity is a stipulation on the centers of all solvent molecules to be excluded from a certain region. A double stipulation for the same region is the same as a single one, hence the equality in (8.83). This relation is also clear on intuitive grounds; the total work of bringing two cavities from infinite separation to zero separation is equal to the work required to eliminate one cavity. This is exactly the content of the third equality on the rhs of (8.83). (Recall the equivalence of the process of transferring a hard-sphere particle and the creation of a cavity, discussed in Sections 3.11 and 7.3.)

Relation (8.83) for a system of hard spheres is known in the form[12]

$$\ln y(0) = \beta \, \Delta \mu_{\text{HS}}^\circ \tag{8.84}$$

which is the same as (8.83), due to equality (8.30). [See, for instance, Meeron and Siegert (1968) and Rowlinson (1968).]

In the following, we elaborate on two generalizations of (8.83), one for two real solutes and the second for any number $M$ of either real or hard-sphere solutes.

Before doing this, it is appropriate to point out a few features of $\delta A^{\text{HI}}(R = 0)$ for hard-sphere solutes. First, this quantity may be viewed as the most fundamental measure of the strength of the HI. We stress this fact in the context of our study of the properties of the *solvent* as a medium for HI. Thus, since hard spheres are the simplest solutes we can envisage, the measure of $\delta A^{\text{HI}}(R = 0)$ would also require the minimum knowledge of the properties of the solutes.

Second, relation (8.83) shows that the HI at zero separation for hard spheres and the corresponding standard free energy of solution provide the same information on the properties of the solvent. This finding brings us back to Kauzmann's idea that $\Delta \mu_S^\circ$ may serve as a measure of the strength of HI. We stressed in Sections 8.2 and 8.6 that there is a fundamental difference in the information we get from the process of transferring a solute from water to another phase and the process of bringing two solutes to close separation in the same solvent. Here, we have a unique example in which exactly the same information is obtained from both processes.

---

[12] Relation (8.83) is slightly more general since it is concerned with two hard spheres in *any* solvent, provided we assign a proper hard-core diameter $\sigma_W$ to the solvent molecules. If the solute molecules are not spherical, relation (8.83) still applies with a reinterpretation of the cavity as a field of force produced by the solute.

## Table 8.8

**Values of $\Delta\mu_{HS}^{\circ}$ Computed from the Scaled Particle Theory[a] for Water and Various Solvents[b]**

| Solvent | $\Delta\mu_{HS}^{\circ}$, kcal/mole | | Solvent diameter, Å |
|---|---|---|---|
| | Solute diameter 3.4 Å | Solute diameter 3.82 Å | |
| Ether | 1.83 | 2.15 | 4.93 |
| Chloroform | 2.68 | 3.17 | 4.78 |
| Carbon disulfide | 3.01 | 3.58 | 4.33 |
| Methanol | 2.09 | 2.47 | 3.32 |
| Chlorobenzene | 3.26 | 3.85 | 5.46 |
| Acetone | 2.32 | 2.75 | 4.47 |
| Water | 5.68 | 6.87 | 2.90 |

[a] See Section 7.3.
[b] Two diameters, 3.4 Å (for argon) and 3.83 Å (for methane) were selected. The effective diameters of the solvent molecules were taken from Reiss (1966) (i.e., the diameters used to fit the surface tension). All values are for 20°C.

Relation (8.83) permits a novel application of the scaled particle theory to the problem of HI.[13] Table 8.8 gives some computed values of $\Delta\mu_{HS}^{\circ}$ for water and various nonaqueous solvents. [See Ben-Naim (1971a).] It is quite clear that the values of $\Delta\mu_{HS}^{\circ}$ are largest in water, which, by virtue of (8.83), means that in water, the HI is the strongest. In spite of some serious reservations that one may have regarding the application of the scaled particle theory to fluids such as water (see Section 7.3), the results of Table 8.8 show the same trend we witnessed in Section 8.6. [Recently, a more detailed examination of the application of the scaled particle theory for this problem has been reported by Wilhelm and Battino (1972). We have discussed here only spherical solutes, for which one needs a spherical cavity. An extension of the scaled particle theory to particles of arbitrary shape has been reported by Gibbons (1969).]

[13] We continue the use of the term hydrophobic interaction even when reference is made to nonaqueous solvents. A more appropriate term for the latter could be "solvophobic interaction."

# Water with Two or More Simple Solutes

We now proceed to obtain two generalizations of (8.83). A quite straightforward one is obtained from (8.67) for $M$ spheres at zero separation, i.e.,

$$\exp[-\beta \,\delta A^{\mathrm{HI}}(\mathbf{R}^M = 0)]$$
$$= \langle\exp[-\beta U(\mathbf{X}^N/\mathbf{R}^M = 0)]\rangle / \langle\exp[-\beta U(\mathbf{X}^N/\mathbf{R}^M = \infty)]\rangle \quad (8.85)$$

$\delta A^{\mathrm{HI}}(\mathbf{R}^M = 0)$ is the total work required to bring $M$ cavities from infinite separation ($\mathbf{R} = \infty$) to zero separation ($\mathbf{R}^M = 0$) (this last indicates that the centers of all the cavities coincide at a single point). Following arguments similar to those of Section 8.7, we can rewrite (8.85) as

$$\delta A^{\mathrm{HI}}(\mathbf{R}^M = 0) = \Delta\mu^\circ_{\mathrm{HS}} - M\,\Delta\mu^\circ_{\mathrm{HS}} = (1 - M)\,\Delta\mu^\circ_{\mathrm{HS}} \quad (8.86)$$

which is a generalization of (8.83) for any $M$.

The second generalization involves real solutes at zero separation. Let us first elaborate on the case of two simple solutes $S$. From (8.82), we get, at $R = |\mathbf{R}_2 - \mathbf{R}_1| = 0$

$$\delta A^{\mathrm{HI}}(R = 0) = -kT \ln\langle\exp[-\beta U(\mathbf{X}^N/R = 0)]\rangle - 2\Delta\mu_S^\circ$$
$$= \Delta\mu^\circ_{2\cdot S} - 2\Delta\mu_S^\circ \quad (8.87)$$

In the first form on the rhs, we recognize the first term as the standard free energy of solution of a hypothetical solute, the field of force of which is "twice" the field of force of a single solute $S$. By "twice" we mean the following: The field of force $U(\mathbf{X}^N/\mathbf{R}_1, \mathbf{R}_2)$ given in (8.82) has the form

$$U(\mathbf{X}^N/\mathbf{R}_1, \mathbf{R}_2) = \sum_{i=3}^{N+2} U(\mathbf{X}_i, \mathbf{R}_1) + \sum_{i=3}^{N+2} U(\mathbf{X}_i, \mathbf{R}_2) \quad (8.88)$$

where each term on the rhs is the field of force produced by a single solute. If $\mathbf{R}_1 = \mathbf{R}_2$, i.e., if the centers of the two fields of force coincide, then we have[14]

$$U(\mathbf{X}^N/R = 0) = 2\sum_{i=3}^{N+2} U(\mathbf{X}_i, \mathbf{R}_1) \quad (8.89)$$

which is the meaning of the field of force assigned to the hypothetical solute, designated in (8.87) by $2\cdot S$. The question arises whether we can find a

---

[14] Note that for hard-sphere solutes, $2\cdot S$ and $S$ are indistinguishable, i.e.,

$$2\sum_{i=3}^{N+2} U(\mathbf{X}_i, \mathbf{R}_1) = \sum_{i=3}^{N+2} U(\mathbf{X}_i, \mathbf{R}_1) \quad \text{(for hard spheres)}$$

real molecule for which the standard free energy of solution is equal to $\Delta\mu_{2.S}^\circ$. In principle, it is possible to find a solute which has almost the same diameter as $S$, but the attractive part of the potential of which is twice as strong as that of $S$.

Let us denote such a solute by $\bar{S}$. Then (8.87) can be written as

$$\delta A^{\mathrm{HI}}(R = 0) = \Delta\mu_{\bar{S}}^\circ - 2\Delta\mu_S^\circ \tag{8.90}$$

A simple way of satisfying a relation similar to (8.90) is to take two solutes, say $a$ and $b$, which differ appreciably in their diameters. For two such solutes at zero separation,

$$\delta A^{\mathrm{HI}}(R = 0) = -kT \ln\left\langle \exp\left\{-\beta \sum_{i=3}^{N+2} [U^{wa}(\mathbf{X}_i, \mathbf{R}) + U^{wb}(\mathbf{X}_i, \mathbf{R})]\right\}\right\rangle$$
$$- \Delta\mu_a^\circ - \Delta\mu_b^\circ \tag{8.91}$$

where we have denoted by $U^{wa}$ and $U^{wb}$ the solvent–solute pair potentials, and $\mathbf{R}$ is any location in the solvent. Now, suppose that $a$ and $b$ are simple spherical molecules such that the diameter of $a$ is much larger than that of $b$. At zero separation, the field of force produced by $a$ will dominate the total field of force produced by the pair of solutes. This is shown schematically in Fig. 8.14. A solvent molecule $W$ approaching the combined field of force will be repelled from the excluded volume of $a$ before it has a chance to approach the excluded region of $b$. Thus, insofar as we are concerned with the hard-core part of the solute–solvent interaction, this is taken care of by the large molecule $a$. Furthermore, since $b$ is buried deeply in the excluded region of $a$, the attractive part of the field of force of $b$ will

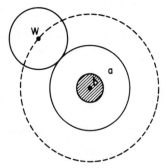

Fig. 8.14. Two solutes $a$ and $b$ of different diameters at zero separation. A water molecule $W$ is shown at the closest distance to the solute $a$.

be very weak in those regions accessible to solvent molecules. Hence, we can neglect this part of the field of force of $b$ as well. Mathematically, the approximation is

$$U^{wa}(\mathbf{X}_i, \mathbf{R}) + U^{wb}(\mathbf{X}_i, \mathbf{R}) \simeq U^{wa}(\mathbf{X}_i, \mathbf{R}) \qquad (8.92)$$

which is valid for all configurations $X_i$ that have nonzero probabilities. Using (8.92) in (8.91), we get

$$\delta A^{\mathrm{HI}}(R = 0) = \Delta\mu_a^\circ - \Delta\mu_a^\circ - \Delta\mu_b^\circ = -\Delta\mu_b^\circ \qquad (8.93)$$

That is, the HI at zero separation between two solutes $a$ and $b$, with $b$ much smaller than $a$, is determined by the standard free energy of solution of the solute $b$.

Let us return to relation (8.90) for two solutes $S$ which are to be replaced by a single solute $\bar{S}$ with a field of force approximately twice as strong as that of $S$. It is claimed that if we find a solute $\bar{S}$ having the same *diameter* as $S$, we already have a reasonable approximation for (8.90). The reason is based on the splitting of the standard free energy of solution into two contributions (see Section 7.3)

$$\Delta\mu_S^\circ = \Delta\mu_S^H + \Delta\mu_S^S \qquad (8.94)$$

where the two terms on the rhs of (8.94) correspond to the hard ($H$) and soft ($S$) parts of the solute–solvent pair potential.

There are reasons to believe that $\Delta\mu_S^H$ is the dominant term in (8.94) [for more details based on the scaled particle theory, see Pierotti (1963) and Wilhelm and Battino (1972)]. If this is true, then an approximation for $\Delta\mu_S^S$ is an approximation of a relatively small term which should have a negligible effect on $\Delta\mu_S^\circ$. Let us rewrite (8.90) after splitting each term according to (8.94):

$$\delta A^{\mathrm{HI}}(R = 0) = \Delta\mu_{\bar{S}}^H + \Delta\mu_{\bar{S}}^S - 2(\Delta\mu_S^H + \Delta\mu_S^S) \qquad (8.95)$$

If $S$ and $\bar{S}$ have the same diameter, then

$$\delta A^{\mathrm{HI}}(R = 0) = -\Delta\mu_S^H + (\Delta\mu_{\bar{S}}^S - 2\Delta\mu_S^S) \qquad (8.96)$$

From (8.96), it is clear that even if the soft part of the field of force of $\bar{S}$ is roughly equal to that of $S$, the second term on the rhs of (8.96) is expected to be small relative to $-\Delta\mu_S^H$. This approximation becomes exact for hard-sphere solutes, and it is expected to be a good one for simple solutes such as noble gas molecules.

## 8.9. HYDROPHOBIC INTERACTION AT MORE REALISTIC DISTANCES

In Section 8.5, we devised a measure of the HI at some distance $R < \sigma$ ($\sigma$ being the solute diameter) which was found useful for a systematic comparison of various solvents. In Section 8.8, we discussed the case $R = 0$. In both cases, it was argued that since our primary interest is the solvent rather than the solute, we can study the problem of HI at any convenient distance for which we get a connection with experimental quantities.

In this section, we turn to a more practical question: What is the strength of the HI at contact, or approximately contact, distance for a pair of solute particles? After all, one should eventually return to the original problem of HI between real solute molecules approaching each other to a distance of $R \approx \sigma$ at most.

We outline here two possible routes for obtaining such information. The full answer to the question is contained in the study of the function $g_{SS}(R)$ [or equivalently $y_{SS}(R)$], a topic which will be mentioned in Section 8.12.

### 8.9.1. Extrapolation from $\delta A^{\mathrm{HI}}(0)$ and $\delta A^{\mathrm{HI}}(\sigma_1)$

We recall the relation (Section 8.3)

$$y(R) = \exp[-\beta\, \delta A^{\mathrm{HI}}(R)] \tag{8.97}$$

[in this section, $y(R)$ is used for $y_{SS}(R)$].

In Section 2.6, we also saw that the function $y(R)$ is a monotonically decreasing function of $R$ in the range $0 \leq R \leq \sigma$. We write here the form of $y(R)$ from the exact solution of the Percus–Yevick equation for hard spheres (Wertheim, 1963; Thiele, 1963; Percus, 1964) in the region $0 \leq R \leq \sigma$:

$$y(R) = -(1-\eta)^{-4}[-(1+2\eta)^2 + 6\eta(1+\tfrac{1}{2}\eta)^2(R/\sigma) \\ - \tfrac{1}{2}\eta(1+2\eta)^2(R/\sigma)^3] \tag{8.98}$$

where $\eta = \pi\varrho\sigma^3/6$, with $\sigma$ the diameter of the particles and $\varrho$ the number density. Figure 8.15 shows the function $y(R^*)$ ($R^* = R/\sigma$) for two densities: (a) the regular close-packed density $\eta = \sqrt{2}\,\pi/6$ and (b) the density of water at 4°C, $\varrho = 0.03346$ Å$^{-3}$. Clearly, the function is almost a linearly

# Water with Two or More Simple Solutes 415

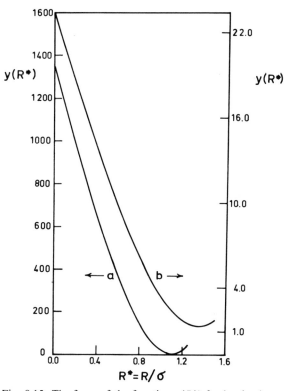

Fig. 8.15. The form of the function $y(R^*)$ for hard spheres [Eq. (8.98)]. Curve $a$ is for hard spheres at the regular close-packed density ($\eta = \sqrt{2}\,\pi/6$). Curve $b$ is the same function computed for hard spheres of diameter 2.82 Å and with the same density as water at 4°C ($\varrho = 0.03345$ Å$^{-5}$). Right and left scales correspond to curves $b$ and $a$, respectively.

decreasing function of $R$. The point we make here is the possibility of utilizing the information on $y(R)$ in the range $0 \leq R \leq \sigma$ to extract information on $y(\sigma)$ [and hence on $\delta A^{\mathrm{HI}}(\sigma)$] from the previously obtained estimates of $y(R)$ at distances $R \leq \sigma$. As an illustration, suppose we can assume that $y(R)$ is linear in the range $0 \leq R \leq \sigma$, i.e.,

$$y(R) = A + BR \tag{8.99}$$

Then, if we know $y(0)$ and $y(\sigma_1)$, where $\sigma_1$ is any value $0 \leq \sigma_1 \leq \sigma$, we can determine $A$ and $B$ from the two equations

$$y(0) = A, \qquad y(\sigma_1) = A + B\sigma_1 \tag{8.100}$$

and thus use (8.99) to calculate $y(\sigma)$. This is the simplest way* of extrapolating the value of $y(\sigma)$ from two values of $y(R)$ at distances shorter than $\sigma$. We will not elaborate further on this method, because, first, we have only approximate values of $y(0)$ and $y(\sigma_1)$, and second, our knowledge of the form of $y(R)$ in this range for simple solutes in complex solvents is very poor. We stress, however, that this method could, in principle, be made useful, since all evidence indicates that the function $y(R)$ in the range $0 \leq R \leq \sigma$ can be approximated by a simple analytical form [for further details, see Ben-Naim (1971a, 1972c,d)].

### 8.9.2. Hydrophobic Interaction between Bulky Molecules

Consider the three pairs of solute particles shown in Fig. 8.16. In the first, we bring two methane molecules to form ethane, in the second, we bring two ethane molecules to form butane (shown schematically as a linear molecule), and in the third, we bring two benzene molecules to form biphenyl. It is evident from the figure that as the molecules become bulkier, the extent of penetration, relative to the molecular size, becomes smaller and smaller (assuming that in all of these cases the molecule on the rhs of the figure is obtained by forming a C–C bond which has almost a constant bond length).

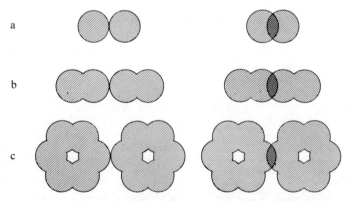

Fig. 8.16. Relative penetration of various molecules. The two molecules at contact distance on the left are compared with the combined molecule $a \cdot b$ on the right. (a) Two methane molecules forming ethane. (b) Two ethane molecules forming butane (depicted schematically in a linear configuration). (c) Two benzene molecules forming diphenyl. It is evident that the relative difference between the two sets of configurations becomes minor as the dimensions of the molecules increase.

Thus, the final configuration of, say, a biphenyl molecule is not much different from the configuration of two benzene molecules at contact distance (as in Fig. 8.16). Hence, the HI for the reaction

$$\text{benzene} + \text{benzene} \to \text{biphenyl} \quad (8.101)$$

gives quite a good estimate of the strength of the HI at contact distance.

Table 8.9 presents some values of $\delta A^{\text{HI}}$ for pairs of identical molecules for which relevant experimental data are available. Note that as the length of the molecule increases, the HI (in the "head-on" configuration) *decreases*. A rough theoretical rationalization of this fact can be given as follows. Let $\sigma_n$ be a measure of the effective diameter of the molecule (or length, in the case of normal paraffin molecules). Let $d_n$ be the distance between the centers of the two molecules at the final configuration, at which the HI has been evaluated. Denote by $l$ the penetration distance (i.e., the distance the two particles penetrate into each other in the repulsive region of the solute–solute pair potential; for methane, $\sigma = 3.82$ Å and the C–C bond is 1.53 Å, hence $l = 3.82 - 1.53 = 2.29$ Å; $l$ is assumed to be constant for a series of paraffin molecules). The reduced distance between the two molecules is then defined as

$$\sigma^* = d_n/\sigma_n = (\sigma_n - l)/\sigma_n \quad (8.102)$$

Now, we expect the function $y(\sigma^*)$ to be a decreasing function of $\sigma^*$ [see Fig. (8.15)]. Therefore, as $n$ increases, $\sigma^*$ increases and approaches unity (for $n \to \infty$), and $y(\sigma^*)$ will reach its lowest value. Translating this in terms of $\delta A^{\text{HI}}$, we conclude that $|\delta A^{\text{HI}}|$ will decrease with $n$, in accord with the results of Table 8.9.

A different way of gaining information on the HI at more realistic distances is provided by molecules such as neopentane (2,2-dimethylpropane). In Section 8.7, we gave the value of the HI for *five* methane molecules brought to the final configuration of neopentane. The "reaction" is

$$5\text{CH}_4 \to \text{neopentane} \quad (8.103)$$

and the corresponding HI is given by

$$\delta A_5^{\text{HI}} = \Delta\mu^\circ(\text{neopentane}) - 5\Delta\mu^\circ(\text{methane}) \quad (8.104)$$

Before turning to a different way of processing the same experimental data, it is worth noting that the validity of the replacement approximation $\varepsilon^g - \varepsilon^l \approx 0$ discussed in Section 5 differs in this case from that in the case

## Table 8.9
### Values of $\delta A^{\mathrm{HI}}$ for Interaction between Two Identical Hydrocarbons

| RH | + RH | → R–R | $-\delta A^{\mathrm{HI}}$, kcal/mole at 25°C |
|---|---|---|---|
| C | + C | → C–C | 2.16 |
| C–C | + C–C | → C–C–C–C | 1.55 |
| C–C–C | + C–C–C | → C–C–C–C–C–C | 1.41 |
| C–C–C–C | + C–C–C–C | → C–C–C–C–C–C–C–C | 1.35 |
| C=C | + C=C | → C=C–C=C | 1.90 |
| C=C–C | + C–C=C | → C=C–C–C–C=C | 1.52 |
| benzene | + benzene | → biphenyl | 1.21 |

of two methane molecules forming an ethane. The reason for this can be appreciated with the help of Fig. 8.17. We have already noted that the replacement approximation involves neglect of the field of force produced by the two inner hydrogens. This field of force is most noticeable in the region denoted by $A$ in the figure, the closest distance a water molecule can approach the two inner hydrogens. In neopentane, we have four C–C bonds, and the replacement approximation involves ignoring the field of force produced by eight hydrogens. However, the region of closest approach between a water molecule and these inner hydrogens—denoted by $B$ in the figure—is farther away as compared with the case of ethane. Therefore, it is expected that the approximation involved in (8.104) is more justified than the corresponding relations of Section 8.5.

We now consider a novel way of extracting information on the HI from the same data. Consider the process of bringing *four* methane molecules from infinity to the final configuration of the four peripheral methyl groups in neopentane, as indicated schematically in Fig. 8.18. The "reaction" is

$$4\mathrm{CH}_4 \rightarrow \text{four peripheral methyl groups of neopentane} \quad (8.105)$$

For this process,

$$\delta A_4^{\mathrm{HI}}(\mathbf{R}^4) = -kT \ln \langle \exp[-\beta U(\mathbf{X}^N/\mathbf{R}^4)] \rangle - 4\Delta\mu°(\text{methane}) \quad (8.106)$$

# Water with Two or More Simple Solutes

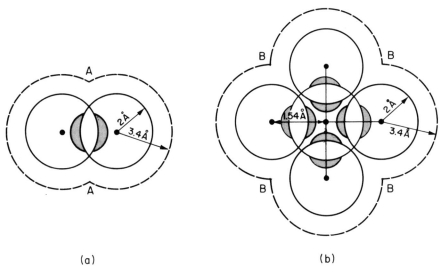

Fig. 8.17. (a) Two methane molecules at a distance of 1.53 Å. The two inner hydrogens are indicated by the dark areas. The replacement approximation essentially involves neglect of the interaction between these two hydrogens and a solvent molecule. The region of closest approach of a water molecule (with radius of about 1.4 Å) to these hydrogens is denoted by $A$. (b) A schematic, two-dimensional description of five methane molecules in the configuration of neopentane. All the inner hydrogens are indicated by dark areas. Because of the partial shielding of the peripheral methane molecules, the region of closest approach of a water molecule is pushed away from the inner hydrogens. These regions are denoted by $B$.

Here, we have denoted by $\mathbf{R}^4$ the configuration of the four methane molecules in the same configuration as the four peripheral methyl groups in neopentane. The field of force that is produced by these molecules is

$$U(\mathbf{X}^N/\mathbf{R}^4) = \sum_{k=1}^{4} \sum_{i=5}^{N+4} U(\mathbf{X}_i, \mathbf{R}_k) \tag{8.107}$$

In (8.107), the configuration $\mathbf{R}^4$ is written more explicitly as $\mathbf{R}_1, \ldots, \mathbf{R}_4$, and we use the indices $i = 5, \ldots, N + 4$ for the solvent molecules.

Clearly, the distance between each pair of methane molecules in this configuration (as shown in Fig. 8.18a) is more realistic than the distance between each pair in the neopentane molecule. (In the latter, we have four C–C bonds at a distance of 1.54 Å and six pairs at a distance of 2.51 Å. In the former, we have only six pairs at a distance of 2.51 Å.)

In order to transform the rhs of (8.106) into a measurable quantity, we must replace the field of force given in (8.107) by a field of force of a

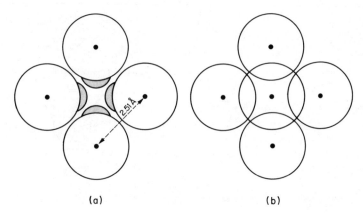

Fig. 8.18. Schematic, two-dimensional description of the replacement process corresponding to "reaction" (8.105). (a) Four methane molecules are brought from infinite separation to the positions of the four peripheral methyl groups of neopentane. The four hydrogens pointing toward the center are indicated by the dark areas. The distance of closest approach between any two of the methane molecules is 2.51 Å (compared to 1.53 Å in the case of ethane). (b) The four methane molecules are replaced by a single neopentane molecule in roughly the same configuration as in (a). There is a new carbon nucleus added which partially compensates for the loss of the four inner hydrogens.

real molecule. To do this, suppose we replace the *four* methane molecules at $\mathbf{R}^4$ by a single neopentane molecule, a replacement that is shown schematically in Fig. 8.18(b). The crucial point to be noted is that the field of force is almost unchanged by this replacement. The reason for this is that the added methyl group is buried deeply in the molecule, and therefore its field of force is hardly noticed by the solvent molecules. [In fact, this approximation might be even better than the one in (8.104). Here, the four hydrogens of the four methanes, indicated by dark areas in Fig. 8.18, are not completely neglected, since they are replaced by the new carbon nucleus which partially compensates for the loss of the field of force produced by these hydrogens.]

Mathematically, the approximation is

$$U(\mathbf{X}^N/\mathbf{R}^4) \approx \sum_{k=1}^{4} \sum_{i=5}^{N+4} U(\mathbf{X}_i, \mathbf{R}_k) + \sum_{i=5}^{N+4} U(\mathbf{X}_i, \mathbf{R}_0)$$
$$= U(\mathbf{X}^N/\mathbf{X}_{\text{neo}}) \qquad (8.108)$$

That is, we add to (8.107) the field of force produced by the nucleus of the carbon at the center of neopentane, denoted by $R_0$. We assume that this addition is negligible for all configurations of the solvent molecules $X^N$ that have nonzero probability. In the second form of the rhs of (8.108), we replace the modified field of force by the field of force produced by a single neopentane, the configuration of which is denoted by $X_{neo}$.

Substituting (8.108) in (8.106) yields

$$\delta A_4^{HI}(R^4) = -kT \ln \langle \exp[-\beta U(X^N/X_{neo})]\rangle - 4\Delta\mu^\circ(\text{methane})$$
$$= \Delta\mu^\circ(\text{neopentane}) - 4\Delta\mu^\circ(\text{methane}) \qquad (8.109)$$

To summarize, we compare (8.109) with (8.104) and note that here we have estimated the HI among *four* methane molecules at more realistic distances than the *five* methane molecules in (8.104). Furthermore, the replacement approximation in (8.109) is expected to be better than the corresponding one in (8.104). Figure 8.19 shows the values of $\delta A_4^{HI}$ and $\delta A_5^{HI}$ as a function of temperature. Clearly, the values of $\delta A_4^{HI}$ are higher than those of $\delta A_5^{HI}$. The reasons are (1) we have HI among four instead of five groups, and (2) the extent of penetration in the latter is larger than in the former case. Note also the values of $\delta A_4^{HI}$ for butane and isobutane.

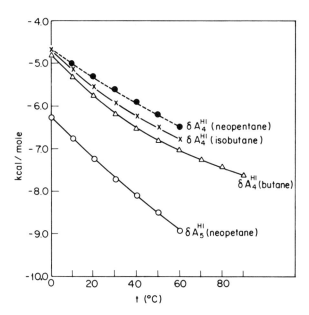

Fig. 8.19. Values of $\delta A_4^{HI}$ for various configurations and $\delta A_5^{HI}$ for neopentane as a function of temperature.

We conclude this section by noting that the arguments given above for neopentane can be extended to bulkier molecules, such as tetraalkylmethane, with large alkyl groups. It is clear that for such molecules, the field of force produced by the central carbon nucleus will be completely negligible; its sole role in the theory is to serve as an agent which holds the four alkane molecules together. Furthermore, the intermolecular distances between the pairs of alkane molecules will involve very slight interpenetration, hence in these cases the HI will refer to realistic distances. The treatment can also be extended to molecules, such as tetraalkylammonium salts, which have very peculiar and interesting properties in aqueous solutions [see Ben-Naim (1972a) and Wen (1972)].

## 8.10. ENTROPY AND ENTHALPY OF HYDROPHOBIC INTERACTION

In this chapter, we have characterized the *strength* of the HI by either $\delta A^{\mathrm{HI}}(R)$ or $\delta G^{\mathrm{HI}}(R)$. We now consider the entropy and the enthalpy changes associated with the process of bringing two solutes from infinite separation to a final distance $R$. We refer to this process as the HI process, and make a distinction between HI processes in $T$, $V$, $N$ and $T$, $P$, $N$ ensembles.

The free energy change for the HI process in the $T$, $V$, $N$ ensemble is denoted by

$$\Delta A^l(R) = A^l_{N+2}(R) - A^l_{N+2}(R = \infty) = U(R) + \delta A^{\mathrm{HI}}(R) \quad (8.110)$$

The entropy change associated with this process is

$$-\Delta S^l(R) = \left(\frac{\partial \Delta A^l(R)}{\partial T}\right)_{V,N} = \left(\frac{\partial \delta A^{\mathrm{HI}}(R)}{\partial T}\right)_{V,N} = -\delta S^{\mathrm{HI}}(R) \quad (8.111)$$

where we have assumed that the direct pair potential $U(R)$ is temperature independent. We can thus obtain $\Delta S^l(R)$ from the temperature dependence of $\delta A^{\mathrm{HI}}(R)$. However, in (8.111), we need a derivative at constant volume, a quantity which is usually not available from experiment.

The energy change associated with the same process is

$$\Delta E^l(R) = \Delta A^l(R) + T\Delta S^l(R)$$
$$= U(R) + \delta A^{\mathrm{HI}}(R) - T[\partial\, \delta A^{\mathrm{HI}}(R)/\partial T] \quad (8.112)$$

Here we demonstrate that this quantity cannot be easily dealt with in terms of ordinary molecular distribution functions. To do this, we write the statistical mechanical expression for $\Delta E^l(R)$ in greater detail.

# Water with Two or More Simple Solutes

For any two fixed positions $\mathbf{R}_1$ and $\mathbf{R}_2$, we define the average potential energy of the system by

$$\langle U_N(\mathbf{R}_1, \mathbf{R}_2) \rangle = \int \cdots \int d\mathbf{X}^N \, P(\mathbf{X}^N/\mathbf{R}_1, \mathbf{R}_2) U(\mathbf{X}^N, \mathbf{R}_1, \mathbf{R}_2) \quad (8.113)$$

Note that the average is taken over all the configurations of the solvent molecules. $P(\mathbf{X}^N/\mathbf{R}_1, \mathbf{R}_2)$ is the conditional probability density of observing a configuration $\mathbf{X}^N$, given that two solute particles are at fixed positions $\mathbf{R}_1, \mathbf{R}_2$. For the following argument, we assume complete pairwise additivity of the total potential energy, i.e.,

$$U(\mathbf{X}^N, \mathbf{R}_1, \mathbf{R}_2) = U(\mathbf{R}_1, \mathbf{R}_2) + \sum_{k=1}^{2} \sum_{i=3}^{N+2} U(\mathbf{X}_i, \mathbf{R}_k) + \sum_{i<j} U(\mathbf{X}_i, \mathbf{X}_j) \quad (8.114)$$

It should be clear from the arguments when we are referring to a solute and when to a solvent molecule. The energy change $\Delta E^l(R)$ is thus

$$\Delta E^l(\mathbf{R}_1, \mathbf{R}_2) = \langle U_N(\mathbf{R}_1, \mathbf{R}_2) \rangle - \langle U_N(\mathbf{R}_1, \mathbf{R}_2; R = \infty) \rangle \quad (8.115)$$

i.e., the difference in the average potential energy for two configurations of the solutes, $(\mathbf{R}_1, \mathbf{R}_2)$ and $R = \infty$, respectively.

It is sufficient to consider the first term on the rhs of (8.115) in order to convey our point. From (8.113) and (8.114), we get

$$\langle U_N(\mathbf{R}_1, \mathbf{R}_2) \rangle = U(\mathbf{R}_1, \mathbf{R}_2) + \sum_k \sum_i \int \cdots \int d\mathbf{X}^N \, P(\mathbf{X}^N/\mathbf{R}_1, \mathbf{R}_2) U(\mathbf{X}_i, \mathbf{R}_k)$$

$$+ \sum_{i<j} \int \cdots \int d\mathbf{X}^N \, P(\mathbf{X}^N/\mathbf{R}_1, \mathbf{R}_2) U(\mathbf{X}_i, \mathbf{X}_j)$$

$$= U(\mathbf{R}_1, \mathbf{R}_2) + \int d\mathbf{X}_3 \, [U(\mathbf{X}_3, \mathbf{R}_1) + U(\mathbf{X}_3, \mathbf{R}_2)] \varrho(\mathbf{X}_3/\mathbf{R}_1, \mathbf{R}_2)$$

$$+ \tfrac{1}{2} \int \int d\mathbf{X}_3 \, d\mathbf{X}_4 \, U(\mathbf{X}_3, \mathbf{X}_4) \varrho(\mathbf{X}_3, \mathbf{X}_4/\mathbf{R}_1, \mathbf{R}_2) \quad (8.116)$$

One sees that an expression for $\Delta E^l(\mathbf{R}_1, \mathbf{R}_2)$ involves molecular distribution functions of up to order four. Therefore, this approach to the problem is impractical at present. In the next section, we pursue a different, conceptually simpler method which involves the mixture-model approach to the solvent.

We now formulate the HI problem in the $T, P, N$ ensemble. All the steps are essentially the same as they are in the $T, V, N$ ensemble, with some minor modifications, as will be noted. (More details on a similar derivation in the $T$, $P$, $N$ ensemble are given in Appendix 9-F.)

The Gibbs free energy change for the HI process is

$$\Delta G^l(\mathbf{R}_1, \mathbf{R}_2) = G^l_{N+2}(\mathbf{R}_1, \mathbf{R}_2) - G^l_{N+2}(R = \infty)$$
$$= U(\mathbf{R}_1, \mathbf{R}_2) + \delta G^{\mathrm{HI}}(\mathbf{R}_1, \mathbf{R}_2) \qquad (8.117)$$

and the statistical mechanical expression for $\delta G^{\mathrm{HI}}(\mathbf{R}_1, \mathbf{R}_2)$ is

$$\delta G^{\mathrm{HI}}(\mathbf{R}_1, \mathbf{R}_2) = -kT \ln\langle \exp[-\beta U(\mathbf{X}^N/\mathbf{R}_1, \mathbf{R}_2)]\rangle - 2\Delta\mu_S^\circ \qquad (8.118)$$

Here, the symbol $\langle\ \rangle$ indicates an average in the $T$, $P$, $N$ ensemble [compare with the corresponding expression, say (8.82), in the $T$, $V$, $N$ ensemble]. $\Delta\mu_S^\circ$ is the Gibbs free energy change for transferring $S$ from a fixed position in the gas to a fixed position in the liquid, the process being carried out at $T$, $P$, and $N$ constant.

Using the same approximation as in Section 8.5, we can write an approximate relation expressing the HI between two methane molecules at the distance $R = \sigma_1 = 1.53$ Å as

$$\delta G^{\mathrm{HI}}(\sigma_1) = \Delta\mu_E^\circ - 2\Delta\mu_M^\circ \qquad (8.119)$$

Here, the subscripts E and M denote ethane and methane, respectively. In (8.119), we view each quantity as a function of $T$ and $P$; hence, the corresponding entropy change for this particular process is

$$-\delta S^{\mathrm{HI}}(\sigma_1) = \left(\frac{\partial\,\delta G^{\mathrm{HI}}(\sigma_1)}{\partial T}\right)_P = \left(\frac{\partial\,\Delta\mu_E^\circ}{\partial T}\right)_P - 2\left(\frac{\partial\,\Delta\mu_M^\circ}{\partial T}\right)_P \qquad (8.120)$$

We now identify the two derivatives on the rhs of (8.120) as the standard entropies of solution for process II (i.e., the transfer from a fixed position in the gas to a fixed position in the liquid; see Section 4.11 for more details). We can thus rewrite (8.120) as

$$\delta S^{\mathrm{HI}}(\sigma_1) = \Delta S_E^\circ(\mathrm{II}) - 2\Delta S_M^\circ(\mathrm{II}) \qquad (8.121)$$

It is very important to stress that the last equality holds only for these particular standard entropies of solution, which is why we have included the symbol II (see Section 4.11) in the notation. We also note that the quantities on the rhs of (8.121) are measurable, since $\Delta\mu_E^\circ$ and $\Delta\mu_M^\circ$ are usually measured as a function of temperature at constant pressure. Thus, (8.121) provides an approximate method of estimating the entropy change for the HI process.

Similarly, the enthalpy change for the same process is obtained from (8.117) by differentiation with respect to temperature:

$$\Delta H^l(\mathbf{R}_1, \mathbf{R}_2) = -T^2 \, \partial[\Delta G^l(\mathbf{R}_1, \mathbf{R}_2)/T]/\partial T = U(\mathbf{R}_1, \mathbf{R}_2) + \delta H^{\mathrm{HI}}(\mathbf{R}_1, \mathbf{R}_2) \tag{8.122}$$

For the particular distance $R = \sigma_1$, we get the approximate expression

$$\delta H^{\mathrm{HI}}(\sigma_1) = -T^2 \, \partial[\delta G^{\mathrm{HI}}(\sigma_1)/T]/\partial T = \Delta H_{\mathrm{E}}^\circ(\mathrm{II}) - 2\Delta H_{\mathrm{M}}^\circ(\mathrm{II}) \tag{8.123}$$

For the volume change, we have

$$\delta V^{\mathrm{HI}}(\sigma_1) = \Delta V_{\mathrm{E}}^\circ(\mathrm{II}) - 2\Delta V_{\mathrm{M}}^\circ(\mathrm{II}) \tag{8.124}$$

Table 8.10 gives some values of $\delta G^{\mathrm{HI}}(\sigma_1)$, $\delta S^{\mathrm{HI}}(\sigma_1)$, $\delta H^{\mathrm{HI}}(\sigma_1)$, and $\delta V^{\mathrm{HI}}(\sigma_1)$ for bringing two methanes to the final configuration of ethane in various solvents. Note that both the entropy and the enthalpy values in water (and heavy water) are positive and large compared with the cor-

### Table 8.10

**Values of $\delta G^{\mathrm{HI}}(\sigma_1)$, $\delta S^{\mathrm{HI}}(\sigma_1)$, $\delta H^{\mathrm{HI}}(\sigma_1)$, and $\delta V^{\mathrm{HI}}(\sigma_1)$ for Bringing Two Methane Molecules to the Distance $\sigma_1 = 1.533$ Å at 10°C[a]**

| Solvent | $\delta G^{\mathrm{HI}}$, kcal/mole | $\delta S^{\mathrm{HI}}$, e.u. | $\delta H^{\mathrm{HI}}$, kcal/mole | $\delta V^{\mathrm{HI}}$, cm³/mole |
|---|---|---|---|---|
| Water | −1.99 | 12 | 1.60 | −23[b] |
| Heavy water | −1.94 | 13 | 1.70 | — |
| Methanol | −1.28 | 0 | −1.4 | — |
| Ethanol | −1.34 | 0 | −1.3 | — |
| 1-Propanol | −1.39 | 2 | −0.6 | — |
| 1-Butanol | −1.44 | −1 | −1.9 | — |
| 1-Pentanol | −1.49 | 0 | −1.3 | — |
| 1-Hexanol | −1.51 | 2 | 0.7 | — |
| 1,4-Dioxane | −1.61 | 3 | −0.8 | — |
| Cyclohexane | −1.36 | 1 | −1.2 | — |

[a] Based on data from Yaacobi and Ben-Naim (1974).
[b] In benzene, the value is −31 cm³/mole.

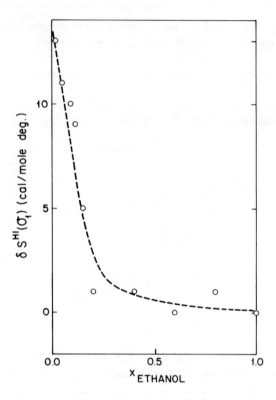

Fig. 8.20. Variation of the entropy $\delta S^{\mathrm{HI}}(\sigma_1)$ associated with the process of hydrophobic interaction as a function of mole fraction of ethanol at 15°C.

responding values in other solvents. Thus, it is tempting to conclude that the behavior of water is unique, even though the comparison has been made only between water and a few other solvents.[15]

It is instructive to estimate the difference between the enthalpy change $\delta H^{\mathrm{HI}}(\sigma_1)$ and the corresponding energy change $\delta E^{\mathrm{HI}}(\sigma_1)$, which, for water at 10°C and 1 atm, is

$$\delta H^{\mathrm{HI}}(\sigma_1) - \delta E^{\mathrm{HI}}(\sigma_1) = P\,\delta V^{\mathrm{HI}}(\sigma_1) = -0.55 \text{ cal/mole} \qquad (8.125)$$

In benzene, the value of $P\,\delta V^{\mathrm{HI}}(\sigma_1)$ is $-0.75$ cal/mole. Thus, the difference

---

[15] Note that $\delta H^{\mathrm{HI}}$ is computed from (8.123), not from $\delta H^{\mathrm{HI}} = \delta G^{\mathrm{HI}} + T\,\delta S^{\mathrm{HI}}$. Because of the relatively large error in computing $\Delta H_{\mathrm{E}}^{\circ}$ and $\Delta H_{\mathrm{M}}^{\circ}$, a discrepancy may arise between the two methods of evaluating $\delta H^{\mathrm{HI}}$.

## Water with Two or More Simple Solutes

between $\delta H^{HI}(\sigma_1)$ and $\delta E^{HI}(\sigma_1)$ is considered to be negligibly small when compared with their individual values.

Figures 8.20 and 8.21 show the variation of $\delta S^{HI}(\sigma_1)$ and $\delta H^{HI}(\sigma_1)$ as a function of the mole fraction of ethanol in water–ethanol mixtures. The most prominent feature of these two curves is the steep decrease in $\delta S^{HI}$ and $\delta H^{HI}$ in the water-rich region ($0 \leq x_{\text{ethanol}} \lesssim 0.2$), followed by an almost constant value in the entire region of composition $x_{\text{ethanol}} \gtrsim 0.2$. An attempt to interpret this behavior is made in the next section.

We conclude this section by pointing out the importance of the temperature dependence of HI in biological systems (Scheraga *et al.*, 1962; Nemethy, 1967). One expects an increase in temperature to lead to a randomization effect. This is revealed in a tendency to unfold a specific conforma-

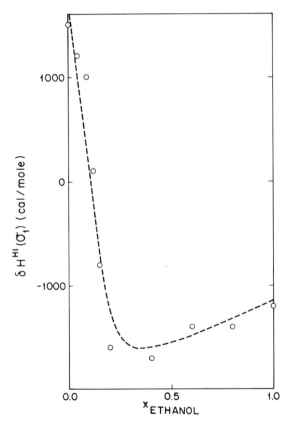

Fig. 8.21. Variation of the enthalpy $\delta H^{HI}(\sigma_1)$ associated with the process of hydrophobic interaction as a function of mole fraction of ethanol at 15°C.

tion of a biopolymer. The fact that the strength of the HI *increases* with temperature counteracts this tendency, and helps to stabilize the particular conformation in spite of the changes in the environmental temperature.

## 8.11. HYDROPHOBIC INTERACTION AND STRUCTURAL CHANGES IN THE SOLVENT

This section is concerned with the interpretation of the large, positive values of $\delta S^{\mathrm{HI}}(\sigma_1)$ and $\delta H^{\mathrm{HI}}(\sigma_1)$ found for water as compared with the corresponding values in other solvents for which relevant experimental results are available.

Consider the two contributions to $\delta G^{\mathrm{HI}}(\sigma_1)$ in water and in ethanol[16] (in kcal/mole at $t = 10°C$)

$$\delta H^{\mathrm{HI}}(\sigma_1) = 1.6, \quad T\,\delta S^{\mathrm{HI}}(\sigma_1) = 3.4, \, \delta G^{\mathrm{HI}}(\sigma_1) = -1.99 \text{ (water)} \quad (8.126\mathrm{a})$$

$$\delta H^{\mathrm{HI}}(\sigma_1) = -1.2, \, T\,\delta S^{\mathrm{HI}}(\sigma_1) \approx 0, \quad \delta G^{\mathrm{HI}}(\sigma_1) = -1.34 \text{ (ethanol)} \quad (8.126\mathrm{b})$$

It is evident that the main contribution to $\delta G^{\mathrm{HI}}(\sigma_1)$ in water comes from the entropy term, whereas in, say, ethanol, it comes from the enthalpy term. This fact provides the basis for the traditional interpretation of the large, negative value of $\delta G^{\mathrm{HI}}$ in water, which is based on the following two statements. (1) The large, positive value of $\delta S^{\mathrm{HI}}$ in water is due to the structural breakdown associated with the HI process. (2) The value of $\delta G^{\mathrm{HI}}$ is determined mainly by the entropy change and not by the enthalpy change. Hence, it is very tempting to conclude that structural changes in the solvent provide the "explanation" for the strong HI in water.[17] We will see that due to compensation effects, contributions to $\delta G^{\mathrm{HI}}$ from structural changes are precluded, and therefore the above interpretation is erroneous.

We stress, however, that the two statements given above are likely to be correct. Later in this section, we shall provide arguments to justify the first one. The second is clear from (8.126a). Nevertheless, the inference made from the combined statements may be misleading.

---

[16] Note that in (8.126a) and (8.126b) $\delta G^{\mathrm{HI}}$ is the most accurate quantity. $\delta H^{\mathrm{HI}}$ and $\delta S^{\mathrm{HI}}$ are determined from (8.121) and (8.123), which are obtained from the temperature dependence of $\Delta\mu_S°$. Therefore these are less accurate quantities, as a result of which the relation $\delta G^{\mathrm{HI}} = \delta H^{\mathrm{HI}} - T\,\delta S^{\mathrm{HI}}$ holds only approximately. (See also Section 8.10.)

[17] This appears in various forms in the literature. In most cases, one refers to $\Delta\mu_t°$ as a measure of the strength of HI rather than referring to $\delta G^{\mathrm{HI}}$. The reasoning given here is the same, however, for both quantities.

## Water with Two or More Simple Solutes

One quick argument can be given on the basis of Eq. (8.119), which relates $\delta G^{\mathrm{HI}}(\sigma_1)$ to the standard free energy changes. We already know from Chapter 7 that these are independent of structural changes in the solvent. Therefore, this must also be true for $\delta G^{\mathrm{HI}}(\sigma_1)$. Because of the importance of this question, we present a direct argument here, valid for any process of bringing two (or more) solutes from infinite separation to a final distance $R$. For simplicity, we use a two-structure model for the solvent. The generalization for any mixture model is quite straightforward.

Consider a two-structure model of $L$-cules and $H$-cules obtained by any exact classification procedure as discussed in Chapter 6. The argument presented below is general, independent of the choice of classification, and valid for any solvent.

Let $N_L$ and $N_H$ be the average number of $L$- and $H$-cules in a system of $N_W$ solvent molecules ($N_W = N_L + N_H$) at given $T$ and $P$, in which two solute particles hold fixed positions at infinite separation $R = \infty$. Consider now the process of bringing the two solutes from infinity to a final configuration of distance $R$. The strength of the HI (i.e., the indirect part of the work) is given by

$$\delta G^{\mathrm{HI}}(R) = G(N_L', N_H', R) - G(N_L, N_H, R = \infty) - U(R) \quad (8.127)$$

where $N_L'$ and $N_H'$ are the new values for the number of $L$- and $H$-cules at the final distance $R$. The same process can be carried out in two steps:

$$\delta G^*(R) = G(N_L, N_H, R) - G(N_L, N_H, R = \infty) - U(R)$$
$$\text{("frozen-in" system)} \quad (8.128)$$

$$\delta G^{\mathrm{r}}(R) = G(N_L', N_H', R) - G(N_L, N_H, R) \quad \text{(relaxation)} \quad (8.129)$$

$$\delta G^{\mathrm{HI}}(R) = \delta G^*(R) + \delta G^{\mathrm{r}}(R) \quad (8.130)$$

That is, we first bring the two solutes to a distance $R$ in a "frozen-in" system, and then let the system relax to its final state. Since the second step is spontaneous, we must have

$$\delta G^{\mathrm{r}}(R) \leq 0 \quad (8.131)$$

Hence, we conclude from (8.130) and (8.131) that

$$\delta G^{\mathrm{HI}}(R) \leq \delta G^*(R) \quad (8.132)$$

which means that the HI in the "frozen-in" system will always be *weaker* (note that the values of $\delta G^{\mathrm{HI}}$ are negative) than the corresponding values in the equilibrated system. This argument can be extended easily to any mixture model.

Next, we consider the order of magnitude of the relaxation term. Expanding (8.129) to second-order terms, we get[18]

$$-\delta G^r(R) = G(N_L, N_H, R) - G(N_L', N_H', R)$$
$$= \mu_L \, dN_L + \mu_H \, dN_H + \tfrac{1}{2}(\mu_{LL} - 2\mu_{LH} + \mu_{HH}) \, dN_L^2 \quad (8.133)$$

Since at equilibrium we have $\mu_L = \mu_H$ (note that $dN_L = -dN_H$), relation (8.133) reduces to

$$-\delta G^r(R) = \tfrac{1}{2}(\mu_{LL} - 2\mu_{LH} + \mu_{HH}) \, dN_L^2 \quad (8.134)$$

Thus, the sign of $\delta G^r(R)$ is always negative, as we found before (see also Section 5.10).

In a similar fashion, we can write, say, the entropy change as

$$\delta S^{\mathrm{HI}}(R) = \delta S^*(R) + \delta S^r(R) \quad (8.135)$$

Expanding $\delta S^r(R)$, we get

$$-\delta S^r = S(N_L, N_H, R) - S(N_L', N_H', R) = (\bar{S}_L - \bar{S}_H) \, dN_L + \cdots \quad (8.136)$$

and similarly,

$$-\delta H^r = H(N_L, N_H, R) - H(N_L', N_H', R) = (\bar{H}_L - \bar{H}_H) \, dN_L + \cdots \quad (8.137)$$

The essential difference between the expansion (8.134), on the one hand, and (8.136) and (8.137) on the other, is that the first has no contribution from the first-order term. This is an important observation. For, if we take the limit of a macroscopic system $N_W \to \infty$ ($P$, $T$ constant), all of the quantities $\mu_{ij}$ in (8.134) will tend to zero (note that these quantities are homogeneous functions of order $-1$). Hence, at this limit $\delta G^r \to 0$, whereas $\delta S^r$ and $\delta H^r$ in (8.136) and (8.137) remain finite.

A different way of stating the same thing is as follows. For any distance $R$, we write

$$\delta G^{\mathrm{HI}} = \delta G^* + \delta G^r$$
$$= \delta H^* - T \, \delta S^* - (\bar{H}_L - \bar{H}_H) \, dN_L + T(\bar{S}_L - \bar{S}_H) \, dN_L + \cdots$$
$$= \delta H^* - T \, \delta S^* - (\mu_L - \mu_H) \, dN_L + \cdots \quad (8.138)$$

---

[18] The expansion is done about the point $(N_L', N_H', R)$, which is an equilibrium state for the system when the two solutes are at a distance $R$.

Because of the equilibrium condition, the first-order relaxation term drops out of this equation.

We can extend the above argument to any structural change in the solvent.[19] Let **N** be any vector that may be used as a quasicomponent distribution function (Chapter 5). For simplicity, we assume that **N** contains discrete components. Let **N'** and **N** be the composition of the solvent when the two solutes are at $R$ and at $R = \infty$, respectively. Then instead of (8.133) we have

$$-\delta G^r(R) = G(\mathbf{N}, R) - G(\mathbf{N'}, R)$$

$$= \sum_{i=0}^{\infty} \frac{\partial G}{\partial N_i} dN_i + \frac{1}{2} \sum_{i,j} \frac{\partial^2 G}{\partial N_i \partial N_j} dN_i \, dN_j + \cdots$$

$$= \frac{1}{2} \sum_{i,j} \frac{\partial^2 G}{\partial N_i \partial N_j} dN_i \, dN_j + \cdots \quad (8.139)$$

The last equality holds because of the equilibrium condition. That is, we have, say, $\mu_i = \mu_0$ (for any $i$), and hence

$$\sum_i \mu_i \, dN_i = \sum_i \mu_0 \, dN_i = \mu_0 \sum_i dN_i = 0 \quad (8.140)$$

Hence, the first-order term drops out of (8.139). We thus conclude that in the limit of macroscopic systems the free energy change per pair of particles for the HI process is the same whether the process is carried out in the "frozen-in" or equilibrated system. On the other hand, the entropy and the enthalpy changes associated with the HI process may get a large contribution due to structural changes in the solvent.

We now proceed to interpret $\delta S^{\mathrm{HI}}$ and $\delta H^{\mathrm{HI}}$ in terms of structural changes in the solvent. We use relations (8.121) and (8.123) for any distance $R$ to obtain

$$\delta S^{\mathrm{HI}}(R) = \Delta S_D{}^\circ(\mathrm{II}) - 2\Delta S_M{}^\circ(\mathrm{II}) = \Delta S_D{}^* - 2\Delta S_M{}^* + \Delta S_D{}^r - 2\Delta S_M{}^r \quad (8.141)$$

$$\delta H^{\mathrm{HI}}(R) = \Delta H_D{}^\circ(\mathrm{II}) - 2\Delta H_M{}^\circ(\mathrm{II}) = \Delta H_D{}^* - 2\Delta H_M{}^* + \Delta H_D{}^r - 2\Delta H_M{}^r \quad (8.142)$$

Here, $M$ and $D$ stand for monomer and dimer, respectively, and the symbol II reminds us that we are using process II (of Chapter 4) for the definition

---

[19] We refer here only to "structural changes" in the sense of Chapters 5 and 7. In principle, one may devise other definitions of the structure of the solvent and structural changes to which the present treatment does not apply.

of the standard entropy and enthalpy of solution. The second equalities on the rhs of (8.141) and (8.142) hold for any mixture-model approach.

For concreteness, we use a two-structure model and express the relaxation term of, say, $\delta H^{\mathrm{HI}}(R)$ as [see also Eq. (7.46)]

$$\Delta H_D^{\mathrm{r}} - 2\Delta H_M^{\mathrm{r}} = (\bar{H}_L - \bar{H}_H)\left(\frac{\partial N_L}{\partial N_D} - 2\frac{\partial N_L}{\partial N_M}\right) \qquad (8.143)$$

In order to determine the sign of the contribution to the enthalpy change associated with the HI process, we need to know the difference in the stabilization effects of $M$ and $D$ on the structure of the solvent.

Using the results of Section 7.6, we rewrite the difference of the stabilization effects in (8.143) as

$$\frac{\partial N_L}{\partial N_D} - 2\frac{\partial N_L}{\partial N_M} = \varrho_W x_L x_H [-(G_{WH} - G_{WL}) + (G_{LD} - G_{HD})$$
$$- 2(G_{LM} - G_{HM})] \qquad (8.144)$$

Relation (8.144) is very general and applies to any two-component system. As in Section 7.6, we now specialize to a particular two-structure model for water, where $L$ and $H$ are the components with low and high local density. With this choice, we immediately have one negative term in (8.144), namely

$$-(G_{WH} - G_{WL}) < 0 \qquad (8.145)$$

The reason for this inequality follows directly from the definition of the two components and has been discussed in detail in Section 7.6.

Next, we consider the two terms $G_{LD} - G_{HD}$ and $G_{LM} - G_{HM}$, both of which are expected to be positive, the arguments being essentially the same as those presented in Section 7.6. We recall that $\varrho_D(G_{LD} - G_{HD})$ can be interpreted as the average excess of dimers $D$ in the neighborhood of $L$ relative to $H$. The primary reason for assuming this excess to be positive is the relative difference in the local densities of $L$ and $H$. Therefore, we conclude that $G_{LD} - G_{HD}$ and $G_{LM} - G_{HM}$ are roughly of the same order of magnitude, and therefore the term $(G_{LD} - G_{HD}) - 2(G_{LM} - G_{HM})$ is expected to be negative.[20] Hence, from (8.144) we have

$$\frac{\partial N_L}{\partial N_D} - 2\frac{\partial N_L}{\partial N_M} < 0 \qquad (8.146)$$

[20] Due to the averaging over all the orientations of $D$, it is possible for $G_{LD} - G_{HD}$ to be slightly larger than $G_{LM} - G_{HM}$. However, it seems unlikely that it will be larger than twice the quantity $G_{LM} - G_{HM}$.

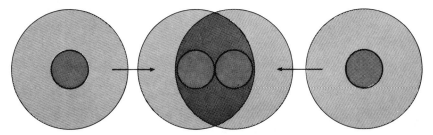

Fig. 8.22. Schematic illustration of the change in the structure of water around the solutes when brought from infinite separation to a close distance. The stabilization of the structure of water induced by two separate solutes is expected to be larger than the corresponding effect for the same pair of solutes at close distance. The large circles surrounding the solute particles designate the regions of influence of the solute on the solvent.

All of the comments of Section 7.6 regarding the magnitude of the stabilization effect also apply here.

The conclusion (8.146) can be summarized with the help of the schematic drawing in Fig. 8.22. We assume that each solute has a "radius of influence" $R_f$ which may be defined as the distance at which the solute–solvent pair correlation function is practically unity. In this region, we know that the concentration of the $L$ form is larger than its bulk concentration, a fact which has been referred to as the stabilization of the $L$ form by the solute. When the two solute particles come close to each other, the spheres of influence of the two solutes must overlap. It is therefore likely that the total excess of $L$-cules in the union of the two spheres of influence is smaller than the excess of $L$-cules in the two spheres when they do not overlap. If we identify the $L$ form as the more structured species, then we conclude that the structure of water decreases as a result of the HI process.

In Section 7.6, we presented some arguments showing that the large stabilization effect of the structure of water is probably a unique phenomenon of liquid water. The same arguments hold here as well. Furthermore, we have stressed that a unique feature of the structure of water is the coupling between low local density and strong binding energy. Therefore, if $L$ is chosen to be the species with the relatively low local density (say, all molecules with coordination numbers less than five), then the quantity $\bar{H}_L - \bar{H}_H$ is expected to be negative.

We can now conclude that as the two solutes are brought from infinite separation to the final close distance $R \approx \sigma$, the net effect is a destabilization of the $L$ form. Furthermore, this destabilization is coupled with a negative

$\bar{H}_L - \bar{H}_H$, so that the whole term in (8.143) becomes positive. It is also likely to be particularly large, for reasons discussed in Section 7.6.

The above arguments form the basis of the interpretation of the large, *positive* enthalpy change associated with the HI process. Because of the compensation law $\Delta H^{\mathrm{r}} = T \Delta S^{\mathrm{r}}$ (for both solutes $M$ and $D$), the same interpretation holds for the corresponding entropy change.

We commenced this section with two statements providing the traditional interpretation of the large, negative value of $\delta G^{\mathrm{HI}}$. We have given arguments showing that, due to structural changes, the value of $\delta S^{\mathrm{HI}}$ is likely to be large and positive. From (8.126a), we see that it is $T \delta S^{\mathrm{HI}}$ that is responsible for the relatively large value of $\delta G^{\mathrm{HI}}$. Nevertheless, because of the compensation effect, structural changes cannot contribute to $\delta G^{\mathrm{HI}}$. In fact, we have

$$\delta G^{\mathrm{HI}} = \delta H^{\mathrm{HI}} - T \delta S^{\mathrm{HI}} = \delta H^* - T \delta S^* \qquad (8.147)$$

i.e., the HI is determined only by the *static* parts of the enthalpy and entropy changes.

Thus, $\delta G^{\mathrm{HI}}$ may depend on the *structure* of water, but it does not receive any contribution from *structural changes*.[21]

It is also appropriate at this stage to apply a continuous mixture-model approach to demonstrate the general aspect of the contribution of structural changes in the solvent. For simplicity, we refer to the energy change associated with the HI process, and apply relation (7.140) for $D$ and $M$ to get

$$\delta E^{\mathrm{HI}} = \Delta E_D^\circ - 2\Delta E_M^\circ$$
$$= (\langle B_D^\circ \rangle - 2\langle B_M^\circ \rangle) + \frac{1}{2} \int v \left[ \frac{\partial N_W(v)}{\partial N_D} - 2 \frac{\partial N_W(v)}{\partial N_M} \right] dv \qquad (8.148)$$

where $\langle B_D^\circ \rangle$ and $\langle B_M^\circ \rangle$ are the average binding energies (at infinite dilution) of $D$ and $M$, respectively, and the second term on the rhs is the difference in the contributions due to structural changes of $D$ and two $M$'s.

It is evident that the average binding energy of $D$ (to the solvent) is weaker than twice the average binding energy of a single monomer $M$.

---

[21] Some care must be exercised in the interpretation of this statement. We saw in Section 7.7 that part of the structural changes (defined in a very broad sense) can be absorbed in the static part of the partial molar entropies and enthalpies of a solute. The above statement applies to any particular definition of the structure which has been adopted for the present purpose.

## Water with Two or More Simple Solutes

Therefore, the first term on the rhs of (8.148) is expected to be positive for any solvent. For hard-sphere solutes, this term is zero, and for simple nonpolar solutes, it is likely to have a small absolute value.

In a "normal" solvent, we expect structural changes in the solvent to be small, for reasons discussed in Chapter 7. Therefore, the second term on the rhs of (8.148) is likely to be small in such solvents. In water, the large, positive energy of the HI process can be attributed to a structural change in the solvent such that the average binding energy of a water molecule *decreases*, an effect which can also be ascribed to a net breaking of hydrogen bonds.

The above discussion, especially the one employing a two-structure model, gives a qualitative interpretation of the large, positive values of $\delta H^{\mathrm{HI}}$ and $\delta S^{\mathrm{HI}}$ in water. Figures 8.20 and 8.21 show that these anomalous values disappear quite rapidly as we add alcohol. In fact, for mole fractions of alcohol larger than $x \approx 0.2$, the values of $\delta H^{\mathrm{HI}}$ and $\delta S^{\mathrm{HI}}$ attain their "normal" values as in pure alcohol.

The temperature dependence of the HI, as revealed through either $\delta S^{\mathrm{HI}}$ or $\delta H^{\mathrm{HI}}$, has been explained qualitatively in terms of structural changes in the solvent. This explanation can be used to correlate some other phenomena involving HI. For any reasonable definition of the "structure of water" (Chapter 6), we expect that the structure will break down as temperature increases. Let $x_{\mathrm{ST}}$ be a measure of the structure of water; we write

$$\text{if } T_2 > T_1, \text{ then } \begin{cases} |\operatorname{HI}(\text{at } T_2)| > |\operatorname{HI}(\text{at } T_1)| \\ x_{\mathrm{ST}}(\text{at } T_2) < x_{\mathrm{ST}}(\text{at } T_1) \end{cases} \quad (8.149)$$

These relations are presumed to hold for temperatures between, say, 0 and 80°C. We can thus correlate an increase in structure with a decrease in the strength of HI. This seems to work for other cases to which the same definition of structure applies. (1) The HI in $H_2O$ is slightly larger than in $D_2O$ (see Table 8.10), which is consistent with the common opinion that heavy water is more "structured" than light water. (2) Addition of a small amount of ethanol is presumed to increase the structure of water (in the sense of Section 7.6). The experimental finding is that for very dilute solutions of ethanol in water, the HI is weakened (see Fig. 8.11). (3) For higher concentrations of alcohol, say $0.03 \lesssim x_{\mathrm{ethanol}} \lesssim 0.2$, the structure of water breaks down due to the dilution of water in the mixture, and the HI appears to strengthen (this effect is particularly clear at 10°C; see Fig. 8.11). (4) Addition of various solutes which are presumed to decrease the structure of water have a strengthening effect on the HI (see Table 8.3).

The above qualitative correlation between changes of structure and changes of HI seems to hold for systems that do not deviate to any appreciable extent from pure water. One certainly could not apply this correlation to a comparison of the HI in water and, say, pure ethanol.

We wind up this section by noting that in spite of the fact that a plausible interpretation of the entropy and the enthalpy of the HI process can be given in terms of structural changes in the solvent, we are still left without any interpretation of the large, negative values of $\delta G^{\text{HI}}$ in water. There have been a few attempts to interpret this phenomenon in terms of the surface tension effect. Such an explanation may be plausible for two cavities at very short distances, where they form a sort of hourglass bubble. Such a bubble is very unstable and will tend to coalesce into a single bubble, thereby reducing the total surface. It is more difficult to see why the HI is large for real solutes at a distance of the order of $\sigma$. Moreover, there is a fundamental difficulty in characterizing the surface area and surface tension for bubbles of microscopic diameter, a difficulty that was recognized long ago by Kirkwood (1939).

In the next section, we outline a possible approach by way of which one can hopefully elucidate the problem of HI and its dependence on the characteristic features of the aqueous medium.

## 8.12. SIMULATIONS

The problem of HI as formulated in Section 8.3 can, in principle, be studied by various simulation techniques. Such an approach encounters some serious difficulties, however, in addition to those listed in Chapter 6, in connection with the study of pure liquid water. It is well known that the available techniques of simulation, such as the Monte Carlo and molecular dynamics methods, are not suitable for the computation of the free energy or the entropy of a system [see, for example, Wood (1968)]. Since the HI problem has been formulated in terms of a difference in the free energy of the system for two configurations of a pair of solutes, its direct computation by these methods is at present unfeasible.

We illustrate one source of the difficulty, using the formulation of the HI problem in terms of the solute–solute pair correlation function $g_{SS}(R)$ (see Section 8.3).

In a pure, one-component system, the function $g(R)$ is computed as an average over *all* pairs of particles. Suppose now that we label each particle and take averages over some finite number of configurations. In

## Water with Two or More Simple Solutes

such a computation, two specific particles, say 1 and 2, need not attain all the possible separations from each other. For instance, if we start with a configuration for which 1 and 2 are at a separation of, say, $R_{12} \approx 4\sigma$, then after a *finite* series of steps, we find these two particles at a distance apart close to $4\sigma$. One cannot expect the entire range of separations $\sigma \leq R \lesssim 5\sigma$ to be realized in such a computation. The fact that we do get the correct $g(R)$ is a result of the averaging process over *many* pairs, each of which attains only a small range of distances $R_{ij}$.

Now, suppose we study a system of $N_W$ waterlike particles and two (or a small number $N_S \ll N_W$) solute particles. Because of the finite series of configurations, one cannot expect to realize all the possible distances between the two solutes. Hence, $g_{SS}(R)$, computed only for one pair of particles, is unlikely to be a faithful representation of the exact limit of $g_{SS}(R)$ at high dilution.

It is also intuitively evident that as the solvent–solvent interaction increases, the greater is the difficulty in computing $g_{SS}(R)$.

A different method of studying the same problem is by applying the set of four Percus–Yevick equations for a two-component system. These can be written by a simple generalization of the one-component equation:

$$y_{\alpha\beta}(1,2) = 1 + \sum_{l=1}^{2} (\varrho_l/2\pi)$$
$$\times \int d(3) y_{\alpha l}(1,3) f_{\alpha l}(1,3) [y_{l\beta}(3,2) f_{l\beta}(3,2) + y_{l\beta}(3,2) - 1] \quad (8.150)$$

where

$$f_{\alpha\beta}(i,j) = \exp[-U_{\alpha\beta}(i,j)/kT] - 1 \quad (8.151)$$

$$y_{\alpha\beta}(i,j) = \{\exp[U_{\alpha\beta}(i,j)/kT]\} g_{\alpha\beta}(i,j) \quad (8.152)$$

The subscripts designate the species, and the arguments of the functions denote the configurations of the corresponding particles.

We present here one result obtained by this method. The system is two dimensional, similar to the system described in Section 6.11. Here, we have waterlike particles at a number density of $\varrho_W = 0.65$ and a simple solute with a number density of $\varrho_S = 0.05$. The other molecular parameters for this particular computation are

$$\varepsilon_H/kT = -3.7, \quad R_H = 1, \quad \sigma_{SS} = 0.75, \quad \sigma_{WW} = 1.05$$
$$\varepsilon_{SS}/kT = \varepsilon_{SW}/kT = \varepsilon_{WS}/kT = \varepsilon_{WW}/kT = 0.185 \quad (8.153)$$

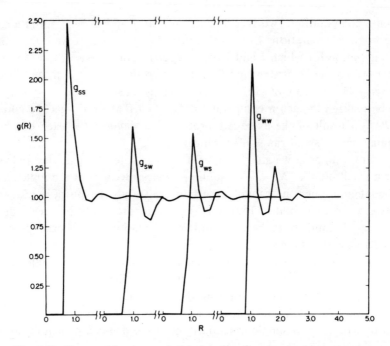

Fig. 8.23. The spatial pair correlation functions for a two-component system in two dimensions. The computations were carried out by solving the four Percus–Yevick equations (8.150) with the parameters listed in (8.153). [For more details see appendix 9-E.]

[For more details on the method of computation and the choice of parameters, see Ben-Naim (1971c, 1972c,d).]

Figure 8.23 shows the pair correlation functions for the four pairs of species. The most remarkable finding is the pronounced peak of $g_{SS}(R)$ at about $R \approx \sigma_{SS}$. A high peak at this distance corresponds to a large probability of finding two solute particles at a distance $R \approx \sigma_{SS}$.

Although this method does not suffer from the drawbacks of the various simulation techniques, it encounters some severe difficulties of its own, which are discussed in detail by Ben-Naim (1972c,d).

It is expected that future work along the routes outlined in this section will eventually lead to better comprehension of the molecular factors responsible for the phenomenon of hydrophobic interaction.

## 8.13. CONCLUSION

The problem of hydrophobic interaction can now be placed in the broader perspective of the phenomena discussed elsewhere in this book, which have a common feature: The process of averaging over part of the degrees of freedom of the system adds new properties to the interaction between a pair of particles. We recall two such examples. (1) The interaction between two ideal dipoles has the characteristic $R^{-3}$ dependence on the separation $R$, and the interaction among a set of ideal dipoles is pairwise additive (see Section 1.7). If we average over all orientations of the dipoles, we get a new behavior of the averaged interaction. The pairwise additivity is lost, and we get an $R^{-6}$ dependence at large distances. (2) Two hard spheres do not attract each other at any separation. The total interaction of a system of hard spheres is additive. If we start with such a system and average over all positions of all but two of the particles, we get the potential of average force. This potential reveals alternating regions of repulsive and attractive interaction (Section 2.5), and the pairwise additivity is also lost.

In the HI problem, we start with two simple solutes, such as argon, methane, or even hard spheres. We specifically note that the attractive forces between two such particles are assumed to be very weak. Putting the same two solutes in water and averaging over all the configurations of the water molecule produces new features in the "solute–solute interaction." The most outstanding of these are the strengthening of the attractive forces at short distances and the anomalous temperature dependence of these forces. The elucidation of the interrelation between the averaging process and the newly created properties will no doubt be the most challenging aspect of the problem of HI.

For the next step in our study of the HI, we must return to the original system from which we borrowed the problem. We recall that the original problem of HI was derived from biopolymer chemistry. There, the HI was only one of many factors which governed the conformational changes of biopolymers (other factors are hydrogen bonding, charge–charge interaction, and so forth). The ultimate aim of our study must be to strive to understand the way in which these factors combine when they operate on the real biopolymer. This task will certainly be difficult, since there is still a great deal of foundation that needs to be laid on the more elementary problems. To this end, elucidation of the HI problem is crucially dependent on our progress in understanding the structure of water around simple solutes. This, in turn, depends on our knowledge of the molecular structure of pure liquid water.

## Chapter 9
# Appendix

## 9-A. ROTATIONAL PARTITION FUNCTION FOR A RIGID ASYMMETRIC MOLECULE

Let $I_1$, $I_2$, and $I_3$ be the three principal moments of inertia of the molecule. The rotational partition function of a single molecule is given by

$$q_r = h^{-3} \int \cdots \int \exp(-\beta E_k)\, d\phi\, d\theta\, d\psi\, dP_\phi\, dP_\theta\, dP_\psi \quad (9.1)$$

where $E_k$ is the rotational kinetic energy of the molecule

$$E_k = (1/2) \sum_{i=1}^{3} I_i w_i^2 \quad (9.2)$$

and the $w_i$ are the components of the angular velocity of the molecule along the principal axes. The integration in (9.1) is carried out over all the orientations (expressed in terms of the three Euler angles) and their conjugate momenta. The latter are defined as the partial derivatives of the Lagrangian of the system, which in this case is the same as $E_k$, namely[1]

$$P_\phi = \partial E_k / \partial \dot\phi, \qquad P_\theta = \partial E_k / \partial \dot\theta, \qquad P_\psi = \partial E_k / \partial \dot\psi \quad (9.3)$$

The evaluation of the integral (9.1) is made convenient by performing a transformation of variables. The components of the angular velocity can be expressed in terms of the three Euler angles and their time derivatives

---
[1] A dot over a letter stands for a time derivative, e.g., $\dot\phi = \partial \phi / \partial t$.

[for more details, see Goldstein (1950)]:

$$w_1 = \dot\phi \sin\theta \sin\psi + \dot\theta \cos\psi$$
$$w_2 = \dot\phi \sin\theta \cos\psi - \dot\theta \sin\psi \quad (9.4)$$
$$w_3 = \dot\phi \cos\theta + \dot\psi$$

From the definitions of the conjugate momenta in (9.3) and from (9.4), we get

$$P_\phi = I_1 w_1 \sin\theta \sin\psi + I_2 w_2 \sin\theta \cos\psi + I_3 w_3 \cos\theta$$
$$P_\theta = I_1 w_1 \cos\psi - I_2 w_2 \sin\psi \quad (9.5)$$
$$P_\psi = I_3 w_3$$

The Jacobian of the transformation from the variables $P_\phi$, $P_\theta$, $P_\psi$ into $w_1$, $w_2$, $w_3$ is

$$\frac{\partial(P_\phi, P_\theta, P_\psi)}{\partial(w_1, w_2, w_3)} = \begin{vmatrix} I_1 \sin\theta \sin\psi & I_2 \sin\theta \cos\psi & I_3 \cos\theta \\ I_1 \cos\psi & -I_2 \sin\psi & 0 \\ 0 & 0 & I_3 \end{vmatrix}$$
$$= -I_1 I_2 I_3 \sin\theta \quad (9.6)$$

The integral (9.1) can now be rewritten as

$$q_r = h^{-3} \int \cdots \int \left[ \exp\left(-\tfrac{1}{2}\beta \sum_{i=1}^{3} I_i w_i^2\right) \right] I_1 I_2 I_3 \sin\theta \, d\phi \, d\theta \, d\psi \, dw_1 \, dw_2 \, dw_3 \quad (9.7)$$

where the integrations over the angular velocities yield

$$\int_{-\infty}^{\infty} \exp(-\beta I_i w_i^2/2) \, dw_i = (2\pi/\beta I_i)^{1/2} \quad (9.8)$$

and also

$$\int_0^{2\pi} d\phi \int_0^{\pi} \sin\theta \, d\theta \int_0^{2\pi} d\psi = 8\pi^2 \quad (9.9)$$

Thus, the integral (9.7) yields

$$q_r = (I_1 I_2 I_3)^{1/2} 8\pi^2 (2\pi kT)^{3/2}/h^3 \quad (9.10)$$

In applying (9.10) for a water molecule, we have to divide by a symmetry factor $\sigma = 2$ to account for the fact that rotation by 180° about the symmetry axis produces an indistinguishable configuration of the molecule.

## 9-B. FUNCTIONAL DERIVATIVE AND FUNCTIONAL TAYLOR EXPANSION

We present here the operations of the functional derivative and functional Taylor expansion in a formal fashion, based on an analogy with the discrete case. For more details on the mathematical aspects, the reader is referred to Volterra (1931).

Consider first the simple function of one variable

$$f(x) = x \tag{9.11}$$

The derivative with respect to $x$ is unity and with respect to any other variable $y$ is zero,

$$df/dx = 1, \quad df/dy = 0 \tag{9.12}$$

Next, consider a function of $n$ independent variables $x_1, \ldots, x_n$, for example,

$$f(x_1, \ldots, x_n) = \sum_{i=1}^{n} a_i x_i \tag{9.13}$$

The partial derivative of $f$ with respect to, say, $x_j$ is

$$\frac{\partial f}{\partial x_j} = \sum_{i=1}^{n} a_i \frac{\partial x_i}{\partial x_j} = \sum_{i=1}^{n} a_i \, \delta_{ij} = a_j \tag{9.14}$$

where $\delta_{ij}$ is the Kronecker delta function.

A further generalization of (9.13) is the case of a vector $\mathbf{y} = (y_1, \ldots, y_n)$ which is a function of a vector $\mathbf{x} = (x_1, \ldots, x_n)$. This connection can be written symbolically as

$$\mathbf{y} = \mathbf{F}\mathbf{x} \tag{9.15}$$

where $\mathbf{F}$ is an operator. A simple example of an operator $\mathbf{F}$ is a matrix.

We now generalize (9.15) as follows. We first write the vector $\mathbf{x}$ in a new notation

$$\mathbf{x} = (x_1, \ldots, x_n) = [x(1), \ldots, x(n)] \tag{9.16}$$

In (9.16), we stress the fact that each component $x(i)$ depends on the discrete variable $i = 1, 2, \ldots, n$. We now let $i$ take any value in a continuous range of real numbers, $a \leq i \leq b$. In this way, we get a vector $\mathbf{x}$ with an infinite number of components $x(i)$. The functional relation (9.15) is then reinterpreted as a relation between a *function* $\mathbf{x}$ and a *function* $\mathbf{y}$.

Now **F** is an operator acting on a functional space rather than on a vector space. A simple relation between two such functions is

$$y(t) = \int_a^b K(s, t) x(s)\, ds \qquad (9.17)$$

That is, for each function **x** [whose components are $x(s)$], we get a new function **y** [whose components are $y(t)$]. The function $K(s, t)$ is presumed to be known.

In the discrete case, for any two components $x_i$ and $x_j$ of the vector **x**, we have the relation

$$\partial x_i / \partial x_j = \delta_{ij} \qquad (9.18)$$

Similarly, viewing $x(t)$ and $x(s)$ as two "components" of the function **x**, we have

$$\delta x(t) / \delta x(s) = \delta(t - s) \qquad (9.19)$$

where the Dirac delta function replaces the Kronecker delta function in (9.18).

In (9.14), the quantity $\partial f / \partial x_j$ can be referred to as the partial derivative of $f$ with respect to the component $x_j$. Similarly, the functional derivative of **y** in (9.17) with respect to the "component" $x(s')$ is[2]

$$\frac{\delta y(t)}{\delta x(s')} = \int K(s, t) \frac{\delta x(s)}{\delta x(s')}\, ds = \int K(s, t)\, \delta(s - s')\, ds = K(s', t) \qquad (9.20)$$

In Section 5.8, we encountered the following example of a functional. The average volume of a system in the $T, P, N$ ensemble is written as

$$V(\mathbf{N}) = \int_0^\infty \phi N(\phi)\, d\phi \qquad (9.21)$$

(Here, we use **N** instead of the more cumbersome notation $N_\psi^{(1)}$ of Chapter 5.)

---

[2] The quantity defined in (9.20) is often called the functional derivative of **y** with respect to the function **x** at the point $s'$ (Bogoliubov, 1962). However, the nomenclature used above is clearer, especially when it is also necessary to specify at which point the derivative is evaluated. For instance, later in this appendix, we shall write

$$\delta f(\mathbf{x}) / \delta x(s')_{\mathbf{x}=0}$$

which we shall refer to as the functional derivative of $f$ with respect to the component $x(s')$, evaluated at the "point" $\mathbf{x} = 0$.

The functional derivative of $V$ with respect to the component $N(\phi')$ is

$$\frac{\delta V(N)}{\delta N(\phi')} = \int_0^\infty \phi \, \frac{\delta N(\phi)}{\delta N(\phi')} \, d\phi = \int_0^\infty \phi \, \delta(\phi - \phi') \, d\phi = \phi' \quad (9.22)$$

As a second example of the application of the functional derivatives, we show that the pair distribution function can be obtained as a functional derivative of the configurational partition function.

For a system of $N$ spherical particles, with pairwise additive potential, we write

$$Z(\mathbf{U}) = \int \cdots \int d\mathbf{R}^N \exp\left[-\beta \sum_{i<j} U(\mathbf{R}_i, \mathbf{R}_j)\right] \quad (9.23)$$

We view $Z$ as a functional of the function $\mathbf{U}$ (i.e., the pair potential, which here is considered as a function of six variables).

The functional derivative of $Z$ with respect to the "component" $U(\mathbf{R}', \mathbf{R}'')$ is

$$\frac{\delta Z(\mathbf{U})}{\delta U(\mathbf{R}', \mathbf{R}'')}$$

$$= \int \cdots \int d\mathbf{R}^N \left[-\beta \sum_{i<j} \frac{\delta U(\mathbf{R}_i, \mathbf{R}_j)}{\delta U(\mathbf{R}', \mathbf{R}'')}\right] \exp\left[-\beta \sum_{i<j} U(\mathbf{R}_i, \mathbf{R}_j)\right]$$

$$= \int \cdots \int d\mathbf{R}^N \left[-\beta \sum_{i<j} \delta(\mathbf{R}_i - \mathbf{R}') \, \delta(\mathbf{R}_j - \mathbf{R}'')\right] \exp\left[-\beta \sum_{i<j} U(\mathbf{R}_i, \mathbf{R}_j)\right]$$

$$= -\beta \varrho^{(2)}(\mathbf{R}', \mathbf{R}'') Z(\mathbf{U})/2 \quad (9.24)$$

where we have used the definition of the pair correlation function (see Section 2.3). Relation (9.24) can be written as

$$\varrho^{(2)}(\mathbf{R}', \mathbf{R}'') = -2kT \, \delta[\ln Z(\mathbf{U})]/\delta U(\mathbf{R}', \mathbf{R}'') \quad (9.25)$$

Before turning to functional Taylor expansion, we note that many operations with ordinary derivatives can be extended to functional derivatives. We note, in particular, the chain rule of differentation.

For functions of one variable $y = f(x)$ and $x = g(t)$, we have[3]

$$\frac{dy}{dx} \frac{dx}{dt} = \frac{dy}{dt} \quad (9.26)$$

---

[3] We assume that all the inverse derivatives exist. In general, this is not guaranteed automatically.

and, in particular,
$$\frac{dy}{dx}\frac{dx}{dy} = 1 \tag{9.27}$$

In the case of functions of $n$ variables, say
$$y_k = f_k(x_1, \ldots, x_n), \qquad k = 1, \ldots, n \tag{9.28}$$
we have
$$\frac{dy_k}{dt} = \sum_{i=1}^{n} \frac{\partial y_k}{\partial x_i} \frac{dx_i}{dt} \tag{9.29}$$
and, in particular,
$$\delta_{kj} = \frac{dy_k}{dy_j} = \sum_{i=1}^{n} \frac{\partial y_k}{\partial x_i} \frac{\partial x_i}{\partial y_j} \tag{9.30}$$

The generalization of (9.30) is straightforward. We view $\mathbf{y} = F(\mathbf{x})$ as a connection between the two functions whose components are $y(t)$ and $x(t)$, respectively, and write
$$\frac{\delta y(t)}{\delta y(v)} = \int \frac{\delta y(t)}{\delta x(s)} \frac{\delta x(s)}{\delta y(v)} ds = \delta(t - v) \tag{9.31}$$
where integration replaces the summation in (9.30).

We now consider the functional Taylor expansion. We start with a one-dimensional function $f(x)$, for which the Taylor expansion about $x = 0$ is
$$f(x) = f(0) + \left.\frac{\partial f}{\partial x}\right|_{x=0} x + \frac{1}{2} \left.\frac{\partial^2 f}{\partial x^2}\right|_{x=0} x^2 + \cdots \tag{9.32}$$

As an example, $f(x) = a + bx$. Then, we have
$$f(0) = a, \qquad \partial f/\partial x = b \tag{9.33}$$
Hence, from (9.32) we get
$$f(x) = a + bx \tag{9.34}$$
i.e., the expansion to first order in $x$ is, in this case, exact for any $x$. In general, if we take the first-order expansion
$$f(x) = f(0) + \left.\frac{\partial f}{\partial x}\right|_{x=0} x \tag{9.35}$$
we get an *approximate* value for $f(x)$. The quality of the approximation depends on $x$ and on the function **f**.

For a function of $n$ variables $f(x_1, \ldots, x_n)$, the Taylor expansion is

$$f(x_1, \ldots, x_n) = f(0, \ldots, 0) + \sum_{i=1}^{n} \frac{\partial f}{\partial x_i}\bigg|_{\mathbf{x}=0} x_i + \cdots \tag{9.36}$$

where all the derivatives are taken at the point $\mathbf{x} = 0$.

The generalization to the continuous case is, by analogy to (9.36),

$$f(\mathbf{x}) = f(0) + \int \frac{\delta f(\mathbf{x})}{\delta x(t)}\bigg|_{\mathbf{x}=0} x(t)\, dt + \cdots \tag{9.37}$$

where the partial derivative has become the functional derivative and the summation over $i$ has become the integration over $t$. We note again that the first-order expansion in (9.37) can be viewed as an approximation to $f(\mathbf{x})$. The quality of the approximation depends both on $\mathbf{x}$ and on the function $\mathbf{f}$. In Appendix 9-D, we present an application of such a first-order Taylor expansion.

As for the nomenclature, the quantity $\partial f/\partial x_i|_{\mathbf{x}=0}$ in (9.36) is referred to as the partial derivative of $f$ with respect to the component $x_i$, evaluated at the point $\mathbf{x} = 0$. Similarly, in (9.37), we have the functional derivative of $f$ with respect to the "component" $x(t)$, at the point $\mathbf{x} = 0$.

## 9-C. THE ORNSTEIN–ZERNIKE RELATION

The original derivation of the Ornstein–Zernike relation (Ornstein and Zernike, 1914) employs arguments on local density fluctuations in the fluid. We present here a different derivation based on the method of functional derivatives (Appendix 9-B). A very thorough discussion of this topic is given by Münster (1969).

Consider the grand partition function of a system of spherical particles exposed to an external potential $\boldsymbol{\psi}$:

$$\Xi(\boldsymbol{\psi}) = \sum_{N=0}^{\infty} (z^N/N!) \int \cdots \int d\mathbf{R}^N \exp[-\beta U(\mathbf{R}^N, \boldsymbol{\psi})] \tag{9.38}$$

where

$$U(\mathbf{R}^N, \boldsymbol{\psi}) = U_N(\mathbf{R}^N) + \sum_{i=1}^{N} \psi(\mathbf{R}_i) \tag{9.39}$$

and

$$z = q[\exp(\beta\mu)]/\Lambda^3 \tag{9.40}$$

Here, $U_N(\mathbf{R}^N)$ is the total potential energy due to interactions among the particles and the second term on the rhs of (9.39) is due to interaction of the system at configuration $\mathbf{R}^N$ with the external potential. As in Appendix 9-B, we use the symbol $\boldsymbol{\psi}$ to designate the whole function whose "components" are $\psi(\mathbf{R}_i)$.

The functional derivative of $\ln \Xi$ with respect to the component $\psi(\mathbf{R}')$ is

$$\frac{\delta \ln \Xi(\boldsymbol{\psi})}{\delta \psi(\mathbf{R}')}$$

$$= \Xi^{-1} \sum_N \left(\frac{z^N}{N!}\right) \int \cdots \int d\mathbf{R}^N \left[-\beta \sum_{i=1}^N \frac{\delta \psi(\mathbf{R}_i)}{\delta \psi(\mathbf{R}')}\right] \exp[-\beta U(\mathbf{R}^N, \boldsymbol{\psi})]$$

$$= -\beta \Xi^{-1} \sum_N \left(\frac{z^N}{N!}\right) \int \cdots \int d\mathbf{R}^N \left[\sum_{i=1}^N \delta(\mathbf{R}_i - \mathbf{R}')\right] \exp[-\beta U(\mathbf{R}^N, \boldsymbol{\psi})]$$

$$= -\beta \varrho^{(1)}(\mathbf{R}'|\boldsymbol{\psi}) \qquad (9.41)$$

where, in the last step on the rhs, we used the definition of the singlet molecular distribution function (Section 2.9) of a system in an external potential $\boldsymbol{\psi}$.

Next, consider the second functional derivative of $\ln \Xi$ with respect to $\psi(\mathbf{R}'')$, which can be obtained from (9.41),

$$\frac{\delta^2 \ln \Xi(\boldsymbol{\psi})}{\delta \psi(\mathbf{R}') \, \delta \psi(\mathbf{R}'')}$$

$$= -\beta \frac{\delta \varrho^{(1)}(\mathbf{R}'|\boldsymbol{\psi})}{\delta \psi(\mathbf{R}'')}$$

$$= \beta^2 \Xi^{-1}(\boldsymbol{\psi}) \Bigg\{ \sum_N \left(\frac{z^N}{N!}\right) \int \cdots \int d\mathbf{R}^N \left[\sum_{i,j=1}^N \delta(\mathbf{R}_i - \mathbf{R}') \, \delta(\mathbf{R}_j - \mathbf{R}'')\right]$$

$$\times \exp[-\beta U(\mathbf{R}^N, \boldsymbol{\psi})] - \varrho^{(1)}(\mathbf{R}'|\boldsymbol{\psi})\varrho^{(1)}(\mathbf{R}''|\boldsymbol{\psi}) \Bigg\}$$

$$= \beta^2 \{\varrho^{(2)}(\mathbf{R}', \mathbf{R}''|\boldsymbol{\psi}) + \varrho^{(1)}(\mathbf{R}'|\boldsymbol{\psi}) \, \delta(\mathbf{R}' - \mathbf{R}'') - \varrho^{(1)}(\mathbf{R}'|\boldsymbol{\psi})\varrho^{(1)}(\mathbf{R}''|\boldsymbol{\psi})\} \qquad (9.42)$$

where $\varrho^{(2)}(\mathbf{R}', \mathbf{R}''|\boldsymbol{\psi})$ is the pair distribution function in the presence of the external potential $\boldsymbol{\psi}$. In the second step on the rhs of (9.42), we have separated the double sum over $i$ and $j$ into two terms, the first containing all terms for which $i \neq j$ and the second all terms with $i = j$.

We define the total correlation function by

$$h(\mathbf{R}', \mathbf{R}'') = g(\mathbf{R}', \mathbf{R}'') - 1 \qquad (9.43)$$

and rewrite (9.42), when evaluated at $\boldsymbol{\psi} = 0$ as

$$-kT \left.\frac{\delta \varrho^{(1)}(\mathbf{R}'|\boldsymbol{\psi})}{\delta \psi(\mathbf{R}'')}\right|_{\psi=0} = \varrho^2 h(\mathbf{R}', \mathbf{R}'') + \varrho^{(1)}(\mathbf{R}') \, \delta(\mathbf{R}' - \mathbf{R}'') \quad (9.44)$$

i.e., the functional derivative of the singlet molecular distribution function at the limit of $\boldsymbol{\psi} = 0$ is "almost" equal to the total correlation function. The singular case arises when $\mathbf{R}' = \mathbf{R}''$.

We now introduce the so-called *direct* correlation function, which is defined in terms of the inverse functional derivative in (9.44). To do this, we must view $\boldsymbol{\psi}$ as a functional of the density, which we write symbolically as $\psi(\mathbf{R}'|\rho^{(1)})$. Now, the "external" potential is produced by preparing a system with an arbitrary local density $\rho^{(1)}$. [It is for this reason that it is preferable to work in the grand ensemble where an arbitrary density change may be envisaged; see Percus (1964).]

The direct correlation function is defined by

$$c(\mathbf{R}', \mathbf{R}'') = \beta \, \frac{\delta \psi(\mathbf{R}'|\rho^{(1)})}{\delta \varrho^{(1)}(\mathbf{R}'')} + \frac{\delta(\mathbf{R}' - \mathbf{R}'')}{\varrho^{(1)}(\mathbf{R}')} \quad (9.45)$$

Now, we apply the chain rule of functional derivatives (see Appendix 9-B), which, for the present case, takes the form

$$\int \frac{\delta \psi(\mathbf{R}'|\rho^{(1)})}{\delta \varrho^{(1)}(\mathbf{R}''')} \left. \frac{\delta \varrho^{(1)}(\mathbf{R}'''|\boldsymbol{\psi})}{\delta \psi(\mathbf{R}'')} \right|_{\psi=0} d\mathbf{R}''' = \delta(\mathbf{R}' - \mathbf{R}'') \quad (9.46)$$

Substituting (9.44) and (9.45) in (9.46), we get

$$-\int [c(\mathbf{R}', \mathbf{R}''') - \varrho^{-1} \delta(\mathbf{R}' - \mathbf{R}''')][\varrho^2 h(\mathbf{R}'', \mathbf{R}''') + \varrho \, \delta(\mathbf{R}'' - \mathbf{R}''')] \, d\mathbf{R}'''$$
$$= \delta(\mathbf{R}' - \mathbf{R}'') \quad (9.47)$$

which yields, upon rearrangement,

$$h(\mathbf{R}', \mathbf{R}'') = c(\mathbf{R}', \mathbf{R}'') + \varrho \int c(\mathbf{R}', \mathbf{R}''') h(\mathbf{R}'', \mathbf{R}''') \, d\mathbf{R}''' \quad (9.48)$$

which is the Ornstein–Zernike relation for a system of spherical particles. By substituting the rhs of (9.48) into the integrand of the rhs of (9.48) and continuing this process, we get

$$h(\mathbf{R}', \mathbf{R}'') = c(\mathbf{R}', \mathbf{R}'') + \varrho \int c(\mathbf{R}', \mathbf{R}''') c(\mathbf{R}'', \mathbf{R}''') \, d\mathbf{R}'''$$
$$+ \varrho^2 \int c(\mathbf{R}', \mathbf{R}''') c(\mathbf{R}''', \mathbf{R}'''') c(\mathbf{R}'''', \mathbf{R}'') \, d\mathbf{R}''' \, d\mathbf{R}'''' + \cdots \quad (9.49)$$

where here the total correlation function is viewed as a sum of "chains" of direct correlation functions between the two points $\mathbf{R}'$ and $\mathbf{R}''$.

## 9-D. THE PERCUS–YEVICK INTEGRAL EQUATION

We present here a derivation of the Percus–Yevick (PY) equation based on the material of Appendices 9-B and 9-C. More details may be found in the work by Percus and Yevick (1958), Percus (1962, 1964), Henderson and Davison (1967), Rushbrooke (1968), and Münster (1969).

As in Appendix 9-C, we consider a system in an external potential $\psi$ which, in the present case, is produced by a *particle* (identical to the other particles of the system) fixed at $\mathbf{R}_0$

$$\psi(\mathbf{R}_i) = U(\mathbf{R}_0, \mathbf{R}_i) \tag{9.50}$$

i.e., the "external" potential at $\mathbf{R}_i$ is equal to the potential produced by putting a particle at $\mathbf{R}_0$.

Consider the singlet density at $\mathbf{R}_1$ in the presence and absence of $\psi$. Clearly, we have (using the notation of Appendix 9-C)

$$\varrho^{(1)}(\mathbf{R}_1 \mid \psi) = \varrho(\mathbf{R}_1/\mathbf{R}_0) \tag{9.51}$$

$$\varrho^{(1)}(\mathbf{R}_1 \mid \psi = 0) = \varrho^{(1)}(\mathbf{R}_1) \tag{9.52}$$

Viewing $\varrho^{(1)}(\mathbf{R}_1 \mid \psi)$ as a functional of $\psi$, we write the functional Taylor expansion

$$\varrho^{(1)}(\mathbf{R}_1|\psi) = \varrho^{(1)}(\mathbf{R}_1|\psi=0) + \int \frac{\delta\varrho^{(1)}(\mathbf{R}_1|\psi)}{\delta\psi(\mathbf{R}_2)}\bigg|_{\psi=0} \psi(\mathbf{R}_2)\, d\mathbf{R}_2 + \cdots \tag{9.53}$$

This particular expansion does not prove to be useful. The reason, as stressed in Appendix 9-B, is that a first-order Taylor expansion is expected to be useful when the increment, here $\psi$, is "small." For instance, in Eq. (9.35) of Appendix 9-B, if $x$ is very large, we cannot expect that a first-order Taylor expansion will lead to a good approximation. In (9.53), $\psi$ replaces $x$ (of the one-dimensional example). Since $\psi(\mathbf{R}') \to \infty$ as $\mathbf{R}' \to \mathbf{R}_0$, the increment $\psi$ cannot be considered to be "small."

Instead, Percus (1962) suggested a different expansion of a functional of a function which is everywhere finite.

Consider the following two functionals of $\psi$:

$$\xi(\mathbf{R}_1 \mid \psi) = \varrho^{(1)}(\mathbf{R}_1 \mid \psi) \exp[\beta\psi(\mathbf{R}_1)] \tag{9.54}$$

$$\eta(\mathbf{R}_1 \mid \psi) = \varrho^{(1)}(\mathbf{R}_1 \mid \psi) \tag{9.55}$$

# Appendix

where we have

$$\xi(\mathbf{R}_1 | \boldsymbol{\psi} = 0) = \varrho^{(1)}(\mathbf{R}_1), \qquad \eta(\mathbf{R}_1 | \boldsymbol{\psi} = 0) = \varrho^{(1)}(\mathbf{R}_1) \qquad (9.56)$$

We now view $\xi$ as a functional of $\eta$ which itself is a funtional of $\boldsymbol{\psi}$. In this way, we avoid the possibility of an infinite increment as in (9.53). Thus, the first-order functional Taylor expansion is

$$\xi(\mathbf{R}_1|\boldsymbol{\psi}) = \xi(\mathbf{R}_1|\boldsymbol{\psi} = 0) + \int \frac{\delta \xi(\mathbf{R}_1)}{\delta \eta(\mathbf{R}_2)} [\eta(\mathbf{R}_2|\boldsymbol{\psi}) - \eta(\mathbf{R}_2|\boldsymbol{\psi} = 0)] \, d\mathbf{R}_2 \quad (9.57)$$

The functional derivative[4] in (9.57) is of $\xi$, viewed as a functional of $\eta$, taken at the point $\boldsymbol{\eta} = \boldsymbol{\rho}^{(1)}$ [i.e., at $\boldsymbol{\psi} = 0$; see (9.56)]. Using (9.51), (9.52), and (9.54)–(9.56), we rewrite (9.57) as

$$\varrho(\mathbf{R}_1/\mathbf{R}_0) \exp[\beta \psi(\mathbf{R}_1)]$$
$$= \varrho^{(1)}(\mathbf{R}_1) + \int [\delta \xi(\mathbf{R}_1)/\delta \eta(\mathbf{R}_2)][\varrho(\mathbf{R}_2/\mathbf{R}_0) - \varrho^{(1)}(\mathbf{R}_2)] \, d\mathbf{R}_2 \quad (9.58)$$

Using the chain rule for the functional derivative (Appendix 9-B), we find

$$\frac{\delta \xi(\mathbf{R}_1)}{\delta \eta(\mathbf{R}_2)} = \int \frac{\delta \xi(\mathbf{R}_1)}{\delta \psi(\mathbf{R}_3)} \frac{\delta \psi(\mathbf{R}_3)}{\delta \eta(\mathbf{R}_2)} \, d\mathbf{R}_3 \qquad (9.59)$$

The two functional derivatives on the rhs can be obtained from (9.54) and (9.55):

$$\frac{\delta \xi(\mathbf{R}_1)}{\delta \psi(\mathbf{R}_3)} = \frac{\delta \varrho^{(1)}(\mathbf{R}_1|\boldsymbol{\psi})}{\delta \psi(\mathbf{R}_3)} \exp[\beta \psi(\mathbf{R}_1)] + \beta \varrho^{(1)}(\mathbf{R}_1|\boldsymbol{\psi}) \exp[\beta \psi(\mathbf{R}_1)] \, \delta(\mathbf{R}_1 - \mathbf{R}_3) \quad (9.60)$$

which, at the point $\boldsymbol{\psi} = 0$, reduces to [see also Eq. (9.44) of Appendix 9-C]

$$\left.\frac{\delta \xi(\mathbf{R}_1)}{\delta \psi(\mathbf{R}_3)}\right|_{\boldsymbol{\psi}=0} = -\beta[\varrho^2 h(\mathbf{R}_1, \mathbf{R}_3) + \varrho^{(1)}(\mathbf{R}_1) \, \delta(\mathbf{R}_1 - \mathbf{R}_3)] + \beta \varrho^{(1)}(\mathbf{R}_1) \, \delta(\mathbf{R}_1 - \mathbf{R}_3) \quad (9.61)$$

Similarly, from (9.55) and from Eq. (9.45) of Appendix 9-C, we get

$$\left.\frac{\delta \psi(\mathbf{R}_3)}{\delta \eta(\mathbf{R}_2)}\right|_{\boldsymbol{\psi}=0} = \left.\frac{\delta \psi(\mathbf{R}_3)}{\delta \varrho^{(1)}(\mathbf{R}_2)}\right|_{\boldsymbol{\psi}=0} = \beta^{-1} \left[ c(\mathbf{R}_2, \mathbf{R}_3) - \frac{\delta(\mathbf{R}_2 - \mathbf{R}_3)}{\varrho^{(1)}(\mathbf{R}_2)} \right] \quad (9.62)$$

---

[4] In (9.57) and the following equation, we use a shorthand notation whenever possible. For instance, the notation $\delta \xi(\mathbf{R}_1)/\delta \eta(\mathbf{R}_2)$ means that we view $\xi$ as a functional of $\eta$ and take the derivative with respect to $\eta(\mathbf{R}_2)$, the derivative being evaluated at the point $\boldsymbol{\eta} = \boldsymbol{\rho}^{(1)}$, corresponding to $\boldsymbol{\psi} = 0$.

Substituting (9.61) and (9.62) into (9.59), we get, after rearrangement,

$$\left.\frac{\delta \xi(\mathbf{R}_1)}{\delta \eta(\mathbf{R}_2)}\right|_{\psi=0} = \varrho h(\mathbf{R}_1, \mathbf{R}_2) - \varrho^2 \int h(\mathbf{R}_1, \mathbf{R}_3) c(\mathbf{R}_2, \mathbf{R}_3) \, d\mathbf{R}_3$$
$$= \varrho c(\mathbf{R}_1, \mathbf{R}_2) \tag{9.63}$$

where in the second step on the rhs we have used the Ornstein–Zernike relation (Appendix 9-C).

Substituting (9.63) into (9.58), we get

$$\varrho g(\mathbf{R}_1, \mathbf{R}_0) \exp[\beta \psi(\mathbf{R}_1)] = \varrho + \varrho^2 \int c(\mathbf{R}_1, \mathbf{R}_2) h(\mathbf{R}_2, \mathbf{R}_0) \, d\mathbf{R}_2 \tag{9.64}$$

Using the functions

$$f(\mathbf{R}_1, \mathbf{R}_2) = \exp[-\beta U(\mathbf{R}_1, \mathbf{R}_2)] - 1 \tag{9.65}$$

$$y(\mathbf{R}_1, \mathbf{R}_2) = g(\mathbf{R}_1, \mathbf{R}_2) \exp[\beta U(\mathbf{R}_1, \mathbf{R}_2)] \tag{9.66}$$

and the Ornstein relation, we can rewrite (9.64) as

$$c(\mathbf{R}_1, \mathbf{R}_2) = y(\mathbf{R}_1, \mathbf{R}_2) f(\mathbf{R}_1, \mathbf{R}_2) \tag{9.67}$$

which is often referred to as the Percus–Yevick approximation. If we use (9.67) in the Ornstein–Zernike relation, we get an integral equation for $y$:

$$y(\mathbf{R}_1, \mathbf{R}_2) = 1 + \varrho \int y(\mathbf{R}_1, \mathbf{R}_3) f(\mathbf{R}_1, \mathbf{R}_3)$$
$$\times [y(\mathbf{R}_2, \mathbf{R}_3) f(\mathbf{R}_2, \mathbf{R}_3) + y(\mathbf{R}_2, \mathbf{R}_3) - 1] \, d\mathbf{R}_3 \tag{9.68}$$

This is the Percus–Yevick integral equation for $y$ as cited in Section 2.7. Another simpler and useful form of this equation is obtained by transforming to bipolar coordinates (see Section 2.5),

$$u = |\mathbf{R}_1 - \mathbf{R}_3|, \quad v = |\mathbf{R}_2 - \mathbf{R}_3|, \quad R = |\mathbf{R}_1 - \mathbf{R}_2| \tag{9.69}$$

The element of volume is

$$d\mathbf{R}_3 = 2\pi uv \, du \, dv / R \tag{9.70}$$

and (9.68) is transformed into

$$y(R) = 1 + 2\pi \varrho R^{-1} \int_0^\infty y(u) f(u) u \, du \int_{|R-u|}^{R+u} [y(v) f(v) + y(v) - 1] v \, dv \tag{9.71}$$

If we define the function

$$z(R) = y(R)R \qquad (9.72)$$

we get from (9.71) an integral equation for $z(R)$ which reads

$$z(R) = R + 2\pi\varrho \int_0^\infty z(u)f(u)\,du \int_{|R-u|}^{R+u} [z(v)f(v) + z(v) - v]\,dv \qquad (9.73)$$

a convenient form for a numerical solution. This is discussed further in Appendix 9-E.

## 9-E. SOLUTION OF THE PERCUS–YEVICK (PY) EQUATION

The exact solution of the PY equation is known for a one-component system of hard spheres (Wertheim, 1963; Thiele, 1963) and for mixtures of hard spheres (Lebowitz, 1964). Numerical solutions of the PY equation (for Lennard-Jones particles) have been carried out by many authors (e.g., Broyles, 1960, 1961; Broyles et al., 1962; Throop and Bearman, 1966; Baxter, 1967; Watts, 1968; Mandel et al., 1970).

We present here a brief account of the numerical procedure employed for the computations of $g(R)$ which we have used for our illustrations.

We start with the integral equation for the function $z(R)$ (see Appendix 9-D),

$$z(R) = R + 2\pi\varrho \int_0^\infty z(u)f(u)\,du \int_{|R-u|}^{R+u} [z(v)f(v) + z(v) - v]\,dv \qquad (9.74)$$

We begin the iterative procedure by substituting the initial function

$$z_0(R) = R \qquad (9.75)$$

in the rhs of (9.74),to obtain

$$z_1(R) = R + 2\pi\varrho \int_0^\infty uf(u)\,du \int_{|R-u|}^{R+u} vf(v)\,dv \qquad (9.76)$$

Next, $z_1(R)$ from (9.76) is substituted in the rhs of (9.74) to obtain $z_2(R)$, and so forth. It turns out that for high densities $\varrho$, such a procedure does not lead to a convergent solution. Instead, one uses a "mixing" parameter $\lambda$, $0 \leq \lambda \leq 1$ (Broyles, 1960; Throop and Bearman, 1966; Ben-Naim, 1972c,d) so that the $(k+1)$th input function is constructed from the $k$th input and the $k$th output functions, as follows:

$$z_{k+1}^{in}(R) = \lambda z_k^{out}(R) + (1-\lambda)z_k^{in}(R) \qquad (9.77)$$

In general, it is found that as $\varrho$ increases, one is compelled to use smaller values of $\lambda$ in (9.77) and large numbers of iterations to get a convergent result. As a criterion of convergence, we can choose the quantity

$$\zeta = N_R^{-1} \sum_{i=1}^{N_R} |z_k(i) - z_{k-1}(i)| \qquad (9.78)$$

where $N_R$ is the number of division points at which the function $z$ is evaluated. $z_k(i)$ denotes the value of the function $z_k$ at the point $R_i$. The iterative procedure is terminated when $\zeta$ falls below a certain small value, say $10^{-5}$ or $10^{-6}$, depending on the required accuracy.

For mixtures of, say, two components, (9.74) is generalized to

$$z_{\alpha\beta}(R) = R + \sum_{\gamma=\alpha,\beta} 2\pi\varrho_\gamma \int_0^\infty z_{\alpha\gamma}(u)f_{\alpha\gamma}(u)\,du \int_{|R-u|}^{R+u} [z_{\gamma\beta}(v)f_{\gamma\beta}(v) + z_{\gamma\beta}(v) - v]\,dv \qquad (9.79)$$

where the sum includes two terms $\gamma = \alpha, \beta$. The numerical procedure is similar to the case for one component. One starts with

$$z_{\alpha\beta}(R) = R \qquad (9.80)$$

for all the four functions $z_{\alpha\beta}(R)$ and proceeds to solve the four integral equations (9.79) by iteration.

## 9-F. THE CHEMICAL POTENTIAL IN THE T, P, N ENSEMBLE

In Section 3.5, we showed that the chemical potential of a one-component system in the $T, V, N$ ensemble can be written as [see Eq. (3.58)]

$$\begin{aligned}\mu &= A(T, V, N+1) - A(T, V, N) \\ &= -kT \ln\langle\exp(-\beta B)\rangle + kT \ln(\varrho \Lambda^3 q^{-1})\end{aligned} \qquad (9.81)$$

We now show that the same expression can be obtained in the $T, P, N$ ensemble provided that we reinterpret $\varrho$ as $N/\langle V\rangle$ [rather than $N/V$ in (9.81)], and the $\langle\ \rangle$ is reinterpreted as is discussed below.

## Appendix

The chemical potential of a one-component system of simple spherical particles in the $T$, $P$, $N$ ensemble is given by

$$\mu = (\partial G/\partial N)_{T,P} = G(T, P, N + 1) - G(T, P, N) \quad (9.82)$$

where the justification of the second equality on the rhs is the same as the one given for Eq. (3.52). The statistical mechanical expression for the chemical potential is

$$\exp(-\beta\mu)$$
$$= \frac{q \int dV \int d\mathbf{R}_0 \int \cdots \int d\mathbf{R}^N \exp[-\beta U_{N+1}(\mathbf{R}_0, \ldots, \mathbf{R}_N) - \beta PV]}{\Lambda^3(N+1) \int dV \int \cdots \int d\mathbf{R}^N \exp[-\beta U_N(\mathbf{R}_1, \ldots, \mathbf{R}_N) - \beta PV]}$$
$$(9.83)$$

For notation, compare with (3.53).

We now use the conditional distribution function of finding a configuration $\mathbf{R}^N$ given a system with volume $V$ [see Eq. (1.56)]

$$P(\mathbf{R}^N/V) = \frac{P(\mathbf{R}^N, V)}{P(V)} = \frac{\exp[-\beta U_N(\mathbf{R}^N)]}{\int \cdots \int d\mathbf{R}^N \exp[-\beta U_N(\mathbf{R}^N)]} \quad (9.84)$$

and rewrite (9.83) as

$$\exp(-\beta\mu) = \frac{q}{\Lambda^3(N+1)}$$
$$= \times \frac{\int dV \int d\mathbf{R}_0 \int \cdots \int d\mathbf{R}^N \exp[-\beta U_N(\mathbf{R}_1, \ldots, \mathbf{R}_N) - \beta B - \beta PV]}{\int dV \int \cdots \int d\mathbf{R}^N \exp[-\beta U_N(\mathbf{R}_1, \ldots, \mathbf{R}_N) - \beta PV]}$$
$$= \frac{q}{\Lambda^3(N+1)} \int dV \int d\mathbf{R}_0 \int \cdots \int d\mathbf{R}^N P(\mathbf{R}^N, V) \exp(-\beta B)$$
$$= \frac{q}{\Lambda^3(N+1)} \int dV\, P(V) \int d\mathbf{R}_0 \int \cdots \int d\mathbf{R}^N P(\mathbf{R}^N/V) \exp(-\beta B)$$
$$(9.85)$$

In (9.85), the quantity $B$ is the same as in Section 3.5. This is the binding energy of the added particle at $\mathbf{R}_0$ with the rest of the system at a specific configuration $\mathbf{R}^N$.

In a macroscopic system, we assume that the probability density $P(V)$ is concentrated very sharply at the average value $\langle V \rangle$. For simplicity, we assume that $P(V)$ is a Dirac delta function

$$P(V) = \delta(V - \langle V \rangle) \quad (9.86)$$

Hence, from (9.85), we get

$$\exp(-\beta\mu) = \frac{q}{\Lambda^3(N+1)}$$
$$\times \int dV\, \delta(V - \langle V \rangle) \int d\mathbf{R}_0 \int \cdots \int d\mathbf{R}^N\, P(\mathbf{R}^N/V) \exp(-\beta B)$$
$$= \frac{q}{\Lambda^3(N+1)} \int_{\langle V \rangle} d\mathbf{R}_0 \int_{\langle V \rangle} \cdots \int d\mathbf{R}^N\, P(\mathbf{R}^N/\langle V \rangle) \exp(-\beta B)$$
$$= \frac{q\langle V \rangle}{\Lambda^3(N+1)} \int_{\langle V \rangle} \cdots \int d\mathbf{R}^N\, P(\mathbf{R}^N/\langle V \rangle) \exp(-\beta B)$$
$$= \frac{q}{\Lambda^3 \varrho} \langle \exp(-\beta B) \rangle \tag{9.87}$$

In the second form on the rhs, we use the basic property of the Dirac delta function. The integration over $\mathbf{R}_0$ can be carried out to obtain the factor $\langle V \rangle$ (for more details, see Section 3.5). In the last form on the rhs of (9.87), we put $\varrho = (N+1)/\langle V \rangle \approx N/\langle V \rangle$, which is the average density of the system. The average of the quantity $\exp(-\beta B)$ is taken over all the configurations of the $N$ particles, using the conditional distribution function[5] $P(\mathbf{R}^N/\langle V \rangle)$.

The final expression for the chemical potential is

$$\mu = -kT \ln\langle \exp(-\beta B) \rangle + kT \ln(\varrho \Lambda^3 q^{-1}) \tag{9.88}$$

which is formally the same as (3.58).

In a very similar way, we can define the pseudo-chemical potential in the $T, P, N$ ensemble as

$$\bar{\mu} = G(T, P, N+1, \mathbf{R}_0) - G(T, P, N) \tag{9.89}$$

(compare with Section 3.6). Going through similar steps, we get

$$\exp(-\beta\bar{\mu}) = q\langle \exp(-\beta B) \rangle \tag{9.90}$$

and hence

$$\mu = \bar{\mu} + kT \ln(\varrho \Lambda^3) \tag{9.91}$$

which is formally the same as (3.92), with the proper reinterpretation of the various quantities.

---

[5] Note that the average sign in (9.87) is used in two senses. $\langle V \rangle$ is a proper average in the $T, P, N$ ensemble, whereas $\langle \exp(-\beta B) \rangle$ is an average with respect to the conditional density $P(\mathbf{R}^N/\langle V \rangle)$. The latter would be equal to an average in the $T, V, N$ ensemble, were the volume chosen as $V = \langle V \rangle$.

## 9-G. THE CHEMICAL AND THE PSEUDO-CHEMICAL POTENTIAL IN A LATTICE MODEL

In Section 3.6, we found the following expression for the chemical potential:

$$\mu = \bar{\mu} + kT \ln(N\Lambda^3/V) \qquad (9.92)$$

The second term on the rhs was interpreted in terms of the three contributions that are "gained" when the particle is released from the fixed position and allowed to wander in the entire volume.

If we add one particle, say $A$, that is different from the other particles, we get, instead of (9.92),

$$\mu_A = \bar{\mu}_A + kT \ln(\Lambda_A^3/V) \qquad (9.93)$$

The reason for this is that the added particle is presumed to be distinguishable from the others whether it is free or restricted to a fixed position.

It is instructive to present an example in which only the "volume" term is gained by releasing the particle. This can be demonstrated by a simple lattice model where particles are devoid of translational energy. Consider a mixture of $N_A$ and $N_B$ molecules of $A$ and $B$ species, respectively. We assume that each molecule occupies a single site of a regular lattice. Each site has $c$ nearest neighbors and all sites are occupied.

The partition function of such a system is

$$Q(T, N_A, N_B) = \sum_E \Omega(E) \exp(-\beta E) \qquad (9.94)$$

where the summation is over all energy levels $E$ of the system. Assuming only nearest-neighbor interaction, the total interaction energy is

$$E = N_{AA} W_{AA} + N_{BB} W_{BB} + N_{AB} W_{AB} \qquad (9.95)$$

where $N_{\alpha\beta}$ is the number of $\alpha\beta$ pairs and $W_{\alpha\beta}$ is the interaction energy between the $\alpha$ and $\beta$ species on adjacent lattice sites.

The general solution of (9.94) is not known [see, for example, Guggenheim (1952), Prigogine (1957)]; however, we are interested here in one simple and solvable example, namely the case of $N_A = 1$, $N_B$, for which we have $N_{AA} = 0$, $N_{AB} = c$, $N_{BB} = \frac{1}{2}c(N_B - 1)$. Hence,

$$Q(T, N_A = 1, N_B) = (N_B + 1) \exp\{-\beta[cW_{AB} + \tfrac{1}{2}cW_{BB}(N_B - 1)]\} \quad (9.96)$$

The degeneracy in this case is $N_B + 1$, which is the number of positions in which the $A$ particle can be placed. The partition function for pure $B$ is

$$Q(T, N_B) = \exp(-\beta c W_{BB} N_B/2) \qquad (9.97)$$

From (9.96) and (9.97), we get the chemical potential of $A$ in the limit of a very dilute solution in $B$, i.e.,

$$\exp(-\beta \mu_A) = \frac{Q(T, N_A = 1, N_B)}{Q(T, N_B)} = (N_B + 1) \exp(-\beta c W_{AB} + \tfrac{1}{2}\beta c W_{BB}) \qquad (9.98)$$

or, equivalently,

$$\mu_A = c W_{AB} - \tfrac{1}{2} c W_{BB} + kT \ln(N_B + 1)^{-1} \qquad (9.99)$$

Repeating the same arguments as above, but requiring that the added particle be confined to a *fixed* lattice position, we get the pseudo-chemical potential

$$\bar{\mu}_A = c W_{AB} - \tfrac{1}{2} c W_{BB} \qquad (9.100)$$

Hence, the chemical potential is written as

$$\mu_A = \bar{\mu}_A + kT \ln(N_B + 1)^{-1} \qquad (9.101)$$

In this case, the release of the constraint on a *fixed* position results only in gaining accessibility to the total number of sites $N_B + 1$, which here replaces the volume in (9.93).

# Glossary of Abbreviations

| | |
|---|---|
| BE | Binding energy |
| CN | Coordination number |
| DI | Dilute ideal |
| GMDF | Generalized molecular distribution function |
| HB | Hydrogen bond |
| HI | Hydrophobic interaction |
| HS | Hard sphere |
| IG | Ideal gas |
| lhs | Left-hand side |
| LJ | Lennard-Jones |
| MC | Monte Carlo |
| MDF | Molecular distribution function |
| MM | Mixture model |
| PF | Partition function |
| PMHC | Partial molar heat capacity |
| PY | Percus–Yevick |
| QCDF | Quasi-component distribution function |
| RDF | Radial distribution function |
| rhs | Right-hand side |
| SI | Symmetric ideal |
| TSM | Two-structure model |
| VP | Voronoi polyhedron |

# References

Alder, B. J., and Hoover, W. G. (1968), in *Physics of Simple Liquids*, edited by H. N. V. Temperley, J. S. Rowlinson, and G. S. Rushbrooke, North-Holland, Amsterdam.
Alder, B. J., and Weinwright, T. E. (1957), *J. Chem. Phys.* **27**, 1208.
Angell, C. A. (1971), *J. Phys. Chem.* **75**, 3698.
Barker, J. A., and Henderson, D. (1972), *Annual Rev. of Phys. Chem.* **23**, 439.
Barker, J. A., and Watts, R. O. (1969), *Chem. Phys. Letters* **3**, 144.
Barker, J. A., Fisher, R. A., and Watts, R. O. (1971), *Mol. Phys.* **21**, 657.
Barone, G., Crescenzi, V., Pispisa, B., and Quadrifoglio, F. (1966), *J. Macromol. Chem.* **1**, 761.
Battino, R., and Clever, H. L. (1966), *Chem. Rev.* **66**, 395.
Baxter, R. J. (1967), *Phys. Rev.* **154**, 170.
Bell, G. M. (1969), *J. Math. Phys.* **10**, 1753.
Bell, G. M., and Lavis, D. A. (1970a), *J. Phys. A: Gen. Phys.* **3**, 427.
Bell, G. M., and Lavis, D. A. (1970b), *J. Phys. A: Gen. Phys.* **3**, 568.
Ben-Naim, A. (1965a), *J. Phys. Chem.* **69**, 3240.
Ben-Naim, A. (1965b), *J. Phys. Chem.* **69**, 3245.
Ben-Naim, A. (1968), *J. Phys. Chem.* **72**, 2998.
Ben-Naim, A. (1970a), *J. Chem. Phys.* **52**, 5531.
Ben-Naim, A. (1970b), *Trans. Faraday Soc.* **66**, 2749.
Ben-Naim, A. (1971a), *J. Chem. Phys.* **54**, 1387.
Ben-Naim, A. (1971b), *J. Chem. Phys.* **54**, 3696.
Ben-Naim, A. (1971c), *J. Chem. Phys.* **54**, 3682.
Ben-Naim, A. (1972a), *J. Chem. Phys.* **57**, 5257.
Ben-Naim, A. (1972b), *J. Chem. Phys.* **57**, 5266.
Ben-Naim, A. (1972c), *Mol. Phys.* **24**, 705.
Ben-Naim, A. (1972d), *Mol. Phys.* **24**, 723.
Ben-Naim, A. (1972e), in *Water and Aqueous Solutions*, edited by R. A. Horne, John Wiley and Sons, New York.
Ben-Naim, A. (1972f), *J. Chem. Phys.* **56**, 2864.
Ben-Naim, A. (1972g), *J. Chem. Phys.* **57**, 3605.
Ben-Naim, A. (1973a), *J. Stat. Phys.* **7**, 3.

Ben-Naim, A. (1973b), *J. Chem. Phys.*, **59**, 6535.
Ben-Naim, A., and Friedman, H. L. (1967), *J. Phys. Chem.* **71**, 448.
Ben-Naim, A., and Stillinger, F. H. (1972), in *Water and Aqueous Solutions*, edited by R. A. Horne, John Wiley and Sons, New York.
Ben-Naim, A., and Yaacobi, M. (1974), *J. Phys. Chem.*, **78**, 170.
Ben-Naim, A., Wilf, J., and Yaacobi, M. (1973), *J. Phys. Chem.* **77**, 95.
Berendsen, H. J. C. (1967), in *Theoretical and Experimental Biophysics*, Vol. I, ed. A. Cole, Edward Arnold, London.
Bernal J. D., and Fowler, R. H. (1933), *J. Chem. Phys.* **1**, 515.
Bernal, J. D., and King S. V. (1968), in *Physics of Simple Liquids*, edited by H. N. V. Temperley, J. S. Rowlinson, and G. S. Rushbrooke, North-Holland, Amsterdam.
Bjerrum, N. (1951), *K. Dan. Vidensk. Selsk. Mat.-Fys. Medd.* **27**, 1.
Bogoliubov, N. N. (1962), in *Studies in Statistical Mechanics*, edited by J. DeBoer and G. E. Uhlenbeck, Vol. I, Amsterdam.
Brown, W. B. (1958), *Mol. Phys.* **1**, 68.
Broyles, A. A. (1960), *J. Chem. Phys.* **33**, 456.
Broyles, A. A. (1961), *J. Chem. Phys.* **35**, 493.
Broyles, A. A., Chung, S. U., and Sahlin, H. L. (1962), *J. Chem. Phys.* **37**, 2462.
Buff, F. P., and Brout, R. (1955), *J. Chem. Phys.* **23**, 458.
Buijs, K., and Choppin, G. R. (1963), *J. Chem. Phys.* **39**, 2035.
Bulsaeva, N. N., and Samoilov, O. Ya. (1963), *Zh. Strukt. Khim.* **4**, 502.
Callen, H. B. (1960), *Thermodynamics*, John Wiley and Sons, New York.
Chadwell, H. M. (1927), *Chem. Rev.* **4**, 375.
Chay, T. R., and Frank, H. S. (1972), *J. Chem. Phys.* **57**, 2910.
Clever, H. L., Battino, R., Saylor, J. S., and Gross, P. M. (1957), *J. Phys. Chem.* **61**, 1078.
Corner, J. (1948), *Trans. Faraday Soc.* **44**, 914.
Coulson, C. A., and Eisenberg, D. (1966), *Proc. Roy. Soc. A* **291**, 445, 454.
Covington, A. K., and Jones P. (editors) (1968), *Hydrogen-Bonded Solvent Systems*, Taylor and Francis, London.
Crowe, R. W., and Santry, D. P. (1973), *Chem. Phys. Letters* **22**, 52.
Danford, M. D., and Levy H. A. (1962), *J. Am. Chem. Soc.* **84**, 3965.
Davis, C. M., and Jarzynski, J. (1967), *Adv. Molecular Relaxation Processes* **1**, 155.
Davis, C. M., and Litovitz, T. A. (1965), *J. Chem. Phys.* **42**, 2563.
Davis, C. M., and Litovitz, T. A. (1966), *J. Chem. Phys.* **45**, 2461.
Del-Bene, J., and Pople, J. A. (1970), *J. Chem. Phys.* **52**, 4858.
Denbigh, K. (1966), *The Principles of Chemical Thermodynamics*, Cambridge University Press.
Dorsey, N. E. (1940), *Properties of Ordinary Water-Substance*, Reinhold, New York.
Drost-Hansen, W. (1967), in *Equilibrium Concepts in Natural Water Systems*, (Adv. Chem. Ser. No. 67), edited by W. Stumm, Washington, D. C.
Drost-Hansen, W. (1971), *Fed. Proc.* **30**, 1539.
Drost-Hansen, W. (1972), in *Chemistry of the Cell Interface*, edited by H. D. Brown, Academic Press, New York.
Dubin, P. L., and Strauss, U. P. (1970), *J. Phys. Chem.* **74**, 2842.
Edsall, J. T., and Wyman, J. (1958), *Biophysical Chemistry*, Academic Press, New York.
Eisenberg, D., and Kauzmann, W. (1969), *The Structure and Properties of Water*, Oxford University Press.

# References

Eley, D. D. (1939), *Trans. Faraday Soc.* **35**, 1281.
Eley, D. D. (1944), *Trans. Faraday Soc.* **40**, 184.
Eliassaf, J. and Silderberg, A. (1959), *J. Polym. Sci.* **41**, 33.
Erlander, S. R. (1968), *J. Macromol. Sci. Chem.* **A2**, 595.
Eyring, H., and Jhon, M. S. (1969), *Significant Liquid Structures*, John Wiley and Sons, New York.
Falk, M. and Ford, T. A. (1966), *Can. J. Chem.* **44**, 1699.
Feller, W. (1960), *An Introduction to Probability Theory and Its Applications*, Vol. I, John Wiley and Sons, New York.
Fine, R. A., and Millero, F. J. (1973), *J. Chem. Phys.* **59**, 5529.
Fisher, I. Z. (1964), *Statistical Theory of Liquids*, Chicago University Press.
Fisher, I. Z., and Adamovich, V. I. (1963), *J. Strukt. Khim.* **4**, 759.
Fletcher, N. H. (1970), *The Chemical Physics of Ice*, Cambridge University Press, Cambridge.
Fletcher, N. H. (1971), *Rep. Prog. Phys.* **34**, 913.
Frank, H. S. (1958), *Proc. Roy. Soc. Lond.* **A247**, 481.
Frank, H. S. (1965), *Fed. Proc.* **24** (Part III), S-1.
Frank, H. S. (1970), *Science* **169**, 635.
Frank, H. S., and Evans, M. W. (1945), *J. Chem. Phys.* **13**, 507.
Frank, H. S., and Franks, F. (1968), *J. Chem. Phys.* **48**, 4746.
Frank, H. S., and Quist, A. S. (1961), *J. Chem. Phys.* **34**, 604.
Frank, H. S., and Wen W. Y. (1957), *Disc. Faraday Soc.* **24**, 133.
Franks, F. (1968), in *Hydrogen-Bonded Solvent Systems*, edited by A. K. Covington and P. Jones, Taylor and Francis, London.
Franks, F. (editor), (1972), *Water, A Comprehensive Treatise*, Vol. I, Plenum Press, New York.
Franks, F. (editor), (1973a), *Water, A Comprehensive Treatise*, Vol. II, Plenum Press, New York.
Franks, F. (editor), (1973b), *Water, A Comprehensive Treatise*, Vol. III, Plenum Press, New York.
Franks, F., and Ives, D. J. G. (1966), *Quart. Rev.* **20**, 1.
Friedman, H. L., and Krishnan, C. V. (1973), in *Water, A Comprehensive Treatise*, Vol. III, edited by F. Franks, Plenum Press, New York.
Friedmann, H. (1962), *Adv. Chem. Phys.* **4**, 225.
Friedmann, H. (1964), *Physica* **30**, 921.
Friend, L. and Adler, S. B. (1957), *Chem. Eng. Prog.* **53**, 452.
Frisch, H. L., and Helfand, E. (1960), *J. Chem. Phys.* **32**, 269.
Frisch, H. L., and Lebowitz, J. L. (1964), *The Equilibrium Theory of Classical Fluids*, Benjamin, New York.
Frisch, H. L., and Wasserman, E. (1961), *J. Am. Chem. Soc.* **83**, 3789.
Gibbons, R. M. (1969), *Mol. Phys.* **17**, 81.
Gjaldbaek, J. C., and Niemann, H. (1958), *Acta Chem. Scand.* **12**, 1015.
Glew, D. N. (1962), *J. Phys. Chem.* **66**, 605.
Glew, D. N. (1968), in *Hydrogen-Bonded Solvent Systems*, edited by A. K. Covington and P. Jones, Taylor and Francis, London.
Goldstein, H. (1950), *Classical Mechanics*, Addison-Wesley, Reading, Massachusetts.
Grigorovich, Z. I., and Samoilov, O. Ya. (1962), *Zh. Strukt. Khim.* **3**, 464.
Grjotheim, K., and Krogh-Moe, J. (1954), *Acta Chem. Scand.* **8**, 1193.

Grundke, E. W., Henderson, D., and Murphy, R. D. (1973), *Canad. J. of Phys.* **51**, 1216.
Grundke, E. W., Ph. D. Thesis, University of Waterloo, Waterloo, Ontario (1972).
Guggenheim, E. A. (1952), *Mixtures*, University Press, Oxford, England.
Guggenheim, E. A. (1959), *Thermodynamics*, North-Holland, Amsterdam.
Gurikov, Yu. V. (1963), *Zh. Strukt. Khim.* **4**, 824.
Gurikov, Yu. V. (1965), *Zh. Strukt. Khim.* **6**, 817.
Gurikov, Yu. V. (1966), *Zh. Strukt. Khim.* **7**, 8.
Gurikov, Yu. V. (1968a), *Zh. Strukt. Khim.* **9**, 599.
Gurikov, Yu. V. (1968b), *Zh. Strukt. Khim.* **9**, 944.
Gurikov, Yu. V. (1969), in *Water in Biological Systems*, edited by L. P. Kayushin (translated from Russian), Consultants Bureau, New York.
Hadzi, D. (editor), (1959), *Hydrogen-Bonding*, Pergamon Press, New York.
Haggis, G. H., Hasted, J. B., and Buchanan, T. J. (1952), *J. Chem. Phys.* **20**, 1452.
Hagler, A. T., Scheraga, H. A., and Némethy, G. (1972), *J. Phys. Chem.* **76**, 3229.
Hall, L. (1948), *Phys. Rev.* **73**, 775.
Hankins, D., Moskowitz, J. W., and Stillinger, F. H. (1970), *J. Chem. Phys.* **53**, 4544.
Hasted, J. B. (1973), in: *Water and Aqueous Solutions*, edited by R. A. Horne, Wiley Interscience, New York.
Helfand, E., Reiss, H., Frisch, H. L., and Lebowitz, J. L. (1960), *J. Chem. Phys.* **33**, 1379.
Henderson, D., and Davison, S. G. (1967), in *Physical Chemistry, An Advanced Treatise*, edited by H. Eyring, D. Henderson, and W. Jost, Academic Press, New York.
Hildebrand, J. H. (1968), *J. Phys. Chem.* **72**, 1841.
Hildebrand, J. H., and Scott, R. L. (1950), *The Solubility of Nonelectrolytes*, Reinhold, New York.
Hildebrand, J. H., Prausnitz, J. M., and Scott, R. L. (1970), *Regular and Related Solutions*, Van Nostrand Reinhold, New York.
Hill, T. L. (1956), *Statistical Mechanics*, McGraw-Hill, New York.
Hill, T. L. (1960), *Introduction to Statistical Thermodynamics*, Addison-Wesley, Reading, Massachusetts.
Himmelblau, D. M. (1959), *J. Phys. Chem.* **63**, 1803.
Hirschfelder, J. O., Curtiss, C. F., and Bird, R. B. (1954), *Molecular Theory of Gases and Liquids*, John Wiley and Sons, New York.
Horiuti, J. (1931), *Sci. Papers Inst. Phys. Chem. Res. (Tokyo)* **17**, 125.
Horne, R. A. (1972), *Water and Aqueous Solutions, Structure, Thermodynamics and Transport Processes*, Wiley—Interscience, New York.
Huang, K. (1963), *Statistical Mechanics*, John Wiley and Sons, New York.
Ives, D. J. G., and Lemon, T. H. (1968), *R. I. C. Reviews* **1**, 62.
Jhon, M. S., Grosh, J., Ree, T., and Eyring, H. (1966), *J. Chem. Phys.* **44**, 1465.
Katchalsky, A., Eisenberg, H., and Lifson, S. (1951), *J. Am. Chem. Soc.* **73**, 5889.
Kauzmann, W. (1954), in *A Symposium on the Mechanism of Enzyme Action*, edited by W. D. McElroy and B. Glass, Johns Hopkins University Press, Baltimore, Maryland.
Kauzmann, W. (1959), *Adv. Protein Chem.* **14**, 1.
Kavanau, J. L. (1964), *Water and Solute–Water Interactions*, Holden-Day, San Francisco, California.
Kayushin, L. P. (editor), (1966), *Water in Biological Systems*, A Symposium on Biophysics, USSR [translated from Russian, Consultants Bureau, New York (1969)].
Kell, G. S. (1972), in *Water and Aqueous Solutions*, edited by R. A. Horne, John Wiley and Sons, New York.

# References

Kell, G. S., McLaurin, G. E., and Whalley, E. (1968), *J. Chem. Phys.* **48**, 3805.
Kirkwood, J. G. (1935), *J. Chem. Phys.* **3**, 300.
Kirkwood, J. G. (1939), *J. Chem. Phys.* **7**, 919.
Kirkwood, J. G. (1954), in *A Symposium on the Mechanism of Enzyme Action*, edited by W. D. McElroy and B. Glass, Johns Hopkins University Press, Baltimore, Maryland.
Kirkwood, J. G., and Buff, F. P. (1951), *J. Chem. Phys.* **19**, 774.
Klotz, I. M. (1958), *Science* **128**, 815.
Klotz, I. M., and Franzen, J. S. (1962), *J. Am. Chem. Soc.* **84**, 3461.
Klotz, I. M., and Luborsky, S. W. (1959), *J. Am. Chem. Soc.* **81**, 5119.
Kollman, P. A. (1972), *J. Am. Chem. Soc.* **94**, 1837.
Kollman, P. A., and Allen, L. C. (1969), *J. Chem. Phys.* **51**, 3286.
Kollman, P. A., and Allen, L. C. (1970), *J. Am. Chem. Soc.* **92**, 753.
Kollman, P. A., and Allen, L. C. (1972), *Chem. Rev.* **72**, 283.
Körösy, F. (1937), *Trans. Faraday Soc.* **33**, 416.
Kozak, J. J., Knight, W. S., and Kauzmann, W. (1968), *J. Chem. Phys.* **48**, 675.
Kresheck, G. C., Schneider, H., and Scheraga, H. A. (1965), *J. Phys. Chem.* **69**, 3132.
Krestov, G. A. (1963), *Zh. Strukt. Khim.* **4**, 18.
Krestov, G. A. (1964), *Zh. Strukt. Khim.* **5**, 909.
Krindel, P., and Eliezer, I. (1971), *Coordination Chem. Rev.* **6**, 217.
Krishnan, C. V., and Friedman, H. L. (1971), *J. Phys. Chem.* **75**, 388.
Kruh, R. F. (1962), *Chem. Rev.* **62**, 319.
Kunimitsu, D. K., Woody, A. Y., Stimson, E. R., and Scheraga, H. A. (1968), *J. Phys. Chem.* **72**, 856.
Lebowitz, J. L. (1964), *Phys. Rev.* **133**, A895.
Lebowitz, J. L. (1968), *Ann. Rev. Phys. Chem.* **19**, 389.
Lebowitz, J. L., and Percus, J. K. (1961), *Phys. Rev.* **122**, 1675.
Lebowitz, J. L., and Percus, J. K. (1963), *J. Math. Phys.* **4**, 116.
Lebowitz, J. L., and Rowlinson, J. S. (1964), *J. Chem. Phys.* **41**, 133.
Lennard-Jones, J., and Pople, J. P. (1951) *Proc. Roy. Soc. Lond.* **A205**, 155.
Lentz, B. R., and Scheraga, H. A. (1973), *J. Chem. Phys.* **58**, 5296.
Lieb, E. H. (1967), *Phys. Rev.* **162**, 162.
Lovett, R. A., and Ben-Naim, A. (1969), *J. Chem. Phys.* **51**, 3108.
Luck, W. A. P., and Ditter, W. (1967), *J. Mol. Struct.* **1**, 339.
Lumry, R., and Rajender, S. (1970), *Biopolymers* **9**, 1125.
Maitland, G. C., and Smith, E. B. (1973), *Chem. Soc. Reviews* (London) **2**, 181.
Malenkov, G. G. (1966), *Zh. Strukt. Khim.* **7**, 331.
Malenkov, G. B., and Samoilov, O. Ya. (1965), *Zh. Strukt. Khim.* **6**, 9.
Mandel, F., Bearman, R. J., and Bearman, M. Y. (1970), *J. Chem. Phys.* **52**, 3315.
Marchi, R. P., and Eyring, H. (1964), *J. Phys. Chem.* **68**, 221.
Mason, E. A., and Monchick, L. (1967), in *Intermolecular Forces, Advances in Chemical Physics*, Vol. 12, p. 329.
Masterton, W. L. (1954), *J. Chem. Phys.* **22**, 1830.
Matyash, I. V., and Yashkichev, V. I. (1964), *Zh. Strukt. Khim.* **5**, 13.
Mayer, J. E., and Mayer, M. G. (1940), *Statistical Mechanics*, John Wiley and Sons, New York.
McAuliffe, C. (1966), *J. Phys. Chem.* **70**, 1267.
McMillan, W. G., and Mayer, J. E. (1945), *J. Chem. Phys.* **13**, 276.

Meeron, E., and Siegert, A. J. F. (1968), *J. Chem. Phys.* **48**, 3139.
Metropolis, N. A., Rosenbluth, A. W., Rosenbluth, M. N., Teller, A. H., and Teller, E. (1953), *J. Chem. Phys.* **21**, 1087.
Mikhailov, V. A. (1961), *Zh. Strukt. Khim.* **2**, 677.
Mikhailov, V. A. (1967), *Zh. Strukt. Khim.* **8**, 189.
Mikhailov, V. A. (1968), *Zh. Strukt. Khim.* **9**, 397.
Mikhailov, V. A., and Ponomareva, L. I. (1968), *Zh. Strukt. Khim.* **9**, 12.
Mikhailov, V. A., Grigoreva, E. F., and Semina, I. I. (1968), *Zh. Strukt. Khim.* **9**, 958.
Moon, A. Y., Poland, D. C., and Scheraga, H. A. (1965), *J. Phys. Chem.* **69**, 2960.
Morokuma, K., and Pederson, L. (1968), *J. Chem. Phys.* **48**, 3275.
Morokuma, K., and Winick, J. R. (1970), *J. Chem. Phys.* **52**, 1301.
Morrell, W. E., and Hildebrand, J. H. (1934), *Science* **80**, 125.
Morrison, T. J., and Billett, F. (1952), *J. Chem. Soc.* 3819.
Mukerjee, P. (1965), *J. Phys. Chem.* **69**, 2821.
Mukerjee, P., Kapauan, P., and Meyer, H. G. (1966), *J. Phys. Chem.* **70**, 783.
Münster, A. (1969), *Statistical Thermodynamics*, Vol. I, Springer-Verlag, Berlin.
Murty, T. S. S. R. (1971), *J. Phys. Chem.* **75**, 1330.
Namiot, A. Yu. (1961), *Zh. Strukt. Khim.* **2**, 408.
Namiot, A. Yu. (1967), *Zh. Strukt. Khim.* **8**, 408.
Narten, A. H. (1970), ONRL Report No. 4578, July 1970.
Narten, A. H., and Levy, H. A. (1971), *J. Chem. Phys.* **55**, 2263.
Narten, A. H., and Levy, H. A. (1972), in *Water, A Comprehensive Treatise*, Vol. I edited by F. Franks, Plenum Press, New York.
Narten, A. H., and Levy, H. A. (1969), *Science* **165**, 447.
Narten, A. H., Danford, M. D., and Levy, H. A. (1967), *Disc. Faraday Soc.* **43**, 97.
Némethy, G. (1965), *Fed. Proc.* **24** (Part III), S-38.
Némethy, G. (1967), *Angew. Chem. (Int.)* **6**, 195.
Némethy, G., and Scheraga, H. A. (1962a), *J. Chem. Phys.* **36**, 3382.
Némethy, G., and Scheraga, H. A. (1962b), *J. Chem. Phys.* **36**, 3401.
Némethy, G., and Scheraga, H. A. (1962c), *J. Phys. Chem.* **66**, 1773.
Némethy, G., and Scheraga, H. A. (1964), *J. Chem. Phys.* **41**, 680.
Némethy, G., Scheraga, H. A., and Kauzmann, W. (1968), *J. Phys. Chem.* **72**, 1842.
Nozaki, Y., and Tanford, C. (1963), *J. Biol. Chem.* **238**, 4074.
Nozaki, Y., and Tanford, C. (1965), *J. Biol. Chem.* **240**, 3568.
Ornstein, L. S., and Zernike, F. (1914), *Proc. Acad. Sci. Amsterdam* **17**, 793.
Papoulis, A. (1965), *Probability, Random Variables, and Stochastic Processes*, McGraw-Hill, New York.
Pauling, L. (1935), *J. Am. Chem. Soc.* **57**, 2680.
Pauling, L. (1960), *The Nature of the Chemical Bond*, 3rd ed., Cornell University Press, New York.
Percus, J. K. (1962), *Phys. Rev. Lett.* **8**, 462.
Percus, J. K. (1964), in *The Equilibrium Theory of Classical Fluids*, edited by H. L. Frish and J. L. Lebowitz, Benjamin, New York.
Percus, J. K., and Yevick, G. L. (1958), *Phys. Rev.* **110**, 1.
Peterson, S. W., and Levy H. A. (1957), *Acta Cryst.* **10**, 70.
Pierotti, R. A. (1963), *J. Phys. Chem.* **67**, 1840.
Pierotti, R. A. (1965), *J. Phys. Chem.* **69**, 281.
Pierotti, R. A. (1967), *J. Phys. Chem.* **71**, 2366.

Pimentel, G. C., and McClellan, A. L. (1960), *The Hydrogen Bond*, W. H. Freeman, San Francisco.
Pings, C. J. (1968), in *Physics of Simple Liquids*, edited by H. N. V. Temperley, J. S. Rowlinson, and G. S. Rushbrooke, North-Holland, Amsterdam.
Popkie, H., Kistenmacher, H., and Clements, E. (1973), *J. Chem. Phys.* **59**, 1325.
Pople, J. A. (1951), *Proc. Roy. Soc. Lond.* **A205**, 163.
Priel, Z., and Silberberg, A. (1970), *J. Polym. Sci. A-2* **8**, 689, 705, 713.
Prigogine, I. (1957), *The Molecular Theory of Solutions*, North-Holland, Amsterdam.
Prigogine, I., and Defay, R. (1954), *Chemical Thermodynamics* (translated by D. H. Everett), Longmans Green, New York.
Rahman, A. (1964), *Phys. Rev.* **136**, A405.
Rahman, A., and Stillinger, F. H. (1971), *J. Chem. Phys.* **55**, 3336.
Rahman, A., and Stillinger, F. H. (1973), *J. Am. Chem. Soc.* **95**, 7943.
Ramachandran, G. N., and Sasisekharan, V. (1970), *Adv. Protein Chem.* **23**, 283.
Rao, N. R. (1972), in *Water, A Comprehensive Treatise*, Vol. I, edited by F. Franks, Plenum Press, New York.
Reiss, H. (1966), *Adv. Chem. Phys.* **9**, 1.
Reiss, H., and Tully-Smith, D. M. (1971), *J. Chem. Phys.* **55**, 1674.
Reiss, H., Frisch, H. L., and Lebowitz, J. L. (1959), *J. Chem. Phys.* **31**, 369.
Reiss, H., Frisch, H. L., Helfand, E., and Lebowitz, J. L. (1960), *J. Chem. Phys.* **32**, 119.
Rice, S. A. and Gray P. (1965), *The Statistical Mechanics of Simple Liquids*, Interscience, John Wiley and Sons, New York.
Röntgen, W. C. (1892), *Ann. Phys.* **45**, 91.
Rowlinson, J. S. (1949), *Trans. Faraday Soc.* **45**, 974.
Rowlinson, J. S. (1951), *Trans. Faraday Soc.* **47**, 120.
Rowlinson, J. S. (1968), in *Physics of Simple Liquids*, edited by H. N. V. Temperley, J. S. Rowlinson, and G. S. Rushbrooke, North-Holland, Amsterdam.
Rowlinson, J. S. (1969), *Liquids and Liquid Mixtures*, Butterworth, London.
Rupley, J. A. (1964), *J. Phys. Chem.* **68**, 2002.
Rushbrooke, G. S. (1968), in *Physics of Simple Liquids*, edited by H. N. V. Temperley, J. S. Rowlinson, and G. S. Rushbrooke, North-Holland, Amsterdam.
Samoilov, O. Ya. (1946), *Zh. Fiz. Khim.* **20**, 1411.
Samoilov, O. Ya. (1963), *Zh. Strukt. Khim.* **4**, 499.
Samoilov, O. Ya. (1957), *Structure of Aqueous Electrolyte Solutions and the Hydration of Ions* (translated by D. J. G. Ives), Consultants Bureau, New York (1965).
Samoilov, O. Ya., and Nosova, T. A. (1965), *Zh. Strukt. Khim.* **6**, 798.
Saylor, J. H. and Battino, R. (1958), *J. Phys. Chem.* **62**, 1334.
Scheraga, H. L., Némethy, G., and Steinberg, I. Z. (1962), *J. Biol. Chem.* **237**, 2506.
Schneider, H., Krescheck, G. C., and Scheraga, H. A. (1965), *J. Phys. Chem.* **69**, 1310.
Schrier, E. E., Pottle, M., and Scheraga, H. A. (1964), *J. Am. Chem. Soc.* **86**, 3444.
Shchukarev, S. A., and Tolmacheva, T. A. (1968), *Zh. Strukt. Khim.* **9**, 21.
Sinanoglu, O. (1967), in *Intermolecular Forces, Advances in Chemical Physics*, Vol. 12, p. 283.
Smith, E. B., and Walkley, J. (1962), *J. Phys. Chem.* **66**, 597.
Stillinger, F. H. (1970), *J. Phys. Chem.* **74**, 3677.
Stillinger, F. H., and Rahman, A. (1972), *J. Chem. Phys.* **57**, 1281.
Stockmayer, W. H. (1941), *J. Chem. Phys.* **9**, 398.
Szkatula, A., and Fulinski, A. (1967), *Physica* **36**, 35.

Tanford, C. (1962), *J. Am. Chem. Soc.* **84**, 4240.
Tanford, C. (1968), *Adv. Protein Chem.* **23**, 121.
Tanford, C. (1970), *Adv. Protein Chem.* **24**, 1.
Temperley, H. N. V., Rowlinson, J. S., and Rushbrooke, G. S. (1968), *Physics of Simple Liquids*, North-Holland, Amsterdam.
Thiele, E. (1963), *J. Chem. Phys.* **39**, 474.
Throop, G. J., and Bearman, R. J. (1965), *J. Chem. Phys.* **42**, 2838.
Throop, G. J., and Bearman, R. J. (1966a), *J. Chem. Phys.* **44**, 1423.
Throop, G. J., and Bearman, R. J. (1966b), *Physica* **32**, 1298.
Tully-Smith, D. M., and Reiss, H. (1970), *J. Chem. Phys.* **53**, 4015.
Uhlenbeck, G. E., and Ford, C. W. (1962), in *Studies in Statistical Mechanics*, Vol. I, edited by J. de Boer and G. E. Uhlenbeck, North-Holland, Amsterdam.
Uhlig, H. H. (1937), *J. Phys. Chem.* **41**, 1215.
Vand, V., and Senior, W. A. (1965), *J. Chem. Phys.* **43**, 1869, 1873, 1878.
Vdovenko, V. M., Gurikov, Yu. V., and Legin, E. K. (1966), *Zh. Strukt. Khim.* **7**, 819.
Vdovenko, V. M., Gurikov, Yu. V., and Legin, E. K. (1967a), *Zh. Strukt. Khim.* **8**, 18.
Vdovenko, V. M., Gurikov, Yu. V., and Legin, E. K. (1967b), *Zh. Strukt. Khim.* **8**, 403.
Volterra, V. (1931), *Theory of Functionals*, Blackie, London.
Wada, G. (1961), *Bull. Chem. Soc. Japan* **34**, 955.
Wada, G., and Umeda, S. (1962a), *Bull. Chem. Soc. Japan* **35**, 646.
Wada, G., and Umeda, S. (1962b), *Bull. Chem. Soc. Japan* **35**, 1797.
Wall, T. T., and Hornig, D. F. (1965), *J. Chem. Phys.* **43**, 2079.
Walrafen, G. E. (1966), *J. Chem. Phys.* **44**, 1546.
Walrafen, G. E. (1968), in *Hydrogen-Bonded Solvent Systems*, edited by A. K. Covington and P. Jones, Taylor and Francis, London.
Wasserman, E. (1962), *Sci. Am.* **207**, 94.
Watts, R. O. (1968), *J. Chem. Phys.* **48**, 50.
Weissmann, M., and Cohan, N. V. (1965), *J. Chem. Phys.* **43**, 119.
Wen, W. Y. (1972), in *Water and Aqueous Solutions*, edited by R. A. Horne, Wiley-Interscience, New York.
Wen, W. Y., and Hertz, H. G. (1972), *J. Solution Chem.* **1**, 17.
Wen, W. Y., and Hung, J. H. (1970), *J. Phys. Chem.* **74**, 170.
Wenograd, J., and Spurr, R. A. (1957), *J. Am. Chem. Soc.* **79**, 5844.
Werthein, M. S. (1963), *Phys. Rev. Lett.* **10**, 321.
Wetlaufer, D. B., Malik, S. K., Stoller, L., and Coffin, R. L. (1964), *J. Am. Chem. Soc.* **86**, 508.
Wicke, E. (1966), *Angew. Chem. (Int.)* **5**, 106.
Wilhelm, E., and Battino, R. (1971), *J. Chem. Phys.* **55**, 4012.
Wilhelm, E., and Battino, R. (1972), *J. Chem. Phys.* **56**, 563.
Wishnia, A. (1962), *Proc. Nat. Acad. Sci.* **48**, 2200.
Wishnia, A. (1963), *J. Phys. Chem.* **67**, 2079.
Wishnia, A. (1969), *Biochem.* **8**, 5064.
Wishnia, A. (1969b), *Biochem.* **8**, 5070.
Wishnia, A., and Pinder, T. (1964), *Biochem.* **3**, 1377.
Wishnia, A., and Pinder, R. (1966), *Biochem.* **5**, 1534.
Wood, W. W. (1968), in *Physics of Simple Liquids*, edited by H. N. V. Temperley, J. S. Rowlinson, and G. S. Rushbrooke, North-Holland, Amsterdam.
Yaacobi, M., and Ben-Naim, A. (1973), *J. Solution Chem.* **2**, 425.

# References

Yaacobi, M., and Ben-Naim, A. (1974), *J. Phys. Chem.* **78**, 175

Yashkichev, V. I., and Samoilov, O. Ya. (1962), *Zh. Strukt. Khim.* **3**, 211.

Yastramskii, P. S. (1963), *Zh. Strukt. Khim.* **4**, 179.

Yvon, J. (1935), *La Théorie Statistique des Fluides et l'Équation d'État*, Actualites Scientifiques et Industrielles, Vol. 203, Hermann et Cie. Paris.

Zeidler, M. (1973), in *Water, A Comprehensive Treatise*, Vol. II, edited by F. Franks, Plenum Press.

# Index

Bold numbers indicate that an extensive discussion of a topic begins on that page.

Activity coefficient
  for ideal gas, 96
  for mixtures, 153
  for various reference systems, 164
Average coordination number, 53, 182
Average values of pairwise quantities, 82

Bernal and Fowler conditions
  (for ice), 228
Binding energy, **183**
  definition, 184
  distribution of, 184, 190, 295
  distribution of, for hard spheres, 191
Bipolar coordinates, 48
Bjerrum model for water, 243
Boltzmann constant, 6

Chemical potential
  density derivatives, 108, 144
  for ideal gas, 16
  for the interstitial model, 258
  in a lattice model, 457
  in a mixture-model approach, 206
  in mixtures, 135
  relation with pair distribution functions, 91
  of solutes in water, 332
  in the T, P, N, ensemble, 454
Coefficient of thermal expansion, **200**
  and generalized molecular distribution functions, 200
  for water, 231, 268
Compressibility equation, **104**

Coordination number
  definition, 53, 180
  distribution of, 181, 188, 294
  of water, 235
Correlation in local densities, 114
Coupling parameter, **94,** 323

Dilute aqueous solutions, 309
  experimental observations, 319
Dilute ideal solutions, 155, 309
  small deviations from, 159
Dimerization reactions, 368
Dipole–dipole interaction, **22**
Distribution functions in classical statistical mechanics, **13**

Effective pair potential for water, 225, 238, 241
Energy of hydrophobic interaction, 422
Entropy
  of ideal gas, 16
  of lattice model for water, 260
Entropy–enthalpy compensation, 334, 341, 434
Entropy of hydrophobic interaction, 422
  in water–ethanol mixtures, 426
Euler angles, 3
Euler theorem, 212, 354
Excess thermodynamic functions, 152

Fluctuations
  energy, 11

471

enthalpy, 11
  in local density, 109
  in mixtures, 139
  number of particles, 12
  volume, 12
Functional derivative, 443
Functional Taylor expansion, 443

Generalized Euler theorem, 212
Generalized molecular distribution functions, **177**
  numerical examples, **187**
  and thermodynamics, 195
Gibbs–Duhem relation, 12, 142
Grand partition function, 8
  for mixture, 139

Hard spheres, **18**
Heat capacity, 102
  and generalized molecular distribution functions, 197, 216, 274
  of lattice model for water, 264
  and molecular distribution functions, 102
  of solutes in water, 335
  of water, 232, 271
Hydrogen-bond, 239
Hydrophobic interaction, **363**
  approximate measure of, 384
  between bulky molecules, 416
  connection with experimental results, 378
  definition, 373
  and dimerization reaction, 368, 383
  effect of solutes on, 397
  integral equation for, 437
  among many particles, 398, 402
  and probability, 376, 393
  and second virial coefficient, 370
  and solubility, 367, 396, 409
  for two hard spheres, 386
  in various solvents, 394
  in water–ethanol mixtures, 396
  at zero separation, 407

Ice structure, **227**
Ideal gas, 15

Ideal solutions
  dilute, 155
  and the mixture model, 272
  symmetric, 145
Integral equations, 68
  for waterlike particles, 288, 301
Internal energy
  and generalized molecular distribution functions, 196, 214
  and molecular distribution functions, 85
Interstitial model, **252**
  for aqueous solutions, 337
Isothermal compressibility, 12
  and generalized molecular distribution, 200
  of ideal gas, 107
  for mixtures, 141, 347
  and pair distribution function, **104**
  of water, 231, 272

Kirkwood–Buff theory of solutions, 137, 345

Lennard-Jones
  particles, **20**
  potential, **19**

Mixture-model approach to liquids, **177**
  application to solutions, 354
  based on local properties of the molecules, 208
  general features, 201
Mixtures of very similar components, 135
Molecular distribution functions, **29**
  in mixtures, **124**
  in open systems, 78
  and thermodynamics, **81**
Molecular dynamics simulation, **74**
  for argon, 305
  for water, 302
Momentum partition function, 7
Monte Carlo simulations, **70**
  for waterlike particles, 202, 299

Nonadditivity of the potential, **22**

# Index

Ornstein–Zernike relation, 447
Ostwald absorption coeffcient, 313

Pair and higher order generalized molecular distribution functions, 194
Pair correlation function, **39**
  for mixtures, 126
Pair distribution function, **36**
  generic, **37**
  for mixtures, 126
  specific, 36
Pair potential, **17**
  hard spheres, 18
  Lennard-Jones, **19**
Partial molar enthalphy, 151
  in ideal solutions, 152
  of solutes in water, 332
Partial molar entropy, 151
  in ideal solutions, 152
  of solutes in water, 332
Partial molar heat capacity
  of gases in water, 318, 343
Partial molar volume, 142, 310
  of gases in aqueous solutions, 317, 320, 346, 360
Partition function
  canonical, 6, 9
  classical, 6
  ideal gas, 15
  of interstitial model, 254
  isothermal–isobaric, 8, 10, 255, 338
  open system, 8, 10
  of water, 223
Percus–Yevick equation, 450
  for hard spheres, 414
  numerical solutions of, 453
Perturbation theory for liquids, 12
Planck constant, 6
Potential energy
  pairwise additivity, **17**, 224
Potential of average force
  definition, 54
  dependence on the density, **57**
  and free energy, **62**
  for hard spheres, 56
  for Lennard-Jones particles, 56
Pressure equation, 88
Pseudo-chemical potential, 99, 312, 322
  and work of cavity formation, 117

Quasicomponents, 203
  distribution of, 209

Radial distribution function
  of argon, 235, 305
  definition, 41
  dependence on the density, 51
  dependence on the energy parameter, 53
  for hard spheres, 49
  for ideal gas, 44
  integral equation for, 68
  for Lennard-Jones particles, 49
  methods of evaluating, 65
  properties of, 43
  simulation methods, 69
  for slightly dense gas, 45
  for very dilute gas, 45
  of water, 233
  of waterlike particles in two dimensions, 289
  x-ray diffraction, 65
Residual entropy of ice, 228
Rotational partition function, 441

Scaled particle theory, 326
Simulation methods, **69**
Singlet distribution function, **30**
  generic, 34
  for mixtures, 126
  specific, 33
Singlet generalized molecular distribution function, **179**, 292
Solubility, 313
  of argon in alcohols, 314
  of argon in organic liquids, 315
  of argon in water, 313
Solutions, **123**
Solvophobic interaction, 376
Standard chemical potential, 311
Standard enthalpy of solution
  of methane in water–ethanol mixtures, 320
Standard enthalpy of transfer, 176
Standard entropy of solution, 315
  of methane in water–ethanol mixtures, 318
Standard entropy of transfer, 174

Standard free energy change
  experimental interpretation, 172
  molecular interpretation, 172
  nonconventional, 173
Standard free energy of solution
  definition, 313
  of methane in water–ethanol mixtures, 319
Standard thermodynamic quantities
  of transfer, **170**
Standard volume of transfer, 176
Stirling approximation, 16
Structural changes in the solvent, 328 339, 343
  and hydrophobic interaction, **428**
Structural temperature, 344
Structure of water, **280**
Superposition approximation, 77
Symmetric ideal solutions, 145
  small deviations from, 153

Thermodynamics and statistical mechanics, **9**
Two-structure model, 217, 265, 328

van der Waals equation, 25
Virial coefficients, **26**, 90
  for water, **245**

Virial expansion, **25**, 245
  for osmotic pressure, 163, 363
Volume
  and generalized molecular distribution functions, 196, 215
  of Voronoi polyhedron, 184
Volume density, 129
Voronoi polyhedron, **184**
  definition, 184
  distribution of volume of, 185

Water, 223
  molecular dimensions, 226
  partition function for, **223**
  survey of properties, 225, 230
  survey of theories, 248
Water with one simple solute, **309**
Waterlike particles
  pair potential for, 286
  in three dimensions, 299
  in two dimensions, 283
Work to form a cavity in a fluid, **114**
  in aqueous solutions, 321, 325
  and free energy change, 114
  and pseudo-chemical potential, 117

X-ray diffraction by liquids, 65